BIODEGRADATION - ENGINEERING AND TECHNOLOGY

Edited by **Rolando Chamy** and **Francisca Rosenkranz**

Biodegradation - Engineering and Technology
http://dx.doi.org/10.5772/50829
Edited by Rolando Chamy and Francisca Rosenkranz

CBS, Edition 2016

Contributors

Magdalena Urbaniak, Mohamed Abdallah, Kevin Kennedy, Abid Ali Khan, Ganapati D Yadav, Jyoti B. Sontakke, Mehdi Hassanshahian, Simone Cappello, Sergio Gonzales, Eva Pellicer, Santiago Suriñach, María Dolors Baró, Jordi Sort, Eric Hallerman, Denise Maria Lenz, Derval Rosa, Zeynep Cetecioglu, Bahar Ince, Orhan Ince, Nazli Gokcek, Samet Azman, Nese Coskun, Stephane Godbout, Matthieu Girard, Valeria Oliveira, Isabel Natalia Sierra-García, Iheoma Mary Adekunle, Augutsine Igbuku, Beatriz Perez, Amparo Mauricio-Gutiérrez, Teresita Jiménez-Salgado, Armando Tapia-Hernández, Angélica Santiesteban-López, Tatyana Poznyak, Jessica Amacosta, S Siles, Alejandro Garcia-González, Isaac Chairez, Barbara Rincon, Edixon Gutierrez, Yaxcelys Caldera

Published by InTech

Janeza Trdine 9, 51000 Rijeka, Croatia

Publishing Process Manager Iva Lipovic

Technical Editor InTech DTP team

Cover InTech Design team

Additional hard copies can be obtained from orders@intechopen.com

Biodegradation - Engineering and Technology, Edited by Rolando Chamy and Francisca Rosenkranz
p. cm.
ISBN 978-953-51-1153-5

INTECH

open science | open minds

Contents

Preface

This book contains a collection of different research activities where several technologies have been applied to the optimization of biodegradation processes.

The book has three main sections classified as:

A) Hydrocarbons Biodegradation This section includes chapters that mention the following topics: Biological and physicochemical systems as efficiency alternatives for the treatment of water from crude oil production; Hydrocarbons degradation capacity and bio-surfactant production by microorganisms isolated from hydrocarbon-contaminated soil as efficient candidates for the recovery of soils; Implementation of culture-independent molecular methods to allow the access to the microbial diversity and metabolic potential of microorganisms and bring novel information about microbial diversity and new pathways involved in biodegradation processes taking place in petroleum deposits; Biodegradation of polychlorinated dibenzo-p-dioxins (PCDDs), polychlorinated dibenzofurans (PCDFs) and polychlorinated biphenyls (PCBs) and the last topic; crude oil biodegradation in marine environments.

B) Biodegradation and anaerobic digestion This section contains the followings topics: Microalgae anaerobic digestion; Monitoring and Control of Anaerobic Digestion in Landfills; Anaerobic digestion post-treatment; Anoxic and anaerobic biodegradability characteristics of sulfamethoxasole; and the last topic: Effects of this compound on microbial communities and different processes involved in manure biodegradation, both the emissions that are produced as well as how biodegradation can be used to treat both the manure and the residues from manure management.

C) Biodegradation and sustainability The last sections contain the following: Fundamental aspects of corrosion of magnesium-based alloys in bodily fluids and review of the various techniques that can be used to tune up their degradation rate. The time-dependent evolution of their mechanical properties during the biodegradation process is also outlined; Nitrogen Budget for a Commercial Recirculating Aquaculture Facility; Application of the treatment sequence build up by ozonation followed by biodegradation. This combined treatment is tested in two cases: the first one uses a model with water samples containing some chlorinated phenol derivatives (4-chlorophenol and 2,4-dichlorophenol) and the second case considers real wastewater samples obtained from the pulp and paper industry (Kraft process in bleaching); Emerging Trend in Natural Resource Utilization for Bioremediation of Oil - Based Drilling Wastes in Nigeria; and the last topic is an overview of the current biodegradable polymer matrices and some of the most used reinforcements are described as well as the properties and applications of the bio-composites obtained.

Rolando Chamy and Francisca Rosenkranz
Biotechnology Center of Curauma
Pontificia Universidad Católica de > Valparaíso, Chilr

School of Biochemical Engineering
Pontificia Universidad Católica de > Valparaíso, Chile

Section 1

Biodegradation of Hydrocarbons

Chapter 1

Biodegradability of Water from Crude Oil Production

Edixon Gutiérrez and Yaxcelys Caldera

Additional information is available at the end of the chapter

http://dx.doi.org/10.5772/56328

1. Introduction

According to Gutiérrez *et al.* (2007) the waters of formation (WOF), are those that are naturally in the rocks and are present before the perforation of the well. Their composition depends on the origin of the water and the modification that could happen as soon as they enter in contact with the environment of the subsoil. WOF must be obtained from the bottom of the well; nevertheless, for costs reason the samples are taken at the surface level, in the head of the well. As they rise in the column from the well up to the surface, their characteristics change due to the changes of pressure, temperature and composition of the gases. For this reason the name adapted for these samples of waters is water associated with crude oil production. Other researchers name these waters as water from petroleum, water from oil field production, oily waters, effluent from the extraction of oil, water from petroleum. In this work they are named waters associated with crude oil production (WCP).

Among the characteristics of WCP are their high content of free and emulsified crude oil and hydrocarbons, suspended solid, H_2S and mercaptans (Gutiérrez *et al.*, 2002), aromatic, poliaromatic and phenols compounds (Rincón *et al.*, 2008), high temperature and high salinity (Guerrero *et al.*, 2005; Li *et al.*, 2005), saturated, aromatics, resins and asphaltenes compounds (SARA) (Díaz *et al.*, 2007), and metal traces Na, Ca, Mg, Fe, Sr, Cr, As and Hg (Gutiérrez *et al.*, 2009). According to García *et al.* (2004) among the pollutants with a major potential impact related to the petroleum industry are polycyclic aromatic hydrocarbons (PAH), voltaic organic compounds (VOC) and total hydrocarbons of the oil (THO). The first ones have high carcinogenic, mutagenic and teratogenic potential in aquatic organisms; the second ones contribute to the greenhouse effect and are involved in the direct ozone formation on the soil level and indirectly on the acid rain, besides some individual compounds are toxic, carcinogenic, mutagenic or bioaccumulative, and the last ones present diverse effects on the flora an fauna.

Gives that the WCP volumes generated in the Ulé tank farm, on the east cost of Maracaibo Lake, Venezuela, belonging to the petroleum industry in Venezuela, would exceed the needs for secondary recovery and the systems of reinjection would be rapidly saturated, different research works were realized to present alternatives to the petroleum industry, to diminish the potential pollutant of WCP.

In this aspect, some proposals for the treatment of WCP are aerobic and anaerobic biological processes, physicochemical treatment and some new technologies as constructed wetlands. Among the anaerobic processes are the batch reactors (BR) and the upflow anaerobic sludge blanket reactors (UASB).

The biological mesophilical and thermophilical anaerobic systems have been successful in the treatment of complex waters, with low, moderate and high organic load (Lettinga, 2001). In the case of UASB, these reactors are outlined by their capacity to retain biomass, to form granular sludge with high properties of sedimentation, to handle high organic load to short hydraulic retention time (HRT), produce biogas and remove high concentration of biodegradable organic matter (Lepisto and Rintala, 1990; Lettinga, 2005).

On the other hand, the aerobic systems have been efficient for the treatment of wastewater containing chemical compounds resistant to be biodegraded. Among these systems are the sequential biological reactors (SBR), which have showed excellent results in the degradation of toxic compounds present in industry effluents (Díaz *et al.*, 2005a; González *et al.*, 2007). As well as, the rotating biological contactor reactors (RBC), which produce good quality effluents including total nitrification, low costs and ease of operation and maintenance (Behling *et al.*, 2003).

Among the physicochemical treatment applied to reduce the pollutants in wastewater are the dissolve air flotation (DAF) and the coagulation. The most applied products to treat natural water and wastewater by coagulation and flocculation are iron and aluminium salts. However, the cationic polymers have demonstrated their efficiency in the removal of oils and phenols from industrial wastewater (Renault *et al.*, 2009; Ahmad *et al.*, 2006).

In this investigation was reviewed a several papers from studies conducted at the Universidad del Zulia during 2002 to 2012, to analyze the efficiency of biological and physicochemical systems BR, UASB, SBR and RBC, and the physicochemical treatment as coagulation and flotation (DAF), which have been evaluated to remove COD, hydrocarbons, SARA and phenols, present in the WCP.

The instrument used was a matrix register of the treatment, considering criteria like WCP type, system of treatments, operation conditions, organic load, retention times, temperature, pollutant contents and dose of coagulant. The efficiency of the treatments was compared considering the parameters COD, phenols, hydrocarbons and SARA.

http://dx.doi.org/10.5772/56328

2. Results

2.1. Origin and composition of the waters associated with the crude oil production

The WCP samples were obtained from the Ulé tank farm, located on the east coast of Maracaibo Lake, Tía Juana, Zulia state, Venezuela (Figure 1). The water samples come from the segregations: Tía Juana light (TJL), Urdaneta heavy (UH), Tía Juana medium (TJM), and the dehydrations of the Punta Gorda tank farm (Rosa medium-RM), Shell Ulé (F-6/h-7) and lacustrine terminal of La Salina (LTLS). These waters were obtained from the separation of the water associated with the extraction of light crude oil (>31.8ºAPI) WCPL, from the water associated with the extraction of medium crude oil (22ºAPI-29.9ºAPI) WCPM, from the water associated with the extraction of heavy crude oil (10ºAPI-21.9ºAPI) WCPH, classified according to the American Petroleum Institute. Also, water samples were taken from the converged point of the three cuts (WCPC).

The Tables 1, 2, 3 and 4 present the principal characteristics of WCPL, WCPM, WCPH and WCPC, respectively. In general, it is observed that the physicochemical characteristics of the WCP are different depending on the contact of these waters with the crude oil associated. They are waters with high pollutant contents and they do not comply with the Venezuelan environmental regulations to be discharged into water bodies (Gaceta Oficial, 1995). On the other hand, the differences in the characteristics reported by the researchers, might be related to the changes that have been given in the productive processes of the petroleum industry in the last years.

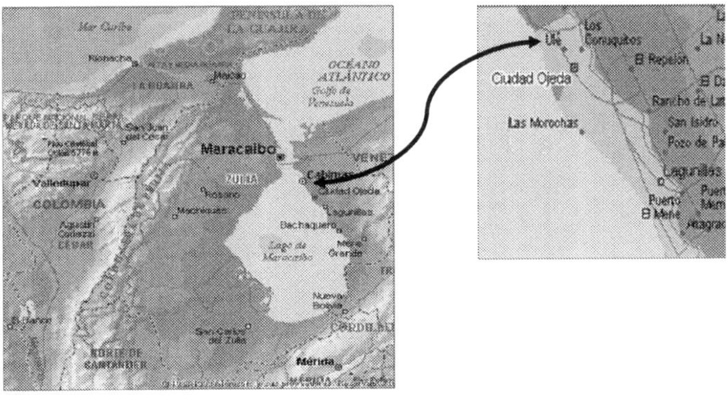

Figure 1. Geographical location of the Ulé tank farm, Tía Juana Zulia state, Venezuela.

Parameters	Díaz *et al.* (2005a)	Díaz *et al.* (2005b)	Gutiérrez et al. (2012)	González et al. (2007)	Rincón et al. (2008)
pH	7.9	8.0	8.3	7.99	NR
Alkalinity (mg $CaCO_3$/L)	2933	2215	2670	2412	NR
COD soluble (mg/L)	1065.2	799	1400	1105	106.2
Phenols (mg/L)	19.36	1.73	NR	16.8	NR
Nitrogen NTK (mg/L)	23.82	28.8	20	21.2	23.82
Phosphorous (mg/L)	1.07	1.0	2.2	1.57	1.07
Hydrocarbons (mg/L)	NR	91	224.2	78.0	NR
Chlorines (mg/L)	NR	NR	NR	NR	NR
TSS (mg/L)	NR	NR	104	NR	NR
VSS (mg/L)	NR	NR	54	NR	NR
O&G (mg/L)	NR	NR	66	100.7	NR
Saturated (mg/L)	NR	NR	76.6*	NR	1.24
Aromatics (mg/L)	NR	NR	7.04*	NR	17.64
Resins (mg/L)	NR	NR	6.34*	NR	8.51
Asphaltenes (mg/L)	NR	NR	7.73*	NR	7.49

*Values in (%), NR: No register

Table 1. Physicochemical parameters of WCPL from tank farm of Ulé

Parameters	Díaz *et al.* (2005a)	Gutiérrez et al. (2012)	Rincón et al. (2008)	Castro et al. (2008)
pH	8.0	8.5	NR	8.04
Alkalinity (mg $CaCO_3$/L)	3440	2800	NR	2906
COD soluble (mg/L)	782.6	933	782.6	880
Phenols (mg/L)	1.40	NR	NR	NR
Nitrogen NTK (mg/L)	39.20	15.1	39.20	NR
Phosphorous (mg/L)	1.05	3.5	1.05	NR
Hydrocarbons (mg/L)	NR	148.7	NR	NR
Chlorines (mg/L)	NR	NR	NR	NR
TSS (mg/L)	NR	NR	NR	82.57
VSS (mg/L)	NR	NR	NR	69.71
Saturated (mg/L)	NR	25.32*	5.73	0.24

Parameters	Díaz *et al.* (2005a)	Gutiérrez et al. (2012)	Rincón et al. (2008)	Castro et al. (2008)
Aromatics (mg/L)	NR	5.86*	9.77	50.34
Resins (mg/L)	NR	6.49*	5.30	33.22
Asphaltenes (mg/L)	NR	5.99*	5.30	16.10

*Values in (%), NR: No register

Table 2. Physicochemical parameters of WCPM from tank farm of Ulé

Parameters	Díaz *et al.* (2005a)	Gutiérrez et al. (2012)	González et al. (2007)	Gutiérrez et al. (2009)	Caldera et al. (2011)
pH	8.0	8.2	8.3	7.08	8.41
Alkalinity (mg $CaCO_3$/L)	885	1000	885	NR	803.33
COD soluble (mg/L)	307	864	320	1029	259.6
Phenols (mg/L)	2.70	NR	2.5	NR	0.83
Nitrogen NTK (mg/L)	10.61	15.7	9.2	8.26	5.60
Phosphorous (mg/L)	2.68	2.0	9.8	0.013	3.01
Hydrocarbons (mg/L)	NR	52.7	78	35.0	123.21
Chlorines (mg/L)	NR	NR	NR	NR	1101.21
TSS (mg/L)	NR	NR	NR	NR	573.33
VSS (mg/L)	NR	NR	NR	NR	220.00
Color (CU)	NR	NR	NR	NR	718.80
Turbidity (NTU)	NR	NR	NR	NR	140.00
Chrome (mg/L)	NR	NR	NR	4.75	NR
Lead (mg/L)	NR	NR	NR	4.35	0.0
Sodium (mg/L)	NR	NR	NR	89.94	NR
Zinc (mg/L)	NR	NR	NR	2.50	0.30
O&G (mg/L)	NR	NR	113.3	NR	NR
Saturated (mg/L)	NR	23.97*	NR	NR	NR
Aromatic (mg/L)	NR	6.15*	NR	NR	NR
Resins (mg/L)	NR	64.7*	NR	NR	NR
Asphaltenes (mg/L)	NR	5.14*	NR	NR	NR

*Values in (%). NR: No register

Table 3. Physicochemical parameters of WCPH from tank farm of Ulé

Parameters	Behling et al. (2003)[a]	Rincón *et al.* (2004)[a]	Rojas *et al.* (2008)[b]	Blanco *et al.* (2008)[c]
pH	7.72	8	7.74	8.03
Alkalinity (mg $CaCO_3$/L)	2460	2238	2477	2635
COD soluble (mg/L)	823	NR	NR	1391.85
COD total (mg/L)	NR	700	NR	NR
Phenols (mg/L)	NR	5	NR	2.14
Nitrogen NTK (mg/L)	12.92	NR	NR	17.55
Phosphorous (mg/L)	1.40	NR	NR	3.67
Hydrocarbons (mg/L)	NR	100	NR	276.68
Chlorine (mg/L)	NR	NR	1802	1404.87
TSS (mg/L)	170	NR	122	550
VSS (mg/L)	50	NR	NR	82.35
Sulfides (mg/L)	NR	NR	NR	7.32
Turbidity (NTU)	NR	NR	480	NR
Chrome (mg/L)	NR	NR	NR	0.31
Lead (mg/L)	NR	NR	NR	0.17
Sodium (mg/L)	NR	NR	NR	8880.32
Nickel (mg/L)	NR	NR	NR	0.20
Zinc (mg/L)	NR	NR	NR	0.32
Copper (mg/L)	NR	NR	NR	0.19
O&G (mg/L)	NR	181	737	NR

[a] Combination of light, medium and heavy crude oil, and exit of the clarifier

[b] Combination of medium and heavy crude oil, API 5.

[c] Combination of light, medium and heavy crude oil, and in of the clarifier

NR: No register

Table 4. Physicochemical parameters of WCPC from tank farm of Ulé

2.2. Treatment of the waters associated with crude oil production

The Tables 5, 6, 7 and 8 show a summary of the methodology used by each researcher, showing the operational conditions for each system. On the other hand, Table 9 and Table 10 compare the different treatments: physicochemical treatments, aerobic and anaerobic biological treatment, and combined treatments.

2.3. Biological treatment applied to the waters associated with crude oil production

The Tables 5 and 6 show a resume of the aerobic and anaerobic biological treatments applied to WCP, and Table 8 shows the operation conditions of the combined system aerobic-anaerobic applied to WCP. Among the aerobic biological systems are the rotating biological contactor reactors (RBC), the sequential biological reactors (SBR) and the continuous flow reactors (CR); and among the anaerobic biological treatments are the batch reactors (BR) and the upflow anaerobic sludge blanket reactors (UASB), working under mesophilic and thermophilic conditions. Likewise, Table 9and Table 10 present a summary of the results of applying these treatments to WCP.

Researcher, year	Kind of WCP	Treatment systems	Characteristics of the experimental equipment	Operation conditions	Parameters evaluated
Behling *et al.* (2003)	WCPC (WCPL, WCPM and WCPH)	RBC	RBC of 9.5 L, with 50 circular disc of PVC, 0.8 cm separation, supported in an axis of carbon steel 3/8 " diameter, rotation speed of 2.5 rpm. The discs were immersed 40 % in the effluent. The area of contact was 2.44 m^2. The water volume was 7.5 L	The RBC worked under mesophilic condition. The organic load average applied was 2.04 ± 0.7 g COD/m^2d and 5.2 mL/min, TRH of 24 h, temperature 27-32°C.	pH COD TSS VSS Total alkalinity
Díaz *et al.* (2005a)	WCPL, WCPM and WCPH	SBR	The SBR of 4 L were constructed in material of plastic and cylindrical form, with a volume of operation of 2 L, in which 600 mL sludge and 1.4 L of WCP. At the bottom of the reactors were located air diffusers connected to a compressor.	After acclimated and stabilized, they worked with HRT of 16 hours with sequence of 15 hours of ventilation, 30 minutes of sedimentation and 30 minutes for capture of sample and recharges of the reactor. The temperature was mesophilic (37 °C). The SBR-1, SBR-2, SBR-3 operated with organic charges of 1.6; 1.17 and 0.46 kg/m^3d for the WCPL, WCPM and WCPH, respectively.	pH Alkalinity COD Phenols
Díaz *et al.* (2005b)	WCPM	SBR	The SBR of 2 L was constructed in material of plastic, with 600 mL of sludge and 1.4 L of WCPM. They gave oxygen to the reactor by means of a compressor.	After acclimated and stabilized, they were operated at the first stage of 15 hours the HRT and time of cellular retention of 15-20 days with sequence of 14 hours for mixed, ½ hour of rest and ½ hour for discharge and load. Whereas in the second stage the HRT was 24 hours with sequence of 23 hours for mixed and ventilation and one hour of discharge and load. The temperature was 37 °C. The	COD Hydrocarbons Phenols

Researcher, year	Kind of WCP	Treatment systems	Characteristics of the experimental equipment	Operation conditions	Parameters evaluated
				organic load applied was between 0.89 and 0.51 kg/m³d	
González *et al.* (2007)	WCPL and WCPH	SBR	The SBR of 2 L was constructed in material of plastic, in cylindrical form, in which they added 600 mL of sludge and 1.4 L of WCP. They gave oxygen to the reactor by a compressor.	HRT of 8 hours and time of cellular retention of 20 days. Nutrients were added. The COD in the inflow was 1105 and 320 mg/L for WCPL and WCPH, respectively.	COD Hydrocarbons Phenols
Castro *et al.* (2008)	WCPM	Batch reactor	The reactor was a receptacle adjusted as Plexiglas of 3 L, provided with a porous circular stone and a hose connected to the tubes for the supply of compressed air. As effective volume of 0.3 L of bacterial suspension and 0.7 L of WCPM.	They used several functional groups and consortiums of bacteria. The systems were operated under mesophilic conditions (27 °C) and HRT of 144 h. The COD of feeding was 880 mg/L.	pH COD TSS VSS Alkalinity

Table 5. Methodology for aerobic treatment of WCPM

Researcher, year	Kind of WCP	Treatment systems	Characteristics of the experimental equipment	Operation conditions	Parameters evaluated
Gutiérrez *et al.* (2007)	WCPL WCPM and WCPH	Batch rectors	They placed four (4) reactors of 500 mL each one, containing 20 % of the useful volume of mesophilic granular sludge proceeding from a beer industry, and 80 % of effluent to treat. The reactors were immersed in a thermal bath that allowed controlling the temperature. The produced biogas was meter by water displacement.	Initially the reactors were loaded, for ten days, with D +glucose on an equivalent concentration in COD of 1500 mg/L and solution of nutrients, for a retention time (RT) of 24 hours. Later they added to three reactors WCPL, WCPM and WCPH with concentrations of 1200-1300 mgCOD/L, 857-960 mgCOD/L and 860-870 mgCOD/L, respectively. The fourth reactor worked with glucose (D+ glucose). To reach the thermophilic conditions (55°C ± 1°C) the temperature was increased from the mesophilic conditions (37°C ± 1°C) at the reason of 1°C/day. The RT in all the cases was 24 hours.	pH COD TSS and VSS Alkalinity VFA Methane

Researcher, year	Kind of WCP	Treatment systems	Characteristics of the experimental equipment	Operation conditions	Parameters evaluated
Gutiérrez *et al.* (2009)	WCPM and WCPH	Batch reactors	They placed three (3) reactors of 500 mL each one, containing 20 % of the useful volume mesophilic granular sludge proceeding from a beer industry, and 80 % of effluent to treat. The reactors were immersed in a thermal bath that allowed controlling the temperature. The produced biogas was meter by water displacement.	Initially the reactors were loaded, for ten days, with D +glucose on an equivalent concentration in COD of 1500 mg/L and solution of nutrients, for a retention time (RT) of 24 hours. Later they added to two reactors WCPM and WCPH with concentrations of 1876.9 and 1029.0 mgCOD/L, respectively. The third reactor worked with glucose (D+glucose). To reach the thermophilic conditions (55°C ± 1°C) the temperature was increased from the mesophilic conditions (37°C ± 1°C) at the reason of 1°C/day. The RT in all the cases was 24 hours.	pH COD TSS and VSS Alkalinity VFA Methane
Rincón *et al.* (2002)	WCPL	UASB reactors	There were employed at a UASB reactor of 4 L, 0.098 m of diameter, 0.67 m high and 0.53 m high of water, inoculated with 30 % of granular sludge from a UASB reactor that treats residual waters of a brewery of the locality.	Initially, the reactor was fed with residual synthetic water that was containing glucose as the only source of carbon (1 g/L) and nutrients. Later, it was operated for 275 days with HRT from 38 to 5 h. The reactors were evaluated for organic loads of 0.78; 1.20; 1.46; 1.64; 1.90; 3.17 and 4.70 kg COD/m^{3}d for HRT of 36, 24, 21, 17, 11, 8 and 6 hours, respectively. They worked under mesophilic conditions (37°C ± 1°C).	pH Alkalinity COD Phenols
Díaz *et al.* (2005a)	WCPL WCPM WCPH	UASB reactors	They worked with 3 UASB reactors of 4 L, inoculated with 1.2 L of granular sludge from an UASB reactor treating residual waters of a brewery of the locality.	Initially, the reactor was fed with residual synthetic water that was containing glucose as the only source of carbon (850 mg/L) and nutrients. Later, the reactors UASB-1, UASB-2 and UASB-3 were fed by WCPL, WCPM and WCPH why organic loads of 1.06; 0.78 and 0.31 kg COD/m^{3}d, respectively. They worked under mesophilic conditions (37°C ± 1°C) during 1 month with HRT of 24 h.	pH Alkalinity COD SARA

Researcher, year	Kind of WCP	Treatment systems	Characteristics of the experimental equipment	Operation conditions	Parameters evaluated
Gutiérrez *et al.* (2006)	WCPL	UASB reactors	They used two UASB reactors constructed in Plexiglas with volumes of 1.7 and 2.5 L, operating under temperatures of 37 ± 1°C and 55 ± 1°C, respectively. The reactors were provided with a jacket, supporting the temperature for recirculation of warm water. Both reactors were inoculated with mesophilic anaerobic granular sludge (20 % of the useful volume) proceeding from a beer industry.	Initially the reactors were load, for two days, with D+glucose on an equivalent concentration in COD of 1500 mg/L and solution of nutrients and TRH of 24 h; then WCPL was added. Later to reach the thermophilic conditions (55°C ± 1°C) in the thermophilic reactor, the temperature was increased from the mesophilic condition (37°C ± 1°C) to a rate of 1°C/day. The reactors were evaluated for organic loads of 1.4, 1.9, 2.8 and 5.6 kg COD/m³d and RTH of 24, 18, 12 and 6 hours, respectively.	pH Alkalinity COD VFA Methane Enzymes
Caldera *et al.* (2007)	WCPL	UASB reactor	They used a UASB reactor constructed in Plexiglas with volume of 2.5 L, inoculated with anaerobic mesophilic granular sludge (30 % of the useful volume) proceeding from a beer industry. The reactor was provided with a jacket, supporting the temperature of 55 ± 1°C for recirculation of warm water.	Initially the reactor was loaded, for two days, with D+glucose on an equivalent concentration in COD of 1500 mg/L and solution of nutrients, and HRT of 24 h; then WCPL was added. Later to reach the thermophilic condition (55°C ± 1°C), in the thermophilic reactor, the temperature was increased from the mesophilic condition (37°C ± 1°C) to a rate of 1°C/day. The reactor was evaluated for 42 days, with HRT of 24 and 12 and organic loads of 1.4 and 2.8 kg COD/m³d, respectively.	pH Alkalinity COD VFA Methane
Rincón *et al.* (2008)	WCPL WCPM	UASB reactors	They used two UASB reactors of 2.5 L, inoculated with 0.75 L of granular sludge from an UASB reactor treating residual waters of a brewery of the locality.	Initially, the reactors were fed with residual synthetic water that was containing glucose as the only source of carbon (850 mg/L) and nutrients. Later, the reactors UASB-1 and UASB-2 were fed with WCPL and WCPM APPL and organic load of 1.06 and 0.78 kg COD/m³d respectively. They worked mesophilic conditions (37°C ± 1°C) during 1 month with HRT of 24 h.	pH Alkalinity COD SARA

Table 6. Methodology for anaerobic treatment of WCP

Researcher, year	Kind of WCP	Treatment systems	Characteristics of the experimental equipment	Operation conditions	Parameters evaluated
Rojas *et al.* (2008)	WCPC (WCPM and WCPH)	Coagulation and DAF	The DAF, consisted of a pressurization cell or saturation camera, constructed in material of transparent plastic of 90 mm of external diameter and 270 mm high. Inside the camera was finding a manual agitator of stainless steel and a filter that worked as diffuser; in addition, a series of connections and valves of the distribution and pressure of the water and air.	They worked with pressures of 30, 40 and 50 psi and recycle of 30%, 40% and 50%, and temperature of 25ºC. They evaluated a cationic flocculants of high molecular weight, in concentration of 0.006 % in volume (3.54 mg/L)	TSS Turbidity O&G
Caldera *et al.* (2009)	WCPH	Coagulation-flocculation	They used a Jar Test model JLT6; adding 1 L of WCP, to each of six precipitation jar of 1000 mL, taking one of these as a control.	They simulated coagulation, flocculation, and sedimentation processes to 100 rpm for rapid agitation for 1 minute and 30 rpm for slow agitation by 20 minutes. The sedimentation was 30 minutes. The initial turbidity of the water was 140 NTU. They used as coagulant commercial chitosan (CCH) (Sigma Chemical Co.) and chitosan obtained in the laboratory (LCH) to 100 ºC dissolved in acetic acid 0.10 M, preparing solutions of 0.6 %. They worked with concentrations of 24, 30, 36, 42 and 48 mg/L of solution of LCH and CCH, respectively.	pH COD TSS VSS Turbidity Color O&G Hydrocarbons
Caldera *et al.* (2011)	WCPH	Coagulation-flocculation	They used a Jar Test model JLT6; adding 1 L of WCP, to each of six precipitation jar of 1000 mL, taking one of these as a control.	They simulated coagulation, flocculation and sedimentation processes to 100 rpm for rapid agitation for 2 minutes, and 100 rpm for slow agitation for 30 minutes. The sedimentation was 30 minutes. The turbidity initial was 52 NTU. As coagulant agent was used commercial chitosan (CCH) dissolved in acetic acid 0.10 M, preparing solutions of 1.0%. The concentrations evaluated were 40,	pH COD TSS VSS Turbidity Color O&G Hydrocarbons

Researcher, year	Kind of WCP	Treatment systems	Characteristics of the experimental equipment	Operation conditions	Parameters evaluated
				42, 44, 46 and 48 mg/L of CCH solution.	

Table 7. Methodology for physicochemical treatment of WCP

Researcher year	Kind of WCP	Treatment systems	Characteristics of the experimental equipment	Operation conditions	Parameters evaluated
Rincón *et al.* (2004)	WCPL WCPC	UASB-SBR system	They used two types of reactors placed in series, a reactor UASB of 2.5 L of useful volume and a SBR. The reactor UASB was inoculated by sludge from an UASB reactor treating residual waters of a brewery. While the SBR reactor was inoculated with aerobic sludge from a wastewater treatment plant.	The system worked 195 days, in two stages. The first was feeding with WCPL from 1100 to 1230 mg COD/L (133 days) and the second one with WCPC of 176 and 264 mg COD/L (66 days). The effluent treated in the UASB was fed in the SBR. The HRT was 24 hours and the temperatures were UASB 37ºC and SBR 28 ºC.	pH Alkalinity COD Hydrocarbons Phenols
Paz *et al.* (2012)	WCPC	Superficial constructed wetlands (SCWFF)	They used two superficial constructed wetlands of free flow (SCWFF) to pilot scale. The support material was gravel and soil, and aquatic emergent plants that counted of support of gravel and soil, and aquatic emergent plants (*Cyperus luzulae y Cyperus ligulari* – SCWFF I, y *Cyperuz feraz, Paspalum sp. y Typha dominguesis* – SCWFF II), and a control (C) without plants.	They placed 30 plants for each species. The depth of the support was 0.25 cm with 7 % of gravel and 93 % of soil, and a water layer of 0.05 m of water. The flow fed was 8 mL/min, with a HRT of 7 days and organic load of 23.5 g COD/m^{2}d. The samples were collected weekly for 80 days.	COD pH Sulphide Phenols TSS VSS DO
Blanco *et al.* (2008)	WCPC	Sub-superficial constructed wetlands (SSCW)	The system SSCW consisted of three polyethelene tray of 1.28 m long for 0.45 m wide and 0.45 m high, one that of them as control (without plants) and the others two with emergent aquatic plants *Cyperus luzulae, Cyperus feraz L.C, Cyperus ligularis L.* y *Typha dominguensis* (SSCW I y SSCW II). The beds of the tray were constituted by 86400 cm^3 of gravel as support and a water level of 1.5 L to simulate a natural system of wetland.	The systems worked to continue flow, without recirculation of the effluent with an organic load of 29.42 g/m^{2}d, a flow of 10 mL/min and HRT of 7 days.	pH Alkalinity COD VSS Hydrocarbons Phenols

Table 8. Methodology for combined treatment of WCP

Researcher year	Treatment systems / WCP	COD (%)	TSS (%)	VSS (%)	Hydro-carbons (%)	O&G (%)	Phenols (%)	Turbidity (%)	pH	Alkalinity (mg$CaCO_3$/L)
Rojas et al. (2008)	Coagulation and DAF WCPC	—	77	—	—	90	—	69	—	—
Caldera et al. (2009)	Coagulation-flocculation WCPH	50.7	—	—	70.1	—	—	90.7	8.0-8.2	—
Caldera et al. (2011)	Coagulation-flocculation WCPH	12.5	55-61	41-63	70-90	39-59	—	76-78	7.9	—
Behling et al. (2003)	RBC WCPC	76.1	< 4	< 3	—	—	—	—	8.9	2343
Díaz et al. (2005a)	SBR WCPL, WCPM and WCPH	88.8 65.2 62.9	—	—	—	—	96.8 89.2 82.8	—	9.0-9.9 9.0-9.6 8.9-9.4	—
Díaz et al. (2005b)	SBR WCPM	65.1[a] 60.9[b]	—	—	76.8 79.5	55.5 62.4	87.5 92	—	—	—
González et al. (2008)	SBR WCPL and WCPH	88 66	—	—	84.4 73.8	—	95.6 79.4	—	—	—
Castro et al. (2008)	Batch reactor WCPM	62.4-89.8	63.3-9.5	—	—	—	—	—	7.4-6.6	—

a and b: different HRT

Table 9. Results of the treatment of WCP

Researcher year	Treatment systems / WCP	COD (%)	VSS (%)	Hydro-carbons (%)	Phenols (%)	pH	Alkalinity (mg$CaCO_3$/L)	SARA (%)	Methane content (%)
Gutiérrez et al. (2007)	Batch reactors WCPL, WCPM and WCPH	70.7 59.9 62.1	—	—		7.6 7.6 7.2	2673.7 2620.0 936.7	—	73.1 51.9 54.1
Gutiérrez et al. (2009)	Batch reactors WCPM and WCPH	68.2-69.2 55.9-50.4	—	—		8.2 7.5	—	—	—

Rincón et al. (2002)	UASB WCPL	23.8-86.1	—	—	10-59	7.6-8.0	2500-2800		24-95
Díaz et al. (2005a)	UASB WCPL, WCPM and WCPH	81.7 23.5 35.7	—	—	55.1 74.7 92.5	7.3-8.6 7.1-8.5 7.2-8.4	—	—	—
Gutiérrez et al. (2006)	UASB WCPL	$40\text{-}80^{M}$ $67\text{-}84^{T}$	42-73 52-67	—	—	7.4-8.5 7.9-8.0	1960-2633 2190-2454	—	53-79 54-80
Caldera et al. (2007)	UASB WCPL	78^{a} 77^{b}	—	—	—	8.0 8.2	2413 2945		87 77
Rincón et al. (2008)	UASB WCPL and WCPM	93 26	—	—	—	7.5 7.8	1955 2520	84 54	—
Rincón et al. (2004)	UASB-SBR WCPL-WCPC	95 79	—	74 82	99.9 90	9 9	2468 2405	—	—
Paz *et al.* (2012)	Superficial constructed wetlands WCPC	-	-		64.3 61.3	9.13-10.5 8.84-9.93	-	-	-
Blanco et al. (2008)	Sub-superficial constructed wetlands WCPC	31.4-65.7	45.2-91.9	77.5	94.7	8.9	2508	—	—

M: Mesophilic T: Thermophilic; a and b : different HRT

Table 10. Results of the treatment of WCP

2.4. Biological treatment of the waters associated with light crude oil production

The waters associated with the production of light crude oil (WCPL) are biodegradable in aerobic biological treatments, in anaerobic biological treatments and in a combination of these treatments. Díaz *et al.* (2005a) report that the COD removal in SBR was 88.8%, and the removal of phenols was 96.8%.

Likewise, the WCPL showed be biodegradable in anaerobic conditions in batch and continuous systems, under mesophilic conditions (37ºC) and thermophilic conditions (55ºC). In batch systems the COD removal reached 70.7% under mesophilic conditions (Gutiérrez *et al.*, 2007), while in UASB reactors under both temperature conditions, the efficiency of COD removal reached over 75%.

In UASB reactors the HRT influenced in the COD removal; so, Rincón *et al.* (2002) reported that under mesophilic conditions the optimal HRT was between 15 and 10 hours, with COD removal above 80%, but for HRT under 10 hours the system did not allow the methanogenic microorganisms to be able to transform volatile fatty acid (VFA), provoking the inhibition of the system. On the other hand, Gutiérrez *et al.* (2006) indicated that for the same temperature conditions, the HRT optimal was 18 hours with COD removal of 80%; they indicate also that for thermophilic conditions, the optimal HRT was 18 hours with COD removal of 84%, maintaining good COD efficient removal for HRT of 6 hours (67%).

When the efficiency of COD removal of WCPL in UASB reactors under mesophilic and thermophilic conditions were compared, major percentages of COD removal under thermophilic conditions for HRT under at 15 hours were observed. This removal of COD can be associated to high temperature accelerate the enzymatic biological systems. Nevertheless, there were not significant differences ($p>0.05$) between the values obtained for mesophilic and thermophilic temperature conditions, for the HRT from 12 to 24 hours.

When combined systems were used, the COD removal of the system was higher than those obtained in each separated system (Rincón *et al.*, 2004).

The maximum COD removal reached for the systems applied to WCPL were between 67% and 95%. In this aspect, the petroleum industry has alternatives to treat the WCPL; however, the final decision will be an economic decision between the temperature, the size of the reactor and energy costs. In the case of thermophilical route, it is necessary to considerate the cost of raising the temperature of the water, because the WCPL is at atmospheric conditions. For aerobic processes, the costs of the energy associated must be considered.

2.5. Biological treatment of the waters associated with medium crude oil production

It is observed in the Table 9 and Table 10 that the WCPM presented lower biodegradability than WCPL, for both aerobic and anaerobic systems. In discontinuous batch aerobic systems, Castro *et al.* (2008) report that the COD removal was between 62.4% and 89.8%; while for SBR reactors was between 60.9% and 65.2% (Díaz *et al.*, 2005 a; Díaz *et al.*, 2005b). On the other hand, in batch anaerobic reactors under thermophilic conditions, the COD removal was between 59.9% and 69.2% (Gutiérrez *et al.* 2007; Gutiérrez *et al.*, 2009). In UASB reactors, the COD removal was between 23.5% (Díaz *et al.*,2005a) and 26% (Rincón *et al.*, 2008).

In the different treatment systems it is observed that the COD removal for WCPM was between 23.5% and 89.8%.

2.6. Biological treatment of the waters associated with heavy crude oil production

In the case of the waters associated with heavy crude oil production (WCPH), the behavior was similar to WCPM. In SBR systems the COD removal was between 62.9% (Díaz *et al.*, 2005a) and 66% (González *et al.*, 2007). While in anaerobic batch reactor systems under thermophilic conditions, the COD removal was between 50.4% and 62.1% (Gutiérrez *et al.*, 2007; Gutiérrez *et al.*, 2009). On the other hand, in UASB reactors under mesophilic conditions, the COD removals were lower than 40% (Díaz *et al.*, 2005a; Rincón *et al.*, 2008).

2.7. Biological treatment of the combination of waters associated with crude oil production

The WCPC represent the combination of the waters in contact with different fractions of crude oil, whether produced in plant or by the researchers. The biodegradability of these waters has been studied in RBC and combined systems UASB-SBR (Table 8).

Behling *et al.* (2003) commented that the COD removal in RBC system used to treat WCPC was 76.1%, while Rincón *et al.* (2004) studied a UASB-SBR system and reported that the COD removal reached 79%, indicating that was important removals of phenols and hydrocarbons were obtained.

3. Discussion

Comparing the biodegradability of WCPL, WCPM and WCPH, it is observed that the WCPL present the major biodegradability in the different treatment systems and operating conditions studied.

The biodegradability of the WCP has been associated to diverse factors as SARA composition, phenols concentration, alkalinity, organic load, metals concentration and temperature.

Some researchers (Rincón *et al.*, 2002; Gutiérrez *et al.*, 2006; Gutiérrez *et al.*, 20007) argue that the WCPL biodegradability is good in anaerobic systems under mesophilic conditions and under thermophilic conditions. The final decision between the temperature used and size of the reactor will be economical, because the WCPL are at atmospheric temperature and the termophilic route implicates to consider the costs associated of warming the water. In the cases of WCPM and WCPH the studies realized up to the moment are not conclusive.

Other researchers (Gutiérrez *et al.*, 2007; Gutiérrez *et al.*, 2001; Gutiérrez *et al.*, 2012) share that the biodegradability of WCP is associated to the SARA composition present in these waters, as product of the contact with the crude oil associated. The difference of composition of these fractions confer characteristics that influence in their biodegradability, because the SARA fractions change in relation to the crude oil that is in contact with the WCP, being the WCPL the waters with the biggest percentages of saturated, considered more biodegradable than WCPM and WCPH.

When the organic fractions present in the WCP are compared, it is observed that WCPM and WCPH present a similar content of organic fractions ($p>0.05$). The opposite case was observed with the WCPL, which organic fractions are different in saturated, aromatics and resins, in comparing to WCPM and WCPH ($p>0.05$).

On the other hand, there is a tendency to increase the saturated fractions in WCP ($r=0.871$) with the increase of the API gravity of the crude oil with the water associated, following the order WCPL>WCPM>WCPH. In relation to the resins, it was observed that it increases with regard to the decrease of the API gravity of the crude oil with the WCP were associated following the order WCPL<WCPM<WCPH.

A study realized by Díaz *et al.* (2007) with WCPM from other tank farm of the Venezuelan petroleum industry, indicated that the SARA fractions can be removed from the WCP using UASB reactors. They obtained removals of 72% of saturated, 91% of resins and 71% of asphaltenes, and did not obtain removals of aromatics. They associated these results with the increases of the aromatic fractions for degradation of the fractions like resins and asphaltenes to aromatics.

Also the researchers have presented biodegradability percentages of different types of WCP under anaerobic conditions. They report values for mesophilic and thermophilic anaerobic systems of 80% and 78%, 45% and 86%, and 20% and 0%, for WCPL, WCPM and WCPP respectively in batch reactors (Gutiérrez and Caldera, 2011; Gutiérrez *et al.*, 2007; Rincón *et al.*, 2006).

In regard to phenols concentration, the studies mention that the consortium of microorganisms developed in mesophilic UASB reactors were influenced by the initial phenols concentrations, indicating that the phenols removal might be associated with the presence of different phenols compounds in the different types of WCP, with varied resistance to degradation and metabolism (aerobic/anaerobic).

Additionally, the studies indicate that the alkalinity values in the WCP were between 900 and 3000 mg $CaCO_3$/L. It has been commented that the difference of COD removal might be due to the acidity-basicity conditions in the WCP. The WCPH presented lower values of alkalinity (642.9-580.4 mg $CaCO_3$/L) and lower COD removal than WCPM. The alkalinity of WCPM was superior to 2000 mg $CaCO_3$/L. As for the pH, the WCP presented basic pH (7-10) for the different treatments.

In other cases, it is mentioned that the presence of metals in the WCP makes the treatment more complex. However, the metals K, Na, Fe, Cr, Pb and Zn can be used by thermophilic microorganisms or can be removed from the WCP and reach to the sludge by diverse mechanisms (Gutiérrez *et al.*, 2009).

In relation to degrading microorganisms present in the WCP, some have been isolated, and identified the genus *Aeromonas, Klebsielle, Xanthomona, Bacteroides* and *Acinetobacter*, as well as a consortium of them, that resulted to be effective in COD decrease (Castro *et al.*, 2008).

The Table 7 shows that WCP has been treated by coagulation-flocculation at laboratory level using chitosane as a coagulating agent in concentrations of 24 to 38 mg/L of solution of commercial chitosane (CCH), and by dissolved air flotation (DAF) using a cationic flocculants of high molecular weight.

Rojas *et al.* (2008) reported that the TSS removal and the turbidity in the WCPC were 77% and 69% respectively. On the other hand, Caldera *et al.* (2009, 2011) commented that the turbidity removal in the WCPH was 90.7%, accompanied of COD removal of 50.7%. In any case, the hydrocarbons removal and oils removal by physicochemical methods were between 70% and 90%, concluding that the cationic polymers represent an alternative to remove oily compounds in the WCP.

Table 8 shows other alternatives applied to treat WCP. In constructed sub-superficial wetlands COD removal of WCPC was between 31.4% and 65.7%, while in constructed superficial wetlands there was no COD removal. Both systems showed efficiency to remove more than 60% of the hydrocarbons present in the WCPC (Paz *et al.*, 2012; Blanco *et al.*, 2008).

The application of ozone also has been proposed to increase the biodegradability of the WCP. According to Gutiérrez *et al.* (2002), the application of ozone improves considerably the biodegradability of the WCP, with an increase of up to 87%. They concluded that the application of doses of ozone to WCP in the order of 30 mg/L of ozone, would affect favorably in the later biological processes applied.

4. Conclusions

The WCP from the different cuts: light (WCPL), medium (WCPM), heavy (WCPH) and combinations of them (WCPC), have different characteristics and their biodegradability or treatment are associated on the SARA compositions, organic matters concentration, hydrocarbons and phenols concentrations, and the operation conditions (HRT and temperature).

The biodegradability of the WCP followed the order WCPL>WCPM>WCPH.

The COD removal in biological systems changed between 67%-95%, 23.5%-89.8% and 35%-66% for WCPL, WCPM and WCPH, respectively.

The physicochemical treatment DAF and coagulation, removed hydrocarbons and oils between 70% and 90%.

Other parameters like phenols, hydrocarbons and SARA fractions, can be removed from the WCP by biological treatments.

It is necessary to analyze other parameters and operating conditions, as well as to conduct an economic evaluation before the treatment selection.

Author details

Edixon Gutiérrez[1] and Yaxcelys Caldera[2]

*Address all correspondence to: egutierr12@gmail.com; yaxcelysc@gmail.com

1 Centro de Investigación del Agua. Facultad de Ingeniería, Universidad del Zulia, Maracaibo, estado Zulia, Venezuela

2 Laboratorio de Investigaciones Ambientales, Núcleo Costa Oriental del Lago, Universidad del Zulia. Cabimas, estado Zulia, Venezuela

References

[1] Ahmad, A, Sumathi, S, & Hameed, B. (2006). Coagulation of residue oil and suspended solid in palm oil milk effluent by chitosan, alum and PAC. Chemical Engineering Journal, 118 (1-2), 99-105.

[2] Behling, E, Marín, J, Gutiérrez, E, & Fernández, N. (2003). Aerobic treatment of two industrial effluents using a rotating biological contactor reactor. Multiciencias, 3 (2), 126-135.

[3] Blanco, E, Gutiérrez, E, Caldera, Y, Núñez, M, & Paz, N. (2008). Tratamiento de aguas de producción a través de humedales construidos de flujo subsuperficial a escala piloto. Memorias del XXXI Congreso Interamericano de Ingeniería Sanitaria y Ambiental, AIDIS, 12 al 15 de octubre, Santiago de Chile, Chile.

[4] Caldera, Y, Rodriguez, Y, Oñate, H, Prato, J, & Gutiérrez, E. (2011). Efficiency of chitosan as coagulant during treatment of low turbidity water associated crude oil production. Revista Tecnocientífica URU, 1 (1), 54-60.

[5] Caldera, Y, Clavel, N, Briceño, D, Nava, A, Gutiérrez, E, & Mármol, Z. (2009). Chitosan as a coagulant during treatment of water from crude oil production. Boletín del Centro de Investigaciones Biológicas, 43 (4), 541-555.

[6] Caldera, Y, Gutiérrez, E, Madueño, P, Griborio, A, & Fernández, N. (2007). Anaerobic biodegradability of industrial effluents in an UASB reactor. Impacto Científico, 2 (1), 11-23.

[7] Castro, F, Fernández, N, & Chávez, M. (2008). Diminution of the COD in formation waters using bacterial stocks. Rev. Téc. Ing. Univ. Zulia, 31 (3), 256-265.

[8] Díaz, A, Rincón, N, López, F, Fernández, N, Chacín, E, & Debellefontaine, H. (2005b). The biological treatment in sequencing batch reactor (SBR) of effluents from the extraction of medium oil production. Multiciencias, 5 (2), 150-156.

[9] Díaz, A, Rincón, N, Marín, J, Behling, E, Chacín, E, & Fernández, N. (2005a). Degradation of total phenols during biological treatment of oilfields produced water. Ciencia, 13 (3), 281-291.

[10] Díaz, V, Bauza, R, Cepeda, N, Behling, E, Díaz, A, Fernández, N, & Rincón, N. (2007). Development of micro-SARA method for organic fractions of crude oil determination on the oil extraction production waters with anaerobic treatment. Ciencia, 15, (1), 95-104.

[11] Gaceta Oficial 5021 de la República de Venezuela ((1995). Caracas 18 de diciembre de 1995. 5021, Extraordinaria. Decreto 883. Normas para la clasificación y el control de las aguas de los cuerpos de agua y vertidos o efluentes líquidos.

[12] García, A, Arreguín, F, Hernández, S, & Lluch, D. (2004). Impacto ecológico de la industria petrolera en sonda de Campeche, México, tras tres décadas de actividad: Una revisión. Interciencia, 29(6), 311-319.

[13] González, Y, Rincón, N, López, F, & Díaz, A. (2007). Organic matter removal from the petroleum effluents by a sequencing batch reactor (SBR). Rev. Téc. Ing. Univ. Zulia, 30 (Edición Especial), 82-89.

[14] Guerrero, C, Escobar, S, & Ramírez, D. (2005). Manejo de la salinidad en aguas asociadas de producción de la industria petrolera. Investigación e Ingeniería, 25 (3), 27-33.

[15] Gutiérrez, E, & Caldera, Y. (2011). Tratamiento de aguas provenientes de la extracción de crudo en sistemas biológicos anaerobios. Intellectus, 1 (1), 11-21.

[16] Gutiérrez, E, Caldera, E, Ruesga, L, Villegas, C, Gutiérrez, R, Paz, N, Blanco, N, & Mármol, Z. (2012). Organic fractions in water from crude oil production. Revista Tecnocientífica URU, 2 (2), 31-37.

[17] Gutiérrez, E, Caldera, Y, Contreras, K, Blanco, E, & Paz, N. (2006). Anaerobic mesophilic and thermophilic degradation of waters from light crude oil production. Boletín del Centro de Investigaciones Biológicas, 40 (3), 242-256.

[18] Gutiérrez, E, Caldera, Y, Fernández, N, Blanco, E, Paz, N, & Mármol, Z. (2007). Thermophilic anaerobic biodegradability of water from crude oil production in batch reactors. Rev. Téc. Ing. Univ. Zulia, 30 (2), 111- 117.

[19] Gutiérrez, E, Caldera, Y, Perez, F, Blanco, E, & Paz, N. (2009). Behavior of metals in water from crude oil production during thermophilic anaerobic treatment. Boletín del Centro de Investigaciones Biológicas, 43 (1), 145-160.

[20] Gutiérrez, E, Fernández, N, Herrera, L, Sepúlveda, J, & Mármol, Z. (2002). The effect of ozone applications on biodegradability in water used for oil production. Multiciencias, 2 (1), 50-54

[21] Lepisto, R, & Rintala, J. (1999). Extreme thermophilic (70ºC), VAF-FED UASB reactor; performance, temperature response, load potential and comparation with 35 and 55ºC UASB reactor. Water Research, 33 (14), 3162-3170.

[22] Lettinga, G. (2005). A good life environment for all through conceptual, technological and social innovations. Memorias del VIII Taller y Simposio Latinoamericano sobre Digestión Anaerobia. Punta del Este, Uruguay, 1-15.

[23] Lettinga, G. (2001). Digestion and degradation, air for life. Water Science Tech., 44 (8), 157-176.

[24] Li, Q, Kang, C, & Zhang, C. (2005). Wastewater produced from an oilfield and continuous treatment with an oil-degrading bacterium. Process Biochemistry, 40 (2), 873-877.

[25] Paz, N, Blanco, E, Gutiérrez, E, Núñez, M, & Caldera, Y. (2012). Pilot scale superficial flow constructed wetlands for sulfide and phenols removals from oil field producer water. Rev. Téc. Ing. Univ. Zulia, 35 (1), 71-79.

[26] Renault, F, Sancey, B, Bodot, P, & Crini, G. (2009). Chitosan for coagulation/flocculation process- An eco-friendly approach. European Polymer Journal, 45 (5), 1337-1348.

[27] Rincón, N, Behling, E, & Cepeda, N. (2004). Combinación de tratamientos anaerobio-aerobio de aguas provenientes de la industria petrolera venezolana. Memorias del

XXXI Congreso Interamericano de Ingeniería Sanitaria y Ambiental, AIDIS, 22 al 27 de agosto, San Juan, Puerto Rico.

[28] Rincón, N, Cepeda, N, Díaz, A, Behling, E, Marín, J, & Bauza, R. (2008). Behavior of organic fraction in water separated from extrated crude oil with anaerobic digestion. Rev. Téc. Ing. Univ. Zulia, 31 (2), 169-176.

[29] Rincón, N, Chacín, E, Marín, J, Moscoso, L, Fernández, L, Torrijos, M, Moletta, R, & Fernández, N. (2002). Optimum time of hydraulic retention for the anaerobic treatment of light oil production wastewater. Rev. Téc. Ing. Univ. Zulia, 25 (2), 90-99.

[30] Rincón, N, Torrijos, M, Colina, G, Behling, E, Chacín, E, Marín, J, & Fernández, N. (2006). Method to determine anaerobic biodegradability of oil production waters. Impacto Científico, 1 (1), 9-20.

[31] Rojas, C, Rincón, N, Díaz, A, Colina, G, Behling, E, Chacín, E, & Fernández, N. (2008). Evaluation of dissolved air flotation unit for oil produced water. Rev. Téc. Ing. Univ. Zulia, 31 (1), 50-57.

Chapter 2

Emulsification of Hydrocarbons Using Biosurfactant Producing Strains Isolated from Contaminated Soil in Puebla, Mexico

Beatriz Pérez-Armendáriz,
Amparo Mauricio-Gutiérrez,
Teresita Jiménez-Salgado,
Armando Tapia-Hernández and
Angélica Santiesteban-López

Additional information is available at the end of the chapter

http://dx.doi.org/10.5772/56143

1. Introduction

Among Mexico's main riches are its oil and the great expanses of land used to grow food. A large number of pipelines pass through Mexico's agricultural region carrying diesel, gasoline or crude oil, however, lack of maintenance of the pipeline installations, fuel theft, vehicle transport and even the topographical, terrain and hydrological conditions of the site cause a high incidence of contamination.

Petrolic activities have generated extensive pollution of soils worldwide, mainly in those regions where petroleum is explored, extracted, and refined. The composition of hydrocarbons on polluted soil varies according to environmental conditions and natural degradation processes. In México there are soil impacted by weathered hydrocarbons, which are predominantly saturated and aromatic, become more recalcitrant if polluted soils are not remediated, affecting the underground water, food chains, and diverse human activities.

Hydrocarbon spills on agricultural soil have direct repercussions on soil quality and its function. Some authors [1] indicate that hydrocarbon contamination reduces food crop growth by preventing water and nutrient absorption through the roots, and reducing the transport of metabolites and respiration rate.

The recovery of hydrocarbon-contaminated agricultural soil in Mexico is a complex theme because the producers harvest the crops for sustenance or sale. A remedy is therefore needed that uses sustainable biological technologies which do not pose a risk for the products of the harvests. The production of biosurfactants to recover agricultural soil used for food production is a viable alternative because of their biodegradability. Furthermore, biosurfactants have been used in the oil industry to recover oils from hydrocarbons, in the emulsification of heavy hydrocarbon fractions and in the degradation of polychlorinated biphenyls [2] and polycyclic aromatic hydrocarbons (PAH's) [3].

2. Approach to the problem

In the agricultural fields of Puebla, Mexico two hydrocarbon spills have been reported due to lack of pipeline maintenance. In 2002, a crude oil spill in the town of Acatzingo, Puebla affected a large expanse of agricultural land (approximately 50 hectares) [4]. And in San Martin Texmelucan, Puebla on December 19, 2010, the explosion caused by a crude oil spill took 30 human lives and greatly affected the agricultural land of the population [5]. The inhabitants of the affected regions still perceive damage to the soil and do not consider the land to be fully recovered [4].

In Mexico, the environmental impact of oil industry activities is rigorously controlled by the authorities (Federal Environmental Protection Agency, *Procuraduría Federal de Protección al Medio Ambiente*, PROFEPA) and therefore recuperation should take only a short time. Bioremediation processes have not given the expected results: expanses of contaminated land are heterogeneous as far as climate, water availability and oxygen availability, and the biostimulation of microbial populations is insufficient due to competing autochthonous microorganisms and inadequate nutritional balance [6, 7].

Mexico relies mainly on micro-encapsulation technology for the restoration of hydrocarbon-contaminated land, according to the National Ecology Institute (*Instituto Nacional de Ecologia*, INE) [6] using chemical substances which encapsulate hydrocarbons and prevent biodegradation. Surfactants have also been used to restore marine sediment with a recovery of 45,000 t [8]. Chemical surfactants, however, are not always environmentally biodegradable [9] and so there is a need to use biosurfactants to recover oil hydrocarbons in impacted soils.

3. Area of application

Biosurfactants are molecules with a polar region and a non-polar region, and are hence considered amphipathic, produced by extracellular or intracellular microorganisms, also can reduce surface tension at the air-water interface between two immiscible liquids or between the solid-water interface [10].

Biosurfactants have other advantages over chemical detergents since they are non-toxic and ecologically acceptable [10]. They are also highly effective at breaking down surface tension

[11]. Several authors have reported bacterial strains isolated from hydrocarbon-contaminated soil and water which present emulsifying activity and which are capable of growing in oil using it as sole carbon source. The reported microorganisms are: *Pseudomonas aeruginosa, P. mendocina, P. aureofasciens, Listonella damsela, Bacillus sphaericus, B. brevis, Enterobacter cloacae, Acinetobacter calcoaceticus* var. *anitratus, Hafnia alvei, Citrobacter freundii, C. amalonaticus, Sphingobacterium multivorum, Staphylococcus* sp, *Neisseria* sp, *Micrococcus* sp, *Serratia rubidae, Alcaligenes, Flavobacterium, Nocardia, Achromobacter, Arthrobacter* [12-16]. There has been a recent rise in the study of biosurfactant for their antimicrobial characteristics as fungicide [17, 18] and as, zoospore inhibitors [19].

The use of biosurfactants for the bioremediation of hydrocarbon contaminated soil has been studied intensely since the last decade [2-3, 20]. Biosurfactants have been used by the oil industry to enhanced oil recovery [21, 22], in the emulsification of heavy hydrocarbon fractions [23], and in the treatment of wastewater with insoluble substances. They have also been used in the degradation of polychlorinated biphenyls [2]. Chemical surfactants have the advantage of being non-toxic, environmentally friendly, and biodegradable and can be produced from agricultural substrates [10].

Biosurfactants can be used as additives to stimulate bioremediation; however, the concentration of these can also be increased by the addition of bioemulsifier-producing bacteria. Bioemulsifier-producing bacteria can participate in the biodegradation of hydrocarbons and, alternatively, function as a family of bacteria that supply emulsifiers to another group of bacteria that degrade the contaminants [24].

A mixture of biosurfactants including cellular lipids produced during the degradation of heavy hydrocarbons, and additives increases solubility and facilitates hydrocarbon degradation. Cellular lipids have excellent surfactant properties and can form micelles at low concentrations, but these surfactants do not release the solubilized organic compounds to degrade them [25]. An increase in the apparent solubility of naphthalene has been observed when the concentration of glycolipids excreted by *Pseudomonas areuginosa* 19SJ exceeds the critical micellar concentration (CMC) [26].

Biosurfactants have different chemical compositions depending on the microorganism that produces them and may be lipopeptides, lipoproteins, fatty acids or phospholipids [27]. The production of biosurfactants depends on physicochemical factors (aeration, pH, substrate availability) and their evaluation will depend on kinetic factors (substrate consumption, product formation, and biomass production). Knowing the kinetics of biosurfactant production will allow the proposal of sustainable oil hydrocarbon recovery technologies for aqueous or solid systems.

Mexico has large areas of soil contaminated by oil activities; especially agricultural soils have few alternatives of sustainable technologies, therefore in this work different microorganisms were isolated from hydrocarbons-contaminated soil and the kinetics of biosurfactant production was studied to generate a proposal for the recovery of oil hydrocarbons as Maya crude oil.

4. Materials and methods

4.1. Isolation of biosurfactant-producing strains

Soil sampling was done in an agricultural area of Acatzingo, Puebla, Mexico with the following geographical coordinates 18° 57' 03.0" N 97° 46' 20.5" W. Biosurfactant-producing strains were isolated using 1 g of soil in 10 mL of pre-sterilized distilled water. The culture medium was composed of (g / L): $(NH_4)_2SO_4$ 7.7, KH_2PO_4 5.7, K_2HPO_4 2, $MgSO_47H_2O$ 2, $CaCl_22H_2O$ 0.005, $FeCl_36H_2O$ 0.0025, agar 15; distilled water 1,000 mL and preadapted to a petroleum environment using the Maya petroleum provided by the Mexican State company (PEMEX). Maya petroleum was added on sterilized filter paper (3 cm^2; with 2 g petroleum) to every lid in order to develop an atmosphere of volatile hydrocarbons inside the petri dish.

The bacteria were then isolated and grown in a liquid mineral medium (g / L): $(NH_4)_2SO_4$ 7, KH_2PO_4 5.7, K_2HPO_4 2, $MgSO_47H_2O$ 2, $CaCl_22H_2O$ 0.005, $FeCl_36H_2O$ 0.0025, Yeast extract 0.1, glucose 20. Strains presenting biosurfactant production were identified as UPAEP 6, UPAEP 8, UPAEP 9, UPAEP 10, UPAEP 12 and UPAEP 15. The following bacteria were also bought *Arthrobacter* sp ATCC 31012, *Bacillus subtilis* ATCC 21332, *Candida petrophilum* ATCC 20226.

4.2. Strain selection

The selected strains were grown in 50 mL of Lebac medium (g / L): $(NH_4)_2SO_4$ 7, KH_2PO_4 5.7, K_2HPO_4 2, $MgSO_47H_2O$ 2, $CaCl_22H_2O$ 0.005, $FeCl_36H_2O$ 0.0025, Yeast extract 0.1, glucose 20, pH 7.0; in 200 mL Erlenmeyer flasks with a 200 µL aliquot of microorganisms. Twenty-four flasks of each strain were placed in an incubator (FELISA) at 37 °C under constant agitation at 200 rpm. Three flasks were removed at each interval over a 44 and 48 h kinetic.

The parameters evaluated over time were: biomass production, pH, emulsification activity on engine oil and glucose consumption.

Biomass production was determined by taking 2 mL of culture medium and passing it through a pre-dried and pre-weighed cellulose nitrate membrane filter (0.22 µm in diameter). The filter with the biomass was then dried at 100°C for 24 h until constant weight was attained; the biomass was reported in g obtained by weight difference.

4.3. Emulsification index

Emulsification activity was determined by placing 6 mL of engine oil and 4 mL of culture medium with the biosurfactant-producing strains in a vortex [28]. They were agitated for 2 minutes and left to rest for 24 h. The percentage of emulsification was estimated according the following expression:

% Emulsifier = ((Total height of the mixture - Height of emulsified oil) / Total height of the mixture) * 100

http://dx.doi.org/10.5772/56143

4.4. Glucose consumption, pH and Critical Micelle Concentration (CMC)

The glucose was determined by the AOAC 969.39 method taking a 2 mL aliquot of culture medium. If necessary it was diluted with distilled water.

The pH was determined with a potentiometer (Conductronic pH 10). In this investigation, pH was maintained close to neutrality by adding 0.1N NaOH.

The Critical Micelle Concentration (CMC) was determined according to [29].

4.5. Statistical analysis

The results were adjusted to a linear model to obtain the rate of substrate consumption (g glucose h^{-1}), the rate of biomass production (g biomass h^{-1}) and emulsification activity (% emulsifier h^{-1}). The slopes (rates) and correlation coefficients were obtained from regression linear model.

In addition, the average initial and final samples of emulsification activity were analyzed by variant analysis to find significant differences and Duncan-Waller multiple comparison tests. The statistical package used was Minitab version 13 (licensed to UPAEP, Mexico).

4.6. Biodegradation tests of maya crude oil

A preculture of selected strains was grown in Banat broth at 30 °C under constant agitation (200 rpm) for 24 h. An aliquot of the selected strains was taken at an absorbance of 70 UK, inoculated in flasks with 50 mL of medium at a pH of 6.5 with 20,000 ppm of crude oil and incubated at 30 °C for 15 days. Following the incubation process, the samples were put in contact with HPLC grade hexane and agitated for 2 minutes. The mixture was then sonicated (Branson 1210 Ultrasonic Cleaner) for 10 minutes before being transferred to a 250 mL separatory funnel leaving the aqueous phase to decant for later use (Figure 1A). The organic phase, in which the hydrocarbons are found, was recovered by means of an asbestos filter and Na_2SO_4 anhydrous as a desiccant in a 50 mL balloon flask. The organic phase was then distilled using a Büchi Rotavapor R11 with operating temperature of 45 °C (Figure 1B).

4.7. Viability of microorganisms

In addition to the hydrocarbon degradation capacity, the viability of the strains was determined at 8, 16 and 24 days of incubation. The organic phase was therefore eliminated by centrifugation (3000 rpm for 5 minutes) and successive serial dilutions made of 10^{-6} and cultivated on plates of Lebac medium. Isolates strains were grown overnight in Lebac broth at 37 °C under constant agitation at 200 rpm. The biochemical characterization was carried out by the API 20 E, API 20 NE and API 50 CH systems (references No. 20160, 20050 and 50300; bioMérieux) following the manufacturer's recommendations. The identification was assessed by APIweb™ identification software (bioMerieux).

4.8. Biosurfactant recovery

The purification of biosurfactant was performed according to a modified technique described in [30]. With the strains with highest percentage of emulsifier. The strains were previously grown in 500 mL of Lebac medium. The biosurfactant was then extracted from the bacteria with isopropanol-ethanol (3:1) analytical grade (Merck, México) in a separatory flask. It was centrifuged at 1200 rpm for 30 minutes (Solbat), and the supernatant was eliminated. The sample was then filtered using cellulose paper grade 101 (Millipore 2.5μ M). The precipitate obtained was dried for 24 h at 60 °C in an oven (FELISA) and stored in an Eppendorf vial to determine the yield.

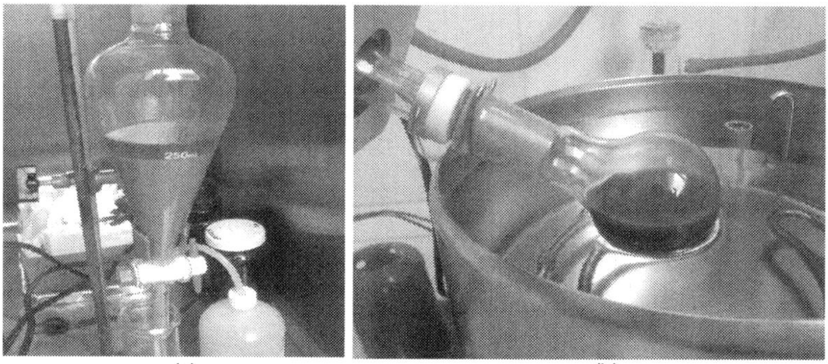

(a) (b)

Figure 1. a) Emulsification of hydrocarbons. (b) Oil recovery.

5. Results

5.1. Presumptive identification of isolated microorganisms

Six microorganisms were isolated and identified according to their morphology. Table 1 shows the results of the presumptive tests for the identification of bacteria and yeasts by API galleries. The strains UPAEP 8 and UPAEP 15 were related to *Klebsiella pneumoniae* (99 and 97.6 % likelihood respectively). UPAEP 6 strain was closely related to *Klebsiella ornithinolytica* (99 %) and UPAEP 9 strain to *Klebsiella* sp (97 %). Whereas UPAEP 10 strain showed high likelihood (99 %) to *Serratia marcescens* and UPAEP 12 strain to *Candida inconspicua* (75 %).

5.2. Glucose consumption and biomass production

The kinetic characteristics of the bacteria showed similar behavior regarding rapid growth, good adaptation to hydrocarbons and rapid glucose consumption.

All strains consumed glucose in a range of 92 to 100 %. However, the glucose consumption percentage of the commercial strains was lower than the isolates studied; with the exception

http://dx.doi.org/10.5772/56143

of *Bacillus subtilis* ATCC 21332 which consumed 93.6 % of glucose (Table 2). Nevertheless, the glucose consumption was inversely proportional to the biomass production during the cell growth (data not shown). The emulsification index was directly proportional at production of biomass, except *Candida petrophilum* ATCC 20226 which showed no relation. Biosurfactant synthesis and biomass production by UPAEP 6, 9, 10, and 15 (Figures 2, 4,5 and 10) strains began during the first few hours (4 to 8) as a response to substrate consumption; UPAEP 8, 12 (Figure 3 and 6) and *Arthrobacter* sp ATCC 31012 (figure 8) strains began at 20, 28 and 50 h. In contrast, *Bacillus subtilis* ATCC 21332 strain biosurfactant production occurred at the end of microbial growth (after of 76 h).

UPAEP 6 strain showed the highest increase in biomass and biosurfactant production at 24 h. Maximum biomass production occurs at 44 h and with a maximum value of 5.3 g L^{-1}. The maximum value of the biosurfactant production (80 %) at 40 h was high considering that crude oil is heavy with a density of 0.92-1.01 g mL^{-1} and an API gravity of 10.1-22.3 and viscosity can reach 10,000 cP [31] (Figure 2).

On the other hand, UPAEP 9, UPAEP 10 and UPAEP 15 strains (Figures 4, 5 and 7) showed maximum biomass production at 28, 24, and 77 h with values of 3.6, 5.3 and 9.5 g L^{-1} respectively. Biosurfactant production started from the first couple of hours and up to 49 h by UPAEP 9 and UPAEP 15 strains reached emulsification of 58 and 69 %; and at 20 h UPAEP 10 strain showed 70 % of biosurfactant production.

UPAEP 8 (Figure 3), UPAEP 12 (Figure 6) and *Arthrobacter* sp ATCC 31012 (Figure 8) strains showed slow biosurfactant production in contrast with the isolated strains. Biosurfactant production started only at 20, 28 and 50 h, and reached a maximum value of 65, 37 and 30 % respectively (at 40, 49 and 72 h). *Arthrobacter* sp ATCC 31012 showed slow growth, the highest biomass production was of 8.5 g L^{-1} at 55 h. Anyhow, UPAEP 12 strain showed a highest increase in biomass production between 28 and 46 h with a final value of 7 g L^{-1} (77 h) and UPAEP 8 strain showed maximum biomass production of 6.6 g L^{-1} at 24 h.

However, *Bacillus subtilis* ATCC 21332 strain (Figure 9) showed maximum biomass production in the first 10 h with 4.6 g L^{-1}. Maximum biosurfactant production (27 %) is observed at the end of the kinetic (70 h).

The *Candida petrophilum* ATCC 20226 strain (Figure 10) showed an important decrease in glucose up to 70 h (data not shown). Biosurfactant production began at 20 h. No relation to substrate consumption or to biomass production was observed. The maximum emulsification percentage obtained was 80 % after 70 h.

The initial pH of the culture medium was 7.0 and lowers during the cellular growth of the studied isolates, therefore was adjusted with NaOH 0.1N to obtain a pH closer to neutrality (data not shown). Thus, the final pH of the culture medium ranged from 6.07 to 7.37 (Table 2). It is interesting to observe, that the drop in pH occurred just before the biosurfactant synthesis, possibly due to a prior synthesis of organic acids as precursors of biosurfactants by UPAEP 6, UPAEP 8, UPAEP 9, UPAEP 10 and UPAEP 15 strains. Yet, the pH was maintained between 6.5 and 6 with few changes during the entire kinetic by UPAEP 12 strain, and *Arthrobacter* sp ATCC 31012 showed only a small drop at 49 h. *Bacillus subtilis* ATCC 21332 and

Candida petrophilum ATCC 20226 strains remained the pH close to neutrality during the entire kinetic.

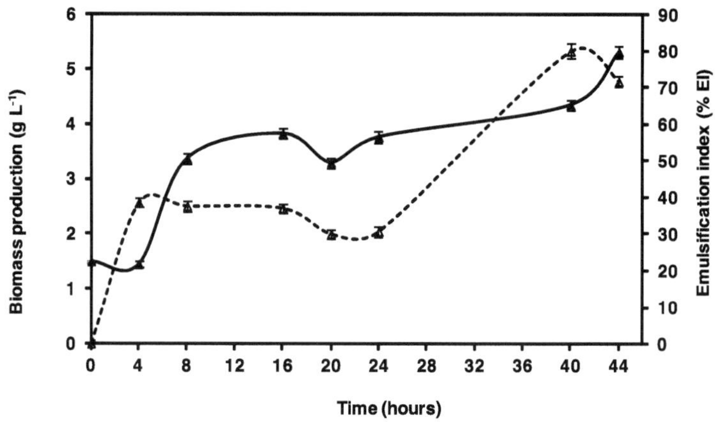

Figure 2. Bacterial growth by bacteria strain UPAEP 6 associated to biomass production (▲), and Emulsification Index EI (%) (Δ). Results are the averages of triplicate experiments ± standard deviation.

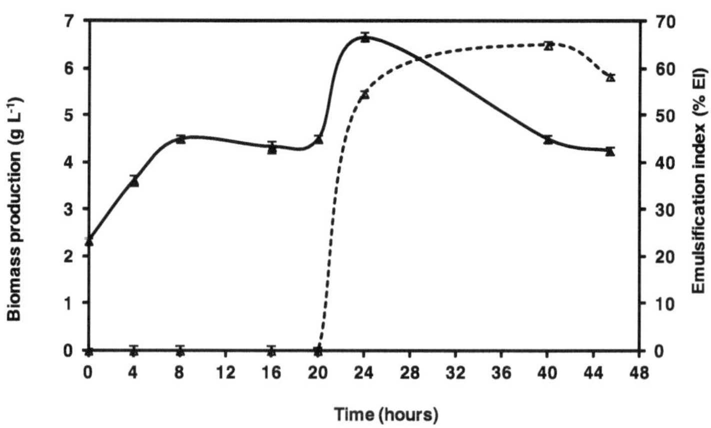

Figure 3. Bacterial growth by bacteria strain UPAEP 8 associated to biomass production (▲), and Emulsification Index EI (%) (Δ). Results are the averages of triplicate experiments ± standard deviation.

http://dx.doi.org/10.5772/56143

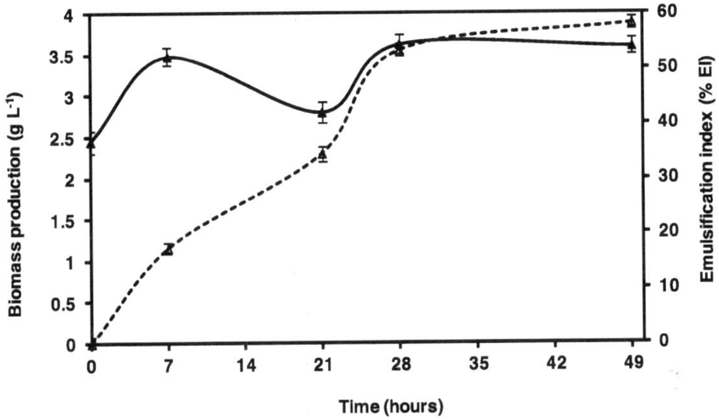

Figure 4. Bacterial growth by bacteria strain UPAEP 9 associated to biomass production (▲), and Emulsification Index EI (%) (Δ). Results are the averages of triplicate experiments ± standard deviation.

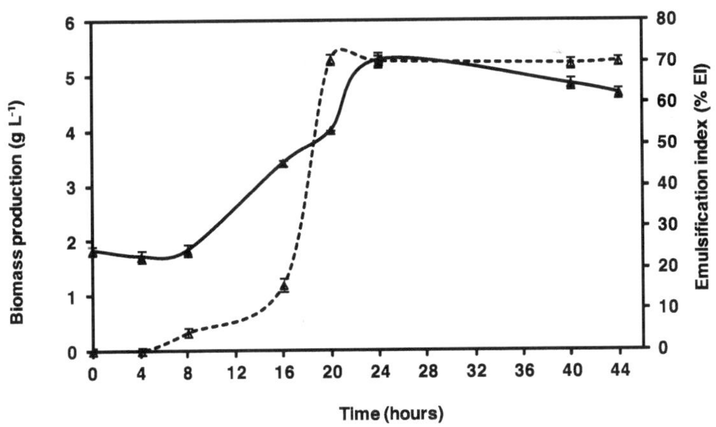

Figure 5. Bacterial growth by bacteria strain UPAEP 10 associated to biomass production (▲), and Emulsification Index EI (%) (Δ). Results are the averages of triplicate experiments ± standard deviation.

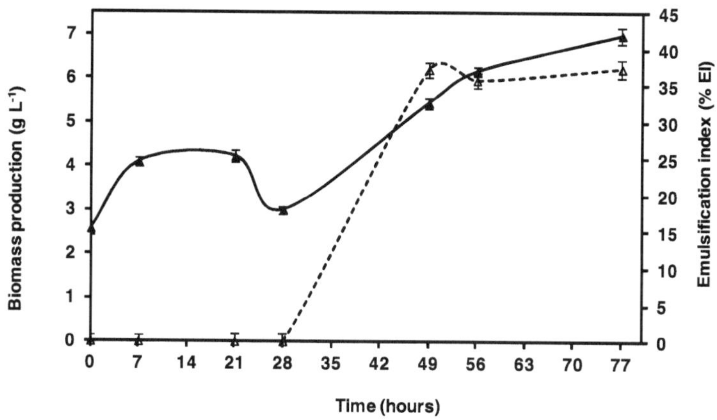

Figure 6. Bacterial growth by bacteria strain UPAEP 12 associated to biomass production (▲), and Emulsification Index EI (%) (Δ). Results are the averages of triplicate experiments ± standard deviation.

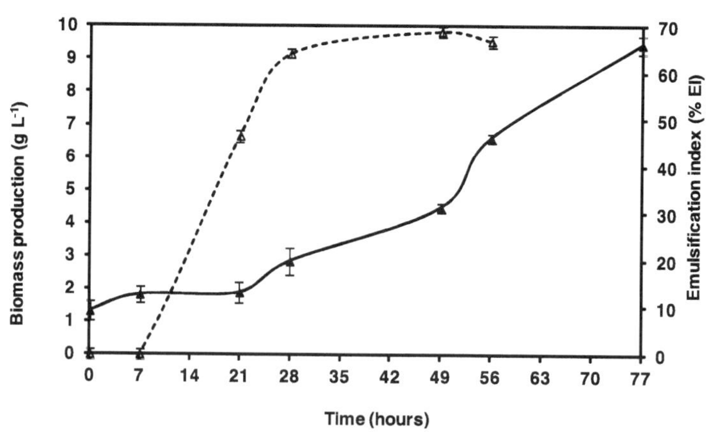

Figure 7. Bacterial growth by bacteria strain UPAEP 15 associated to biomass production (▲), and Emulsification Index EI (%) (Δ). Results are the averages of triplicate experiments ± standard deviation.

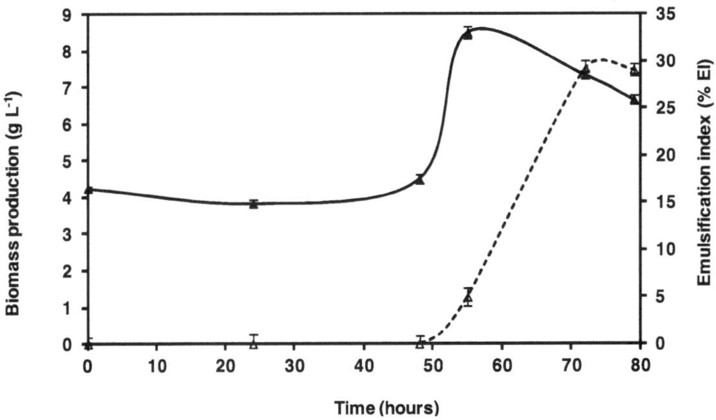

Figure 8. Bacterial growth by bacteria strain commercial *Arthrobacter* sp ATCC 31012 associated to biomass production (▲), and Emulsification Index EI (%) (Δ). Results are the averages of triplicate experiments ± standard deviation.

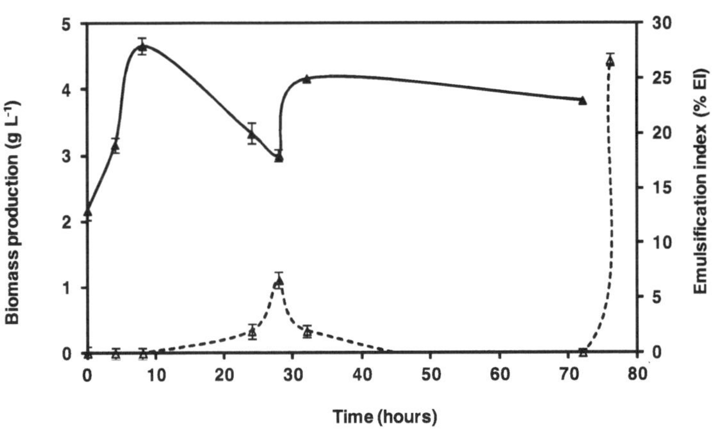

Figure 9. Bacterial growth by bacteria strain commercial *Bacillus subtilis* ATCC 21332 associated to biomass production (▲), and Emulsification Index EI (%) (Δ). Results are the averages of triplicate experiments ± standard deviation.

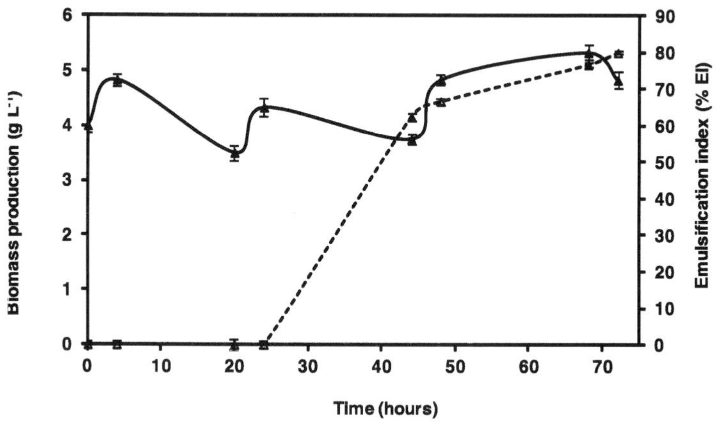

Figure 10. Bacterial growth by bacteria strain commercial *Candida petrophilum* ATCC 20226 associated to biomass production (▲), and Emulsification Index EI (%) (Δ). Results are the averages of triplicate experiments ± standard deviation.

5.3. Production rates

Table 3 shows the results of the estimated rates. The UPAEP 6 strain showed the highest biomass production rate with 0.178 g h^{-1}. The strains with best biosurfactant production rates were UPAEP 10 and UPAEP 8 with 2.5 and 2.39 % h^{-1}, respectively. Significant differences were found in the variance analysis of the emulsification final values with 70% (*Serratia marcescens*) and 80% (*Klebsiella pneumonia*). The highest rates of emulsification were for UPAEP 8 and the yeast *Candida petrophilum* ATCC 20226 (80%). CMC results of the selected strains are similar to that reported for Tergitol (0.0149 mg L^{-1}) and 10 times less than *Serratia marcescens* subsp. *marcescens*.

The capacity of these bacteria to degrade toxic compounds depends on the contact time with the compound, the environmental conditions in which they develop and their physiological versatility.

5.4. Biodegradation tests of maya crude oil

Once the strains had been evaluated, the next step was to evaluate the removal percentage of Maya crude oil (20,000 ppm) using UPAEP 8 (*Klebsiella pneumoniae*) and UPAEP 10 (*Serratia marcescens*). These two bacteria showed a greater than 80 % degradation for Maya crude oil (Figure 11,12 and 13).

http://dx.doi.org/10.5772/56143

Bacterial strain UPAEP	Classification	% likelihood
6	*Klebsiella ornithinolytica*	99
8	*Klebsiella pneumoniae*	99
9	*Klebsiella* sp	75
10	*Serratia marcescens*	99
12	*Candida inconspicua*	75
15	*Klebsiella pneumoniae*	97.6

Table 1. Identification of the bacterial strains was by the API galleries.

Strain UAPEP	Initial pH		Final pH *		% Glucose consumption **	
6	7.0	± 0.2	7.37	± 0.09	99.7	± 0.9
8	7.0	± 0.1	6.65	± 0.11	99.9	± 0.9
9	7.0	± 0.1	6.25	± 0.14	95.8	± 1.0
10	7.0	± 0.1	6.64	± 0.06	99.7	± 0.9
12	7.0	± 0.1	6.07	± 0.14	92.0	± 0.5
15	7.0	± 0.2	6.86	± 0.24	100	± 0.1
Strain ATCC						
31012	7.0	± 0.1	6.22	± 0.15	66.96	± 0.6
20226	7.0	± 0.1	6.45	± 0.12	76.48	± 0.5
21332	7.0	± 0.1	6.64	± 0.13	93.61	± 0.5

* pH values for isolates incubates in Lebac medium for 44 and 48 h at 37°C under constant agitation at 200 rpm (see Methods); each value represents the average of three replicates ± standard deviation.

* * Glucose consumption percentage is the difference between initial and final glucose concentration; each value represents the average of three replicates ± standard deviation.

Table 2. Changes of pH and Glucose consumption by Biosurfactants-producing bacterial strains during the bacterial growth.

Figure 11. Maya oil Bioemulsification. Experiment with 20,000 ppm of petroleum and biosurfactan-producing microorganisms.

Strain UAPEP	Rate Biomass production (g h^{-1})	R^2	Emulsification Activity (% h^{-1})	R^2	Rate substrate consumption (g glucose h^{-1})	R^2	Emulsification Index Final value * (%)	CMC (mg L^{-1})
6	0.178	0.76	1.72	0.51	0.86	87.1	65^{b,c}	0.0016
8	0.074	0.68	2.39	0.93	0.277	88.5	80^{a}	0.0047
9	0.018	0.80	1.13	0.86	0.336	70.0	49^{c}	0.0014
10	0.074	0.81	2.5	0.82	N.D.**	N.D	70^{b}	0.0014
12	0.05	0.72	0.01	0.41	0.218	87.0	58^{c}	0.0010
15	0.100	0.86	1.39	0.64	0.404	78.0	70^{b}	0.062
Strain ATCC								
31012	0.071	0.83	1.16	0.66	0.428	84.2	40^{c}	0.005
20226	0	0.21	1.32	0.88	0.390	97.6	80^{a}	0.005
21332	0.031	0.78	0.19	0.74	0.380	80.0	27^{d}	0.0015

* Final value Means with different letters are significantly different ($P<0.05$).

* * It was not determined.

Table 3. Biosurfactants-producing bacterial strains isolated from polluted soil with hydrocarbons.

http://dx.doi.org/10.5772/56143

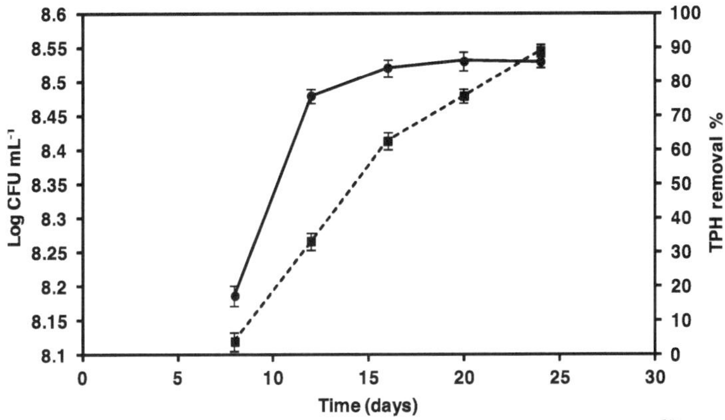

Figure 12. Removal of TPH by bacteria *Klebsiella pneumoniae* (UPAEP 8 strain) isolated from contaminated soil. Strain was grown at 30 °C, and 20000 ppm of mayan crude oil. Removal of TPH (■). Cell growth of strain with 20000 ppm of mayan crude oil (●). Results are the average of triplicate experiments ± standard deviation.

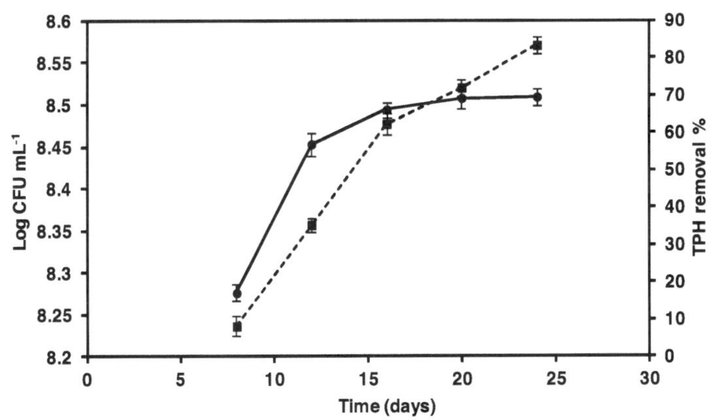

Figure 13. Removal of TPH by bacteria *Serratia marcescens* (UPAEP 10 strain) isolated from contaminated soil. Strain was grown at 30 °C, and 20000 ppm of mayan crude oil. Removal of TPH (■). Cell growth with 20000 ppm of mayan crude oil (●). Results are the average of triplicate experiments ± standard deviation

6. Discussion

Serratia genus have been reported by other authors as biosurfactants-producing bacterial capable degrader oily compounds [32, 33]. According to [34] bacteria with high capacity to produce biosurfactant promising remain very still, because many companies wish to replace chemical biological to chemical surfactants. The biosurfactant production rate for *Serratia marcescens and Klebsiella pneumonia* 2.39 and 2.5 (% h-1) respectively show the significant potential for industrialization of the strains. Biosurfactants-production remains a topic of industrial interest [35] emulsified 20% of 1500 mg / L of octadecane, while the present work with the best strains emulsified 80 and 90% of Mayan crude oil at an initial concentration of 2000 mg/L. According to [34] states that the genus *Pseudomonas* is the most promising from the industrial point of view, among other things because of the chemical nature of the rhapnolipids, in work [35] are employed *Pseudomonas aeruginosa* ATCC 9027, however the strains studied in this work were even better at the emulsification even using oil that is more complex relative to octadecano.

All the selected strains presented emulsifying activity, the majority associated with the growth of microorganisms and a decrease in pH. Some authors [19] reported that for the *Pseudomonas* species, an association has been found in the synthesis of different metabolites (fatty acids, lipopeptides, peptides and amino acids), which can be used for cellular synthesis and biosurfactant production. Although this work is focused on the degradation of recalcitrant hydrocarbons such as Maya crude oil, there is wide interest in biosurfactant production due to its applications in various fields. Other authors [32] performed a chemical and antimicrobial characterization of pseudofactin II, a biosurfactant secreted by *Pseudomonas fluorescens* BD 5 identified as a new cyclic lipopeptide with broad-spectrum bactericidal activity.

The bacteria used the Maya crude oil as sole carbon source, associated with high biomass content and a very high capacity to emulsify hydrocarbon compounds in relatively short operating times (15, 17 and 24 days) compared to those reported by other authors [36-38]. The values of the production kinetics of are very important considering of the scaling the process, *Klebsiella pneumoniae* showed up to 90 % removal and is a promising strain for future biodegradation studies.

The results will allow the use of these cultures as possible inoculants, in real bioremediation experiences where large quantities of inoculants are required. Crude oil biodegradation has been studied extensively because of the high variability of crude oil amount, incubation times and methodologies used to quantify degradation.

7. Future work

In Mexico, particularly on agricultural land, biological techniques which leave no chemical residue and with low-toxicity are required to recover impacted soil. The impact on agricultural

http://dx.doi.org/10.5772/56143

soil and its recovery for farmers is a major problem. Sustainable biological techniques may be an alternative and raise the expectations of farmers hoping to plant their crops without risk. Biosurfactants have shown their potential in bioremediation of contaminated soil and water with oil and its derivatives. Because of its low toxicity and biodegradability these are considered as an accepted alternative and environmentally friendly.

However, the *in situ* production of these compounds by microorganisms in natural environments are link to many factors including the type of contaminant, nitrogenous compounds content, interaction with native microorganisms and some others. It is important to perform tests on real soil before the scaling tests since several studies have reported inconsistent results. Therefore the use of microorganisms producing biosurfactants in bioaugmentation processes requires a careful study; new research on the scaling processes to optimize biosurfactants production must be conducted.

The rhamnolipids produced by *Pseudomonas auriginosa* biosurfactants have been extensively studied, but there are other organisms that produce substances with emulsifier, such as those produced by the serrawettin by *Serratia marcescens* this it is a bacteria which has been described as plant growth promoting rhizobacteria (PGPR), which refers to the promotion of growth when plants are inoculated, because it has the ability to produce indole-3-acetic acid (IAA). Due to the activities of the oil industry in Mexico, agricultural soils are contaminated with hydrocarbons, leading to impairment of soil properties and the consequent decline in agricultural production. Technologies should be applied for the recovery of the ground with the least environmental impact. The plant-assisted bioremediation (phytoremediation) is an alternative for the *in situ* treatment of soil contaminated with hydrocarbons. The UPAEP 10 strain of *S. marcescens* is capable of producing biosurfactants and degrades crude oil which is needed for investigating the ability of promoting plant growth in order to develop rhizoremediation technologies.

8. Conclusions

This study showed microorganism isolated of contaminated soils with high capacity of degrading recalcitrant compounds. In México there is a great need to develop clean technologies due to oil spill accidents in agricultural soils. Biosurfactant production by native strains as *Klebsiella pneumoniae* (UPAEP 8 strain) and *Serratia marcescens* (UPAEP 10 strain) showed emulsification rates of up to 80 %, and CMC values were similar than commercial detergents; therefore may be a promising way for recovery of weathered soils with heavy hydrocarbon particles. The implementation of clean technologies will allow farmers to continue producing their products of the harvests harmless and safe for sale and consumption.

Author details

Beatriz Pérez-Armendáriz[1*], Amparo Mauricio-Gutiérrez[1], Teresita Jiménez-Salgado[2], Armando Tapia-Hernández[2] and Angélica Santiesteban-López[1]

*Address all correspondence to: beatriz.perez@upaep.mx

1 Universidad Popular Autónoma del Estado de Puebla, Interdisciplinary Center for Postgraduate Studies, Research and Consulting, Santiago. CP, Puebla, Mexico

2 Benemérita Universidad Autónoma de Puebla. ICUAP, Centro de Investigaciones en Ciencias Microbiológicas, Laboratorio de Microbiología del Suelo. Edificio J 1er. Piso, C.U., Puebla, Mexico

References

[1] Xu JGRJohnson L. 1995. Root growth, microbial activity and phosphatase activity in oil contaminated, remediated and uncontaminated soil planted to barley and field pea. Plant and Soil (1995)., 173(1), 3-10.

[2] Robinson, K, Ghosh, M, & Shi, Z. Mineralization enhancement of non-aqueous phase and soil-bound PCB using biosurfactant. Water Science &Technology (1996).

[3] Deschenes, P, Lafrance, J. P, Villeneuve, R, & Samson, R. Adding sodium dodecyl sulfate and *Pseudomonas aeruginosa* UG2 biosurfactants inhibits polycyclic aromatic hydrocarbon biodegradation in a weathered creosote-contaminated soil. Applied Microbiology & Biotechnology (1996).

[4] Rivera-pineda, F, Ramírez-valverde, B, Juárez-sánchez, J. P, Pérez-armendáriz, B, Estrella-chulim, N, Escobedo-castillo, F, & Ramírez-valverde, G. Implicaciones en la agricultura por el derrame de hidrocarburos en Acatzingo, México. In Recursos naturales y contaminación ambiental. Series Ciencias Ambientales; (2012)., 203-218.

[5] PEMEX Petróleos MexicanosFuga de crudo en el oleoducto nuevo Teapa-Venta de Carpio-Tula. Boletín (2002). http://www.pemex.com/index.cfm?action=news&sectionID=8&catID=40&contentID=269accesed 12 july 2012).(19)

[6] Ortínez-brito, O, Ize-lema, I, & Gavilán-garcía, A. La restauración de suelos contaminados con hidrocarburos en México. Instituto Nacional de Ecología. http://www2.ine.gob.mx/publicaciones/gacetas/422/restauracion.htmlaccesed 14 july (2012).

[7] Pérez-armendáriz, B, Martínez-carrera, D, Calixto-mosqueda, M, Alba, J, & Rodríguez-vázquez, R. Filamentous fungi remove weathered hydrocarbons from polluted

soil of tropical México. Revista Internacional de contaminación ambiental (2010). , 26(3), 193-99.

[8] Saval, S. La biorremediación como alternativa para la limpieza de sitios contaminados con hidrocarburos. In: seminario internacional sobre restauración de sitios contaminados. Instituto Nacional de Ecología- SERMANAP, agencia de cooperación Internacional del Japón y centro nacional de investigación y Capacitación Ambiental. México (1997).

[9] Pérez-armendáriz, B, Castañeda-antonio, D, Castellanos, G, Jiménez-salgado, T, Tapia-hernández, A, & Martínez-carrera, D. Efecto del antraceno en la estimulación del maíz y el frijol. Terra Latinoamericana (2011). , 29(1), 95-102.

[10] Yu, H, & Huang, G. H. Isolation and Characterization of Biosurfactant- and Bioemulsifier-Producing Bacteria from Petroleum Contaminated Sites in Western Canada. Soil and Sediment Contamination (2011). , 20(3), 274-88.

[11] Salihu, A, Abdulkadir, I, & Almustapha, M. N. An investigation for potential development on biosurfactants. Biotechnology and Molecular Biology Reviews (2009). , 3(5), 111-17.

[12] Arenas, S. Aislamiento y Caracterización de Bacterias de Ambientes Contaminados por Petróleo en la Refinería "La Pampilla". Collegethesis. PhD Thesis. Universidad Nacional Mayor de San Marcos; (1999).

[13] Rentería, A, & Miranda, H. Aislamiento y Selección Primaria de Microorganismos Capaces de Utilizar Petróleo Como Única Fuente de Carbono. Abstract Book: proceedings of the I Congreso Peruano de Biotecnología y Bioingeniería, ICA, November 1998, Trujillo, Perú. Perú; (1998). , 12-15.

[14] Tantalean, J, & Altamirano, R. Aislamiento y Evaluación del Crecimiento de *Pseudomonas spp.* hidrocarburoclásticas en Petróleo Diesel Abstract Book: proceedings of the I Congreso Peruano de Biotecnología y Bioingeniería, ICA, November 1998, Trujillo, Perú. Perú; (1998). , 12-15.

[15] Merino, F. Estudio de Microorganismos Nativos Productores de Emulsificantes de Petróleo. Masters Degree thesis. Universidad Nacional Mayor de San Marcos; (1998).

[16] Leahy, J. G, & Colwell, R. R. Microbial Degradation of Hydrocarbons in the Environment. Microbiological Reviews (1990). , 54(3), 305-15.

[17] Kiran, T, Nalini, S, Sistla, R, & Sadanandam, M. Surface solid dispersion of glimepiride for enhancement of dissolution rate. International Journal of PharmTech Research (2009). , 1(3), 822-31.

[18] Mukherjee, S, Das, P, Sivapathasekaran, C, & Sen, R. Enhanced production of biosurfactant by a marine bacterium on statistical screening of nutritional parameters. Biochemical Engineering Journal (2008). , 42(3), 254-60.

[19] Tran, H, Kruijt, M, & Raaijmakers, J. M. Diversity and acivity of biosurfactant-producing *Pseudomonas* in the rhizosphere of black pepper in Vietnam. Journal of Applied Microbiology (2008)., 104(3), 839-51.

[20] Banat, I. M, Samarah, N, Murad, M, Horne, R, & Banerjee, S. Biosurfactant Production and Use in Oil Tank Clean- Up. World Journal of Microbiology and Biotechnology (1991)., 7(1), 80-8.

[21] Henry, N. D, Robinson, L, Johnson, E, Cherrier, J, & Abazinge, M. Phenanthrene Emulsification and Biodegradation Using Rhamnolipid Biosurfactants and *Acinetobacter calcoaceticus* In Vitro. Bioremediation Journal (2011)., 15(2), 109-20.

[22] Zheng, C, Luo, Z, Yu, L, Huang, L, & Bai, X. The Utilization of Lipid Waste for Biosurfactant Production and Its Application in Enhancing Oil Recovery. Petroleum Science and Technology (2011)., 29(3), 282-89.

[23] Thavasi, R, Jayalakshmi, S, Balasubramanian, T, & Banat, I. M. Production and characterization of a glycolipid biosurfactant from *Bacillus megaterium* using economically cheaper sources. World Journal of Microbiology and Biotechnology (2008). , 24(7), 917-25.

[24] Ron, E. Z, & Rosenberg, E. Biosurfactants and oil bioremediation. Current Opinion in Biotechnology (2002)., 13(3), 249-52.

[25] Falatko, D. M, & Novak, J. T. Effects of biologically produced surfactants on the mobility and biodegradation of petroleum hydrocarbons. Water Environment Research (1992)., 64(2), 163-9.

[26] Deziel, E, Paquette, G, Villemur, R, Lepine, F, & Bisaillon, J. Biosurfactant Production by a Soil *Pseudomonas* Strain Growing on Polycyclic Aromatic Hydrocarbons. Applied and Environmental Microbiology (1996)., 62(6), 1908-12.

[27] Das, K, & Mukherjee, A. K. Crude petroleum-oil biodegradation efficiency of *Bacillus subtilis* and *Pseudomonas aeruginosa* strains isolated from a petroleum-oil contaminated soil from North-East India. Bioresource Technology (2007)., 98(7), 1339-45.

[28] Cooper, D. G, & Goldenberg, D. G. Surface-active agent from two Bacillus species. Applied and Environmental Microbiology (1987)., 28, 224-229.

[29] Margaritis, A, Kennedy, K, Zajic, J, & Gerson, D. Biosurfactant production *by Nocardia erythropolis*. Dev. Ind. Microbiol. (1979)., 20, 623-630.

[30] Samadi, N, Abadian, N, Akhavan, A, & Fazeli, M. R. Biosurfactant production by the strain isolated from contaminated soil. Journal of Biological Sciences (2007). , 7(7), 1266-9.

[31] Batzle, M, & Wang, Z. Seismic properties of pore fluids. Geophysics (1992). , 57(11), 1396-1408.

[32] Rajasekar, A, Ganesh-babu, T, Maruthamuthu, S, Karutha-pandian, S T, Mohanan, S, & Palaniswamy, N. Biodegradation and corrosion behaviour of *Serratia marcescens*

ACE2 isolated from an Indian diesel-transporting pipeline. Journal of Microbiology and Biotechnology. (2007). DOIs11274-006-9332-0, 23(8), 1065-1074.

[33] Janek, T, Lukaszewicz, M, & Krasowska, A. Antiadhesive activity of the biosurfactant pseudofactin II secreted by the Arctic bacterium *Pseudomonas fluorescens* BD5. BMC Microbiology (2012). DOI:10.1186/1471-2180-12-24., 12(24), 1-9.

[34] Marchant, R, & Banat, I. M. Microbial biosurfactants: challenges and opportunities for future exploitation. Trends in Biotechnology (2012). , 30(11), 558-565.

[35] Zhang, Y, & Miller, R M. Enhanced octadecane dispersion and biodegradation by a Pseudomonas rhamnolipid surfactant (biosurfactant). Applied and Environmental Microbiology.(1992). , 58(10), 3276-3282.

[36] Belloso, C, Carrario, J, & Viduzzi, D. Biodegradación de hidrocarburos en suelos contenidos en terrarios: conference proceedings, November 1-5, 1998, XXVI Congreso Interamericano de Ingeniería Sanitaria y Ambiental,Lima, Perú. REPIDISCA: 45167; (1998).

[37] Díaz, I, Favela, E, Gallegos, M, & Gutiérrez, M. Biodegradación de hidrocarburos por un consorcio microbiano de la rizósfera de *Cyperus laxus* lam. Memories proceedings of the VIII Congreso Nacional de Biotecnología y Bioingeniería y IV Congreso Latinoamericano de Biotecnología y Bioingeniería, 12-17 Sep 1999, Huatulco, Oaxaca, México; (1999). http://astonjournals.com/manuscripts/GEBJ-3_Vol2010.pdfaccesed 10july 2012).

[38] Ifeanychukwu Atagana HHaynes R J, Wallis F M. Fungal Bioremediation of Creosote-Contaminated Soil: A Laboratory Scale Bioremediation Study Using Indigenous Soil Fungi. Water, Air and Soil pollution. (2006). 172(1-4):201-219DOI 10.1007/s11270-005-9074-x

Chapter 3

Microbial Hydrocarbon Degradation: Efforts to Understand Biodegradation in Petroleum Reservoirs

Isabel Natalia Sierra-Garcia and
Valéria Maia de Oliveira

Additional information is available at the end of the chapter

http://dx.doi.org/10.5772/55920

1. Introduction

The understanding of the phylogenetic diversity, metabolic capabilities, ecological roles, and community dynamics taking place in oil reservoir microbial communities is far from complete. The interest in studying microbial diversity and metabolism in petroleum reservoirs lies mainly but not only on providing a better comprehension of biodegradation of crude oils, since it represents a worldwide problem for petroleum industry. Generally, biodegradation of oil affects physical and chemical properties of the petroleum, resulting in a decrease of its hydrocarbon content and an increase in oil density, sulphur content, acidity and viscosity, leading to a negative economic consequence for oil production and refining operations [1,2]. Another important point for studying biodegradation lies on its important role in the global carbon cycle and the direct impact on bioremediation of polluted ecosystems. Furthermore, many of the enzymes involved in the degradation pathways are considered key catalysts in industrial biotechnology [3].

Despite these motivations and long recognition of petroleum as a the most important "primary energy" source, at present, microorganisms and factors involved in biodegradation of crude oil hydrocarbons in petroleum reservoirs are still not fully understood. The inaccessibility and complex microbiological sampling of petroleum reservoirs as well as the inherent limitations of the traditional culturing methods conventionally employed can explain this fact. Culture-based techniques have traditionally been the primary tools utilized for studying the microbiology of terrestrial and subsurface environments [4], which allowed the recovery and documentation of a large collection of bacteria capable of hydrocarbon utilization. Studies of numerous aerobic and anaerobic bacterial isolates have revealed mechanisms, which allow them to degrade specific classes of the highly diverse range of hydrocarbon compounds.

Therefore, all we know about the degradation of petroleum compounds has come from studying isolated microorganisms. Here, we provide an overview of what is currently known about the mechanisms of aerobic and anaerobic degradation of hydrocarbons, as a result from biochemical and genomic approaches, we give a perspective of the petroleum microbial diversity unraveled so far, and finally we discuss the common oil reservoir characteristics that can be used to predict the most probable mechanism of degradation into deep petroleum reservoirs.

It is well known that microbial diversity in environment is several orders of magnitude higher than the one assumed based on previous cultivation methods [5]. A particularly large number of novel techniques have been developed, which now allow the determination of the *in situ* microbial diversity and activity on a particular site, screening for a particular gene or activity of interest, gene quantification, and DNA and mRNA sequencing and analysis from total communities. This book chapter will address how the implementation of such culture-independent molecular methods allow the access to the microbial diversity and metabolic potential of microorganisms and bring novel information about microbial diversity and new pathways involved in biodegradation processes taking place in petroleum reservoirs. This information will certainly contribute to a broader perspective of the biodegradation processes and corroborate with previous findings that degradation of pollutants in many cases is carried out by microbial consortia rather than a single species [6], where key species and catabolic genes are often not identical to those that have been isolated and described in the laboratory [7, 8].

2. Microbial diversity in oil reservoirs

Recognition of indigenous microbiota harbored by oil reservoirs has been discussed for a long time. Actually, determining the nature of isolated microorganisms from oil reservoirs (indigenous or nonindigenous) is a difficult issue concerning petroleum microbiologists. The reasons for this controversy rely mainly on the difficulty of aseptic sampling in deep oil reservoirs. This means that microorganisms observed in oil field fluids conceivably could be contaminants introduced during drilling operations and/or during sample retrieval, or could be material sloughed from biofilms growing in installed pipes. Another reason for skepticism is the commonplace practice of ''water- flooding'' (injection of surface waters or re-injection of natural formation waters to maintain reservoir pressure for oil production); since in this case microbes would be introduced during injection and therefore would not necessarily represent indigenous species [9].

In addition to this controversy, there is the fact that petroleum reservoirs are considered extreme environments where *in situ* conditions, like high pressure, temperature, salinity and anaerobic conditions, are considered as inhospitable to microbial activity. In fact, perception of deep subsurface as a sterile environment has only changed during the past two decades with the increasing awareness of the ability of microbes to colonize extreme environments. Actually, with the use of more sophisticated and appropriate sampling and cultivation techniques, as well as the application of molecular biological techniques to oil field fluids, the dogma of the sterile deep subsurface has been dispelled [9]. Rather, it has become clear that many oil reservoirs do harbor indigenous microbes (*e.g.* the genera *Geotoga* and *Petrotoga* isolated only from oil reservoirs) [10]. Nowadays it is clear that worldwide petroleum reserves are dominated by deposits that have been microbially degraded over geological time and biodegraded petroleum reservoirs represent the most dramatic manifestation of the deep biosphere [11]

In spite of the polemics on which micro-organisms would actually be native and which would be contaminants in oil reservoirs, a wide range of microbial taxonomic groups have been identified in oil reservoirs geographically distant using traditional techniques adapted to *in situ* conditions, as described by L'Haridon et al. [12], Grassia et al. [13] and reviewed by Magot et al [14], or combined with cultivation-independent molecular methods, as reported by Orphan et al. [15]. Table 1 summarizes the various physiological and taxonomical groups and species that have been isolated from oil reservoirs.

3. Aspects from oil reservoir determining microbial degradation

For a long time, the mechanism considered to be prevalent for oil degradation in petroleum reservoirs was the well documented aerobic microbial metabolism and it has long been thought that the flow of oxygen through meteoric waters was necessary for in-reservoir petroleum biodegradation [16]. This mechanism has been widely accepted despite the fact that oxygen would likely be consumed by oxidation of organic matter in near surface sediments and therefore, would be very unlikely for oxygen to reach deep petroleum reservoirs [11].

Recently, the discovery of the ability of microorganisms to degrade anaerobically hydrocarbon oil components and the detection of metabolites characteristic of anaerobic hydrocarbon degradation in oil samples from biodegraded reservoirs, but not in non-degraded reservoirs or aerobically degraded oils [11], have provided valuable information to determine the processes involved in the degradation of oil reservoirs. Nowadays, evidences of such degradation through anaerobic rather than aerobic processes are becoming more substantial and compelling [17].

It is known that microorganisms in anaerobic conditions can use a variety of final electron acceptors, including nitrate, iron, sulfate, manganese and, more recently, chlorate. Anaerobic degradation has also been coupled to methanogenesis, fermentation and phototrophic metabolism but growth of these microorganisms and, therefore, biodegradation rates are significantly lower compared to aerobic degraders. These anaerobic processes have been demonstrated in surface sediments and pure cultures or enrichments in laboratories [18] and all of them potentially play a role in oil biodegradation in anoxic petroleum reservoirs [11]. However, nitrate, like oxygen, is highly reactive and would likely be completely consumed before it could reach the oil reservoir [17]. In deep reservoirs, the supply of large amounts of Fe(III) or manganese(IV) via meteoric water influx are unlikely due to poor solubility and slow water recharge rates in subterranean cycles. Therefore, iron and manganese, which could be used as electro acceptors for oil oxidation, are unlikely to be responsible for significant compositional changes in the oil, considering their limited availability in the reservoir. Accordingly, oil degradation linked to sulfate reduction and methanogenic would therefore explain the consistent hydrocarbon compositional patterns seen in degraded oils worldwide [17]. Sulfate arises from geological sources, such as evaporitic sediments and limestone, or from the injection of seawater for pressure stabilization, and may lead to significant oil degradation and increased residual-oil sulfur content. Methanogenic oil degradation, on the other hand, does not require external electron acceptors and leads to less overall souring of the oil reservoir. Several studies have described *in vitro* methanogenic degradation of crude oil related compounds [19, 20] Jones et al., 2008), including n-alkanes [21, 20] and aromatic hydrocarbons [17].

Organism	Taxonomical group	Metabolism	Origin	Reference
Thermodesulforhabdus norvegicus	Deltaproteobacteria	Sulfate-reducer	Oil field in Norway	[22]
Desulfacinum infernum	Deltaproteobacteria	Sulfate-reducer	North see petroleum reservoir near Scotland	[23]
Desulfomicrobium norvegicum	Deltaproteobacteria	Sulfate reducer	Petroleum reservoir in Canada	[24]
Desulfovibrio sp.	Deltaproteobacteria	Sulfate reducer	Petroleum reservoir in Canada	[24]
Dethiosulfovibrio peptidovorans	Bacteria, Synergistetes	Sulfate reducer	Oil well in the Emeraude oilfield in Congo, Central Africa,	[25]
Desulfotomaculum thermocisternum	Bacteria, Firmicutes	Sulfate reducer	Oil reservoir in the North sea	[26]
Deferribacter sp.	Bacteria, Deferribacteres	Sulfate reducer	California oil fields	[15]
Halanaerobium congolense	Bacteria, Firmicutes	Thiosulfate- and sulfur-reducing bacterium	African oil field	[27]
Thauera phenylacetica	Betaproteobacteria	Nitrate reducer	Petroleum reservoir in Canada	[24]
Pseudomonas stutzeri	Gammaproteobacteria	Nitrate reducer	Petroleum reservoir in Canada	[24]
Garciella nitratireducens	Bacteria, Firmicutes	Nitrate reducer	Oil field in Tabasco, Gulf of Mexico	[28]
Geobacillus subterraneus, Geobacillus uzenensis	Bacteria, Firmicutes	Nitrate reducer	Petroleum reservoir in China	[29]

Organism	Taxonomical group	Metabolism	Origin	Reference
Lactosphaera pasteurii	Bacteria, Firmicutes	Fermentative	Petroleum reservoir in Canada	[24]
Propionicimonas paludicola	Bacteria, Firmicutes	Fermentative	Petroleum reservoir in Canada	[24]
Anaerobaculum	Bacteria, Synergistetes	Fermentative	California oil fields	[15]
Thermococcus sp.	Archaea, Euryarchaeota	Fermentative	California oil fields	[15]
Thermococcus sibericus	Archaea, Euryarchaeota	Fermentative	Petroleum reservoir in Western Siberia	[30]
Petrotoga sp.	Bacteria, Thermotogae	Fermentative	California oil fields	[15]
Petrotoga olearia; P. siberica	Bacteria, Thermotogae	Fermentative	Petroleum reservoir in Western Siberia	[12]
Thermoanaerobacter	Bacteria, Firmicutes	Fermentative	California oil fields	[15]
Thermotoga sp.	Bacteria, Thermotogae	Fermentative	California oil fields	[15]
Thermosipho geolei	Bacteria, Thermotogae	Fermentative	Petroleum reservoir in Western Siberia	[12]
Anaerobaculum thermoterrenum	Bacteria, Synergistetes	Fermentative	Oil well in Utah	[23]
Fusibacter paucivorans	Bacteria, Firmicutes	Fermentative	Oil well in the Emeraude oilfield in Congo, Central Africa	[31]
Thermovirga lienii	Bacteria, Synergistetes	Fermentative	Oil reservoir in the North sea	[32]
Methanococcus	Archaea, Euryarchaeota	Methanogen	California oil fields	[15]
Methanococcus thermolithotrophicus	Archaea, Euryarchaeota	Methanogen	North sea old field in Norway	[33]
Methanoculleus	Archaea, Euryarchaeota	Methanogen	California oil fields	[15]
Methanobacterium	Archaea, Euryarchaeota	Methanogen	California oil fields	[15]

Table 1. Summary of bacteria isolated from oil reservoirs worldwide.

Deep subsurface environments such as petroleum reservoirs are logistically much more difficult to study than contaminated shallow subsurface environments [17]. Since in many biodegraded petroleum reservoirs most biodegradation occurs close to the oil water transition zone, it has been proposed that the oil–water transition zone (OWTZ) provides suitable physical and chemical conditions for microbial activity [17].

There are other physical and chemical parameters influencing *in situ* biodegradation. Temperature is one of the main factors which limits oil degradation in reservoir, and, empirically, it has been repeatedly observed that biodegradation does not occur in oil reservoirs with *in situ* temperatures >80-90°C [34]. Salinity is another factor that affects in-reservoir oil biodegradation, especially in combination with temperature [13]. Typically, reservoirs with highly saline waters show limited oil biodegradation [11]. This is consistent with the observations that it has not been possible to cultivate microorganisms from reservoir waters with salinity greater than 100 g/L [13]. Pressure seems to be a less limiting factor, except that it may select for certain physiological types and influences the pH of pore waters by increasing dissolution of CO_2 [9]. The availability of electron donors and acceptors governs the type of bacterial metabolic activities within oil field environments [14]. The potential electron donors include CO_2, hydrocarbons, H_2 and numerous organic molecules. Availability of fixed nitrogen is unlikely to limit microbial activity in reservoirs. However, the availability of water-soluble nutrients, like phosphorus and/ or oxidants (terminal electron acceptors such as ferrous iron, sulfate or CO2), is more likely to limit *in situ* microbial activity [9]. Nonetheless, physiological characteristics of microorganisms indigenous to petroleum reservoirs shed light on the conditions under which petroleum degradation may occur and the potential degradation mechanisms.

4. Hydrocarbon degradation

Hydrocarbons are understood as the compounds that consist exclusively of carbon and hydrogen. Because of the lack of functional groups, hydrocarbons are largely apolar and exhibit low chemical reactivity at room temperature. Differences in their reactivities are primarily determined by the occurrence, type and arrangement of unsaturated bonds. Therefore, in this chapter, we will use the common way to classify hydrocarbons according to their bonding features: i) aliphatic group, which includes straight-chain (n-alkanes), branched-chain and cyclic compounds and ii) aromatic group which includes mono or polycyclic hydrocarbons an many important compounds which also contain aliphatic hydrocarbon chains (*e. g.*, alkylbenzenes).

Already a century ago, bacterial isolates had been reported to use aliphatic and aromatic hydrocarbons as sole carbon and energy sources [35]. Since then, numerous aerobic, and also anaerobic, bacterial isolates have been studied in order to understand the mechanisms which allow them to degrade specific members of the highly diverse aliphatic and aromatic compounds. Degradation by such isolates has been investigated thoroughly and results have revealed that they can completely degrade most classes of hydrocarbons, including alkanes, alkenes, alkynes and aromatic compounds. Such degradation can occur aerobically, with oxygen, or anaerobically, with nitrate, ferric iron, sulfate or other electron acceptors [36].

Efforts to overview the metabolism of hydrocarbons in microorganisms are confronted with the chemical diversity of such compounds and their reactivities, as well as with various microbial life styles [36]. The study of biodegradation is conventionally treated in separate areas: aliphatic vs. aromatic hydrocarbons, aerobic vs. anaerobic degradation pathways, physiology and overall metabolic pathways vs. enzymatic mechanisms and structures, often

http://dx.doi.org/10.5772/55920

with limited knowledge and data exchange. Nonetheless, each of these study areas deals with the same central point that is the "metabolic challenge" to guide an apolar, unreactive compound composed only of carbon and hydrogen into the metabolism [36]. The hydrocarbon must be first functionalized and currently it has been recognized that there is a surprisingly diversity of reactions of activation that had evolved in microorganisms (Table 2).

Mechanisms for hydrocarbon activation		
	Aerobic	**Anaerobic**
Short-Chain non-methane alkanes C2-C10	• Non-heme iron monooxygenase similar to sMMO (C2-C9) • Copper-containing monooxygenase similar to pMMO (C2-C9) • Heme-iron monooxygenases (also refered as soluble Cytochrome P450 (C5-C12)	• Fumarate addition
Long-Chain alkanes >C10	• Heme-Monooxygenase (P450 type) • [Fe2]-Monooxygenase • Non-heme iron monooxygenase (AlkB-related) (C3-C13 or C10-C20) • Flavin-binding monooxygenase (AlmA) (C20- C36) • Thermophilic flavin-dependent monooxygenase (LadA) (C10-C30)	• Fumarate addition • Carboxylation
Aromatic hydrocarbons	• [Fe]-Dioxygenase • [Fe2]-Monooxygenase • [Flavin]-Monooxygenase	• Fumarate addition • Hydroxylation • Carboxylation

Table 2. Overview of aerobic and anaerobic mechanisms for hydrocarbon activation in bacteria.

Mechanisms for hydrocarbon activation are basically different in aerobic and anaerobic microorganisms. Under oxic conditions, hydrocarbon metabolism is always initiated using molecular oxygen as a co-substrate in mono- or dioxygenase reactions that enable the terminal or sub-terminal hydroxylation of aliphatic alkane chains or the mono or dihydroxylation of aromatic rings [37]. In the hydrocarbon activation under anoxic conditions, some proposed reactions comprise: (1) addition to fumarate by glycyl-radical enzymes, (2) methylation of unsubstituted aromatics, (3) hydroxylation with water by molybdenum cofactor containing enzymes of an alkyl substituent via dehydrogenase, and (4) carboxylation catalyzed by yet-uncharacterized enzymes which may actually represent a combination of reaction (2) followed by reaction (1) [38; 37]. Although all these mechanisms of hydrocarbon anaerobic activation have been proposed, the required signature metabolites and enzymes involved have been characterized only for (1) addition to fumarate (demonstrated for toluene, xylene, ethylbenzene, methylnaphthalene, alkanes and alicyclic alkanes); for (3) hydroxylation (demonstrated for ethylbenzene); and for (4) carboxylation (demonstrated for benzene and naphtalene) [39].

5. Biochemical and genetic pathways of microbial hydrocarbon degradation

The enzymatic reactions involved in the aerobic degradation of hydrocarbons by bacteria have been extensively studied for several decades [37]. Genes encoding enzymes for degradation are relatively well understood for aerobic and easily cultivable microorganisms, particularly for a *Pseudomonas* strain, known as *P. putida* GPo1, as well as for the strains *Acinetobacter* sp. ADP1 and *Mycobacterium tuberculosis* H37Rv [39, 40]. On the other hand, the anaerobic hydrocarbon degradation has gained more attention since is supposed to be the predominant mechanism occurring in several polluted environments and oil reservoirs. However, its study is an incipient area because of the peculiarities of the reservoir environment and difficulties that arise from attempts to characterize these communities. Nevertheless, several bacteria from other environments able to use alkanes as carbon source in the absence of oxygen have been described in the last few years [41], but anaerobic bacteria able to degrade hydrocarbons under conditions found in deep petroleum reservoirs have not been isolated so far [2]. Figure 1 represents an overview of the main mechanisms and pathways used by microorganisms to degrade hydrocarbon compounds under aerobic and anaerobic conditions.

5.1. Aerobic degradation

5.1.1. Aliphatic hydrocarbons

In most degradation pathways described, the substrate n-alkane is oxidized to the corresponding alcohol by substrate-specific terminal monooxygenases/hydroxylases. The alcohol is then oxidized to the corresponding aldehyde, and finally converted into a fatty acid. Fatty acids are conjugated to CoA and subsequently processed by β – oxidation to generate acetyl-CoA [42, 40]. Subterminal oxidation has also been described for both short and long-chain alkanes [40]. Both terminal and sub-terminal oxidation can coexist in some microorganisms [41]. Initial terminal hydroxylation of n-alkanes in bacteria can be carried out by enzymes belonging to different classes, named: (1) propane monooxygenase (C3), (2) different classes of butane monooxygenase (C2-C9), (3) CYP153 monooxygenases (C5-C12), (4) AlkB-related non-heme iron monooxigenase (C3-C10 or C10-C20), (5) flavin-binding monooxigenase AlmA (C20-C36), (6) flavin-dependent monooxygenase LadA (C10-C30), (7) copper flavin-dependent dioxygenase (C10-C30) [43].

Among all the alkane activating enzymes, the integral membrane non-heme iron monooxygenase (AlkB) is the best characterized one. Microorganisms degrading medium (C5-C11) and long (>C12)-length alkanes have been frequently related to the presence of *alk*B genes and that is why the presence of such genes have been widely used as functional biomarker for the characterization of aerobic alkane-degrading bacterial populations in several environmental

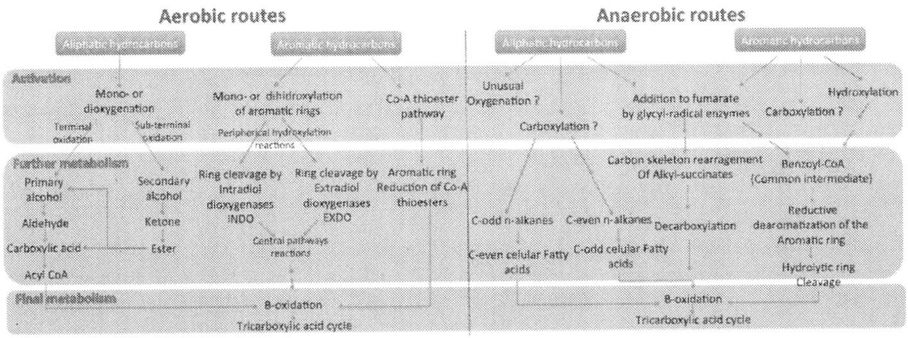

Figure 1. Pathways for aerobic and anaerobic bacterial degradation of hydrocarbon compounds. Two arrows represent more than one reaction.

samples [44, 45] and in bioremediation experiments [46, 47]. The degradation pathway of the *alk* system was first described in *Pseudomonas putida* GPo1 (formerly identified as *P. oleovorans* GPo1), where it is located on the OCT plasmid. In this model system, OCT plasmid contains two operons: *alk*BFGHJKL and *alk*ST [48]. The first operon encodes two components of the *alk* system, a particulate non-heme integral membrane alkane monooxygenase (AlkB) and the soluble protein rubredoxin (AlkG), as well as other enzymes involved in further steps. The second operon encodes for a rubredoxin reductase (AlkT and AlkS), which regulates the expression of the *alk*BFGHJKL operon [48, 49]. Since this system was described, AlkB homologous have been found in many alkane-degrading α- β – and γ –Proteobacteria and high G + C content Gram-positive bacteria (Actinobacteria) [39] and an increasing collection of alkane hydroxylase gene sequences has allowed the diversity analysis of hydrocarbon-degrading microbial populations in different ecosystems. However, comparisons of cloned *alk*B genes or gene fragments have showed that sequence diversity is very high, even among *alk*B genes within the same species [50].

In despite of the relevance of *alkB* genes as a functional biomarker of alkane-degrading bacterial communities, knowledge on the presence and diversity of *alk*B genes in oil reservoirs is scarce. Tourova et al. [51] analysed *alk*B diversity in thermophilic bacterial strains of the genus *Geobacillus* isolated from oil reservoirs or hot springs. They detected, for the first time, sets of *alk*B gene homologous in thermophilic bacteria, and some strains showed different homologous within the same genome. This fact was explained by the occurrence of horizontal gene transfer among these bacteria. Recently, Li et al. [52] aimed to evaluate *alk*B gene diversity and distribution in production water from 3 oilfields in China through a specific PCR-DGGE method. Results showed that sequences found in the water samples were similar to *alk*B genes from other corresponding alkane-degrading strains. But at the same time, they showed the presence of a considerable genetic diversity of alkB genes in the wastewater as evidenced by a total of 13 unique DNA bands detected. Studies on the degradation of alkanes in oil reservoirs are currently in a start point, but in the future they certainly will help to understand the process of degradation in oil reservoir.

In comparison to the few efforts in studying *alk*B system in oil reservoirs, much less is known about the presence of the other enzymatic systems previously listed, which have been described for aerobic degradation of n-alkanes in isolated bacteria or laboratory microcosms. For the most recent elucidated systems for alkane oxidation, named *almA* and *ladA* genes, nothing is known about the environmental distribution of these type of genes in petroleum contaminated sites [53] or oil fields, although the LadA complete degradation pathway has been characterized through genome and proteome analysis of *Geobacillus thermodenitrificans* NG80-2, a thermophilic strain isolated from a deep oil reservoir in Northern China [54]. Currently, it is believed that there are enzyme systems for alkane degradation which have still not been characterized and that may include new proteins unrelated to those already known [41]. Moreover, in many alkane degraders more than one alkane oxidation system have been observed, which have been reported exhibiting overlapping substrate ranges [39, 40]. These observations point out that in order to characterize and explore metabolic diversity and functions involved in alkane degradation one should take into consideration the high diversity of enzymes capable of initiating such metabolism.

5.1.2. Aromatic hydrocarbons

The aerobic bacterial catabolism of aromatic compounds involves a wide variety of peripheral pathways that activate structurally diverse substrates into a limited number of common intermediates that are further cleaved and processed by a few central pathways to the central metabolism of the cell [55]. Metabolic pathways and encoding genes responsible for the degradation of specific members of a highly diverse range of aromatic compounds have been characterized for many isolated bacterial strains, predominantly from the Proteobacteria and Actinobacteria phyla [56]. Degradation by such isolates is typically initiated by members of one of the three superfamilies: the Rieske non-heme iron oxygenases (RNHO), the flavoprotein monooxygenases (FPM) and the soluble diiron multicomponent monooxygenases (SDM). Further metabolism is achieved through di- or trihydroxylated aromatic intermediates. Alternatively, activation is mediated by CoA ligases where the formed CoA derivates are subjected to selective hydroxylation [58, 53]. In the case of hydrophobic pollutants, such as benzene, toluene, naphthalene, biphenyl or polycyclic aromatics, aerobic degradation is usually initiated by activation of the aromatic ring through oxygenation reactions catalyzed by RNHO enzymes or, as intensively described for toluene degradation, through members of SDM enzymes [56].

Further intermediates can be catalyzed by two kinds of enzyme, intradiol and extradiol dioxygenases, which represent two classes of phylogenetically unrelated proteins [58]. These enzymes are key enzymes in the degradation of aromatic compounds, and many of such proteins and their encoding sequences have been described, purified and characterized in the last decades [56]. While all intradiol dioxygenases described so far belong to the same superfamily, the extradiol dioxygenases include at least three members of different families. Type I extradiol dioxygenases (e.g. catechol 2,3-dioxygenases and 1,2-dioxygenases) belong to the vicinal oxygen chelate superfamily enzymes. Type II extradiol dioxygenases are related to LigB superfamily (e.g. protocatechuate 4,5-dioxygenases) and the type III enzymes belongs to the cupin superfamily (e.g. gentisate dioxygenases) [53]. However, members of novel superfamilies performing crucial steps in aromatic metabolic pathways are still being discovered [56, 53].

The knowledge of metabolic properties of isolates has allowed the monitoring of the ability of microorganisms to mineralize aromatic hydrocarbons in soils. Typically, these studies have used primers designed based on conserved gene regions and focused on RNHO or SDM as targets for initiating degradation, or on Extradiol dioxygenases (EXDO) cleaving the aromatic ring [59]. These studies range from those searching for a narrow range of genes similar or identical to those observed in type strains using non-degenerated primers to those searching for subfamilies of homologous genes using degenerated primers [59]. However, due to the immense heterogeneity of such enzymes [57], there will never be a pair of primers that will reliably cover the huge diversity of a catabolic gene family in nature [53].

5.2. Anaerobic degradation

5.2.1. Aromatic hydrocarbons

We have already described the main mechanism for degradation of aromatic compounds in aerobic conditions, where oxygen is not only the final electron acceptor but also co-substrate of two key processes: hydroxylation and cleavage of the aromatic ring by oxygenases. In contrast, in the absence of oxygen, microorganisms use a complete different pathway, based in reductive reactions to attack the aromatic ring [61].

The biochemistry of some anaerobic degradation pathways of aromatic compounds has been studied to some extent; however, the genetic determinants of all these processes and the mechanisms involved in their regulation are much less studied [55]. Recent advances in genome sequencing have led to the complete genetic information for six bacterial strains that are able to anaerobically degrade aromatic compounds using different electron acceptors and that belong to different taxonomic groups of bacteria: denitrifying betaproteobacteria, *Thauera aromatica* and *Azoarcus* sp. EbN1, two alphaproteobacteria, the phototroph *Rhodopseudomonas palustris* strain CGA009 and the denitrifying *Magnetospirillum magneticum* strain AMB-1, and two obligate anaerobic deltaproteobacteria, the iron reducer *Geobacillus metallireducens* GS-15 and the fermenter *Syntrophus aciditrophicus* strain SB [55]. It is worth remembering that, in recent years, important inferences and generalizations have been made about the genetics involved in hydrocarbon metabolism based on these isolated bacteria under conventional laboratory conditions. However, potential novel genes, enzymes and metabolic pathways responsible for degradation processes are probably harbored by yet uncultivated bacteria.

The best understood and apparently the most widespread of these anaerobic mechanisms is the radical-catalyzed addition of fumarate to hydrocarbons, yielding substituted succinate derivatives. This reaction has been recognized for the activation of several alkyl-substituted benzenes as well for n-alkanes [62]. However, understanding of this fumarate-dependent hydrocarbon activation is most advanced in the case of toluene. The key enzyme in this process is the enzyme benzylsuccinate synthase. All enzymes required for β-oxidation of benzylsuccinate are encoded by the *bbs* operon. Subsequent degradation of benzoyl-CoA proceeds via reductive dearomatization, hydrolytic ring cleavage, β-oxidation to acetyl-CoA units and terminal oxidation to Co_2 [63]. In contrast to the anaerobic metabolism of toluene, degradation of ethylbenzene (and probably other alkylbenzenes with carbon chain of at least 2) is entirely different, despite the chemical and structural similarities between the two compounds, and involves a direct oxidation of the methylene carbon via (S)-1-phenylethanol to acetophenone [55]. Ethylbenzene is anaerobically hydroxylated and dehydrogenated to acetophone, which is then carboxyled and converted to benzoylCoA as the first common intermediate of the two pathways [62].

Genetics of the enzymatic system have been only characterized for these two mechanisms for anaerobic hydrocarbon activation. Genes encoding pathways that involve fumarate addition are typically organized in two operons. One operon includes the three structural genes of the protein catalyzing fumarate addition and the other includes genes required for converting succinate derivates to benzoyl-CoA [64]. Gene sequences and organization are relatively conserved among nitrate-reducing bacteria but differ somewhat from those of the iron reducer *G. metallireducens* [64] and substantially from those of the hexane-degrading nitrate reducer strain HxN1 [65]. Hydrocarbon dehydrogenation pathway is also organized in two operons. One operon contains the structural genes for the first two reactions (ethylbenzene dehydrogenase and 1-phenylethanol dehydrogenase) and the other contains the structural genes for acetophone carboxylase [64].

Kane et al. [66] developed the first real-time polymerase chain reaction (PCR) method to quantify hydrocarbon utilizers based on *bss*A genes of nitrate-reducing Betaproteobacteria. Since then, there have been several additional studies investigating the presence and/or distribution of anaerobic hydrocarbon utilizers in anaerobic environments via functional gene surveys of *bss*A, extending the range of detectable hydrocarbon-degrading microbes to iron and sulfate-reducing Deltaproteobacteria and revealing partially novel, site specific degrader populations [67, 68]. Other *bss*A-based detection studies in impacted environments, as well as studies that combine field metabolomics and molecular tools, are described by other authors [69, 70, 71]. Despite of the role of benzylsuccinate synthase in aromatic hydrocarbon degradation and its use as a biomarker are well documented, there is no study on the presence of this gene in oil reservoirs.

5.2.2. Aliphatic hydrocarbons

Anaerobic degradation of alkanes has not been extensively studied as for some aromatic compounds. The presumable reasons include the greater attention given to BTEX compounds (benzene, toluene, ethylbenzene and xylenes) because of their classification as priority pollutants [71], also the fact that anaerobic growth with n-alkanes is even slower than that with the alkylbenzenes, and finally the fact that long chain alkanes are poorly soluble and often prevents the cultivation of cells homogeneously in the medium [72]. However, anaerobic degradation of alkanes is equally relevant, since alkanes are quantitatively the most important hydrocarbon components of petroleum, and some are acutely toxic and difficult to remediate [71]. Several anaerobic bacteria capable of degrading n-alkanes with 6 or more carbons in length, particularly hexadecane (C16), using sulfate or nitrate as electron acceptors have been isolated [72, 73].

The two main mechanisms of anaerobic degradation of n-alkanes described involve unprecedented biochemical reactions that differ completely from those employed in aerobic hydrocarbon metabolism [73]. The first involves activation at the subterminal carbon of the alkane by the addition of fumarate, analogously to the formation of benzyl succinate during anaerobic degradation of toluene, however further reactions are completely different involving dehydrogenation and hydration [72]. Studies conducted with established axenic cultures have indicated that anaerobic metabolism of oil allkanes predominantly proceeds via addition of fumarate to the double bound [72]. Although alkylsuccinate metabolites have rarely been detected in oil reservoir fluids [74, 75], they have been reported in oil-contaminated environments as well as in oilfield facilities, where their detection is indicative of *in situ* microbial degradation of oil alkanes [71, 75]. Alkylsuccinic acids as intermediates of anaerobic alkane oxidation were first studied by Gieg and Suflita [76] when surveying these metabolites in aquifers contaminated with condensate gas, natural gas liquids, gasoline, diesel, alkanes and BTEX. They found alkylsuccinates originating from C3 to C11 alkanes, as well as putative metabolites originating from compounds with one degree of unsaturation, such as alkenes or alicyclic alkanes. Since this report, other studies have detected alkylsuccinate derivates in petroleum contaminated groundwater systems [76], coal beds [70] and oil fields [74, 77]. The formation of alkylsuccinates is catalyzed by a strictly anaerobic glycyl radical enzyme which has been termed as alkylsuccinate synthase or (1-methyl-alkyl)succinate synthase (Ass or Mas). The genes encoding Ass have recently been identified in the alkane degrading sulfidogenic bacteria *D. alkenivoras* AK-01 [78] and *Desulfoglaeba alkanedexens* ALDC^T [71], as well as in nitrate reducing strains HxN1 [65] and OcN1 [79], all affiliated to the Proteobacteria phylum [80]. Recently, Callaghan et al. [71] detected *ass*A genes in a propane-utilizing mixed culture and in a paraffin-degrading enrichment culture maintained under sulfate-reducing conditions. Despite of no genes for benzyl-and alkylsuccinate synthase were found when environmental metagenome datasets of uncontaminated sites were analyzed in Callaghan et al [71], the authors consider that *ass*A gene could be a useful biomarker for anaerobic alkane metabolism.

The second mechanism for alkane anaerobic degradation is the carboxylation, mainly developed from the growth pattern of the sulfate-reducing strain Hxd3 [81], tentatively named as *Desulfococcus oleovorans*. This strain differs from other alkane degraders for converting C-even alkanes into C-odd cellular fatty acids whereas growth on C-odd alkanes resulted in C-even cellular fatty acids [81, 72]. More recently, Callaghan et al. [82] suggested that a carboxylation-like mechanism analogous to the activation strategy previously proposed by So et al. [81] was the probable route for the anaerobic biodegradation of hexadecane in an alkane-degrading, nitrate-reducing consortium. However, in both cases, the hypothetical fatty acid intermediate (2-ethylalkanoate) that should result from the incorporation of inorganic carbon at C-3 of the alkane has never been detected. There is an on-going debate about this initial activation mechanism. From an energetic point of view, the carboxylation of alkanes is not feasible under physiological conditions, unless the concentration of the fatty acid (2-ethylalkanoate) is in the micromolar order of magnitude or less [80].

Other alternative activation mechanisms are proposed for the anaerobic degradation of alkanes. For instance, the mechanism referred as "unusual oxygenation" is used by the strain *Pseudomonas chloritidismutans* AW-1^T, that is assumed to produce its own oxygen via chlorate respiration used for subsequent metabolism of alkanes [60]. Other alternative mechanism considers that activation in the anaerobic methanogenic system may be initiated by an anaerobic hydroxylation reaction [83].

6. Mechanisms involved in oil biodegradation in petroleum reservoirs

From those microorganisms studied in oilfields, methanogens have received particular attention since they have been isolated and molecularly detected in both low- and high-temperature reservoirs [88, 89]. Their physiological characteristics and potential activity possibly involved in methanogenesis occurring in oil reservoirs have been demonstrated [90]. Furthermore, recently, Jones et al. [20] provided evidence that the patterns of hydrocarbon degradation observed in biodegraded petroleum reservoirs were the result of methanogenic processes. Therefore, microbiological and biogeochemical investigations have indicated that methanogenesis is a widely distributed process in petroleum reservoirs, although still poorly understood [90]. Methanogenesis is the terminal process of biomass degradation. Acetate and hydrogen are the most important immediate precursors for methanogenesis, and are converted into methane by acetoclastic and hydrogenotrophic methanogens, respectively [91]. Acetate can also be a precursor for methanogenesis through syntrophic acetate oxidation coupled to hydrogenotrophic methanogenesis, which is mediated by syntrophic bacteria and methanogenic archaea [92, 93, 94, 95]. Interestingly, acetate is generally abundant in many petroleum reservoirs, at concentrations ranging between 0.3 and 20 mM [96] hence, acetate metabolism is considered an important methane production process in those environments [90].

Cultivation-dependent and -independent approaches have shown the presence of acetoclastic and hydrogenotrophic methanogens and putative syntrophic acetate-oxidizing bacteria in reservoirs [88, 89, 102], indicating that there should be two different pathways of acetate metabolism in the environment, namely acetoclastic methanogenesis and syntrophic acetate oxidation coupled with hydrogenotrophic methanogenesis. Some previous studies suggested that syntrophic acetate oxidation was most likely to occur in petroleum reservoirs, based on molecular biological analysis [89] and thermodynamic calculations [98]. In Jones et al. [20], the composition of oil in microcosms exhibiting methanogenic oil degradation is compared to patterns observed in biodegraded oils from the Gullfaks field in the North Sea. Analysis of the methanogenic communities from oil-degrading microcosms revealed a strong selection for CO_2-reducing methanogens against acetoclastic methanogens, and gas isotope modeling also revealed that to match the d13C of methane and carbon dioxide from biodegraded petroleum reservoirs 75–92% of methanogenesis should be via the CO_2 reduction pathway [20, 11].

The reason why syntrophic acetate oxidation predominates over acetoclastic methanogenesis in oil reservoirs remains unclear. There is evidence from studies of oil contaminated aquifers that crude oil can have a detrimental effect on acetoclastic methanogenesis and, in situations where acetoclastic methanogenesis is inhibited, methanogenic alkane degradation via syntrophic acetate oxidation may be thermodynamically the most favorable alternative pathway [11]. Nonetheless, one recent report suggests that acetoclastic methanogenesis may predominate in some methanogenic oil-degrading systems [19]. Although there is currently great interest in how much each of the two pathways contributes to methane production in petroleum reservoirs, no studies are being conducted to address this question [90].

7. Metagenomics as a tool for a better comprehension of biodegradation

As stated previously, cultivation-based methods have traditionally been utilized for studying the microbiology in oil fields and have yielded valuable information about microbial interactions and their relations with hydrocarbons [42]. However, nowadays, it is known that only a small fraction of the microbial diversity in nature (1-10%) can be grown in the laboratory [84, 85, 86]. Therefore, it is assumed that the ecological functions of the majority of microorganisms in nature and their potential applications in biotechnology remain obscure [87].

In metagenomics, total DNA is extracted from appropriately chosen environmental samples, propagated in the laboratory by cloning techniques, submitted to sequence or function-based screenings and/or subjected to large-scale sequence analysis (Fig. 2). Functional screening of metagenomic libraries offer the advantage that it does not rely on sequence homology to known genes, and for this reason, has allowed the isolation of different enzyme classes from several environments. The probability (hit rate) of identifying a certain gene depends on multiple factors that are intrinsically linked to each other: the host–vector system, size of the target gene, its abundance in the source metagenome, the assay method, and the efficiency of heterologous gene expression in a surrogate host [99].

One of the first studies using metagenomics to study microbial degradation of aromatic compounds was performed by Suenaga and colleagues [100], who constructed a metagenomic library from activated sludge for industrial wastewater. The library was functionally screened for extradiol dioxygenase activities (enzymes for aromatic degradation) and 38 clones were subjected to sequencing analysis [101]. As a result, various types of gene subsets were identified that were not similar to the previously reported pathways performing complete degradation. Moreover, the authors discussed the fact that aromatic compounds in the environment may be degraded through the concerted action of various fragmented pathways. Sierra-Garcia [101] reported the organization of hydrocarbon degradation genes of selected metagenomic fosmid clones derived from a metagenomic library from Brazilian petroleum reservoir and functional screening for hydrocarbon degradation activities. The author found many putative proteins of different aerobic and anaerobic well described catabolic pathways, however the complete catabolic pathways described for hydrocarbon degradation in previous studies were absent in the fosmid clones. Instead, the metagenomic fragments comprised genes belonging to different pathways, showing novel gene arrangements where hydrocarbon compounds were degraded through the concerted actions of these fragmented pathways. These results suggest that there are marked differences between the degradation genes found

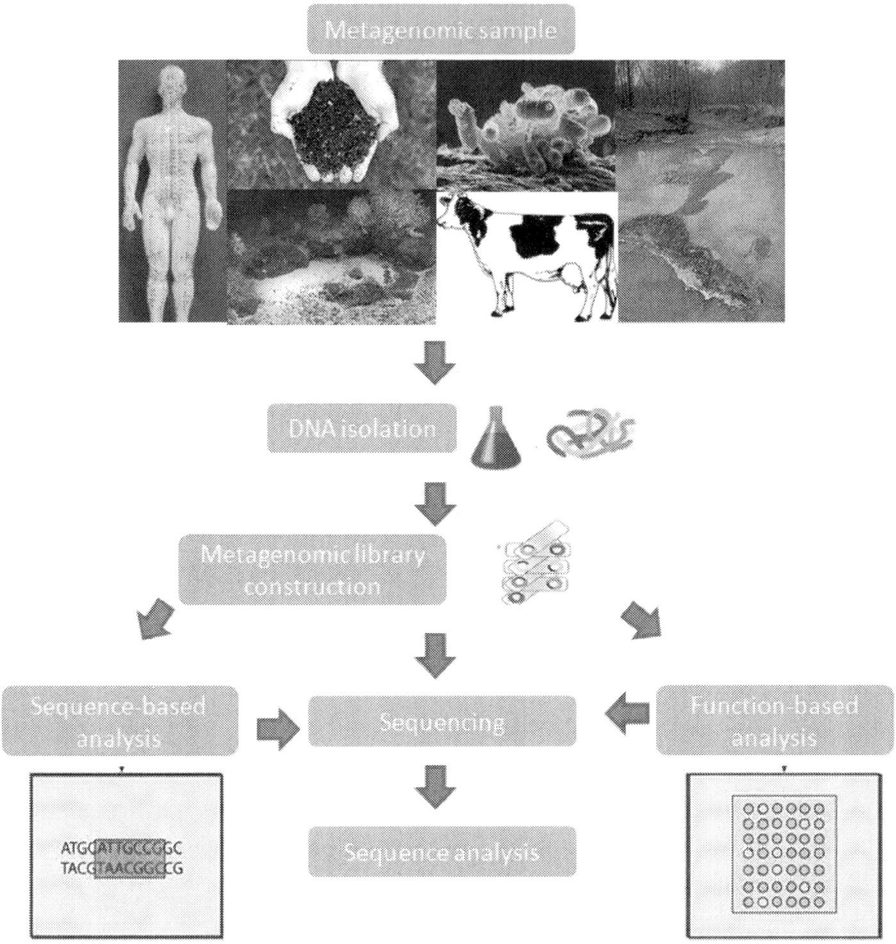

Figure 2. Schematic representation of the different steps for metagenomic analysis.

in microbial communities derived from enrichments of oil reservoir sample and those that have been previously identified in bacteria isolated from contaminated or pristine environments.

However, function-based screening of metagenomic libraries for xenobiotic degradation genes is often considered problematic because of insufficient and biased expression of the heterologous genes in the host *Escherichia coli* [99]. Only a few efforts have been made to solve these problems. In Uchiyama et al. [103], a novel method for function-driven screening is described, which was termed substrate-induced gene expression screening (SIGEX). This high-through-put screening approach employs an operon trap gfp expression vector in combination with fluorescence-activated cell sorting. The screening is based on the fact that catabolic-gene expression is induced mainly by specific substrates and is often controlled by regulatory elements located close to catabolic genes [103]. Using this approach, Uchiyama et al. [103] isolated aromatic-hydrocarbon-induced genes from a metagenomic library derived from groundwater. In Ono et al. [104] another screening strategy was based on functional complementation of a *Pseudomonas putida* host strain containing a naphthalene degrading pathway devoid of the naphthalene dioxygenase (NDO) encoding gene. Two clones were able to restore the ability of the host strain to use naphthalene as a sole carbon source and their genes were similar but no identical to already known operons. The authors refer to the use of other host strains for the construction of metagenomic libraries instead of the well-established *E. coli* as a simpler and economical way to perform function-driven screening in comparison to other reported systems such as SIGEX [103].

In the context of this chapter, several aspects of the hydrocarbon degradation need to be studied to obtain a comprehensive overview of the biodegradation processes that take place in oil reservoirs or petroleum impacted environments. These studies should take into consideration the high diversity of enzymes capable of initiating such metabolism as well as the implementation of integrated studies combining culture and molecular techniques, linking with metabolomics or compound-specific isotope analysis and microcosm studies for a better resolution of in situ microbial activity in petroleum reservoirs.

8. Conclusions and research needs

The understanding about biodegraded petroleum reservoirs have advanced considerably in recent years, but the organisms responsible for the *in situ* activity and a quantitative understanding of the factors which control in-reservoir oil biodegradation remain far from complete. The inaccessibility of petroleum reservoirs and inherent difficulties of microbiological sampling from commercially operating oil wells have required a multidisciplinary approach to delineating the study of subsurface petroleum biodegradation, and to date there are still prevailing paradigms relating to hydrocarbon biodegradation processes. This multidisciplinary approach to study *in situ* petroleum degradation should consider molecular biology, microbiology, and geological and geochemical parameters in order to establish the key organisms, biochemical reactions and mechanisms involved in such complex associations. Indeed, the isolation of anaerobic microorganisms capable of utilizing hydrocarbons is essential for a comprehensive understanding of their role and behavior in anoxic habitats and their complex interactions within methanogenic hydrocarbon-degrading communities. In addition, novel approaches, combining functional metagenomics, transcriptomics, metabolomics and other molecular surveys in microcosms are urgently required to better allow access to a more realistic phylogenetic and metabolic diversity governing oil biodegradation in petroleum reservoirs.

Author details

Isabel Natalia Sierra-Garcia and Valéria Maia de Oliveira*

*Address all correspondence to: vmaia@cpqba.unicamp.br

Microbial Resources Division, Research Center for Chemistry, Biology and Agriculture (CPQBA), University of Campinas, Campinas, Sao Paulo, Brazil

References

[1] Roling, W. (2003). The microbiology of hydrocarbon degradation in subsurface petroleum reservoirs: perspectives and prospects. *Res Microbiol.* 154(5), 321-328.

[2] Head, I. M, Jones, D. M, & Larter, S. R. (2003). Biological activity in the deep subsurface and the origin of heavy oil. *Nature,* 426(6964), 344-52.

[3] Ismail, W, & Gescher, J. (2012). Epoxy coenzyme a thioester pathways for degradation of aromatic compounds. *Appl Environ Microbiol.* , 78(15), 5043-51.

[4] Chandler, D. P, Li, S. M, Spadoni, C. M, Drake, G. R, Balkwill, D. L, Fredrickson, J. K, & Brockman, F. J. (1997). A molecular comparison of culturable aerobic heterotrophic bacteria and 16S rDNA clones derived from a deep subsurface sediment. *FEMS Microbiol Ecol.* , 23, 131-144.

[5] Leigh, M. B, Pellizari, V. H, Uhlik, O, Sutka, R, Rodrigues, J, & Ostrom, N. E. (2007). Biphenyl-utilizing bacteria and their functional genes in a pine root zone contaminated with polychlorinated biphenyls (PCBs). *ISME J 1,* 134-148.

[6] De Lorenzo, V. (2008). Systems biology approaches to bioremediation. *Curr Opin Biotechnol 19,* 579-589.

[7] Jeon, C, Park, W, Padmanabhan, P, Derito, C, Snape, J, & Madsen, E. (2003). Discovery of a bacterium, with distinctive dioxygenase, that is responsible for *in situ* biodegradation in contaminated sediment. *Proc Natl Acad Sci USA 100,* 13591-13596.

[8] Witzig, R, Junca, H, Hecht, H. J, & Pieper, D. H. (2006). Assessment of toluene/ biphenyl dioxygenase gene diversity in benzene-polluted soils: links between benzene biodegradation and genes similar to those encoding isopropylbenzene dioxygenases. *Appl Environ Microbiol 72,* 3504-3514.

[9] Foght, J. (2010). Microbial comminities in oil shales, biodegraded and heavy oil reservoirs, and bitumen deposits. In: K. N. Timmis (Ed.) *Handbook of Hydrocarbon and Lipid Microbiology.* Berlin, Heidelberg: Springer Berlin Heidelberg.

[10] Birkeland, N. K. (2004). The microbial diversity of deep subsurface oil reservoirs. *Stud Surface Sci Catal 151*, 385-403.

[11] Head, I. M, Aitken, C. M, Gray, N. D, Sherry, A, Adams, J. J, Jones, D. M, Rowan, A. K, et al. (2010). Hydrocarbon degradation in petroleum reservoirs. In: K. N. Timmis (Ed.) *Handbook of Hydrocarbon and Lipid Microbiology*. Berlin, Heidelberg: Springer Berlin Heidelberg.

[12] Haridon, L, Reysenbach, S, Glenat, A. L, Prieur, P, & Jeanthon, D. C. ((1995). Hot subterranean biosphere in a continental oil reservoir. *Nature* , 377, 223-224.

[13] Grassia, G. S, Mclean, K. M, Glenat, P, Bauld, J, & Sheehy, A. J. (1996). A systematic survey for thermophilic fermentative bacteria and archaea in high temperature petroleum reservoirs. *FEMS Microbiol* Ecol , 21, 47-58.

[14] Magot, M, Ollivier, B, & Patel, B. K. C. (2000). Microbiology of petroleum reservoirs. *Antonie van Leeuwenhoek.* , 77(2), 103-116.

[15] Orphan, V. J, Taylor, L. T, Hafenbradl, D, & Delong, E. F. (2000). Culture-dependent and culture-independent characterization of microbial assemblages associated with high-temperature petroleum reservoirs. *Appl Environ Microbiol.* , 66(2), 700-11.

[16] Aitken, C. M, Jones, D. M, & Larter, S. R. (2004). Anaerobic hydrocarbon biodegradation in deep subsurface oil reservoirs. *Nature,* , 431(7006), 291-4.

[17] Gray, N. D, Sherry, A, Hubert, C, Dolfing, J, & Head, I. M. (2010). Methanogenic degradation of petroleum hydrocarbons in subsurface environments remediation, heavy oil formation, and energy recovery. *Adv Appl Microbiol.* , 72, 137-61.

[18] Widdel, F, & Rabus, R. (2001). Anaerobic biodegradation of saturated and aromatic hydrocarbons. *Curr Opin Biotechnol 12*, 259-276.

[19] Gieg, L. M, Duncan, K. E, & Suflita, J. M. (2008). Bioenergy production via microbial conversion of residual oil to natural gas. Appl Environ Microbiol , 74, 3022-3029.

[20] Jones, D, Head, I, Gray, N, Adams, J, Rowan, A, Aitken, C, Bennett, B, et al. (2007). Crude-oil biodegradation via methanogenesis in subsurface petroleum reservoirs. *Nature*, 451(7175), 176-180.

[21] Zengler, K, Richnow, H. H, Rossello-mora, R, Michaelis, W, & Widdel, F. (1999). Methane formation from long chain alkanes by anaerobic microorganisms. *Nature* , 401, 266-269.

[22] Beeder, J, Torsvik, T, & Lien, T. (1995). *Thermodesulforhabdus norvegicus* gen. nov., sp. nov., a novel thermophilic sulfate-reducing bacterium from oil field water. *Arch. Microbiol* , 164, 331-336.

[23] Rees, G. N, Grassia, G. S, Sheehy, A. J, Dwivedi, P. P, & Patel, B. K. C. (1995). *Desulfacinum infernum* gen. nov., sp. nov., a thermophilic sulfate-reducing bacterium from a petroleum reservoir. *Int. J. Syst. Bacteriol* , 45, 85-89.

[24] [24] Grabowski, A, Nercessian, O, Fayolle, F, Blanchet, D, & Jeanthon, C. (2005). Microbial diversity in production waters of a low-temperature biodegraded oil reservoir. *FEMS microbiology ecology*, 54(3), 427-43.

[25] Magot, M, Ravot, G, Campaignolle, X, Ollivier, B, Patel, B. K, Fardeau, M. L, Thomas, P, Crolet, J. L, & Garcia, J. L. (1997). Dethiosulfovibrio peptidovorans gen. nov., sp. nov., a new anaerobic, slightly halophilic, thiosulfate-reducing bacterium from corroding offshore oil wells. *Int. J. Syst. Bacteriol.* , 47, 818-824.

[26] Nilsen, R. K, Torsvik, T, & Lien, T. (1996). Desulfotomaculum thermocisternum sp. nov., a sulfate reducer isolated from a hot North Sea oil reservoir. *Int. J. Syst. Bacteriol.* , 46, 397-402.

[27] Ravot, G, Magot, M, Ollivier, B, Patel, B. K. C, Ageron, E, Grimont, P. A. D, Thomas, P, & Garcia, J. L. (1997). *Haloanaerobium congolense* sp. nov., an anaerobic, moderately halophilic, thiosulfate- and sulfur-reducing bacterium from an African oil field. *FEMS Microbiol. Lett.* , 147, 81-88.

[28] Miranda-tello, E, Fardeau, M. L, Fernandez, L, Ramirez, F, Cayol, J. L, Thomas, P, Garcia, J. L, & Ollivier, B. (2003). *Desulfovibrio capillatus* sp. nov., a novel sulfatereducing bacterium isolated from an oil field separator located in the Gulf of Mexico. *Anaerobe* , 9, 97-103.

[29] Nazina, T. N, Tourova, T. P, Poltaraus, A. B, Novikova, E. V, Grigoryan, A. A, Ivanova, A. E, et al. (2001). Taxonomic study of aerobic thermophilic bacilli: Descriptions of *Geobacillus subterraneus* gen. nov., sp. nov. and *Geobacillus uzenensis* sp. nov. from petroleum reservoirs and transfer of *Bacillus stearothermophilus, Bacillus hermocatenulatus, Bacillus thermoleovorans, Bacillus kaustophilus, Bacillus thermoglucosidasius* and *Bacillus thermodenitrificans* to *Geobacillus* as the new combinations *G. stearothermophilus, G. thermocatenulatus, G. thermoleovorans, G. kaustophilus, G. thermoglucosidasius* and *G. thermodenitrificans. Int. J. Syst. Evol. Microbiol.* , 51, 433-446.

[30] Miroshnichenko, M. L, Hippe, H, Stackebrandt, E, Kostrikina, N. A, Chernyh, N. A, Jeanthon, C, Nazina, T. N, Belyaev, S. S, & Bonch-osmolovskaya, E. A. (2001). Isolation and characterization of *Thermococcus sibiricus* sp. nov. from a Western Siberia high-temperature oil reservoir. *Extremophiles.* , 5, 85-91.

[31] Ravot, G, Magot, M, Fardeau, M. L, Patel, B. K. C, Thomas, P, Garcia, J. L, & Ollivier, B. (1999). *Fusibacter paucivorans* gen. nov., sp. nov., an anaerobic, thiosulfate-reducing bacterium from an oil-producing well. *Int. J. Syst. Bacteriol.* , 49, 1141-1147.

[32] Dahle, H, & Birkeland, N. K. (2006). *Thermovirga lienii* gen. nov., sp. nov., a novel moderately thermophilic, anaerobic, amino-acid-degrading bacterium isolated from a North Sea oil well. *Int. J. Syst. Evol. Microbiol.* , 56, 1539-1545.

[33] Nilsen, R. K, & Torsvik, T. (1996). *Methanococcus thermolithotrophicus* isolated from North sea oil field reservoir water. *Appl. Environ. Microbiol.* , 62, 728-731.

[34] Magot, M. (2005). Indigenous microbial communities in oil fields. In B. Ollivier and M. Magot, (Eds.) Petroleum microbiology. ASM, Washington, DC., 21-34.

[35] Söhngen, N. L. (1913). Benzin, Petroleum, Paraffinöl und Paraffin als Kohlenstoff- und Energiequelle für Mikroben. *Zentr Bacteriol Parasitenk Abt II 37*, 595-609.

[36] Widdel, F, & Musat, F. (2010). Diversity and common principles in enzymatic activation of hydrocarbons. In: K. N. Timmis (Ed.) *Handbook of Hydrocarbon and Lipid Microbiology*. Berlin, Heidelberg: Springer Berlin Heidelberg.

[37] Boll, M, & Heider, J. (2010). Anaerobic Degradation of Hydrocarbons: Mechanisms of C-H-Bond activation in the absence of oxygen. In: K. N. Timmis (Ed.) *Handbook of Hydrocarbon and Lipid Microbiology*. Berlin, Heidelberg: Springer Berlin Heidelberg.

[38] Foght, J. (2008). Anaerobic biodegradation of aromatic hydrocarbons: pathways and prospects. *J Mol Microbiol Biotechnol.* 15(2-3): 93-120.

[39] Van Beilen, J. B, & Funhoff, E. G. (2007). Alkane hydroxylases involved in microbial alkane degradation. *Appl Microbiol Biotechnol.* , 74(1), 13-21.

[40] Wentzel, A, Ellingsen, T. E, Kotlar, H. K, Zotchev, S. B, & Throne-holst, M. (2007). Bacterial metabolism of long-chain n-alkanes. *Appl Microbiol Biotechnol.* , 76(6), 1209-1221.

[41] Rojo, F. (2009). *Degradation of alkanes by bacteria. Environmental microbiology.*

[42] Van Hamme, J. D, Singh, A, & Ward, O. P. (2003). Recent advances in petroleum microbiology. *Microbiol Mol Biol Rev.* , 67(4), 503-549.

[43] Rojo, F. (2010). Enzymes for Aerobic Degradation of Alkanes. In K. N. Timmis (Ed.), *Handbook of Hydrocarbon and Lipid Microbiology* (Berlin, Heidelberg: Springer Berlin Heidelberg., 781.

[44] Margesin, R, Labbe, D, Schinner, F, Greer, C, & Whyte, L. (2003). Characterization of hydrocarbon-degrading microbial populations in contaminated and pristine alpine soils. *Appl Environ Microbiol.* , 69(6), 3085-3092.

[45] Kuhn, E, Bellicanta, G. S, & Pellizari, V. H. (2009). New alk genes detected in Antarctic marine sediments. *Environ Microbiol.* , 11(3), 669-673.

[46] Salminen, J. M, Tuomi, P. M, & Jorgensen, K. S. (2008). Functional gene abundances (*nahAc, alkB, xylE*) in the assessment of the efficacy of bioremediation. *Appl Biochem Biotechnol 151*, 638-652.

[47] Hamamura, N, Fukui, M, Ward, D. M, & Inskeep, W. P. (2008). Assessing soil microbial populations responding to crude-oil amendment at different temperatures using phylogenetic, functional gene (*alkB*) and physiological analyses. *Environ Sci Technol 42*, 7580-7586.

[48] Van Beilen, J. B, Wubbolts, M. G, & Witholt, B. (1994). Genetics of alkane oxidation by *Pseudomonas oleovorans*. *Biodegradation 5*, 61-174.

[49] Marchant, R, Sharkey, F. H, Banat, I. M, Rahman, T. J, & Perfumo, A. (2006). The degradation of n-hexadecane in soil by thermophilic geobacilli. *FEMS Microbiol Ecol.* , 56(1), 44-4.

[50] Van Beilen, J. B, Li, Z, Duetz, W. A, Smits, T. H. M, & Witholt, B. (2003). Diversity of Alkane Hydroxylase Systems in the Environment. *Oil Gas Sci Technol.* , 58(4), 427-440.

[51] Tourova, T. P, Nazina, T. N, Mikhailova, E. M, Rodionova, T. A, Ekimov, A. N, Mashukova, A. V, & Poltaraus, A. B. (2008). alkB homologs in thermophilic bacteria of the genus *Geobacillus*. *Mol Biol.* , 42(2), 217-226.

[52] Li, W, Wang, L. Y, Duan, R. Y, Liu, J. F, Gu, J. D, & Mu, B. Z. (2012). Microbial community characteristics of petroleum reservoir production water amended with n-alkanes and incubated under nitrate-, sulfate-reducing and methanogenic conditions. *Inter Biodeterior Biodegradation.* , 69, 87-96.

[53] Vilchez-vargas, R, Junca, H, & Pieper, D. H. (2010). Metabolic networks, microbial ecology and "omics" technologies: towards understanding in situ biodegradation processes. *Environ Microbiol.* , 12, 3089-3104.

[54] Feng, L, Wang, W, Cheng, J, Ren, Y, Zhao, G, Gao, C, Tang, Y, et al. (2007). Genome and proteome of long-chain alkane degrading Geobacillus thermodenitrificans NG80-2 isolated from a deep-subsurface oil reservoir. *Proc Natl Acad Sci U S A.* , 104(13), 5602-7.

[55] Carmona, M, Zamarro, M, Blazquez, B, Durante-rodriguez, G, Juarez, J, & Valderrama, J. (2009). Anaerobic catabolism of aromatic compounds: a genetic and genomic view. *Microbiol Mol Biol Rev. 73*, 71-133.

[56] Brennerova, M. V, Josefiova, J, Brenner, V, Pieper, D. H, & Junca, H. (2009). Metagenomics reveals diversity and abundance of meta-cleavage pathways in microbial communities from soil highly contaminated with jet fuel under air-sparging bioremediation. *Environ Microbiol.* , 11(9), 2216-27.

[57] Pérez-pantoja, D, González, B, & Pieper, D. H. (2010). Aerobic degradation of aromatic hydrocarbons. In: K. N. Timmis (Ed.) *Handbook of Hydrocarbon and Lipid Microbiology*. Berlin, Heidelberg: Springer Berlin Heidelberg.

[58] Jouanneau, Y. (2010). Oxidative inactivation of ring cleavage extradiol dioxigenases: mechanism and ferredoxin mediated reactivation. In: K. N. Timmis (Ed.) *Handbook of Hydrocarbon and Lipid Microbiology*. Berlin, Heidelberg: Springer Berlin Heidelberg.

[59] Junca, H, & Pieper, D. H. (2003). Functional gene diversity analysis in BTEX contaminated soils by means of PCR-SSCP DNA fingerprinting: comparative diversity assessment against bacterial isolates and PCR-DNA clone libraries. *Environ Microbiol.* , 6(2), 95-110.

[60] Mehboob, F, Junca, H, Schraa, G, & Stams, A. J. M. (2009). Growth of Pseudomonas chloritidismutans AW-1(T) on n-alkanes with chlorate as electron acceptor. *Appl Microbiol Biotechnol.*, 83(4), 739-47.

[61] Fuchs, G. (2008). Anaerobic metabolism of aromatic compounds. *Ann N Y Acad Sci.*, 1125, 82-99.

[62] Kube, M, Heider, J, Amann, J, Hufnagel, P, Kühner, S, Beck, A, Reinhardt, R, et al. (2004). Genes involved in the anaerobic degradation of toluene in a denitrifying bacterium, strain EbN1. *Arch Microbiol.*, 181(3), 182-94.

[63] Boll, M, Fuchs, G, & Heider, J. (2002). Anaerobic oxidation of aromatic compounds and hydrocarbons. *Curr Opin Chem Biol.*, 6(5), 604-11.

[64] Kaser, f. M, & Coates, J. D. (2010). Nitrate, Perchlorate and Metal respirers. In: K. N. Timmis (Ed.) *Handbook of Hydrocarbon and Lipid Microbiology*. Berlin, Heidelberg: Springer Berlin Heidelberg.

[65] Grundmann, O, Behrends, A, Rabus, R, Amann, J, Halder, T, Heider, J, & Widdel, F. (2008). Genes encoding the candidate enzyme for anaerobic activation of n-alkanes in the denitrifying bacterium, strain HxN1. *Environ Microbiol.*, 10(2), 376-85.

[66] Kane, S. R, Beller, H. R, Legler, T. C, & Anderson, R. T. (2002). Biochemical and genetic evidence of benzylsuccinate synthase in toluene-degrading, ferric iron-reducing *Geobacter metallireducens. Biodegradation,*, 13(2), 149-54.

[67] Winderl, C, Schaefer, S, & Lueders, T. (2007). Detection of anaerobic toluene and hydrocarbon degraders in contaminated aquifers using benzylsuccinate synthase (bssA) genes as a functional marker. *Environ Microb*iol, 9, 1035-1046.

[68] Winderl, C, Anneser, B, Griebler, C, Meckenstock, R. U, & Lueders, T. (2008). Depth resolved quantification of anaerobic toluene degraders and aquifer microbial community patterns in distinct redox zones of a tar oil contaminant plume. *Appl Environ Microbiol*, 74, 792-801.

[69] Staats, M, Braster, M, & Roling, W. F. M. (2011). Molecular diversity and distribution of aromatic hydrocarbon-degrading anaerobes across a landfill leachate plume. *Environ Microbiol*, 13, 1216-1227.

[70] Wawrik, B, Mendivelso, M, Parisi, V. A, Suflita, J. M, Davidova, I. A, Marks, C. R, Van Nostrand, J. D, Liang, Y, Zhou, J, Huizinga, B. J, et al. (2012). Field and laboratory studies on the bioconversion of coal to methane in the San Juan Basin. *FEMS Microbiol Ecol.*, 81, 26-42.

[71] Callaghan, A. V, Davidova, I. A, Savage-ashlock, K, Parisi, V. A, Gieg, L. M, Suflita, J. M, Kukor, J. J, et al. (2010). Diversity of benzyl- and alkylsuccinate synthase genes in hydrocarbon-impacted environments and enrichment cultures. *Environ Sci Technol.*, 44(19), 7287-94.

[72] Widdel, F, & Grundmann, O. (2010). Biochemistry of the anaerobic degradation of non-methane alkanes. In: K. N. Timmis (Ed.) *Handbook of Hydrocarbon and Lipid Microbiology*. Berlin, Heidelberg: Springer Berlin Heidelberg.

[73] Grossi, V, Cravolaureau, C, Guyoneaud, R, Ranchoupeyruse, A, & Hirschlerrea, A. (2008). Metabolism of n-alkanes and n-alkenes by anaerobic bacteria: A summary. *Org Geochem.* , 39(8), 1197-1203.

[74] Gieg, L. M, Davidova, I. A, Duncan, K. E, & Suflita, J. M. (2010). Methanogenesis, sulfate reduction and crude oil biodegradation in hot Alaskan oilfields. *Environ Microbiol.* , 12(11), 3074-86.

[75] Mbadinga, S. M, Li, K. P, Zhou, L, Wang, L. Y, Yang, S, Liu, Z, Gu, J. F, et al. (2012). Analysis of alkane-dependent methanogenic community derived from production water of a high-temperature petroleum reservoir. *Appl Microbiol Biotechnol.* , 96(2), 531-42.

[76] Gieg, L. M, & Suflita, J. M. (2002). Detection of anaerobic metabolites of saturated and aromatic hydrocarbons in petroleum-contaminated aquifers. *Environ. Sci. Technol.* , 36(17), 3755-3762.

[77] Duncan, K. E, Gieg, L. M, Parisi, V. A, Tanner, R. S, & Suflita, J. M. Green Tringe, S., Bristow, J. ((2009). Biocorrosive thermophilic microbial communities in Alaskan North Slope oil facilities. *Environ Sci Technol 43*, 7977-7984.

[78] Callaghan, A. V, & Wawrik, B. NıChadhain, S.M., Young, L.Y., Zylstra, G.J. ((2008). Anaerobic alkane-degrading strain AK-01 contains two alkylsuccinate synthase genes. *Biochem Biophys Res Commun.* , 366, 142-148.

[79] Zedelius, J, Rabus, R, Grundmann, O, Werner, I, Brodkorb, D, Schreiber, F, Ehrenreich, P, Behrends, A, Wilkes, H, Kube, M, Reinhardt, R, & Widdel, F. (2010). Alkane degradation under anoxic conditions by a nitrate-reducing bacterium with possible involvement of the electron acceptor in substrate activation. *Environ Microbiol Rep.* *3*(1), 125-135.

[80] Mbadinga, S. M, Wang, L. Y, Zhou, L, Liu, J. F, Gu, J. D, & Mu, B. Z. (2011). Microbial communities involved in anaerobic degradation of alkanes. *Inter Biodeterior Biodegradation.* , 65(1), 1-13.

[81] So, C, Phelps, C, & Young, L. (2003). Anaerobic transformation of alkanes to fatty acids by a sulfate-reducing bacterium, strain Hxd3. *Appl Environ. 69*(7), 3892-3900.

[82] Callaghan, A. V, Tierney, M, Phelps, C. D, & Young, L. Y. (2009). Anaerobic biodegradation of n-hexadecane by a nitrate-reducing consortium. *Appl Environ Microb*iol , 75, 1339-1344.

[83] (Head, I., Gray, N., Aitken, C., Sherry, A., Jones, M., Larter, S. (2010). Hydrocarbon activation under sulfate-reducing and methanogenic conditions proceeds by different mechanisms. Geophysical Research Abstracts 12 (EGU General Assembly 2010).

[84] Torsvik, V, Goksoyr, J, & Daae, F. L. (1990). High diversity in DNA of soil bacteria. *Appl Environ Microbiol* , 56, 782-787.

[85] Amann, R. I, Ludwig, W, & Schleifer, K. H. (1995). Phylogenetic identification and in situ detection of individual microbial cells without cultivation. *Microbiol Rev* , 59, 143-169.

[86] Torsvik, V, Daae, F. L, Sandaa, R. A, & Øvreås, L. (1998). Novel techniques for analyzing microbial diversity in natural and perturbed environments. *J Biotechnol* , 64, 53-62.

[87] Kellenberger, E. (2001). Exploring the unknown: the silent revolution of microbiology. *EMBO reports*, 2(1), 2-5.

[88] Orphan, V. J, Goffredi, S. K, Delong, E. F, & Boles, J. R. (2003). Geochemical influence on diversity and microbial processes in high temperature oil reservoirs. *Geomicrobiol J 20*, 295-311.

[89] Nazina, T. N, & Shestakova, N. M. Grigor'yan, A.A., Mikhailova, E.M., Tourova, T.P., Poltaraus, A.B., *et al.* ((2006). Phylogenetic diversity and activity of anaerobic microorganisms of high-temperature horizons of the Dagang oil field (P.R. China). *Microbiology 75*, 55-65.

[90] Mayumi, D, Mochimaru, H, Yoshioka, H, Sakata, S, Maeda, H, Miyagawa, Y, Ikarashi, M, et al. (2011). Evidence for syntrophic acetate oxidation coupled to hydrogenotrophic methanogenesis in the high-temperature petroleum reservoir of Yabase oil field (Japan). *Environ Microbiol.* , 13(8), 1995-2006.

[91] Garcia, J. L, Patel, B. K, & Ollivier, B. (2000). Taxonomic, phylogenetic, and ecological diversity of methanogenic Archaea. *Anaerobe 6*, 205-226.

[92] Zinder, S. H, & Koch, M. (1984). Non-acetoclastic methanogenesis from acetate: acetate oxidation by a thermophilic syntrophic coculture. *Arch Microbiol 138*, 263-272.

[93] Schnurer, A, Houwen, F. P, & Svensson, B. H. (1994). Mesophilic syntrophic acetate oxidation during methane formation by a triculture at high ammonium concentration. *Arch Microbiol 162*, 70-74.

[94] Hattori, S, Kamagata, Y, Hanada, S, & Shoun, H. (2000). *Thermacetogenium phaeum* gen. nov., sp. nov., a strictly anaerobic, thermophilic, syntrophic acetate-oxidizing bacterium. *Int J Syst Evol Microbiol 50*, 1601-1609.

[95] Balk, M, Weijma, J, & Stams, A. J. (2002). *Thermotoga lettingae* sp. nov., a novel thermophilic, methanoldegrading bacterium isolated from a thermophilic anaerobic reactor. *Int J Syst Evol Microbiol 52*, 1361-1368.

[96] Barth, T. (1991). Organic-acids and inorganic-ions in waters from petroleum reservoirs, Norwegian continental-shelf: a multivariate statistical-analysis and comparison with American reservoir formation waters. *Appl Geochem 6*, 1-15.

[97] Silva, T. R, & Verde, L. C. L. Santos Neto, E.V., Oliveira, V.M. ((2012). Diversity analyses of microbial communities in petroleum samples from Brazilian oil fields. Inter Biodeterior Biodegradation doi:10.1016/j.ibiod.2012.05.005.

[98] Dolfing, J, Larter, S. R, & Head, I. M. (2008). Thermodynamic constraints on methanogenic crude oil biodegradation. *ISME J 2*, 442-452.

[99] Uchiyama, T, & Miyazaki, K. (2009). Functional metagenomics for enzyme discovery: challenges to efficient screening. *Curr Opin Biotechnol.*, 20(6), 616-622.

[100] Suenaga, H, Ohnuki, T, & Miyazaki, K. (2007). Functional screening of a metagenomic library for genes involved in microbial degradation of aromatic compounds. *Environ Microbiol.*, 9(9), 2289-2297.

[101] Suenaga, H, Koyama, Y, Miyakoshi, M, Miyazaki, R, Yano, H, Sota, M, Ohtsubo, Y, et al. (2009). Novel organization of aromatic degradation pathway genes in a microbial community as revealed by metagenomic analysis. *ISME J.*, 3(12), 1335-48.

[102] Sierra-garcia, I. N. Caracterização estrutural e funcional de genes de degradação de hidrocarbonetos originados de metagenoma microbiano de reservatório de petróleo. M SC. Thesis. Universidade Estadual de Campinas; (2011).

[103] Uchiyama, T, Abe, T, Ikemura, T, & Watanabe, K. (2005). Substrate-induced gene-expression screening of environmental metagenome libraries for isolation of catabolic genes. *Nat Biotechnol.*, 23(1), 88-93.

[104] Ono, A, Miyazaki, R, Sota, M, Ohtsubo, Y, Nagata, Y, & Tsuda, M. (2007). Isolation and characterization of naphthalene-catabolic genes and plasmids from oil-contaminated soil by using two cultivation-independent approaches. *Appl Microbiol Biotechnol.*, 74(2), 501-10.

Chapter 4

Biodegradation of PCDDs/PCDFs and PCBs

Magdalena Urbaniak

Additional information is available at the end of the chapter

http://dx.doi.org/10.5772/56018

1. Introduction

As a consequence of the rapid development of modern society during the 20th century, a significant amount of organic chemicals has been dispersed into the environment. Many of them have been used as pesticides, insecticides, defoliants and industrial chemicals or produced as undesirable industrial by-products. A large amount of them show several metabolic and toxic activities including mutagenic, immunotoxic and carcinogenic effects. From this group of substances, the organochlorine compounds include polychlorinated dibenzo-*p*-dioxins (PCDDs), polychlorinated dibenzofurans (PCDFs) and polychlorinated biphenyls (PCBs), which have received the most attention according to their persistence in the environment, bioaccumulation and hazard for biota [1].

PCDDs/PCDFs

PCDDs and PCDFs are a group of organic chemicals that contain 75 structurally related individual congeners widely distributed in the environment. They were present on Earth for a long time before humans, as they are formed as a result of forest fires and volcanic explosions. They are also manufactured as unwanted by-products in a range of processes, such as municipal waste incineration, metal smelting, chlorine bleaching in the pulp and paper industry, and vehicular emissions. Such a variety of PCDD/PCDF sources causes their widespread occurrence in the environment. They have been detected in soil, surface water, sediments, plants and animal tissue in all regions of the Earth [2,3].

Chlorinated dioxin's precursor is dibenzo-*p*-dioxin, which consists of two benzene rings bridged by oxygen [4-8] (Fig. 1).

Polychlorinated dibezofurans are similar to polychlorinated dibenzo-*p*-dioxins, in terms of chemical structure and biological activity (Fig. 2).

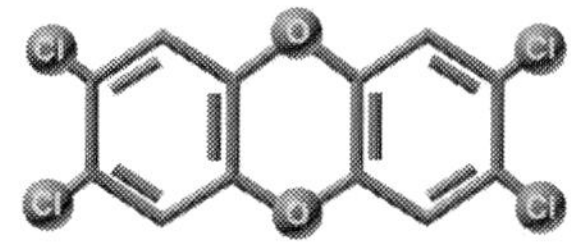

Figure 1. The structural formula of 2,3,7,8-tetrachlorodibenzo-*p*-dioxin [9, changed].

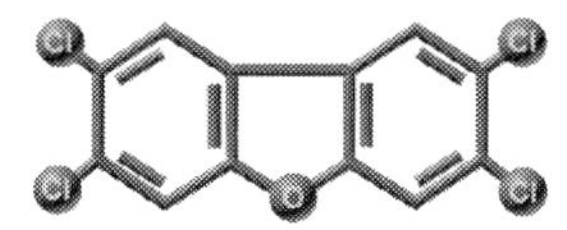

Figure 2. The structural formula 2,3,7,8-tetrachlorodibenzofuran (9, changed].

The physical and chemical properties of toxic congeners of PCDD and PCDF are depicted in Table 1 and 2, respectively.

Compound	Melting point (25°C)	Solubility in water in mg/l (25°C)	Vapour pressure (Pa) in 25°C	Log K_{ow}
2,3,7,8-TCDD	305-306	1.93×10^{-3}	2.0×10^{-7}	6.8
1,2,3,7,8-PeCDD	240-241	1.93×10^{-3}	5.8×10^{-8}	6.64
1,2,3,4,7,8-HxCDD	273-275	4.42×10^{-6}	5.1×10^{-9}	7.8
1,2,3,6,7,8-HxCDD	283-286	4.42×10^{-6}	4.8×10^{-9}	7.8
1,2,3,7,8,9-HxCDD	243-244	4.42×10^{-6}	6.5×10^{-9}	7.8
1,2,3,4,6,7,8-HpCDD	264-265	2.4×10^{-6}	7.5×10^{-10}	8.0
OCDD	325-326	0.75×10^{-7}	$1.1 \times 10^{6,8}$	8.2

Table 1. Physical and chemical properties of PCDDs [10, changed].

PCBs, in turn, due to their stable properties such as low dielectric constant, chemical inertness, non-flammability, high heat capacity, high electrical resistivity and low acute toxicity, were found to be ideal for industrial applications and thus were produced and used in many countries including the United States, Russia, Japan, France and Czechoslovakia. Global PCBs use is estimated to be 1.2 to 1.5 million tonnes. Although the production and use of PCBs was banned almost all over the world more than 30 years ago due to their toxic effects on humans and biota, they are still detected in many ecosystem compartments [11-14]. The PCB molecule consists of two phenyl rings, in which the chlorine atoms are substituted in place of hydrogen atoms. Theoretically, there could be 209 individual PCB congeners (Fig. 3).

Compound	Melting point (25°C)	Solubility in water in mg/l (22.7°C)	Vapour pressure (Pa) in 25°C	Log K_{ow}
2,3,7,8-TCDF	227-228	4.19×10^{-4}	2.0×10^{-6}	6.53
1,2,3,7,8-PeCDF	225-227	4.19×10^{-4}	2.3×10^{-7}	6.79
2,3,4,7,8-PeCDF	196-196.5	2.36×10^{-4}	3.5×10^{-7}	6.92
1,2,3,4,7,8-HxCDF	225.5-226.5	8.25×10^{-6}	3.2×10^{-8}	6.92
1,2,3,6,7,8-HxCDF	232-234	1.77×10^{-6}	2.9×10^{-8}	6.92
1,2,3,7,8,9-HxCDF	246-249	1.77×10^{-6}	2.4×10^{-8}	6.92
2,3,4,6,7,8-HxCDF	239-240	1.77×10^{-6}	2.6×10^{-8}	6.92
1,2,3,4,6,7,8-HpCDF	236-237	1.35×10^{-6}	4.7×10^{-9}	7.92
1,2,3,4,7,8,9-HpCDD	221-223	1.35×10^{-6}	6.2×10^{-9}	7.92
OCDF	258-260	1.16×10^{-6} (in 25 °C)	5×10^{-9}	8.78

Table 2. Physical and chemical properties of PCDFs [10, changed].

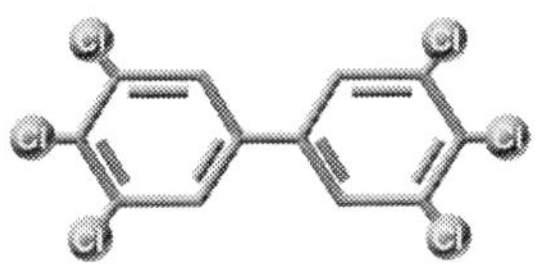

Figure 3. The structural formula of 2,2 ', 3,3', 4,4 '-hexachlorobiphenyl [9, changed].

PCBs have been produced under several trade names, e.g., Clophen (Bayer, Germany), Aroclor (Monsanto, USA), Kanechlor (Kanegafuchi, Japan), Santothrem (Mitsubishi, Japan), Phenoclor and Pyralene (Prodolec, France) (Table 3).

Apirolio	Diaclor	No-Flamol
Areclor	Duconol	Pydraul
Aroclor	Dykanol	Pyralene
Arubren	Elemex	Pyranol
Asbestol	Euracel	Pyroclor
Askarel	Fenchlor	Phenoclor
Bakola	Hivar	Saf-T-Khul
Biclor	Hydol	Santotherm
Chlorextol	Inclor	Santovac
Chlorinol	Iterteen	Siclonyl
Chlorphen	Kennechlor	Solvol
Clophen	Montar	Sovol
Delor	Nepolin	Therminol

Table 3. Major trade names of PCBs [15].

Commercial PCBs are complex mixtures of 30–60 congeners, which are the major PCB components of most environmental extracts. Each individual compound shows a unique combination of physico-chemical and biological properties dependent on the degree of chlorination (Table 4).

Aroclor compound	Water solubility (mg/l) 25°C	Vapour pressure 25°C	Density 25°C [g/cm3]	Appearance	Boiling point [°C]
Aroclor 1016	0.4200	$4.0{\times}10^{-4}$	1.33	Clear oil	325–356
Aroclor 1221	0.5900	$6.7{\times}10^{-3}$	1.15	Clear oil	275–320
Aroclor 1232	0.4500	$4.1{\times}10^{-3}$	1.24	Clear oil	290–325
Aroclor 1242	0.2400	$4.1{\times}10^{-3}$	1.35	Clear oil	325–366
Aroclor 1248	0.0540	$4.9{\times}10^{-4}$	1.41	Clear oil	340–375
Aroclor 1254	0.0210	$7.7{\times}10^{-5}$	1.50	Light, yellow, viscous oil	365–390
Aroclor 1260	0.0027	$4.0{\times}10^{-5}$	1.58	Light, yellow, viscous oil	385–420

Table 4. Physical and chemical properties of selected Aroclors [15, after 16].

Currently, many countries impose strict controls on the use and release of PCDDs/PCDFs and PCBs. As a result their input into the environment has decreased significantly. Nevertheless, their release from contaminated sites and their redistribution on a global scale is still observed [17-18]. Their slow decomposition in the environment and the hazards they pose for living organisms makes PCDDs/PCDFs and PCBs large-scale environmental degraders, especially because their toxicity can be further enhanced by their ability to accumulate in the soil and sediments and their bioaccumulation and biomagnification within aquatic and land food chains (Fig. 4).

It should also be underlined that PCDDs/PCDFs and PCBs also pose a risk to human health. They have been shown to produce toxic responses similar to those caused by 2,3,7,8-TCDD, the most potent congener within this group. Studies on animals demonstrate that PCDDs/PCDFs and PCBs are implicated in mutagenic and carcinogenic effects such as liver damage, malignant melanoma and preneoplastic and neoplastic changes [1, 19]. Other manifestations related to PCDDs/PCDFs and PCBs are gastrointestinal (gastric hyperplasia, ulceration, necrosis), respiratory (chronic bronchitis and coughs), dermal (chloracne, oedema, alopecia, hyperkeratosis of epithelium), neurotoxic (impaired behavioural responses, depressed motor activity, developmental deficits, numbness) and immunotoxic (lymphoid tissue atrophy, leukocyte and lymphocyte reduction, suppressed antibody responses), hepatotoxic (hepatomegaly, hyperplasia of the bile duct, necrosis, fatty degeneration, porphyria) and reproductive problems (decreased sperm motility and number, increased miscarriages, decreased survival and mating success) [1, 19].

http://dx.doi.org/10.5772/56018

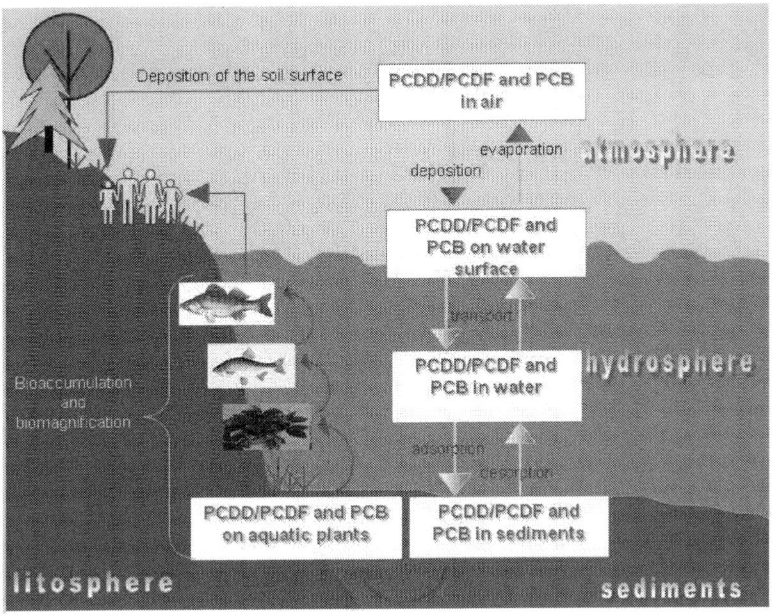

Figure 4. Transport and circulation of PCDDs/PCDFs and PCBs in the environment.

2. Microbiological transformation of PCDDs/PCDFs and PCBs

The degradation of PCDDs/PCDFs and PCBs is classified into two sections: biological transformation by microorganism activity and physico-chemical transformation.

The first group includes anaerobic, aerobic and sequential anaerobic-aerobic transformation. The latter can be classified into photochemical and thermal degradation.

Microbiological transformation depends on enzymes produced by microorganisms which enable modification of toxic compounds into less toxic forms. Biological degradation can carry on as mineralization when microorganisms use the organic compound as a source of carbon and energy, or as co-metabolism where microorganisms need other sources of carbon and energy and the transformation of pollutants occurs as a concurrent process. Products of this process can be further mineralized, otherwise incomplete degradation occurs, leading to the formation and accumulation of more toxic metabolites than parent substrates.

The effectiveness of degradation rates varies depending on the conditions present in the environment and comprises: 1) input of pollutants, 2) physical parameters (oxygen content, temperature, light intensity, pH, conductivity) and 3) biological parameters (presence of microorganisms able to degrade a given pollutant and the availability of carbon and/or other sources of energy). All of the above variables determine the rate of biological and physical transformation of analysed compounds.

2.1. Aerobic conditions

Bacterial cometabolism

Aerobic transformation occurs in environments that are rich in oxygen and involves the use of microbial molecules, such as mono- and dichlorinated PCDDs/PCDFs and PCBs, as a source of carbon and energy. It should be noted that in about 90% of cases, the process takes place as co-metabolism, which means that the microorganisms need an additional source of carbon apart from PCDDs/PCDFs or PCBs.

Data from the literature confirms the aerobic biodegradation of PCDD/PCDF and PCB compounds and the rate of this process increases with the reduction of PCDD/PCDF and PCB chlorination [20-23]. Thus, for example, molecules containing five or more chlorine atoms are not susceptible to the effects of aerobic microorganisms.

PCDDs/PCDFs

In the case of PCDDs and PCDFs the research conducted over the last 30 years has widely described their aerobic biodegradation [19-22, 24]. Worldwide studies have demonstrated that many isolated strains of bacteria, such as *Rhodococus opacus* SAO101, *Beijerinckia* sp. B8/36, *Psudomonas veronii* PH-03, *Psudomonas* sp. HH69, CA10, EE41, *Bacillus megaterium* AL4V, *Sphingomonas* sp. RWI and HL7, are capable of the biodegradation of slightly chlorinated PCDDs/PCDFs under aerobic conditions [21, 24-29]. To increase the rate of aerobic biodegradation of PCDDs/PCDFs and PCBs an additional source of carbon, for example a small amount of un-substituted PCDD or biphenyls [20], carbazole [30], o-dichlorobenzene [25] or benzoic acid or 3-methoxybenzoic [30] can be used.

PCBs

The first data on the aerobic degradation of PCBs was reported by Ahmed and Focht [31] in 1973 and the respective study was devoted to the degradation of biphenyl and monochlorobiphenyl to chlorobenzoic acid by two species of *Achromobacter*. Furukawa et al. [32] demonstrated that a species of *Acinetobacter* and *Alcaligenes* can rapidly adsorb 2,5,2′ trichlorobiphenyl onto the cell surface, then metabolize and release metabolic compounds from the cell. Since then numerous investigations have focused on the occurrence and distribution of PCB-degrading microorganisms and their capability to biodegrade PCBs. For example, Clark et al. [33] reported that *Alcalegenes denitrificants* and *A. odorans* can degradate Aroclor 1242 (a mixture of PCB containing 42% chlorine) by co-metabolism. A study by Novakova et al. [34] showed the results of the degradation of Delor 103 by *Psudomonas* sp. P2 and *Alcaligenes eutropha.* Optimal PCB degradation was obtained by the addition of biphenyl, saccharose, agar or an amino acid mixture as the source of carbon. A reduction of degradation efficiency was observed by the addition of glycerol or pyruvate. To completely degrade PCBs by aerobic bacteria, various microbial strains with specific congener preferences are required.

Bacterial mineralization

According to data described by Field and Sierra-Alvarez [35] there are few well documented examples of chlorinated PCDDs/PCDFs and PCBs serving as the sole source of carbon and energy for pure bacterial strains. This is shown by the research of Hong et al. [28] wherein the *Pseudomonas veronii* PH-03 has been used to utilize 1-CDD and 2-CDD growing on aliphatic acids generated from ring cleavage. The mentioned strain of *Pseudomonas veronii* accumulated the dead products 3-chlorocatchol and 4-chlorocatchol from the chlorinated rings. Similar results were also obtained by Arfmann et al. [36] by using a *Sphingomonas* sp. strain RW1 growing on 4CDF. The substrate of carbon and energy was a 5-carbon aliphatic acid and a 2-hydroxypenta-2,4dienoate released from the ring cleavage and the dead-end products were 3-chlorosalicylic acid.

The complete mineralization of PCDDs/PCDFs was also achieved by using co-cultures including a PCDD/PCDF-degrader and a 3-chlorosalicylic acid-degrader. For example, a study by Wittich et al. [37] showed that use of *Sphingomonas* sp RW16 and *Pseudomonas* sp. RW10 enabled the complete degradation of 2-CDF and 3CDF. The co-culture mixture combined with *Sphingomonas* sp. RW1 and *Burkholderia* sp. JWS was shown to completely degrade 4-CDF [36]. The above research demonstrates that *Sphingomonas* sp. RW16 and *Sphingomonas* sp. RW1 were capable of degrading the CDF and the *Pseudomonas* sp. RW10. *Burkholderia* sp. JWS utilized the 3-chlorosalicylic acid as the released as dead-end product.

Fungal cometabolism

It should also be mentioned that fungi, similarly to bacteria, are capable of PCDD/PCDF degradation in aerobic conditions, in both mineralization and the co-metabolism process.

Fungi use enzymes (lignin peroxidase or manganese peroxidase) to oxidise the molecule of the compound. The first described case of use of the fungal biodegradation is the work of Bumpus et al. [38], in which the authors documented the mineralization of [^{14}C] 2,3,7,8-TCDD to $^{14}CO_2$ within 30 days by the fungi of *Phanerochaete chrysosporium. P. chrysosporium* has also been successfully used to degrade 2,7-DCDD [39].

The biodegradation activity of fungi is not limited to less chlorinated congeners. There is evidence that *P. chrysosporium* is able to remove 34% and 48% of a mixture of congeners containing from 5 to 8 chlorine atoms in the molecule during 7 and 14 days [40].

2.2. Anaerobic conditions

Anaerobic microorganisms are well adapted to pollutants with a high carbon concentration due to the diffusional limitation of oxygen. Anaerobic transformations of PCDDs/PCDFs and PCBs include reductive dehalogenation using PCDDs/PCDFs and PCBs as electron acceptors. During this process a substituent chlorine atom is replaced with a hydrogen atom.

Reductive dehalogenation occurs in soils and sediments, where different microorganisms possessing dehalogenation enzymes responsible for dechlorination and dehalogenation processes exist. The rate, extent and route of dechlorination are dependent on environmental factors, such as carbon availability, electron donors, presence of electron acceptors other than PCDDs/PCDFs and PCBs, temperature and pH. All of these factors influence the composition of a microorganism's community and their activity.

PCDDs/PCDFs

The first evidence of degradation of PCDDs/PCDFs under anaerobic conditions was obtained by spiking sediment microcosms with highly chlorinated congeners of HpCDD, HxCDD and PeCDD [40].The rate of removal of those compounds in biologically active sediments was from 19% to 56% higher in comparison to heat-killed sediments. The products of such biodegradation processes were TCDD and TCDF congeners [40, 41]. The main microorganisms capable of efficient degradation of these compounds were mainly bacteria of the genus *Dehalococcoides* [43-45]. Experiments with the use of OCDD (8 chlorine atoms) at a concentration of 5.3 ml/L applied into sediment microcosms, showed that after 7 months the congener was distributed into forms that contain only 1 to 3 chlorine atoms [46-47].

PCBs

The first evidence of anaerobic degradation of PCBs was reported based on the observed modification of Hudson River and Silver Lakes sediments contaminated by commercially produced PCBs. The increase of low-chlorinated PCBs in comparison to the high-chlorinated congeners was consistent with reductive dechlorination [48]. Furukawa et al. [49] demonstrated that species of *Acinetobacter* and *Alcaligenes* may rapidly adsorb 2,5,2′-trichlorobiphenyl onto the cell surface and then metabolise and release metabolic compounds from the cell. From that time many of investigations were devoted to the occurrence and distribution of PCB-degrading microorganisms and their capability to biodegrade PCBs.

Master et al. [48] showed that many commercial PCB mixtures can be reductively dechlorinated under anaerobic conditions, for example, Aroclor was dechlorinated at rates of 3 μg Cl/g of sediment per week. The dechlorination occurs at temperatures of 12°C and PCB concentrations of 100–1000ppm [49]. Fava et al. [50] described the degradation of Aroclor 1242 by three strains: *Comamonas testosteroni, Rhodococcus rhodochrus* and *Psudomonas putida* with total losses of 13.8%, 19.1% and 24.6%, respectively. In both experiments, the favoured positions for dechlorination were (in order) meta>para>ortho and preference was shown for "open" sites 2 and 3, indicative of the action of 2,3-dioxygenase enzymes [50]. Fava et al. [50] reported that the dechlorination of Fenclor 54 primarily occurred from the meta- and para positions, while ortho-substituted congeners accumulated in the medium. Other studies showed an inability of anaerobic microorganisms to degrade the low chlorinated biphenyls. The occurrence of diortho- and monoorthochlorobiphenyls, as well as the biphenyl rings, was identified even after a one year incubation [31].

2.3. Sequential anaerobic-aerobic conditions

Laboratory experiments showed that microbial degradation of lower chlorinated PCDDs/PCDFs and PCBs occurs at a faster rate than in higher chlorinated ones. Lower chlorinated congeners produced by dechlorination can be readily degraded by indigenous bacteria, which in consequence, reduces the potential bioconcentration risk and the exposure to PCDDs/PCDFs and PCBs by conversion to congeners with a low bioaccumulation potential in the food chain [35, 51]. The lightly chlorinated PCDDs/PCDFs and PCBs congeners produced during the anaerobic dechlorination may then be substrates for oxidative destruction by aerobic microorganisms, which leads to the production of chlorobenzoic acid, which is easily degraded by bacteria.

The findings described above indicate that a complete degradation of PCDDs/PCDFs and PCBs can be achieved by sequential exposure to anaerobic and aerobic biodegradation. Highly chlorinated congeners can be transformed to compounds of lower chlorination during reductive dechlorination under anaerobic conditions. Lightly chlorinated congeners, produced during anaerobic dechlorination, might then become substrates for oxidative destruction by aerobic microorganisms, which can lead to the production of chlorobenzoic acid, which is further easily degraded by bacteria [34, 51].

3. Physical transformation of PCDDs/PCDFs and PCBs

There is also a division of degradation processes that takes into account the physicochemical degradation of PCDD/PCDF and PCB compounds.

3.1. Photochemical degradation

Photochemical degradation called photolysis also depends on the degree of chlorination, the position of chlorine atoms in the biphenyl ring and the solvent used for PCDD/PCDF and PCB dissolution. The primary process in photoreaction is reductive dechlorination, but examples of photo-induced isomerization and condensation of individual chlorobiphenyls have been also reported.

The first laboratory experiments on photolysis were conducted with mercury lamps as the UV source, with a wavelength of about 254nm, which results in the dechlorination of PCBs. Later, sunlight simulating lamps were used, which also confirmed the degradation of the chlorinated compounds [52-54].

It should also be mentioned that the higher chlorinated biphenyls undergo photolysis faster than less chlorinated ones. For example, the exposure of PCB to a 310nm wavelength causes of reduction of about 70% tetra-, 96% of hexa- and 99% of octachlorobiphenyl. Experiments with tetrachlorobiphenyls showed that the major products after irradiation at 300nm are di- and trichlorinated biphenyls [52]. Bunce et al. [53] reported intensified photodegradation with increased irradiation time.

Photolysis is regarded as one of the major processes reducing PCDDs/PCDFs and PCBs in the environment. Bunce at al. [53] estimated the loss of PCBs in natural waters at the magnitude of 10 to 1000g/Km^{-2}/year. In shallow water bodies at least one chlorine atom from mostly chlorinated PCB molecules is photodegradated per year. Zepp et al. [54] reported that humic acids and suspended materials may induce and accelerate PCB photodegradation.

Several researchers described accelerated in-situ photolysis by the addition of various organics, such as isooctane, hexane and cyclohexane, on the surface of contaminated soil [56-58]. Doughtery et al. [59] found that solar-induced photolysis reactions can be a principal mechanism for the transformation of PCDD/PCDF to less toxic forms.

3.2. Thermal degradation

The last group of PCDD/PCDF and PCB transformations is thermal degradation, leading to the complete destruction of toxic substances at temperatures above 700°C or producing more toxic congeners such as TCDD at temperatures below 700°C. This kind of PCDD/PCDF and PCB destruction is well adapted on an industrial scale for the safe disposal of waste products containing PCDDs/PCDFs and PCBs.

4. Environmental biodegradation of PCDD/PCDF and PCB

PCDDs/PCDFs and PCBs are substances that are created during different types of natural and industrial processes. Their appearance in the environment and in consequence in food products creates a serious threat to human health and ecosystem functioning as far as their genotoxic and toxic effects on living organisms are concerned [59]. Therefore, natural transformation of PCDDs/PCDFs and PCBs is a critical event in determining their fate in the environment.

4.1. Phytoremediation

Phytoremediation is defined, according to Macek et al. [61], after Cunningham and Betri [62] and Cunningham et al. [63], as the use of green plants to remove, contain, or render harmless environmental contaminants. According to other authors, phytotechnology is a set of technologies that use plants to remediate contaminated sites [64-68].

Phytoremediation uses living plants for the remediation of contaminated mediums, such as soil, sediment, sludge and water (in situ as well as ex situ) by the removal, degradation or stabilization of a given contaminant [64].

According to Macek et al. [61], after Salt et al. [69], phytoremediation is currently divided into several subtypes:

- phytoextraction
- phytodegradation
- rhizofiltration
- phytostabilization
- phytovolatilization

These techniques are an alternative to the widely used methods of physical, physico-chemical and thermal remediation. Their advantages include the possibility of application ex-situ and in-situ, low investment and operating costs with high effectiveness and non-invasiveness in the environment [70-72].

The main problem with the use of phytoremediation techniques is their long operational time and the fact that many of the bioremediation techniques are still in the experimental stage [70-72].

The genesis of the phytoremediation process was observed by the rate of degradation of organic chemicals in the soil with and without vegetation cover. On the basis of the obtained results it was stated that vegetation cover promotes the reduction of organic compounds in soil. Currently, a variety of research indicates the positive effects of using higher plants to degrade organic compounds [73-81].

Siciliano et al. [73] demonstrated the reduction of organochlorine compounds by about 30% during 2 years of plant cultivation; whereas on the soil without plants, the reduction was 2 times lower. Nedunuri et al. [74] reported the reduction of aromatic compounds by about 42% and 50% by using fibre flax (*Lolium annual*) and St. Augustine grass (*Stenotaphrum secundatum*), respectively, over a period of 21 months. Other examples showed remediation of soil contaminated by crude oil using a combination of grass and fertilizers [74-77]. Vervaeke et al. [78] reported a 57% reduction of aromatic compounds and mineral oils during 1.5 years of willow (*Salix viminalis*) cultivation. Pradham et al. [79] demonstrated the usage of phytoremediation as a primary remediation technology and as a final step for treatment of soil contaminated with PAHs. The authors recorded a 57% reduction in PAHs after 6 months of alfalfa (*Medicago sativa*), switch grass (*Panicum virgatum*) and little bluestem grass (*Schizachyrian scoparium*) growth.

A study by Gregor and Fletcher [80] demonstrated the ability of plant cells to metabolize PCBs. While, research by Jou et al. [81] showed the uptake of PCDDs/PCDFs by *Boussonetia papyrifera* growing on highly contaminated soil. The authors reported similar concentrations and distributions of PCDD/PCDF and PCB congeners in plant tissues and soils. Other research demonstrated that several plants of the genus *Cucurbita* (e.g., courgette, pumpkin and squash) can readily take up PCDD and PCDF from soil and translocate them to leaves and fruits [82-84]. It was also found that *Cucurbita* plants can phytoextract PCBs from soil and translocate some quantities to aerial tissues [85, 86]. This confirms that the PCDD/PCDF and PCB contents in plants may closely relate to the surrounding environments where plants grow [81]. Nevertheless, Uegaki et al. [87] reported no concentration differences in brown rice grown in three different soils: dioxin-contaminated soil, paddy soil and upland soil. The authors assumed that growing rice in soil contaminated with high concentrations of dioxins has no influence of the PCDD/PCDF levels in rice tissue [87].

4.2. Rhizoremediation

Rhizoremediation of organic micropollutants is one of the most effective remediation processes due to existing interactions in the rhizosphere between plant roots, plant exudates, soil and microorganisms. Mackova et al. [64] reported that plants support bioremediation by the release of exudates and enzymes that stimulate both microbial and biochemical activity in the surrounding soil and mineralization in the rhizosphere. Plants can also accelerate bioremediation in surface soils by stimulating the growth and metabolism of soil microorganisms through the release of nutrients and the transport of oxygen to their roots [61-62, 67]. Moreover, the fact that up to 40% of carbohydrates, amino acids and other photosynthesis products are stored in the plant rhizosphere, plays an important role in the availability of carbon used by microorganisms in the co-metabolism process.

A study by Whipps [88] demonstrated that 1g of rhizosphere soil contains a 10^{12} higher amount of microorganisms in comparison to non-planted soil. Microorganisms settling in the rhizosphere also play a role in the protection of plants against pathogens and stress induced by too high a concentration of contaminants and facilitate nutrient uptake by a given plant [89-93].

Bacteria present in the rhizosphere soil serve remediation functions by secreting the appropriate enzymes (e.g., peroxidase, phosphatase, dioxygenase, P450 monooxygenase, dehalogenaza, nitrylases and nitroreductase) involved in the degradation of organic pollutants. Such enzymes are also found in plants and fungi that colonize plant roots. This led to a thesis on the interaction of plants and microorganisms in order to completely destroy a given pollutant [93-99]. This process is called rhizodegradation and is defined as the degradation of pollutants in the root zones of plants (rhizosphere).

The effectiveness of rhizosphere biodegradation depends on the ability of microorganisms to adapt to a given pollution concentration and the effectiveness of root colonization [97]. The interactions between plants, soil and rhizosphere microorganisms are multifaceted and according to Macek et al. [61] can give mutual benefit to both organisms. This mutualistic relationship is responsible for the accelerated degradation of soil contaminants in the presence of plants [101]. Research on this issue is ongoing. Already existing publications confirm the validity of the use of rhizoremediation to reduce PCDDs/PCDFs and dl-PCBs. For example, an article by Kuiper et al. [98] demonstrated that naturally occurring rhizosphere biodegradation can be enhanced by the addition of microorganisms to the rhizosphere.

The important group of substances present in the rhizosphere are complexes of aromatic compounds such as flavonoids and coumarins. These compounds are used by bacterial microflora as a source of carbon and nitrogen [73, 98-99, 102-103]. They are structurally similar to organic compounds such as PCBs and PAHs. This indicates the potential of using such evolutionary established metabolic pathways of rhizosphere microorganisms for the remediation of organic pollutants [104]. Thus, many researchers are interested in the ability of microorganisms inhabiting the rhizosphere to degrade organochlorine pollutants and the role of flavonoids and coumarins produced by plants [99, 103, 105-108].

Worldwide studies describe many kinds of pollutants including PCBs, PAH, petroleum hydrocarbons, chlorinated pesticides like Pentachlorophenol and 2,4-Dichlorophenoxyacetic acid, which were more rapidly degraded in the rhizosphere compared to the bulk soil [64, 109-111]. Research by Betts [112] conducted on soil contaminated by petroleum hydrocarbons showed its considerable improvement by using several plants species such as Bermuda grass, rye grass, white clover and tall fescue. A study by Burken and Schnoor [113] described the positive role of root exudates on atrazine uptake by plants (poplar trees). The research also showed that phenolics, flavonoids and terpenes present in root exudates can induce the bacterial degradation of PCBs [61, 114—115]. A study by Mackova et al. [116] showed the effect of tobacco, nightshade, alfalfa and horseradish on PCB removal from contaminated soil. The obtained results showed 6% to 33.7% removal of PCBs during 6 months of experimentation. The authors also underline the role of the studied plants as a source of bacterial consortia capable of PCB degradation.

5. Perspectives in environmental biodegradation of PCDDs/PCDFs and PCBs

PCDDs/PCDFs and PCBs are compounds that occur in all types and structures of ecosystems. Their transfer takes place through biogeochemical cycles, but it is their long half-life in the environment, their accumulation and biomagnification in aquatic and terrestrial food chains and their toxicity that determine their long-term and large-scale threat to the environment and humans. As a result, one of the priority tasks of recent research on PCDDs/PCDFs and PCBs is to characterize the processes that determine their transport and deposition in ecosystems, in order to regulate their allocation and diminish their concentration. Reversing ecosystem degradation and reducing PCDD/PCDF and PCB concentrations in the environment requires solutions based on integrative problem-solving science, such as ecological engineering and ecohydrology [117].

A key element of the ecohydrology theory is the assumption that an excess amount of pollutants including PCDDs/PCDFs and dl-PCBs and their negative effects on the environment can be limited by so-called "dual regulation". Until now, the above methodology was used to reduce the occurrence of toxic cyanobacterial blooms resulting from excessive inflow of phosphorus into water. This concept involves the use of biological and hydrological processes to control the amount and allocation of phosphorus in the ecosystem through increasing biofiltration and by the formation of ecosystem biota [118-120].

Similarly, in order to diminish the concentration of PCDDs/PCDFs and PCBs in the environment there is a need to not only reduce the pollutant load from point and non-point sources but also to develop and apply in-situ bioremediation strategies [72, 117-123]. The application of bioremediation technologies should focus on the possibilities of exploiting and strengthening the functioning of the given ecosystem to reduce the recorded concentrations of PCDDs/PCDFs and PCBs.

The phyto-and rhizoremediation techniques described above are examples of the use of the natural properties of the ecosystem to reduce the environmental PCDD/PCDF and PCB contamination.

Currently, in order to improve the rate and efficiency of such remediation processes a number of advantages have been developed and applied. Some of them are focused on the stimulation of growth and activity in microbial communities in order to accelerate remediation efficiency and diminish the concentration of PCDDs/PCDFs and PCBs in environment.

It should be underlined that there are two main types of microorganism: indigenous and exogenous. Indigenous ones are those that are found already living at a given site. To stimulate the growth of these indigenous microorganisms, the proper soil temperature, oxygen and nutrient content may need to be provided. If the biological activity needed to degrade a particular contaminant is not present in the soil at the site, microorganisms from other locations, whose effectiveness has been tested, can be added to the contaminated soil. These are called exogenous microorganisms [56]. Research has shown that the stimulation of an indigenous microbial population, by injecting methanol and acetate as an electron donor, enhances the removal of tetrachloroethane (PCE) to ethane [124]. Nevertheless until now, scientists have been faced with the problem of the application of isolated microorganisms in situ, as they are often unable to adapt and compete with microorganisms naturally occurring at contaminated sites. This is mainly due to the inability to grow a culture of microorganisms below a certain depth, the lack of sufficient amounts of nitrogen, phosphorus and carbon in the environment, the low bioavailability of pollutants and the preferential use of carbon from non-toxic substrates rather than toxic. An important role is played by the presence of contaminants that inhibit the growth of microorganisms. Currently, in order to avoid such a situation the analogues of the natural soil contaminant are added to the remediated soil. This stimulates the micropollutants' degradation pathways in the microorganisms' cells [99,105,125].

Another problem with bioremediation is the availability of the contaminant to the degrading organisms. To solve this problem research has been conducted on the use of surfactants as potential agents for enhancing solubility and removing contaminants from soil and sediments [126-128]. As reported by Nakajima et al. [129], the addition of sodium dodecyl sulphate, Triton-100 and sodium taurocholate increases the bioavailability of PCBs and PAHs.

Bioaugmentation is another method used in order to improve the microbial degradation of pollutants. This process is based on the introduction of appropriate species for the degradation of specific contaminants. The efficacy of bioaugmentation is contradictory, as far as both positive and negative results have been obtained. A successful bioaugmentation was observed for the remediation of PAHs in sediments [124]. Nevertheless, other studies have achieved no positive results [130].

On the basis of the above data, contemporary bioremediation strategies should be implemented in combination, for example phytoremediation and biostimulation or rhizoremediation and bioaugmentation. This would accelerate the usage of plants and enhance the activity of degrading microorganisms in order to minimize the risk played by PCDDs/PCDFs and PCBs.

It is also possible to remediate soil by using transgenic organisms. Currently, most of the research into the use of transgenic organisms is carried out on a laboratory scale. These experiments are mainly concerned with the introduction of genes encoding biosynthetic pathways of biosurfactants (in order to increase the bioavailability of contaminants), the introduction of genes that enable increased resistance to given contaminants in microbial communities or genes encoding the enzymes' degradative pathways (e.g., cytochrome P450) [131-136].

The latest research by Lan Chun et al. [136] demonstrated the positive role of the electrical stimulation of microbial PCB degradation. The authors found a 40-60% reduction in total PCB concentration in weathered sediments exposed to electric currents, while no significant decrease in PCB concentration was observed in control sediments.

The techniques described above and their advantages, such as biostimulation and bioaugmentation, can be adopted and used in large-scale remediation processes. Examples of such an approach include the utilization of wetlands and biofilters.

Wetlands are often described as "the kidneys of the landscape" owing to their the intrinsic function to transform and store organic matter and nutrients [138] and associated micropollutants such as PCDDs/PCDFs and PCBs. This ability has been exploited for water quality improvement [138]. Constructed wetlands were first used for wastewater treatment in the 1950s. In recent years constructed wetlands have been widely used for urban and agricultural runoff treatment. They utilize natural processes to purify water in a sustainable, cost and energy effective way with minimal operation and maintenance cost [140]. Furthermore, the usage of constructed wetlands as tools in the treatment of polluted waters, has been gaining popularity as an ecological engineering alternative over conventional, chemical based methods [141-142]. Several scholars have shown successful utilizations of constructed wetlands for the treatment of a wide variety of wastewaters including industrial effluents [142-144], urban storm water, agricultural runoff [146-147], domestic wastewater [148] and animal wastewaters [149]. Schulz and Peall [150] determined the effectiveness of constructed wetlands in retaining agricultural pesticide pollution as 89% during runoff. Several researchers have proven the ability of constructed wetlands to mitigate pesticide pollution derived from various agricultural nonpoint sources [151-155].Considering the above, it appears that the use of constructed wetlands to purify water from organochlorine compounds is a promising challenge.

Furthermore, the use of land-water ecotones constructed in a river valley with different kinds of plants and microorganisms may partially purify the inflowing surface- and groundwater contamination by PCDDs/PCDFs and PCBs [156-157]. Such structures may capture, immobilize and/or degrade PCDDs/PCDFs and dl-PCBs [61, 96, 103].

The other promising solution involves the use of biofilters for the purification of inflowing water, wastewater, leachate etc. Such biofilters combined with areas of intensive sedimentation, which enable the deposition of matter, nutrients and micropollutants and their further biodegradation by existing microbial consortia and areas of macrophyte growth, wherein intensive phytodegradation processes occur, are considered to be one of the most effective solutions for pollutant removal. Results obtained by Urbaniak et al. [158] in the Asella Demonstration Project demonstrated changes in the Toxic Equivalent (TEQ) of PCDDs/PCDFs in the sediments of the Asella river and lake taken before and after biofilter construction. Authors showed a 70% reduction in sediment toxicity after one year of biofilter implementation. This indicates the positive role of biofiltration in the quality of lake ecosystems and in consequence on human health. The implementation of such biofiltration system enabled a reduction in the input of PCDDs/PCDFs into the lake through sedimentation and due to acceleration of photo- and biodegradation processes the quality of the whole river-lake system was improved.

6. Conclusions

PCDDs/PCDFs and PCBs pose one of the most challenging problems in environmental science and technology. Their fate, transport and biodegradation in the environment occur via complex networks, involving complicated interactions with other contaminants and with various physiological, chemical and biological processes. Those processes can be used and modified in order to diminish their environmental concentration. The promising results of such activities performed by researchers worldwide were described in this chapter. Nevertheless, the still existing challenge is to develop a bioremediation strategy that involves and integrates different types of solutions, on the scale of the whole ecosystem, in order to optimize the effectiveness of pollutant removal.

Acknowledgements

This chapter has been carried out as a part of the following projects:

- "Innovative resources and effective methods of safety improvement and durability of buildings and transport infrastructure in the sustainable development" financed by the European Union, from the European Fund of Regional Development based on the Operational Programme of the Innovative Economy, POIG.01.01.02-10-106/09
- The Polish Ministry of Science and Higher Education, Project: N N305 365738 "Analysis of point source pollution of nutrients, dioxins and dioxin-like compounds in the Pilica River catchment and draw up of reclamation methods";
- Ministry of Foreign Affairs of the Republic of Poland within the Polish Aid Programme 2012, project no. 62/2012: "Implementation of Ecohydrology – a transdisciplinary science for integrated water management and sustainable development in Ethiopia".

Author details

Magdalena Urbaniak[1,2*]

1 European Regional Centre for Ecohydrology under the auspices of UNESCO, Łódź, Poland

2 University of Łódź, Department of Applied Ecology, Łódź, Poland

References

[1] Schecter A, Birnbaum L,. Ryan JJ, Constable JD, Dioxins. An overview. Environmental Research 2006, 101 419–428.

[2] Im SH, Kannan K, Matsuda M, Giesy JP, Wakimoto T. Sources and distribution of polychlorinated dibenzo-p-dioxins and dibenzofurans in sediment from Masay Bay, Korea. Environmental Toxicology and Chemistry 2002; 21 245-252.

[3] Hilscherova K, Kannan K, Nakata H, Yamashita N, Bradley P, Maccabe J.M, Taylor A. B, Giesy J.P, Polychlorinated dibenzo-p-dioxin and dibenzofuran concentration profiles in sediments and flood-plain soils of the Tittabawssee River, Michigan. Environmental Science Technology 2003; 37 468-474.

[4] Grochowalski A. Sources of dioxins and ways of their entering into the environment. Problems of waste combustion. I Symposium "Dioxin-man-environment" 22-23.09.1994. Cracow University of Technology 1994.

[5] Makles Z, Świątkowski A, Grybowska S. Hazardous dioxins. Arkady Publisher Warsaw 2001.

[6] Martinez D, Muller RK. Gifte in unsere. Hand, II AUFL. Urania Verlag Leipzig, Jen Berlin 1988.

[7] Rappe C. Dioxin chemistry – on overview. Herbicides in war. The long-term ecological and human consequences. Westing A.H, SIPRI, Stockholm, Taylor and Francis. London, Philadelphia 1984.

[8] Sokołowski M, Śliwakowski M. Sources of dioxin formation outside the combustion process. I National Symposium - Dioxin-Man-Environment. Cracow University of Technology. Krakow 22-23.09 1994.

[9] Kołodziejak-Nieckuła E. Poison targeted Wiedza i życie. 2001; 6 (in Polish).

[10] Wasiela T, Tam I, Krajewski J, Tarkowski S. Environmental health risks, Dioxins. IMP, Łódź 1999.

[11] Novell LH, Capel PD, Dilenis PD. Pesticides in stream sediment and aquatic biota. Distribution, trends, and governing factors. Pesticides in the hydrologic system series. CRC Press, Boca Raton, FL. 1999.

[12] Brasner AMD, Wolff RH. Relationships between land use and organochlorine pesticides, PCBs, and semivolatile organic compounds in streambed sediment and fish on the Island of Oahu, Hawaii. Archives of Environmental Contamination Toxicology 2004; 46, 385-398.

[13] Smith JA, Witkowski PJ, Fusillo TV. Manmade organic compounds in the surface waters of the United States – a review of current understanding. U.S. Geological Survey Circular 1988; 1007 92.

[14] Lulek J. Polychlorinated biphenyls in Poland: history, fate, and occurrence. In: R.L. Lipnick, J.L.M. Hermens, K.C. Jones and D.C. Muir, (Eds), ACS Symposium Series 772 2001; 85.

[15] Urbaniak M. Polychlorinated biphenyls: sources, distribution and transportation in the environment – a literature review, Acta Toxicologica 2007; 15(2) 83-93.

[16] WHO/EURO, PCBs, PCDDs, PCDFs: Prevention and control of accidental and environmental exposures. Environmental Health Series 23. Copenhagen: World Health Organization, Regional Office for Europe 1987.

[17] Bletchly JD. Polychlorinated biphenyls. Production, current use and possible rates of further disposal in OECD member countries. Barres MC, Koeman H, Visser R, [Eds.] Proceedings of PCB seminar. Amsterdam: Ministry of Housing, Physical Planning, and Environment 1984.

[18] Hansen LG. Environmental toxicology of polychlorinated biphenyls. Safe S, Hutzinger O, (Eds.) Environmental Toxin Series. New York: Springer-Verlag 1987; 15–48.

[19] De Vito M, Birnbaum LS 1994.Toxicology of dioxins and related chemicals. In: Schecter A, Dioxins and Health (Ed), New York: Plenum Press 1994; 139-162

[20] Parsons JR, Storms MCM. Biodegradation of chlorinated dibenzo-para-dioxins in batch and continuous cultures of strain JB1. Chemosphere 1989; 19 1297–1308.

[21] Wilkes H, Wittich RM, Timmis KN, Fortnagel P, Francke W. Degradation of chlorinated dibenzofurans and dibenzo-p-dioxins by Sphingomonas sp. strain RW1. Applied Environmental Microbiology 1996; 62 367–371.

[22] Schreiner G, Wiedmann T, Schimmel H, Ballschmiter K. Influence of the substitution pattern on the microbial degradation of mono- to tetrachlorinated dibenzo-p-dioxins and dibenzofurans. Chemosphere 1997; 34 1315–1331.

[23] Keim T, Francke W, Schmidt S, Fortnagel P. Catabolism of 2,7-dichloro- and 2,4,8-trichlorodibenzofuran by Sphingomonas sp. strain RW1. Journal of Industrial Microbiology and Biotechnology 1999; 23 359–363.

[24] Klecka GM, Gibson DT. Metabolism of dibenzo-para-dioxin and chlorinated dibenzo-para-dioxins by a *Beijerinckia* species. Applied Environmental Microbiology 1980; 39 288–296.

[25] Du XY, Zhu NK, Xia XJ, Bao ZC, Xu XB. Enhancement of biodegradability of polychlorinated dibenzo-p-dioxins. Journal Environmental Science Health Part A-Toxic/ Hazard. Subst. Environmental Engineering 2001; 36 1589–1595.

[26] Kimura N, Urushigawa Y. Metabolism of dibenzo-p-dioxin and chlorinated dibenzo-p-dioxin by a gram-positive bacterium, *Rhodococcus opacus* SAO 101. Journal of Bioscience and Bioengineering 2001; 92 138–143.

[27] Habe H, Chung JS, Lee JH, Kasuga K, Yoshida T, Nojiri H, Omori T. Degradation of chlorinated bibenzofurans and dibenzo-p-dioxins by two types of bacteria having angular dioxygenases with different features. Applied Environment microbiology 2001; 67 3610-3617.

[28] Hong HB, Nam IH, Murugesan K, Kim YM, Chang YS. Biodegradation of dibenzo-p-dioxin, dibenzofuran, and chlorodibenzo-p-dioxins by Pseudomonas veronii PH-03. Biodegradation 2004; 15 303–313.

[29] Sulistyaningdyah WT, Ogawa J, Li QS, Shinkyo R, Sakaki T, Inouye K, Schmid RD, Shimizu S. Metabolism of polychlorinated dibenzo-p-dioxins by cytochrome P450BM-3 and its mutant. Journal of Biotechnological Letters 2004; 26 1857–1860.

[30] Habe H, Ashikawa Y, Saiki Y, Yoshida T, Nojiri H, Omori T. *Sphingomonas* sp. strain KA1, carrying a carbazole dioxygenase gene homologue, degrades chlorinated dibenzo-p-dioxins in soil. FEMS Microbiological Letters 2002; 211 43–49.

[31] Ahmed M, Focht DD. Degradation of polychlorinated biphenyls by two species of *Achromobacter*. Canadian Journal of Microbiology 1973; 19 47–52.

[32] Furukawa K, Tonomura K, Kamibayashi A. Effect of chlorine substitution on the biodegradability of polychlorinated biphenyls. Applied and Environmental Microbiology 1978; 35 223–7.

[33] Clark RR, Chian ESK, Griffin RA. Degradation of polychlorinated biphenyls by mixed microbial cultures. Applied and Environmental Microbiology 1979; 37 680–688.

[34] Novakova H, Vosahlikova M, Pazlarova J, Mackova M, Burkhard J, Demnerova K. PCB metabolism by *Peudomonas* sp. P2. *Intern Biodeterior Biodegrad.* 2002, 50, 47–54.

[35] Field JA, Sierra-Alvarez R Microbial degradation of chlorinated dioxins. Chemosphere 2008; 71 1005-1018.

[36] Arfmann HA, Timmis KN, Wittich RM. Mineralization of 4-chlorodibenzofuran by a consortium consisting of *Sphingomonas* sp. strain RW1 and *Burkholderia* sp. strain JWS. Applied Environmental Microbiology 1997; 63 3458–3462.

[37] Wittich RM, Strompl C, Moore ERB, Blasco R, Timmis KN. Interactions of sphingomonas and Pseudomonas strains in the degradation of chlorinated dibenzofurans. Journal of Industrial Microbiology and Biotechnology 1999; 23 353-358.

[38] Bumpus M, Tien D, Wright SD. Oxidation of persistent environmental-pollutants by a white root fungi. Science 1985; 228 1434–1436.

[39] Valli K, Wariishi H, Gold MH. Degradation of 2,7-dichlorodibenzo-para-dioxin by the lignin-degrading basidiomycete *Phanerochaete chrysosporium.* Journal of Bacteriology 1992; 174 2131–2137.

[40] Takada S, Nakamura M, Matsueda T, Kondo R, Sakai K. Degradation of polychlorinated dibenzo-p-dioxins and polychlorinated dibenzofurans by the white root fungus *Phanerochaete sordida* YK-624. Applied Environmental Microbiology 1996; 62 4323–4328.

[41] Adriaens P, Grabic-Galic D. Reductive dechlorination of PCDD/F by anaerobic cultures and sediments. Chemosphere 1994; 29 2253–2259.

[42] Adriaens P, Fu QZ. Grabic-Galic D. Bioavailability and transformation of highly chlorinated dibenzo-p-dioxins and dibenzofurans in anaerobic soils and sediments. Environmental Science and Technology 1995; 29 2252-2260.

[43] Bungie M, Ballerstedt H, Lechner U. Regiospecific dechlorination of spiked tetra- and trichlorodibenzo-p-dioxins by anaerobic bacteria from PCDD/F contaminated Spittelwasser sediments. Chemosphere 2001; 43 675-681.

[44] Bungie M, Adrian L, Kraus A, Opel M, Lorenz WG, Anderssen JR, Gorish H, Lechner U. Reductive dehalogenation of chlorinated dioxins by an anaerobic bacterium. Nature 2003; 421 357-360.

[45] Fennell DE, Nijenhuis I, Wilson SF, Zinder SH, Haggblom MM. *Dehalococcoids ethenogenes* strain 195 reductively dechlorinated diverse chlorinated aromatic pollutants. Environmental Science and Technology 2004; 38 2075-2081.

[46] Barkovskii AL, Adriaens P. Microbial dechlorination of historically present and freshly spiked chlorinated dioxins and diversity of dioxin-dechlorinating populations. Applied Environmental Microbiology 1996; 62 4556-4562.

[47] Barkovskii AL, Adriaens P. Impact of humic constituents on microbial dechlorination of polychlorinated dioxins. Environmental Toxicology and Chemistry 1998; 17 1013-1020.

[48] Master ER, Lai VW, Kuipers B, Cullen WR, Mohn WW. Sequential anaerobic-aerobic treatment of soil contaminated by weathered Aroclor 1260. Environmental Science and Technology 2002; 36 100–3.

[49] Boyle AW, Silvin CJ, Hassett JP, Nakas JP, Tanenbaum SW. Bacterial PCB biodegradation. Biodegradation 1992; 3 285–98.

[50] Fava F, di Gioia D, Cinti S, Marchetti 40. L, Quattroni G. Degradation and dechlorination of low-chlorinated biphenyls by a three-membered bacterial co-culture. Appl Microbiol Biotechnol 1994; 41 117–23.

[51] Bunge M, Lechner U. Anaerobic reductive dehalogenation of polychlorinated dioxins. Appl. Microbial Biotechnol. 2009; 84 429-444.

[52] Ruzo LO, Zabik MJ, Schuetz RD. Photochemistry of bioactive compounds: photoproducts and kinetics of polychlorinated biphenyls. Journal of Agricultural and Food Chemistry 1974; 22 199–202.

[53] Bunce NJ, Kumar Y, Brownlee BG. An assessment of the impact of solar degradation of polychlorinated biphenyls in the aquatic environment. Chemosphere 1978; 7 155–64.

http://dx.doi.org/10.5772/56018

[54] Zepp RG, Baughman GL, Scholtzhauer PF. Comparison of photochemical behavior of various humic substances in water: sunlight induced reactions of aquatic pollutants photosensitized by humic substances. Chemosphere 1981; 10 109–17.

[55] Buekens A, Huang H. Comparative evaluation of techniques for controlling the formation and emission of chlorinated dioxins/furans in municipal waste incineration. Journal of Hazardous Material 1999; 62 1-33.

[56] Kulkarni PS, Crespo JG, Afonso CAM. Dioxins sources and current remediation technologies – a review. Environmental International Journal 2008; 34 139-153.

[57] Balmer ME, Goss KU, Schwarzenbach RP. Photolysis transformation of organic pollutants on soil surfaces – an experimental approach. Environmental Science and Technology 2000; 34 1240-1245.

[58] Goncalves C, Dimou A, Sakkas V, Alpendurda MF, Albanis TA. Photolitic degradation of quinalphos in natural waters and on soil matrices under simulated solar irradiation. Chemosphere 2006; 64 1375-1382

[59] Doughtery EJ, McPeters AL, Overcash MR, Carbonell RG. Theoretical analysis of a method for in situ decontamination of soil containing 2,3,7,8-tetrachlorodibezno-p-dioxin. Environmental Science Technology 1993; 27 505-515.

[60] Van den Berg M, Birnbaum L, Denison M, Farland W. The 2005 World Health Organization reevaluation of human and mammalian toxic equivalency factors for dioxins and dioxin-like compounds. Toxicology Science 2006; 93 223–241.

[61] Macek T, Mackova M, Kas J. Exploitation of plants for the removal of organics in environmental remediation. Biotechnology Advances 2000; 18 23–34.

[62] Cunningham SD, Berti WR. Remediation of contaminated soils with green plants: an overview. In vitro Cell Development Biology 1993; 29 207-212. Cunningham SD, Berti WR, Huang JW. Phytoremediation of contaminated soils. Tibtech Journal 1995; 13 393-397.

[63] Cunningham SD, Anderson TA, Schwab AP, Hsu FC. Phytoremediation of soils contaminated with organic pollutants. Sparks DL (ed.) Advances in Agronomy, Academic Press San Diego Ca. 1996, 56 55-114

[64] Mackova M, Vrchotova B, Francova K, Sylvestre M, Tomaniova M, Lovecka P, Demnerova K, Macek M. Biotransformation of PCBs by plants and bacteria -consequences of plant-microbe interactions .European Journal of Soil Biology 2007; 43 233-241.

[65] Macek T, Mackova M, Kucerova P, Chroma L, Burkhard J, Demnerova K. Phytoremediation,. S.N. Agathos, W. Reineke (Eds.), Biotechnology for the Environment: Soil Remediation, Kluwer Academic Publishers, Brussels 115-137, 2002.

[66] Macek T, Francova K, Kochankova L, Lovecka P, Ryslava E, Rezek J, Sura M, Triska J, Demnerova K, Mackova M. Phytoremediation: biological cleaning of a polluted environment. Reviews on Environmental Health 2004; 19 63-82.

[67] Schnoor JL, Licht L.A, McCutcheon SC, Wolfe NL, Carreira LH, Phytoremediation of organic contaminants, Environmental Science and Technology 1995; 29 318- 323.

[68] Schnoor JL, Phytoremediation of Soil and Ground-water, GWRT Series, E-Series: TE-02-01 2002; 1-45.

[69] Salt DE, Smith RD, Raskin I. Phytoremediation. Ann Rev Plant Physiol Plant Mol Biol 1998; 49 643–68.

[70] Buczkowski R, Kondzielski I, Szymański T. Metody remediacji gleb zanieczyszczonych metalami ciężkimi. Uniwersytet Mikołaja Kopernika w Toruniu; 2002.

[71] Newman LA, Reynolds ChM. Phytodegradation of organic compounds. Current. Opinion in Microbiology 2004; 15 225-230.

[72] Gerhard KE, Huang X-D, Glick BR, Greenberg BM. Phytoremediation and rhizoremediation of organic soil contaminants: Potential and challenges. Plant Science 2009; 176 20-30.

[73] Siciliano SD, Germida JJ, Banks K, Greer CW. Changes in microbial community composition and function during a polyaromatic hydrocarbon phytoremediation field trial. Applied Environmental. Microbiology 2003; 69 483-489.

[74] Nedunuri KV, Govindaraju RS, Banks MK, Schwab AP, Chen Z. Evaluation of phytoteremediation for field-scale degradation of total petroleum hydrocarbons. Journal of Environmental Engineering 2000; 126 483-490.

[75] Robinson SL, Novak JT, Widdowsen MA, Crosswell SB, Fetterolf GJ. Field and laboratory evaluation of the impact of tall fescue on polyaromatic hydrocarbon degradation in aged creosote-contaminated surface oil. Journal of Environmental Engineering 2002; 129 232-240.

[76] White PM Jr, Wolf DC, Thoma GJ, Reynolds CM. Phytoremediation of alkylated polycyclic aromatic hydrocarbons in a crude oil-contaminated soil. Water Air Soil Pollution 2006; 169 207–220.

[77] Banks MK, Kulakow P, Schwab AP, Chen Z, Rathbone K. Degradation of crude oil in the rizosphere of sorghum bicolor. International Journal of Phytoremediation 2003; 5 225-234.

[78] Vervaeke P, Luyssaert S, Mertens J, Meers E, Tack FM, Lust N.. Phytoremediation prospects of willow stands on contaminated sediments: a field trial. Environmental Pollution 2003; 126 27-282.

[79] Pradhan SP, Conrad JR, Paterek JR, Srivastava VJ. Potential of phytoremediation for treatment of PAHs, In: Rainey PB Adaptation of Pseudomonas fluorescens to the plant rhizosphere. Environmental Microbiology 1999; 1 243-257.

[80] Gregor AW, Fletcher JS. The influence of increasing chlorine content on the accumulation and metabolism of polychlorinated biphenyls by Pau's Scarlet Rose cells. Plant Cell Response 1988; 7 329-332.

[81] Jou JJ, Chung JC, Weng YM, Liawc SL, Wang MK: Identification of dioxin and dioxin-like polychlorbiphenyls in plant tissues and contaminated soils. Journal of Hazardous Material 2007; 149174–179.

[82] Hülster A, Marschner H. Transfer of PCDD/PCDF from contaminated soils to food and fodder crop plants. Chemosphere 1993; 27 439–446.

[83] Hülster A, Mueller JF, Marschner H.. Soil–plant transfer of polychlorinated dibenzo-p-dioxins and dibenzofurans to vegetables of the cucumber family (Cucurbitaceae). Environmental Science and Technology 1994; 28 1110–1115.

[84] Engwall M, Hjelm K. Uptake of dioxin-like compounds from sewage sludge into various plant species – assessment of levels using a sensitive bioassay. Chemosphere 2000; 40 1189–1195.

[85] White JC, Parrish ZD, Isleyen M, Gent MP, Iannucci-Berger W, Eitzer BD, Kelsey JW, Mattina MI. Influence of citric acid amendments on the availability of weathered PCBs to plant and earthworm species. International Journal of Phytoremediation 2005; 8 63–79.

[86] Inui H, Wakai T, Gion K, Kim YS, Eun H. Differential uptake for dioxin-like compounds by zucchini subspecies. Chemosphere 2008; 73 1602–1607.

[87] Uegaki R, Seike N, Otani T. Polychlorinated dibenzo-p-dioxins, dibenzofurans and dioxin-like polychlorinated biphenyls in rice plants: possible contaminated pathways. Chemosphere 2006; 65 1537–1543.

[88] Whipps JM. Carbon economy. Lynch JM. (ed.) The rhizosphere. Wiley, New York, 1990; p59–97.

[89] Rainey PB. Adaptation of Pseudomonas fluorescens to the plant rhizosphere. Environmental Microbiology 1999; 1 243-257.

[90] Lugtenberg BJJ, Dekkers L, Bloemberg GV. Molecular determinants of rhizosphere colonization by Pseudomonas. Annual Review Phytopathology 2001; 39 461–490

[91] Gianfreda L, Rao MA, Potential of extra cellular enzymes in remediation of polluted soils: a review. Enzyme Microbiology Technology Journal 2004; 35 339-354.

[92] Liu L, Jiang C-Y, Liu X-Y, Wu J-F, Han J-G, Liu S-J. Plant–microbe association for rhizoremediation of chloronitroaromatic pollutants with Comamonas sp. strain CNB-1. Environmental Microbiology 2007; 9 465–473.

[93] Dams RI, Paton GI, Killham K. Rhizoremediation of pentachlorophenol by Sphinogobium chlorophenolicum ATCC 39723. Chemosphere 2007; 68 864-870.

[94] Macek T, Mackova M, Brkhar J, Demnerova K. Introduction of green plants for the control of metals andorganics I environmental remediation. Holm FW, (ed) Effluents from alternative demilitaryzation technologies. NATO PS Series 1998; 71-85.

[95] Lamoureux GL, Flear DS. Pesticide metabolism in higher plants: In vitro enzyme studies. Paulson GD, Frear DS, Marks EP (eds.). Xenobiotic metabolism. In vitro methods. American chemical Society Symposium Series, 97, Washington DC, ASC 1979; 263-266.

[96] Susarla S, Medina VF, McCutcheon SC. Phytoremediation: an ecological solution to organic chemical contamination. Ecological Engineering 2002; 18 647–658.

[97] Singer AC. The chemical ecology of pollutant biodegradation. Bioremediation and phytoremediation from mechanistic and ecological perspectives Mackova M, Dowling D, Macek T. (eds). Phytoremediation and rhizoremediation. Theoretical background. focus on biotechnology Springer, Dordrecht, 2004; 5-21.

[98] Kuiper I, Lagendijk EL, Bloemberg GV, Lugtenberg BJJ. Rhizoremediation: a beneficial plant–microbe interaction. Molecular Plant Microbe Interactions 2004; 17 6–15.

[99] Chaudhry Q, Blom-Zandstra M, Gupta S, Joner EJ. Utilizing the synergy between plants and rhizosphere microorganisms to enhance breakdown of organic pollutants in the environment. Environmental Science Pollution Researches 2005; 12 34–48.

[100] Yateem A., Al-Sharrah T., Bin-Haji A. Investigation of microbes in the rhizosphere of selected grasses for rhizoremediation of hydrocarbon-contaminated soils. Soil Sed. Contam, 2007; 16 269–280.

[101] Shimp JF, Tracy JC, Davis LC, Lee E, Huang W, Erickson LE, Schnoor JL. Beneficial effects of plants in the remediation of soil and groundwater contaminated with organic materials. Critical Reviews Environmental Science and Technology 1993; 23 41–77.

[102] Leigh MB, Fletcher JS, Fu X, Schmitz FJ. Root turnover: an important source of microbial substrates in rhizosphere remediation of recalcitrant contaminants. Environmental Science and Technology 2002; 36 1579–1583.

[103] Yateem A, Al-Sharrah T, Bin-Haji A. Investigation of microbes in the rhizosphere of selected grasses for rhizoremediation of hydrocarbon-contaminated soils. Soil and Sedimentation Contamination 2007; 16 269–280.

[104] Holden PA, Firestone MK. Soil microorganisms in soil cleanup: how can we improve our understanding? Journal Environmental Quality 1997; 26 32-40.

[105] Ferro AM, Rock SA, Kennedy J, Herrick JJ, Turner DL. Phytoremediation of soils contaminated with wood preservatives: greenhouse and field evaluations. International Journal Phytoremediation 1999; 1 289–306.

[106] Thoma GJ, Lam TB, Wolf DC. A mathematical model of phytoremediation for petroleum contaminated soil: sensitivity analysis. International Journal of Phytoremedation 2003; 5 125–136.

[107] Pillai BVS, Swarup S. Elucidation of the flavonoid catabolism pathway in *Pseudomonas putida* PML2 by comparative metabolic profiling. Appl Environ Microbiol. 2002; 68143–151.

[108] Leigh MB, Prouzova P, Mackova M, Macek T, Nagle DP, Fletcher JS. Polychlorinated biphenyl (PCB)-degrading bacteria associated with trees in a PCB contaminated site. Applied Environmental Microbiology 2006; 72 2331–2342.

[109] Mackova M, Macek T, Ocenaskova J, Burkhard J, Demnerova K, Pazlarova J. Selection of the potential plant degraders of PCB. Chemické Listy 1996; 90 712–3.

[110] Mackova M, Macek T, Kucerova P, Burkhard J, Tiska J, Demnerova K. Plant tissue cultures in model studies of transformation of polychlorinated biphenyls. Chemical Papers 1998; 52 599–600.

[111] Nichols TD, Wolf DC, Rogers HB, Beyrouty CA, Reynolds CM. Rhizosphere microbial populations in contaminated soils. Water, air, Soil Pollution 1997; 95:165-178.

[112] Betts KS. TPH soil cleanup aided by ground cover. Environmental Science Technology 1997; 31 214A.

[113] Burken JG, Schnoor JL. Phytoremediation: plant uptake of atrazine and role of root exudates. Journal Environmental Engineering 1996; 122 968-963.

[114] Donelly PK, Fletcher JS. Potential use of fungi as bioremediation agents. In: Anderson TA. (ed.) Bioremediation through rhizosphere technology. ACS Symposium Series no. 563. American Chemical Society 1994; 93-99.

[115] Fletcher JS, Donnelly PK, Hegde RS. Biostimulation of PCB-degrading bacteria by compounds released from plant toots. Hinche RE, Anderson DB, Hoeppel RE. (eds.) Bioremediation of recalcitrant organics. Battelle Press. Columbus 1995; 131-136.

[116] Mackova M, Pruzova P, Stursa P, Ryslava E, Uhlik O, Beranova K, Rezek J, Kurzawova V, Demnerova K, Macek T. Phyto/rhizoremediation studies using long-term PCB-contaminated soil. Environ. Sci. Pollut. Res. 2009; 16 817-829.

[117] Zalewski M. Ecohydrology for implementation of the Water Framework Directive. Water Management 2011; 164 WM1 1-12.

[118] Zalewski M, Janauer, GA, Jolankaj G. Ecohydrology: a new paradigm for the sustainable use of aquatic resources. Conceptual Background, Working Hypothesis, Ration-

ale and Scientific Guidelines for the Implementation of the IHP-V Projects 2.3:2.4. UNESCO, Paris Technical Documents in Hydrology 1997; 7.

[119] Zalewski M., editor. Guidelines for the Integrated Management of the Watershed-Phytotechnotogy and Ecohydrology. UNEP/UNESCO. UNEP IETC Freshwater Management. Series No 5 2002

[120] Zalewski M, Wagner-Lotkowska I, Robarts RD. Integrated Watershed Management - Ecohydrology and Phytotechnology - Manual. Venice Osaka, Shiga, Warsaw, Lodz 2004.

[121] Zalewski M. Ecohydrology for compensation of global change. Brazilian Journal of Biology 2010; 70(3) 689-695.

[122] Zalewski M, Bis B, Łapinska M, Frankiewicz P, Puchalski W. The importance of the riparian ecotone and river hydraulics for sustainable basin-scale restoration scenarios. Aquatic Conservation: Marine and Freshwater Ecosystems 1998; 8 287-307.

[123] Zalewski M. Ecohydrology - The scientific background to use ecosystem properties as management tools toward sustainability of water resources. Guest Editorial, Ecological Engineering 2000; 16 1-8.

[124] Major DW, McMaster ML, Cox EE, Edwards EA, Dworatzek SM, Hendrokson ER, Starr MG, Payne JA, Buonamici lW. Field demonstration of successful bioaugmentation to achieve dechloriantion of tetrachloroethane to ethane. Environmental Science and Technology 2002; 36 5106-5116.

[125] Brunner W, Sutherland FH, Focht DD. Enhanced biodegradation of polychlorinated biphenyls in soil by analogue enrichment and bacterial inoculation. Journal Environmental Quality 1985; 14 324–328

[126] Yeong SW. Evaluation of the use of capillary numbers for quantifying he removal of DNAPL trapped in a porous medium by surfactant and surfactant foam floods. Journal of Colloid Interface Science 2005; 282 182-187

[127] Johnson DN, Pedit JA, Miller CT. Efficient near-complete removal of DNAPL from three-dimensional, heterogeneous porous media using a novel combination of treatment technologies. Environmental Science and Technology 2004; 38 5149-5156.

[128] West CC, Harwell JH. Surfactants and subsurface remediation. Environmental Science and Technology 1992; 26 2324-2330.

[129] Nakajima F, Baun A, Ledin A, Mikkelsen PS. A novel method for evaluating bioavailability of polycyclic aromatic hydrocarbons in sediments of an urban stream. Water Science and Technology 2005; 51275-281.

[130] Tam NFY, Wong YS. Efectivness of bacterial inoculums and mangrove plants on remediation of sediment contaminated with polycyclic aromatic hydrocarbons. Marine Pollution Bulletin 2008; 57 716-726.

[131] Doty SL, James CA, Moore AL, Vajzovic A, Singleton GL. Ma C, Khan Z, Xin Shang TQ, Wilson AM, Tangen J, Westergeen AD, Newman LA, Strand SE, Gordon MP. Enhanced metabolism of halogenated hydrocarbons in transgenic plants containing mammalian cytochrome P450 2E1. PNAS 97, 2000; 6287-6291.

[132] Dua M, Singh A, Sethunathan N, Johri A. Biotechnology and bioremediation: successes and limitations. Applied Microbiology Biotechnology 2002; 59143 -152.

[133] Lovely DR, Cleaning up with genomics: applying molecular biology to bioremediation. Nat. Rev 2003; 1 35-44.

[134] Kawahigashi H, Hirose S, Ohkawa H, Ohkawa Y.Transgenic rice plants expressing human CYP1A1 exude herbicide metabolites from their roots. Plant Science 2003; 165 373–381.

[135] Cherian S, Oliveira MM. Transgenic plants in phytoremediation: recent advances and new possibilities. Environmental Science and Technology 2005; 39 9377-9390.

[136] Kawahigashi H, Hirose S, Ohkawa H, Ohkawa Y. Phytoremediation of the herbicides atrazine and metolachlor by transgenic rice plants expressing human CYP1A1, CYP2B6 and CYP2C19. Journal of Agricultural and Food Chemistry 2006; 54 2985–2991.

[137] Lan Chun Ch, Payne RB, Sowers KR, May HD. Electrical stimulation of microbial PCB degradation in sediment. Water Research Journal 2013; 24 141-151.

[138] Mitsch WJ, Gosselink JG. Wetlands, 4th edn. John Wiley & Sons, New York 2007.

[139] Brix H. Use of Constructed Wetlands In Water Pollution Control: Historical Development, Present Status and Future Perspectives. Water Science and Technology Journal 1994; 30(8) 209-223.

[140] Zhang D, Richard MG and Tan SK. Constructed wetlands in China. Ecological Engineering 2009; 35 1367–1378.

[141] Mitsch WJ, Jorgensen SE. Ecological Engineering and Ecosystem Restoration. John Wiley & Sons, Inc, New York 2004

[142] Kadlec RH, Knight RL.. Treatment Wetlands. Boca Raton (USA). Lewis. 1996

[143] Chen H. Surface-Flow Constructed Treatment Wetlands for Pollutant Removal: Applications and Perspectives. Wetlands 2011; 31 805–814.

[144] Cheng S, Grosse W, Karrenbrock F, Thoennessen M. Efficiency of constructed wetlands in decontamination of water polluted by heavy metals. Ecological Engineering 2002; 18 317–325.

[145] Cronk JK. Constructed wetlands to treat wastewater from dairy and swine operations: a review. Agriculture, Ecosystems and Environment 1996; 58 97-114.

[146] Fenta BG. Constructed Wetland System for Domestic Wastewater Treatment: A Case Study in Addis Ababa, Ethiopia A thesis submitted to the School of Graduate Studies

of the Addis Ababa University in Partial Fulfilment of the Requirements for the Degree of Master of Science in Environmental Science 2007

[147] Koskiaho J, Ekholm P, Raty M. , Riihimaki J. Puustinen M. Retaining agricultural nutrients in constructed wetlands experiences under boreal conditions. Ecological Engineering 2003; 20 89- 103.

[148] Koukia S, M'hirib F., Saidia N, Belaïd S, Hassen A. Performances of a constructed wetland treating domestic wastewaters during a macrophytes life cycle. Ecological Engineering 2000; 15 77–90.

[149] Cronk JK. Constructed wetlands to treat wastewater from dairy and swine operations: a review. Agriculture, Ecosystems and Environment 1996; 58 97-114.

[150] Schulz R, Peall SKC. Effectiveness of a Constructed Wetland for Retention of Non-point-Source Pesticide Pollution in the Lourens River Catchment, South Africa. Environmental Science and Technology 2001; 35 422-426.

[151] Scholz M, Lee B-H. 2008. Constructed wetlands: a review. International Jour Environ Stud 2005; 62(4) 421–47.

[152] Schulz R. Field studies on exposure, effects and risk mitigation of aquatic nonpoint-source insecticide pollution: a review. Journal Environmental Quality 2004; 33 419–48.

[153] Destandau F, Martin E. and Rozan A. Potential of artificial wetlands for removing pesticides from water in a cost-effective framework, Working Paper No 5. 2011.

[154] Budd R, O'Geen A, Goh KS, Gan J, Efficacy Of Constructed Wetlands In Pesticide Removal From Tailwaters In The Central Valley, Kalifornia. Environmental Science and Technology 2009; 43 2925–2930.

[155] Tournebize J.E, Passeport C, Chaumont C. Fesneau A, Guenne, B. Pesticide decontamination of surface waters as a wetland ecosystem service in agricultural landscapes. Ecological Engineering 2012.

[156] Naiman, RJ & Decamps, H. (eds.) The Ecology and Management of Aquatic–Terrestrial Ecotones. UNESCO, MAB, Parthenon, Paris 1990.

[157] Schiemer, F, Zalewski, M. & Thorpe, JE (Eds) The Importance of Aquatic–Terrestrial Ecotones for Freshwater Fish. Developments in Hydrobiology, 105.Kluwer Academic Publisher, Dordrecht, Boston, London 1995.

[158] Urbaniak M, Zerihun Negussie Y, Zalewski, M, The ecohydrological biotechnology (SBFS) for reduction of dioxin-induced toxicity in Asella lake, Ethiopia. Geophysical Research Abstracts 2012; 14 EGU2012-14431-1.

Chapter 5

Crude Oil Biodegradation in the Marine Environments

Mehdi Hassanshahian and Simone Cappello

Additional information is available at the end of the chapter

http://dx.doi.org/10.5772/55554

1. Introduction

Petroleum is a viscous liquid mixture that contains thousands of compounds mainly consisting of carbon and hydrogen. Oil fields are not uniformly distributed around the globe, but being in limited areas such as the Persian Gulf region. The world production of crude oil is more than three billion tons per year, and about the half of this is transported by sea. Consequently, the international transport of petroleum by tankers is frequent. All tankers take on ballast water which contaminates the marine environment when it is subsequently discharged. More importantly, tanker accidents exemplified by that of the Exxon Valdez in Prince William Sound, Alaska, severely affect the local marine environment. Off-shore drilling is now common to explore new oil resources and this constitutes another source of petroleum pollution. However, the largest source of marine contamination by petroleum seems to be the runoff from land. Annually, more than two million tons of petroleum is estimated to end up in the sea. Fortunately, petroleum introduced to the sea seems to be degraded either biologically or abiotically.

2. The composition of crude oil

Petroleum has been known for several years to occur in the surface seepage and was first obtained in pre-Christian times by the Chinese. The modern petroleum industry had its beginning in Romania and in a well-sunk in Pennsylvania by Colonel E. A. Drake in 1859 [1]. The principal early use of the product of the petroleum industry was for the replacement of expensive whale oil for lighting. Today, its consumption as a fuel and its dominance in the world market as a source of chemicals has diversified tremendously.

Petroleum is defined as any mixture of natural gas, condensate, and crude oil. Crude oil which is a heterogeneous liquid consisting of hydrocarbons comprised almost entirely of the elements hydrogen and carbon in the ratio of about 2 hydrogen atoms to 1 carbon atom. It also contains elements such as nitrogen, sulfur and oxygen, all of which constitute less than 3% (v/v).

There are also trace constituents, comprising less than1% (v/v), including phosphorus and heavy metals such as vanadium and nickel. Crude oils could be classified according to their respective distillation residues as paraffins, naphthenes or aromatics and based on the relative proportions of the heavy molecular weight constituents as light, medium or heavy. Also, the composition of crudes may vary with the location and age of an oil field, and may even be depth dependent within an individual well. About 85% of the components of all types of crude oil can be classified as either asphalt base, paraffin base or mixed base. Asphalt base contain little paraffin wax and an asphaltic residue [2].The sulfur, oxygen and nitrogen contents are often relatively higher in comparison with paraffin base crudes, which contain little or no asphaltic materials. Mixed crude oil contains considerable amount of oxides of nitrogen and asphalt [2].

Crude oil is perhaps the most complex mixture of organic compounds that occurs on earth. Recent advances in ultra-high-resolution mass spectrometry have allowed the identification of more than 17,000 distinct chemical components, and the term petroleomics has been coined to express this newly uncovered complexity [3]. Furthermore, crude oil is not a homogeneous mat erial, and different crude oils have a range of chemical and physical properties that affect their susceptibility to biodegradation and their environmental fate. Within this complexity, however, crude oil can be classified into four main operationally defined groups of chemicals: the saturated hydrocarbons and the aromatic hydrocarbons, and the more polar, non-hydrocarbon components the resins and the asphaltenes. Light oils are typically high in saturated and aromatic hydrocarbons, with a smaller proportion of resins and asphaltenes. Heavy oils, which result from the biodegradation of crude oil under anoxic conditions *in situ* in petroleum reservoirs, have a much lower content of saturated and aromatic hydrocarbons and a higher proportion of the more polar chemicals, the resins and asphaltenes [4] (figure 1). Biodegradation of crude oil in surface environments results in similar changes in crude oil composition and the loss of saturated and aromatic hydrocarbons, together with an increase in the relative abundance of the polar fractions (which are more resistant to biodegradation), is a characteristic signature of crude-oil biodegradation. Because saturated hydrocarbons constitute the largest fraction of crude oil by mass, the biodegradation of saturated hydrocarbons is quantitatively the most important process in the removal of crude oil from the environment. Nevertheless, the aromatic hydrocarbons and polar fractions, which are more toxic and persistent, could be of greater long-term environmental significance [5].

3. Oil pollution as an environmental problem

It is no exaggeration that oil fuels the world's economy, and it is used on a staggering scale. World production was some 80 Mbbl (11 Mt/day) by the end of 2000, and this is expected to increase by 1.9% year in the next decade [6, 7]. Approximately 40% of the world's oil travels by water at some time between its production and final consumption, and again the volumes

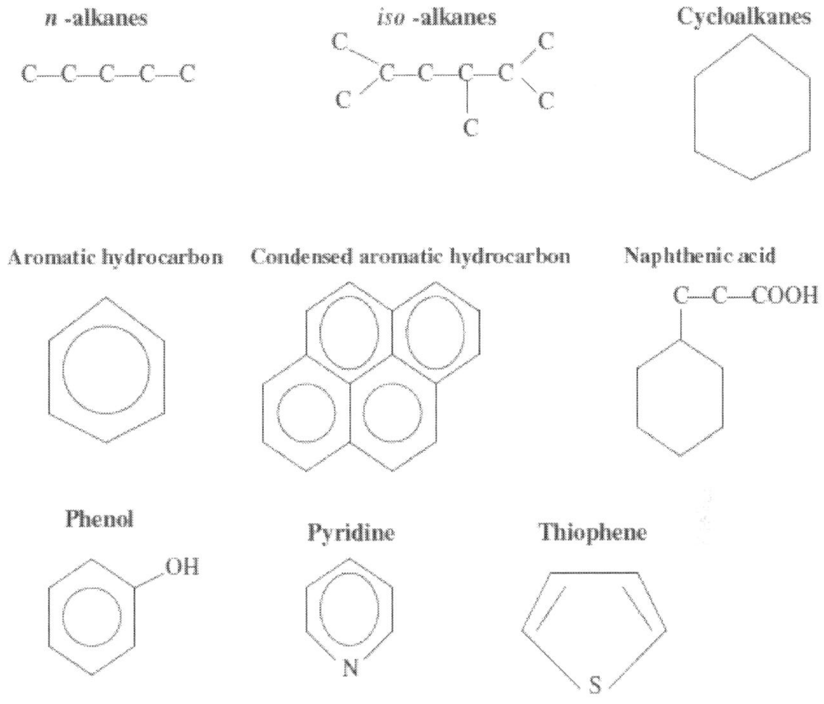

Figure 1. Structural classification of some crude oil components [1].

are staggering. For example, the US imported 350 000 t of oil per day from the Middle East alone in 1999 [7]. Unfortunately, despite the best efforts of the major part of the petroleum industry, a small amount is inevitably spilled. Fortunately this is only a tiny fraction of that transported, and there has been a general improvement in oil spill statistics in the last two decades [7, 8]. Massive releases from pipelines, wells and tankers receive the most public attention, but in fact these account for only a relatively small proportion of the total petroleum entering the environment. The National Research Council has recently updated its classic oil in the sea [7] and now estimates that the total input of petroleum into the sea from all sources is approximately 1.3 Mt/year. Almost 50% comes from natural seeps, and less than 9% emanates from catastrophic releases. Consumption, principally due to non-tanker operational discharges and urban run-off, is responsible for almost 40% of the input (figure 2) skimmers and adsorbents is generally the first priority of responders, but this is neither rarely easy, nor very effective after a large spill. There is therefore a continuing search for alternative and additional responses. Amongst the most promising are those that aim to stimulate the natural process of oil biodegradation [9].

The marine environment is subject to contamination by organic pollutants from a variety of sources. Organic contamination results from uncontrolled releases from manufacturing and refining installations, spillages during transportation, direct discharge from effluent treatment plants and run-off from terrestrial sources.

In quantitative terms, crude oil is one of the most important organic pollutants in marine environments and it has been estimated that worldwide some where between 1.7 and 8.8 ′ 106 tons of petroleum hydrocarbons impact marine waters and estuaries annually [7]. Large oil spills, such as the Exxon Valdez and Sea Empress incidents, invariably capture media attention but such events are relatively rare; however, a substantial number of smaller releases of petroleum hydrocarbons occur regularly in coastal waters. Around the coast of the UK alone, between the years of 1986 and 1996, 6,845 oil spills were reported. Of these, 1,497 occurred in environmentally sensitive areas or were of sufficient magnitude to require clean-up (23). As a consequence of the importance of oil spills relative to other sources of organic contaminants in the marine environment, there is a large body of research on oil-spill bioremediation. Furthermore, studies of oiled shorelines have been far more numerous than open water studies, which have often been equivocal [11, 12].

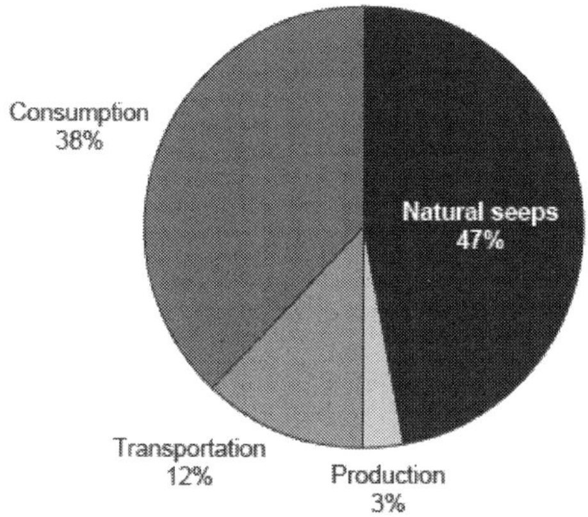

Figure 2. Sources of oil into the sea.

4. The fate of oil in the marine environment

The fate of petroleum in marine ecosystems has been intensively studied [5]. Crude oil and petroleum distillate products introduced to the marine environment are immediately subject to a variety of physical and chemical, as well as biological, changes (figure 3) [13].

Abiological weathering processes include evaporation, dissolution, dispersion, photochemical oxidation, water-in-oil emulsification, adsorption onto suspended particulate material, sinking, and sedimentation. Biological processes include ingestion by organisms as well as microbial degradation [1]. These processes occur simultaneously and cause important changes in the chemical composition and physical properties of the original pollutant, which in turn may affect the rate or effectiveness of biodegradation. The most important weathering process during the first 48 hours of a spill is usually evaporation, the process by which low to medium-weight crude oil components with low boiling points volatilize into the atmosphere. Evaporation can be responsible for the loss of one to two-thirds of an oil spill's mass during this period, with the loss rate decreasing rapidly over time [13].

Roughly one-third of the oil spilled from the Amoco Cadiz, for example, evaporated within the frost 3 days. Evaporative loss is controlled by the composition of the oil, its surface area and physical properties, wind velocity, air and sea temperatures, sea state, and the intensity of solar radiation [14]. The material left behind is richer in metals (mainly nickel and vanadium), waxes, and asphaltenes than the original oil [15]. With evaporation, the specific gravity and viscosity of the original oil also increase. For instance, after several days, spilled crude oil may begin to resemble Bunker C (heavy) oil in composition.

None of the other abiological weathering processes accounts for as significant a proportion of the losses from a spill. For example, the dissolving, or dissolution, of oil in the water column is a much less important process than evaporation from the perspective of mass lost from a spill; dissolution of even a few percent of a spill's mass is unlikely. Dissolution is important, however, because some water soluble fractions of crude oil (e.g., the light aromatic compounds) are acutely toxic to various marine organisms (including microorganisms that may be able to degrade other fractions of oil), and their impact on the marine environment is greater than mass balance considerations might imply [14, 15).

Dispersion, the breakup of oil and its transport as small particles from the surface to the water column extremely important process in the disappearance of a surface slick [15]. Dispersion is controlled largely by sea surface turbulence: the more turbulence, the more dispersion. Chemical dispersants have been formulated to enhance this process. Such dispersants are intended as a first-line defense against oil spills that threaten beaches and sensitive habitats such as salt marshes and mangrove swamps although used widely in other countries, dispersants have had trouble being accepted in the United States. The National Research Council has generally approved their use, but effectiveness and, to a lesser degree, toxicity remain concerns. Dispersed oil particles are more susceptible to biological attack than undispersed ones because they have a greater exposed surface area. Hence, dispersants may enhance the rate of natural biodegradation Water-in-oil emulsions, often termed "mousses are formed when seawater, through heavy wave action, becomes entrained with the insoluble components of oil. Such emulsions can form quickly in turbulent conditions and may contain 30 to 80 percent water [16].

Heavier or weathered crudes with high viscosities form the most stable mousses. Mousse will eventually disperse in the water column and/or be biodegraded, but may first sink or become stranded on beaches. A water-in-oil emulsion is more difficult for microorganisms to degrade than oil alone [17].

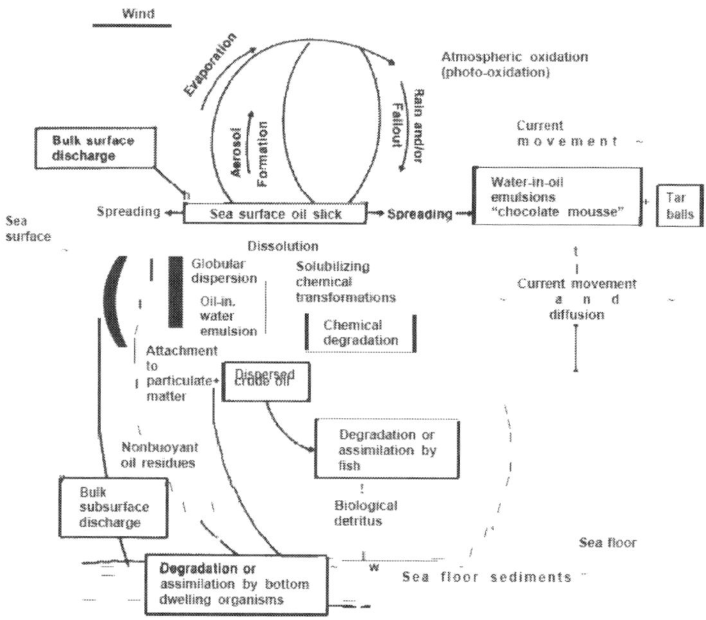

Figure 3. the fate of oil in the marine environment [7].

Mousse formation, for example, has been suggested as a major limiting factor in petroleum biodegradation of the Ixtoc I and Metula spills, probably because of the low surface area of the mousse and the low flux of oxygen and mineral nutrients to the oil-degrading microorganisms within it [17]. Natural biodegradation is ultimately one of the most important means by which oil is removed from the marine environment, especially the nonvolatile components of crude or refined petroleum.

In general, it is the process whereby microorganisms (especially bacteria, but yeasts, fungi, and some other organisms as well) chemically transform compounds such as petroleum hydrocarbons into simpler products. Although some products can actually be more complex, ideally hydrocarbons would be converted to carbon dioxide (i.e., mineralized), nontoxic water-soluble products, and new microbial biomass. The mere disappearance of oil (e.g., through emulsification by living cells) technically is not biodegradation if the oil has not actually been chemically transformed by microbes [17].

The ideal may be difficult to reach, particularly in a reasonably short time, given the recalcitrance of some petroleum fractions to biodegradation (discussed below) and the many variables that affect its rate and extent. Man-made bioremediation technologies are intended to improve the effectiveness of natural biodegradation [17].

5. Response of marine microbial community to oil pollution

Hydrocarbon-degrading microorganisms usually exist in very low abundance in marine environments. Pollution by petroleum hydrocarbons, however, may stimulate the growth of such organisms and cause changes in the structure of microbial communities in the contaminated area [18]. For example Hassanshahian et al (2010) show that oil contamination can induce major changes in marine microbial communities at Persian Gulf and Caspian Sea, that when the pollution occur the number of crude oil degrading bacteria increased and also inhibit some catalytic enzymes [19].

Identification of the key organisms that play roles in pollutant biodegradation is important for understanding, evaluating and developing in situ bioremediation strategies. For this reason, many efforts have been made to characterize bacterial communities, to identify responsible degraders, and to elucidate the catalytic potential of these degraders. In a natural marine environment, the amounts of nutrients, especially those of nitrogen and phosphorus, are insufficient to support the microbial requirements for growth, especially after a sudden increase in the hydrocarbon level associated with an oil spill. Therefore, nitrogen and phosphorus nutrients are added to a contaminated environment to stimulate the growth of hydrocarbon degrading microorganisms and, hence, to increase the rate of biodegradation of the polluting hydrocarbons [20, 21].

6. Crude oil degrading microorganisms

Hydrocarbon-degrading bacteria were first isolated almost a century ago [22] and a recent review lists 79 bacterial genera that can use hydrocarbons as a sole source of carbon and energy, as well as 9 cyanobacterial genera, 103 fungal genera and 14 algal genera that are known to degrade or transform hydrocarbons (Table 1) [23, 24].

Despite the difficulty of degrading certain fractions, some hydrocarbons are among the most easily biodegradable naturally occurring compounds. Many more as-yet-unidentified strains are likely to occur in nature [25]. Moreover, these genera are distributed worldwide. All marine and freshwater ecosystems contain some oil-degrading bacteria. No one species of microorganism, however, is capable of degrading all the components of given oil. Hence, many different species are usually required for significant overall degradation. Both the quantity and the diversity of microbes are greater in chronically polluted areas. In waters that have not been polluted by hydrocarbons, hydrocarbon-degrading bacteria typically make up less than 1 percent of the bacterial population, whereas in most chronically polluted systems (harbors, for example) they constitute 10 percent or more of the total population [26].

Hydrocarbon degrading bacteria and fungi are widely distributed in marine, freshwater, and soil habitats. Similarly, hydrocarbon degrading cyanobacteria have been reported [27, 28] although contrasting reports indicated that growth of mats built by cyanobacteria in the Saudi coast led to preservation of oil residues [29]. Typical bacterial groups already known for their capacity to degrade hydrocarbons include *Pseudomonas, Marinobacter, Alcanivorax, Microbulbifer, Sphingomonas,Micrococcus, Cellulomonas, Dietzia,* and *Gordonia* groups [30]. Molds belonging to the genera *Aspergillus, Penicillium, Fusarium, Amorphoteca, Neosartorya, Paecilomyces, Talaromyces, Graphium* and the yeasts *Candida, Yarrowia* and *Pichia* have been implicated in hydrocarbon degradation [27, 31]. However, reports in literature on the actual numbers of hydrocarbon utilizes are at variance with one another because of the methodological differences used to enumerate petroleum-degrading microorganisms.

Diverse petroleum-degrading bacteria inhabit marine environments. They have often been isolated as degraders of alkanes or of such aromatic compounds as toluene, naphthalene and phenanthrene. Several marine bacteria capable of degrading petroleum hydrocarbons have been newly isolated. These are bacteria of the genera *Alcanivorax* [32], *Cycloclasticus* [33], *Marinobacter* [34], *Neptunomonas* [25], *Oleiphilus* [35] and *Oleispira* [36] within the γ-Proteobacteria, and of the genus *Planococcus* within Gram-positive bacteria [37]. These bacteria, with the possible exception of *Marinobacter* and *Neptunomonas*, use limited carbon sources with a preference for petroleum hydrocarbons and are thus 'professional hydrocarbonoclastic' bacteria. For example, *Alcanivorax* strains grow on n-alkanes and branched alkanes, but cannot use any sugars or amino acids as carbon sources. Similarly, *Cycloclasticus* strains grow on the aromatic hydrocarbons, naphthalene, phenanthrene and anthracene, whereas *Oleiphilus* and *Oleispira* strains grow on the aliphatic hydrocarbons, alkanoles and alkanoates. Many 'non-professional' hydrocarbonoclastic bacteria have been isolated: for example, *Vibrio, Pseudoalteromonas, Marinomonas* and *Halomonas* have been isolated as marine bacteria capable of degrading phenanthrene or chrysene [38].

Some hydrocarbon-degrading bacteria isolated from marine environments have been classified into several genera that include terrestrial hydrocarbon degrading bacteria: namely, naphthalene-degrading *Staphylococcus* and *Micrococcus* [39], 2-methylphenanthrene-degrading *Sphingomonas* [40] and alkane-degrading *Geobacillus* [41]. Although some *Cycloclasticus* strains have been isolated using the extinction culturing method, other strains were isolated by conventional enrichment techniques with petroleum hydrocarbons used as the sources of carbon and energy. Therefore, a greater variety of hydrocarbon-degrading marine bacteria are likely to be isolated if hydrocarbon enrichment is done in combination with the specific resuscitation techniques already described.

7. Pathway for biodegradation of some compartment of crude oil

7.1. Fundamental reactions of aerobic degradation

The fundamental reactions of the aerobic hydrocarbon decomposition have been well known for several decades. Suitable surveys are contained in the books of [42, 43]. Even though many details have been published since, such as the degradation of aliphatic alkenes [44], the fundamental steps are still valid and enable us to understand the dependence of the processes on environmental conditions (Figures 4 and 5). Experiments on the laboratory scale as well as

http://dx.doi.org/10.5772/55554

Bacteria	Yeast	Fungi
Achromobacter	*Candida*	*Aspergillus*
Acinetobacter	*Cryptococcus*	*Cladosporium*
Alcanivorax	*Debaryomyces*	*Corollasporium*
Alcaligenes	*Hamsenula*	*Cunninghamella*
Bacillus	*Pichia*	*Dendryphiella*
Brevibacterium	*Rhodotorula*	*Fusarium*
Burkholderia	*Saccharomyces*	*Gliocladium*
Corynebacterium	*Sporobolomyces*	*Luhworthia*
Flavobacterium	*Torulopsis*	*Penicillium*
Mycobscterium	*Trichosporon*	*Varicospora*
Nocardia	*Yarrowia*	*Verticillium*
Pseudomonas		
Rhodococcus		
Sphingomonas		
Streptomyces		

Table 1. Crude-oil degrading microorganisms

observation of polluted sites have made it possible to estimate the impact of oil degradation on sediment.

The key step of hydrocarbon degradation is the addition of one oxygen atom, in some cases, two oxygen atoms, to the hydrocarbon molecule, which is then converted to an alkanol (in the case of aliphatic hydrocarbons) or to a phenol (in the case of aromatic molecules). In some species, an epoxide is the first intermediate. This activation makes the hydrocarbon more soluble in water, marks a reactive site, and introduces a reactive site for the next reactions. The reaction requires energy, which is typically generated via the oxidation of a reduced biological intermediate such as NADH, which itself is reoxidized by an electron acceptor. For the degradation of alkanes, different enzyme systems are known which carry out the primary attack. An omega-hydroxylase system consisting of three proteins (the rubredoxin reductase, a rubredoxin and an omega-hydroxylase) was isolated and characterized from *Pseudomonas* [45]. In some bacterial or fungal species as well as in mammalian cells, there are enzyme systems which depend on cytochrome P450 acting as a terminal oxidase. The main intermediates of the alkane degradation are fatty acids, which are produced from the alkanols via aldehydes. These acids can be further decomposed by the pathway typical of physiologica carboxylic acid degradation, in which the molecule is shortened stepwise. However, fatty acids can also be excreted by the cells and accumulate in the environment.

Once released, they can produce ambiguous effects. On the one hand, fatty acids can serve as a carbon source for bacteria of a community, thus enhancing the hydrocarbon degradation. On the other hand, fatty acids (chain length 14 C) can inhibit growth and hydrocarbon metabolism because they interfere with the cell membrane [47]. This provokes a toxic effect and reduces growth. Different degradative pathways have been demonstrated for aromatic substrates. The choice of the pathway depends on the type of the organism and/or on the type of the aromatic molecule, especially on its substituents and (in the case of polyaromatic molecules, PAH) on the number of rings [48]. For an overview of the fundamental possibilities of PAH biodegradation, three different metabolic routes considered to be the main pathways are summarized here.

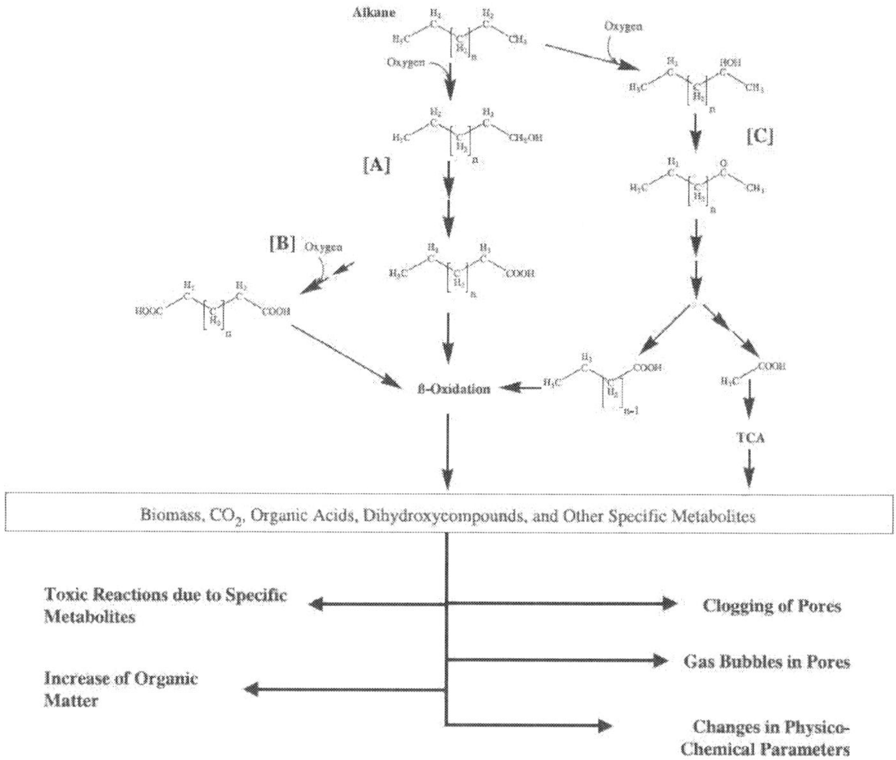

Figure 4. Aerobic degradation of crude oil hydrocarbons with its environmental impact. Biodegradation of n-alkanes: metabolism begins with the activity of a monooxygenase which introduces a hydroxyl group into the aliphatic chain. [A]-monoterminal oxidation, [B]-biterminal oxidation, [C]- subterminal oxidation); TCA-tricarboxylic acid cycle [44]

http://dx.doi.org/10.5772/55554

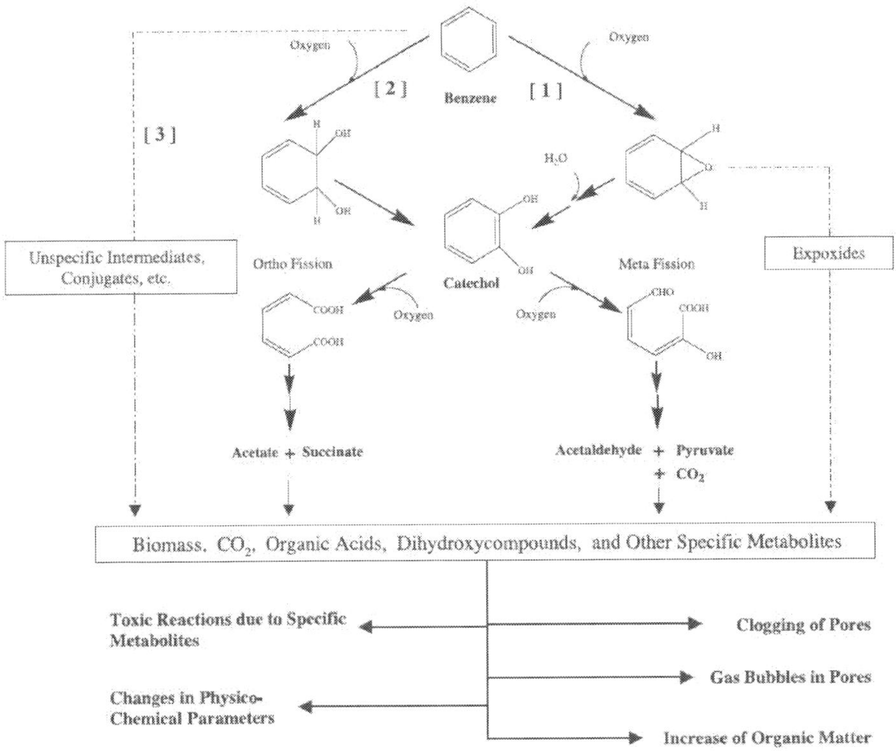

Figure 5. Biodegradation of aromatic hydrocarbons: metabolism begins with the activity of a monooxygenase [1] or a dioxygenase [2] which introduce one or two atoms of oxygen; it can also begin with unspecific reactions [3] [48].

7.2. Complete mineralization or the dioxygenase pathway

This pathway is taken mainly by bacteria. The monoaromatic molecule or one ring of the polyaromatic system is attacked by a dioxygenase, and the molecule is oxidized stepwise via formation of a diol and subsequent ring cleavage. Pyruvate is one of the main intermediates of the pathway. The main products are biomass and carbon dioxide. An accumulation of dead-end products is rare and occurs mostly when cells are deficient in their degradation pathway. The disadvantage of this pathway is that only ring systems of up to four rings are mineralized. Systems with a higher number of rings seem to be recalcitrant [49].

7.3. Cometabolic transformation or the monooxygenase pathway

This pathway has been mainly demonstrated for yeasts and fungi, but it also occurs in bacteria and in some algae. The respective PAH-degrading species can only perform the degradation if a compound is available which can serve as a source of carbon and energy. The characteristic enzymes which perform ring cleavage are monooxygenases (e.g., Cyt P450). The monooxygenase activity results in the formation of an epoxide which is highly reactive, resulting in toxic or mutagenic activity. Epoxides may also be transformed to trans-dihydrodiols. The latter have not been metabolized further in pure cultures in the laboratory and have to be regarded as dead-end products. However, no such metabolites have been detected in soil or in sediment [50].

7.4. Unspecific oxidation via radical reactions

The wood-destroying white rot fungi, e.g., have been shown to destroy the structure of lignin via the activity of extracellular peroxidases and phenol oxidases. They attack the phenolic molecule structure by a nonspecific action, thus also attacking other aromatic structures such as PAH. The type of cleavage product is not predictable. Frequent metabolites of PAHs are quinones, quinoles, and ring systems with a ring number lower than that of the original substance. These compounds may be incorporated into sediments and alter the sediment structure [51].

7.5. Anaerobic hydrocarbon degradation

For many decades, it was assumed that hydrocarbons undergo biodegradation only in the presence of molecular oxygen. However, in 1988 Evans and Fuchs [50] published a review paper on the anaerobic degradation of aromatic compounds, and Aeckersberg et al. (1991) [52] reported on a sulphate-reducing bacterium able to anaerobically mineralize hexadecane. Since that time, a great deal of work has been done on the anaerobic degradation of aliphatic and aromatic hydrocarbons. It has been demonstrated that anaerobic hydrocarbon degradation is not uncommon in nature although, in most cases it is considerably slower than aerobic degradation. Denitrifying, sulfate-reducing, and iron (III)-reducing strains collected at different sites (terrestrial, aquifers, fresh-water and marine systems) are able to anaerobically metabolize hydrocarbons. The same has been demonstrated for the phototrophic bacterium *Blastochloris sulfoviridis* strain ToP1, which uses light as an energy source [53]. Even methanogenic consortia have been shown to degrade hydrocarbons [54, 55]. The metabolic routes of alkane degradation seem to function differently and are not completely understood yet. Several authors have discussed a terminal or sub terminal addition of a one-carbon moiety or a fumarate molecule to the alkane as an activation mechanism [56, 57] (Figure 6). For aromatic molecules, it has been demonstrated that alkyl benzenes which have a methyl group as a side chain undergo an enzymes addition of fumarate, most likely via a radical mechanism. This was demonstrated for toluene. Alkyl benzenes with side chains of two or more carbon atoms are activated by dehydrogenation of the side chain.

This has been shown for ethyl- and propylbenzene [53]. A scheme of the anaerobic degradation is shown in Figure (7).

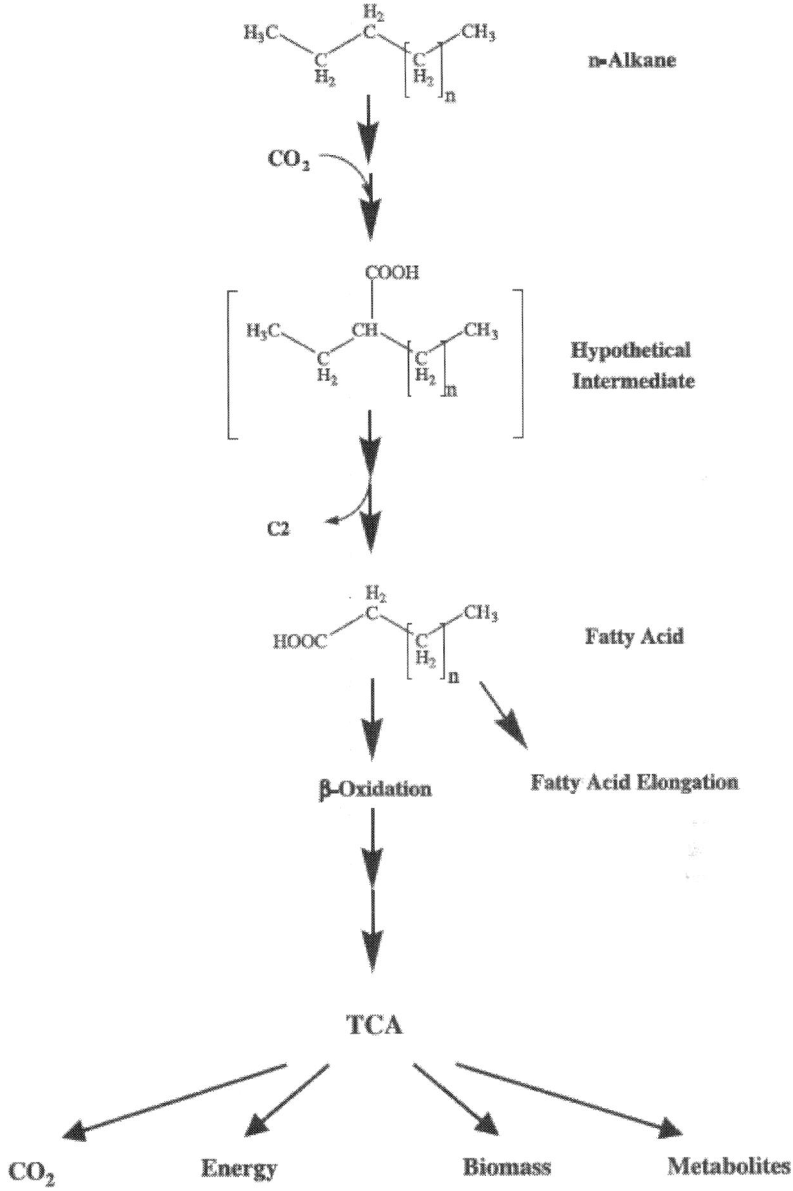

Figure 6. Proposed pathway for anaerobic degradation of n-alkanes; activation via addition of a C1-moiety (subterminal carboxylation at C3). Pathway according to So et al. (2003); TCA tricarboxylic acid cycle [55].

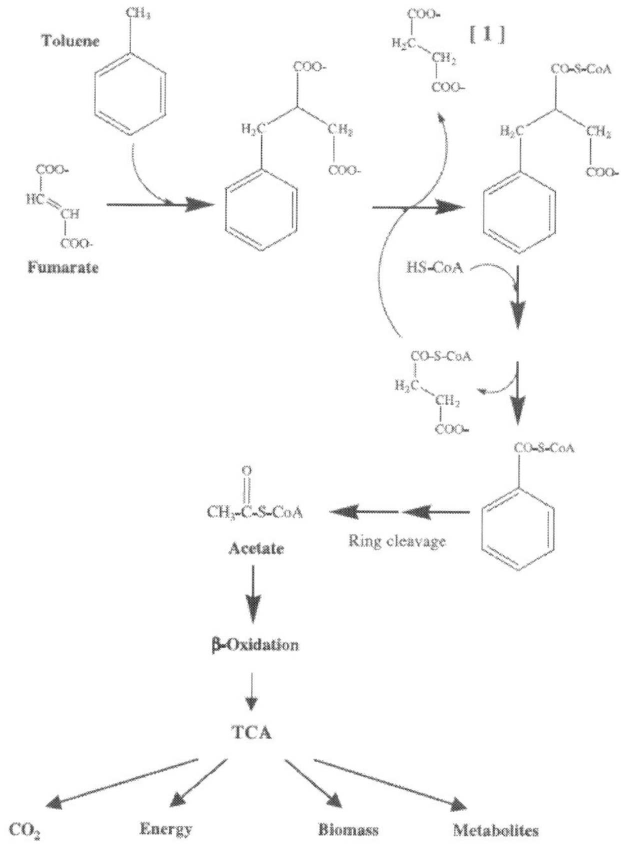

Figure 7. Proposed pathways of anaerobic degradation of aromatic hydrocarbons; activation via addition of fumarate, [1]—succinate. Pathways according to Spormann and Widdel (2000), and Wilkes et al. (2002); TCA—tricarboxylic acid cycle [55].

7.6. Competing processes

The ideal preconditions for biodegradation cited above occur only rarely, e.g., in the case of a rough and nutrient-rich sea or on energy-rich tidal flats. Mostly, however, the reality of oil spills is very different. The ideal steps are rendered difficult, slowed down, or made impossible by competing processes. Such influences are exemplified by the case studies. Heavy oils or heavy oil products such as heavy fuel oil or bunker oil C behave very differently from the light oils described above. Heavy oils incorporate suspended matter, debris, biomass, and even garbage, which increases their viscosity and decreases their biodegradability. Due to their viscosity, the energy needed to emulsify heavy oils is very great. Solar irradiation causes the evaporation of the light components and photodecomposition, resulting in unpredictable compounds. Oil carpets are formed. Where they meet the coastline, beaches are covered. Their removal by natural forces is very slow or even impossible, and technical purification is expensive and troublesome. Biological degradation is extremely slow because the low oil surface to volume ratio limits the bioavailability of the oil.

Oil biodegradation works well on the open sea but proceeds differently on beaches. Vast areas of tidal beaches can be covered by oil when there is wind onshore during the ebb tide. If this oil cover is subjected to strong sun irradiation, the oil does not float up during the next flood because the light components have evaporated. The sediment is soaked with the sticky oil. Tides and wind add further sediment, and the initially liquid, later viscous, pollutant becomes more and more solidified [58].

This solidified material is only slowly attacked by waves, hampering biodegradation because the available surface is too small. Irradiation and the catalyzing capacity of particle surfaces help to convert a part of the original mixture of small molecules into high molecular mass material of low solubility, forming tar and finally asphalt. Such products appear as geological rather than organic matter. Experience has shown that it is difficult for organisms to settle on oil layers. The Persian Gulf spill presented a new experience in so far as thick and vital cyanobacterial mats developed on oil covers within a few months, introducing biomass as well as Aeolian and hydrodynamic sediments fixed by the growing mats. This observation was welcomed initially [59] but then turned out to be disappointing because biodegradation was not favored [60]. In some cases, colonization opened the oil crusts; in other cases, it formed stable covers which prevented the access of oxygen to deeper layers, helping to preserve the pollution. The latter clearly transformed the polluted system, resulting in geo-biological matter that had never been present before. This geo-biological matter dominated the sites after the spill. In the upper eulittoral and the lower supra tidal zone, calcareous incrustation and solid salt supported the conversion of the oil into rock-like matter with a life span of 10 or more years [61].

8. Factors affected crude oil biodegradation in marine environment

Environmental variables can also greatly influence the rate and extent of biodegradation. Variables such as oxygen and nutrient availability can often be manipulated at spill sites to enhance natural biodegradation (i.e., using bioremediation). Other variables, such as salinity, are not usually controllable. The great extent to which a given environment can influence biodegradation accounts for some of difficulty in accurately predicting the success of bioremediation efforts. Lack of sufficient knowledge about the effect of various environmental factors on the rate and extent of biodegradation is another source of uncertainty [19, 62].

8.1. Oxygen

Oxygen is one of the most important requirements for microbial degradation of hydrocarbons. However, its availability is rarely a rate-limiting factor in the biodegradation of marine oil spills. Microorganisms employ oxygen-incorporating enzymes to initiate attack on hydrocarbons. Anaerobic degradation of certain hydrocarbons (i.e., degradation in the absence of oxygen) also occurs, but usually at negligible rates. Such degradation follows different chemical paths, and its ecological significance is generally considered minor. For example, studies of sediments impacted by the Amoco Cadiz spill found that, at best, anaerobic biodegradation is several orders of magnitude slower than aerobic biodegradation. Oxygen is generally necessary for the initial breakdown of hydrocarbons, and subsequent reactions may also require direct incorporation of oxygen. Requirements can be substantial; 3 to 4 parts of dissolved oxygen are necessary to completely oxidize 1 part of hydrocarbon into carbon dioxide and water. Oxygen is usually not a factor limiting the rate of biodegradation on or near the surface of the ocean, where it is plentiful and where oil can spread out to provide a large, exposed surface area. Oxygen is also generally plentiful on and just below the surface of beaches where wave and tide action constantly assist aeration. When oxygen is less available, however, the rates of biodegradation decrease. Thus, oil that has sunk to the sea floor and been covered by sediment takes much longer to degrade. Oxygen availability there is determined by depth in the sediment, height of the water column, and turbulence (some oxygen may also become available as the burrowing of bottom-dwelling organisms helps aeration) [63, 64]. Low-energy beaches and fine-grained sediments may also be depleted in oxygen; thus, the rate of biodegradation may be limited in these areas. Pools of oil are a problem because oxygen is less available below their surfaces. Thus, it may be preferable to remove large pools of oil on beaches, as was done in Alaska, before attempting bioremediation [18, 65].

8.2. Nutrients

Nutrients such as nitrogen, phosphorus, and iron play a much more critical role than oxygen in limiting the rate of biodegradation in marine waters. Several studies have shown that an inadequate supply of these nutrients may result in a slow rate of biodegradation [52]. Although petroleum is rich in the carbon required by microorganisms, it is deficient in the mineral nutrients necessary to support microbial growth [53]. Marine and other ecosystems are often deficient in these substances because non-oil degrading microorganisms (including phytoplankton) consume them in competition with the oil degrading species. Also, phosphorus precipitates as calcium phosphate at the pH of seawater. Lack of nitrogen and phosphorus is most likely to limit biodegradation, but lack of iron or other trace minerals may sometimes be important. Iron, for instance, is more limited in clear offshore waters than in sediment-rich coastal waters Scientists have attempted to adjust nutrient levels (e.g., by adding nitrogen- and phosphorus-rich fertilizers) to stimulate biodegradation of petroleum hydrocarbons. This is the experimental bioremediation approach used recently on about 110 miles of beaches in Prince William Sound, Alaska. Researchers have also experimented with alternative methods of applying nutrients. Given the necessity of keeping nutrients in contact with oil, the method of application is itself likely to be an important factor in the success of bioremediation [65, 66].

8.3. Temperature

The temperature of most seawater is between –2 and 35 °C (55). Biodegradation has been observed in this entire temperature range, and thus in water temperatures as different as those of Prince William Sound and the Persian Gulf. The rates of biodegradation are fastest at the higher end of this range and usually decrease—sometimes dramatically in very cold climates-with decreasing temperature. One experiment showed that a temperature drop from 25 to 5 °C caused a tenfold decrease in response [56]. At low temperature, the rate of hydrocarbon metabolism by microorganisms decreases [57]. Also, lighter fractions of petroleum become less volatile, thereby leaving the petroleum constituents that are toxic to microbes in the water for a longer time and depressing microbial activity. Petroleum also becomes more viscous at low temperature. Hence, less spreading occurs and less surface area is available for colonization by microorganisms. In temperate regions, seasonal changes in water temperature affect the rate of biodegradation, but the process continues year-round.

8.4. Other factors

Several variables, including pressure, salinity, and pH may also have important effects on biodegradation rates. Increasing pressure has been correlated with decreasing rates of biodegradation; therefore, pressure may be very important in the deep ocean [67,68]. Oil reaching great ocean depths degrades very slowly and, although probably of little concern, is likely to persist for a long time [59]. Microorganisms are typically well adapted to cope with the range of salinities common in the world's oceans. Estuaries may present a special case because salinity values, as well as oxygen and nutrient levels, are quite different from those in coastal or ocean areas. However, there is little evidence to suggest that microorganisms are adversely affected by other than hyper saline environments. Extremes in pH affect a microbe's ability to degrade hydrocarbons. However, like salinity, pH does not fluctuate much in the oceans it remains between 7.6 and 8. l and does not appear to have an important effect on biodegradation rates in most marine environments. In salt marshes, however, the pH maybe as low as 5.0, and thus may slow the rate of biodegradation in these habitats [69, 70].

9. Biodegradation strategy for crude oil removal from marine environment (biostimulation and bioaugmentation)

Bioremediation technologies for responding to marine oil spills may be divided into three discrete categories:

1. Nutrient enrichment (Biostimulation)
2. Seeding with naturally occurring microorganisms (Bioaugmentation)
3. Seeding with genetically engineered microorganisms (Bioaugmentation with GEMs)

9.1. Nutrient enrichment (biostimulation)

Of all the factors that potentially limit the rate of petroleum biodegradation in marine environments, lack of an adequate supply of nutrients, such as nitrogen and phosphorus, is probably the most important and perhaps the most easily modified. Nutrient enrichment (sometimes called nutrition) also has been more thoroughly studied than the other two approaches, especially now that EPA, Exxon, and the State of Alaska have carried out extensive nutrient enrichment testing on beaches polluted by oil from the Exxon Valdez [71]. In part for these reasons, many scientists currently view nutrient enrichment as the most promising of the three approaches for those oil spill situations in which bioremediation could be appropriate. This approach involves the addition of those nutrients that limit biodegradation rates (but not any additional microorganisms) to a spill site and conceptually is not much different than fertilizing a lawn [71]. The rationale behind the approach is that oil-degrading microorganisms are usually plentiful in marine environments and well adapted to resisting local environmental stresses. However, when oil is released in large quantities, microorganisms are limited in their ability to degrade petroleum by the lack of sufficient nutrients. The addition of nitrogen, phosphorus, and other nutrients is intended to overcome these deficits and allow petroleum biodegradation to proceed at the optimal rate. Experiments dating to at least 1973 have demonstrated the potential of this approach. Researchers, for example, have tested nutrient enrichment in near shore areas off the coast of New Jersey, in Prudhoe Bay, and in several ponds near Barrow, Alaska. In each case, the addition of fertilizer was found to stimulate biodegradation by naturally occurring microbial populations. The recent nutrient enrichment experiments in Alaska provided a wealth of experimental data about bioremediation in an open environment (box B) [72]. Since previous research findings had already demonstrated the general value of this approach, the experiments were intended to determine for one type of environment how much enhancement of natural biodegradation could be expected and to evaluate the most effective methods of application. The results provided additional evidence that application of nutrients could significantly enhance the natural rate of biodegradation on and below the surface of some beaches. As a result, Exxon was authorized by the Coast Guard on-scene coordinator, in concurrence with the Alaska Regional Response Team, to apply fertilizers to the oiled beaches in Prince William Sound [73]. To date, about 110 miles of shoreline have been treated with nutrients, and a monitoring program has been established. Without additional research, however, it is premature to conclude that nutrient enrichment will be effective under all conditions or that it will always be more effective than other bioremediation approaches, other oil spill response technologies, or merely the operation of natural processes. The results of the Alaska experiments were influenced by the beach characteristics (mostly rocky beaches, well-washed by wave and tide action), the water temperature (cold), the kind of oil (Prudhoe Bay crude), and the type and quantity of indigenous microorganisms in Prince William Sound. Few detailed analyses or performance data are yet available for different sets of circumstances. One smaller-scale test using the same fertilizer as in Alaska was recently conducted on beaches in Madeira polluted by the Spanish tanker Aragon. Results in this very different setting and with a different type of oil were not especially encouraging. Researchers speculated that the unsatisfactory results could have been due to differences in the type of oil, the concentration of fertilizer used the lower initial bacterial

http://dx.doi.org/10.5772/55554

activity, and/or different climatic conditions. At the same time, Exxon recently used what it learned in Alaska to help degrade subsurface no. 2 heating oil spilled in a wildlife refuge bordering the Arthur Kill at Prall's Island, New Jersey. An innovative aspect of this application was the use of two trenches parallel to the beach in which to distribute fertilizer. Nutrients were dissolved with the incoming tide and pulled down the beach with the ebb tide, enabling a more even distribution than point sources of fertilizer. Exxon reports those 3 months after applying fertilizers, the oil in the treated zone had been reduced substantially relative to that in an untreated control zone [74, 75, 76].

9.2. Seeding with naturally occurring microorganisms (bioaugmentation)

Seeding (also called inoculation) is the addition of microorganisms to a polluted environment to promote increased rates of biodegradation. The inoculums maybe a blend of non indigenous microbes from various polluted environments, specially selected and cultivated for their oil-degrading characteristics, or it may be a mix of oil-degrading microbes selected from the site to be remediate and mass-cultured in the laboratory or in on-site bioreactors. Nutrients would usually also accompany the seed culture. The rationale for adding microorganisms to a spill site is that indigenous microbial populations may not include the diversity or density of oil-degraders needed to efficiently degrade the many components of a spill. Some companies that advocate seeding with microorganisms also claim that commercial bacterial blends can be custom-tailored for different types of oil in advance of a spill, that the nutritional needs and limitations of seed cultures are well understood, that microbes can easily be produced in large quantities for emergency situations, and that seed cultures can be stored, ready for use, for up to 3 years.

The value of introducing nonindigenous microorganisms to marine environments is still being evaluated. With some exceptions, the scientilc community has not been encouraging about the promise of seeding marine oil spills. Controlled studies have not been conducted in such settings, so no data are available to evaluate the effectiveness of this approach. Many scientists question the necessity of adding microbes to a spill site because most locales have sufficient indigenous oil-degrading microbes, and in most environments biodegradation is limited more by lack of nutrients than by lack of microbes [72]. At many spill sites, a very low level of oil is often present as "chronic" input, inducing oil-degrading capability in naturally occurring microorganisms. Moreover, the requirements for successful seeding are more demanding than those for nutrient enrichment. Not only would introduced microbes have to degrade petroleum hydrocarbons better than indigenous microbes, they would also have to compete for survival against a mixed population of indigenous organisms well adapted to their environment. They would have to cope with physical conditions (such as local water temperature, chemistry, and salinity) and predation by other species, factors to which the native organisms are likely to be well adapted [77].

The time required for introduced microbes to begin metabolizing hydrocarbons is also important. If a seed culture can stimulate the rapid onset of biodegradation, it would have an advantage over relying on indigenous microbes that may take time to adapt. Despite some claims, seed cultures have not yet demonstrated such an advantage over indigenous microbial communities. Seed cultures are typically freeze-dried (and therefore dormant) and require time before they become active [73].

Seed cultures also must be genetically stable, must not be pathogenic, and must not produce toxic metabelites. Some laboratory and small-scale experiments in controlled environments have demonstrated that seeding can promote biodegradation [75].

However, it is exceedingly difficult to extrapolate the results of such tests to open water where many more variables enter the picture. Results of experimental seeding of oil spills in the field have thus far been inconclusive. As noted in box B, recent EPA tests of two commercial products applied to contaminated beaches in Alaska concluded that, during the period of testing, there was no advantage from their use [77]. In a well-publicized attempt to demonstrate seeding at sea, one company applied microorganisms to oil from the 1990 Mega Borg spill in the Gulf of Mexico [78]. Although the experiment aroused some interest, the results were inconclusive and illustrated the difficulty of conducting a controlled bioremediation experiment at sea and measuring the results. Although there were changes observed in the seeded oil, in the absence of controls the experiment could not tell whether they were due to biodegradation or bioemulsification (the process in which microbes assist the dispersal of surface oil), or were unrelated to the seeding. (Even if bioemulstilcation rather than biodegradation was the process at work in this experiment, it may be of potential interest for oil spill response and could be investigated further.) An attempt has been made to apply a seed culture to a polluted salt marsh [78, 79]. In July 1990 the Greek tanker Shinoussa collided with three barges in the Houston Ship Channel, resulting in a spill of about 700,000 gallons of catalytic feed stock, partially refined oil. Some of this oil impacted neighboring Marrow Marsh. Microbes were applied to experimental areas within the marsh, and control areas were established. Visual observations made by the scientific support coordinator who monitored the application for the National Oceanic and Atmospheric Administration (NOAA) indicated that treated oil changed color within a few minutes to a few hours after treatment, but that after several days there were no significant visual differences between treated and untreated plots.

More importantly, chemical analyses indicated "no apparent chemical differences in petroleum hydrocarbon patterns between treated and untreated plots several days after treatment [70, 80]. Not all of the monitoring data have been analyzed yet, so a final determination of effectiveness has not been made. Seed cultures may be most appropriate for situations in which native organisms are either present as slow growers or unable to degrade a particular hydrocarbon. Especially difficult-to-degrade petroleum components, such as polynuclear aromatic hydrocarbons, might be appropriate candidates for seeding [80]. In other cases, if a time advantage can be realized; there may be some utility in seeding with a culture consisting of indigenous organisms [81]. Thus, the potential environmental adaptation problems of nonindigenous organisms might be avoided. In many cases, fertilizers would also have to be added. Seeding may offer promise in environments where conditions can be more or less controlled. In such cases one would have to consider the proper choice of bacteria, a suitable method of application, and suitable site engineering. Arrangements would have to be made for keeping cells moist and in contact with the oil; for protecting them from excess ultraviolet light; for providing adequate nutrients; and for controlling temperature, pH, and salinity.

However, before claims about the utility of seeding marine oil spills can be proved (or disproved) additional research [82, 83, 84].

9.3. Seeding with genetically engineered microorganisms (bioaugmentation with GEMs)

Although it was not demonstrably superior to indigenous organisms and has never been tested in the field, the frost organism ever patented was a microorganism genetically engineered to degrade oil [82]. The rationale for creating such organisms is that they might possibly be designed either to be more efficient than naturally occurring species or to have the ability to degrade fractions of petroleum not degradable by naturally occurring species. To be effective, such microorganisms would have to overcome all of the problems related to seeding a spill with nonindigenous microbes. EPA has not yet conducted any GEM product reviews for commercial applications, although at least two companies are considering using genetically engineered products for remediating hazardous waste [84, 85]. Since the development and use of GEMs are still limited by scientific, economic, regulatory, and public perception obstacles, the imminent use of bioengineered microorganisms for environmental cleanup is unlikely. Lack of a strong research infrastructure, the predominance of small companies in the bioremediation field, lack of data sharing, and regulatory hurdles are all barriers to the commercial use of genetically engineered organisms [83]. The development of GEMs for application to marine oil spills does not have high priority. Many individuals, including EPA officials, believe that we are so far away from realizing the potential of naturally occurring microorganisms to degrade marine oil spills that the increased problems associated with GEMs render them unnecessary at this time [86, 87, 88].

10. Field evaluation of marine oil spill

There have been several oil spill incidents in which bioremediation products have been used in an attempt to enhance oil biodegradation. In some cases, the response authorities have allowed products to be used for experimental purposes [89]. However, in general, it is difficult to draw valid conclusions from many of these efforts because of the time constraints in planning experiments with appropriate controls after a major spill. Moreover, many of the results are reported second hand with little reliable quantitative information. Despite these limitations, some of these spills have been given as examples of bioremediation success and therefore qualify for scientific appraisal [90, 91]. One notable exception is the work carried out in the aftermath of the Exxon Valdez spill. The assessments of bioremediation products and techniques are based on experiments carried out with considerable scientific rigor, and the work after the Exxon Valdez incident is therefore given prominence in this section. The scientific results of this research have been only recently published in primary publications and conference proceedings. A majority of the papers were not peer reviewed prior to publication in the scientific literature (a fact that applies to much work conducted after oil spill incidents), and thus the results from these studies should be assessed with caution [92, 93].

Also, it is important to emphasize that even in this case, there were significant limitations in the scope of the work. For example, the studies concentrated on North Slope crude oil on cobble shorelines in a high-latitude environment. During the early 1990s, there was an increase in bioremediation field trials associated with accidental spills, largely as a result of the perceived success of the bioremediation program following the Exxon Valdez incident [93]. These are mentioned herein, but many are characterized by having been carried out over a short period and, in some cases, with products in the early stage of development [94].

10.1. Amoco Cadiz

On 16 March 1978, the tanker Amoco Cadiz containing 223,000 tones of Arabian Light and Iranian Light crude oil was wrecked off the coast of France. Rough sea conditions resulted in rapid emulsification of the spilled oil, resulting in an increase in the volume of pollutant. Despite efforts to treat the oil at sea, extensive contamination of the shoreline occurred. Most of the beach cleanup effort focused on pumping and mechanical recovery, particularly during the first few weeks of the operation when there was a thick emulsion on the sand and rocks and in the crevices between the rocks. These operations caused some oil to penetrate the sand. In some places, oily sand was overlaid with clean sand deposited as a result of natural coastal processes. Repeated ploughing and harrowing were used to clean the intertidal zone, and four different products were tested to assess the possibility of promoting the biodegradation of oil trapped in sand [95]: (i) a commercial cleaning compound containing nutrients especially adapted to restore oiled soils; (ii) a mixture of lyophilized adapted bacteria, dispersant, and nutrient; (iii) a chemical fertilizer used in agriculture; and (iv) a talc treated with 0.1% of surfactant. The approaching tourist season seems to have prevented extended experimentation, and other techniques were used to complete the cleanup operations. Hence, the limited results were inconclusive [95, 96]. Some changes in oil content were found in these experiments, but it was not clear if the removal was physically or biologically mediated.

10.2. Apex barge

On 28 July 1990, the Greek tanker Shinoussa collided with two Apex tank barges in the Houston Ship Channel, Galveston Bay, Tex., causing a release of approximately 3,000 m^3 of partially refined catalytic feedstock oil over 2 days, which spread onto the surrounding coastline. Alpha BioSea (Alpha Environmental, Houston, Tex.), a product composed of a lyophilized bacterial mixture and inorganic phosphorus and nitrogen nutrients, was applied 8 days after the spill in selected areas of Pelican Island and Marrow Marsh [97, 98]. Two plots on the beach were treated, and two were left untreated as controls. The 15-m diameter experimental plots (separated by 45 to 75 m) were sampled on a routine basis [99]. The results of the detailed chemical analysis showed that there were no significant differences between pre- and post treatment samples after 96 h of treatment with any of the selected methods. Although visual signs indicated that the condition of the marsh areas improved after treatment [100], there was no conclusive evidence to show significant degradation of the oil within the 4-day monitoring period. Numerous compromises in the experimental design of this study have been identified [99]. For example, the separation of treated and untreated plots and the booming methods used to isolate them may not have prevented mixing and cross-contamination Furthermore, our knowledge from previous laboratory studies and field trials suggests that the 96-h duration of the experiment was insufficient for a definitive test of bioremediation. Unfortunately, no attempt was made to establish which factor (if any) was limiting biodegradation and what the most appropriate bioremediation strategy might be.

10.3. Mega Borg

On 8 June 1990, the Norwegian tanker Mega Borg was carrying out a lightering operation with the Italian tanker Fraqmura about 57 miles off the Texas coast. Following an explosion and fire, the Fraqmura carried out an emergency breakaway operation from the Mega Borg, which resulted in the release of approximately 45 m3 of Angolan Palanca crude oil [101]. The next day, further oil was lost before the situation was controlled. While it was initially predicted that no oil would reach the shoreline, the Louisiana coast was littered with tiny tar balls 16 days after the accident (55). In terms of bioremediation strategies, the On-Scene Coordinator granted permission to conduct a field trial 1 day after the accident occurred. Two portions of the slick were treated with a product containing Alpha BioSea [102]. A 16-hectare patch of slick located about 5 km from the *Mega Borg* was treated 7 days after the accident with 50 kg of microbial agent (Alpha BioSea) which had been rehydrated with seawater. The product was applied with the standard shipboard fire-hose system. The equipment and treatment preparation time of approximately 1 h (105) indicates that very little rehydration time was given to the product. Four traverses of the treatment area were made over a 30-min period. Following large-scale application of the product at sea, visual observations indicated that the treated oil changed from a continuous film of brown oil and sheen to discrete areas of mottled brown and yellow material and sheen. An aerial reconnaissance 16 h after treatment was not able to detect oil in the area. However, there is considerable uncertainty about the fate of the treated oil [102]. The measurements on water samples from the treated slick showed no evidence of acute toxicity to marine life or significantly elevated levels of nutrients or total hydrocarbons. Attempts to assess the effect of the microbial agent from measurements of oil content in the emulsion samples were unsuccessful because of sample variability. By 8 h after treatment, the slick had largely broken up and dissipated. Although little change was observed in the control area, conclusive evidence of bioremediation effectiveness was not achieved because of limitations in the sampling strategy and the chemical evidence obtained. This study demonstrated the potential problems with the application of bioremediation products at sea, including difficulties with uniform product application, representative sampling, and uncertainties about the ultimate fate of the oil. The short periods over which monitoring are often possible may not be sufficient to validate the presence and activity of oil-degrading bacteria or the effectiveness of bioremediation treatments. The observed visual effects may well have been caused by physical or chemical processes such as surfactant action associated with the treatment [102].

10.4. Prall's Island

In January 1990, fuel oil from a pipeline failure spilled into the Arthur Kill waterway in New Jersey and contaminated a gravel beach on the Prall's Island bird sanctuary. Mechanical methods were used to remove the bulk of the oil. Cleanup was suspended in March 1990 to minimize possible adverse effects on migrating birds. However, Exxon was granted permission to carry out a bioremediation experiment on part of a contaminated beach. Two shallow trenches were dug in the intertidal zone to bury bags of beach substrate containing known concentrations of oil and to help overcome possible problems of variable distribution of oil on the beach. A slow-release fertilizer (Customblen, Sierra Chemicals) was placed in the trenches to encourage biodegradation. Over a 92-day period, sub samples were periodically taken from the oiled bags, together with beach samples and water samples for analysis of total petroleum hydrocarbons, GC-MS detection of hydrocarbons, microbial counts, and water quality (nitrogen, phosphorus, ammonia, and dissolved oxygen) determination. No clear trends of increased biodegradation from the fertilized plots could be identified during the experiment, and there was high variability in the levels of total petroleum hydrocarbons, which may have masked any effects of the treatment [89].

10.5. Seal Beach

On 31 October 1990, a well blowout off Seal Beach, Calif., resulted in the release of approximately 2 m^3 of crude oil that contaminated 8,000 to 12,000 m^2 of marsh grassland in the Seal Beach National Wildlife Refuge. One week after the incident, the marsh was hand sprayed with a combination of a microbial product used in sewage treatment plants (INOC 8162) and a commercial fertilizer (Miracle Gro 30-6-6). Two weeks later, the fertilizer alone was applied. Oiled, oiled and treated, and unoiled samples were collected and analyzed for oil content by GC-MS [89].

Measurements were also made of the microbial mineralization of the phenanthrene, microbial respiration, and biomass. The results of a 35-day monitoring effort showed no differences between the treated and untreated oil plots. Subsequently, laboratory tests were carried out with the microbial product and Prudhoe Bay crude oil to compare the performance of the microbial product with nutrient-only controls. After 16 days of incubation, little or no difference was found between treated and control flasks. It was concluded that the microbial product was not effective in accelerating biodegradation of oil under controlled laboratory conditions [89].

Moreover, the salt marsh environment may be difficult to bioremediate simply by adding sources of nitrogen and phosphorus. Oxygen depletion may have been a significant factor in the inhibition of oil biodegradation [103].

10.6. Exxon Valdez

The tanker Exxon Valdez ran aground on Bligh Reef in the Gulf of Alaska on 24 March 1989, spilling approximately 41,000 m^3 of Alaskan North Slope crude oil (primarily Prudhoe Bay crude oil). A major response effort was mounted at sea to recover the oil, but the prevailing conditions and circumstances resulted in the contamination of about 2,090 km of coastline [104]. Some beaches were heavily oiled, particularly those on islands in Prince William Sound that were directly in the path of the slick. Many techniques were adopted in a massive effort to clean up the shoreline of the Sound (72% rock face, 24% mixed boulder and cobble, 3.5% mixed cobble and pebble, and 0.5% fine-grain sand/mud or marsh). These included cold- and warm-water washing, steam cleaning, and manual oil recovery techniques. Initially, the main aim was to remove the heaviest concentrations of oil to minimize the impact on wildlife and fisheries [104, 105]. A bioremediation option based on nutrient enrichment was proposed shortly after the spill. However, it was thought necessary to carry out some research first to establish the potential for effective and safe use of this technique. The limited success of the initial field tests led to the approval of full-scale application in August 1989, and 119 km of shoreline was subsequently treated that year. By 1990, the previous cleanup efforts and winter storms had greatly reduced the extent of shoreline oiling [106] and natural recovery processes were already well advanced [7, 8]. The National Oceanic and Atmospheric Administration applied the concept of net environmental benefit analysis in an evaluation of the main alternative to bioremediation at this time, namely, excavation and rock-washing treatment [107].

It was concluded that this technique would be particularly damaging to the environment. Bioremediation was therefore adopted as a prime cleanup strategy. In 1990 and 1991, bioremediation was used in combination with storm berm relocation, tilling, and manual pickup. On 12 June 1992, the U.S. Coast Guard and the State of Alaska declared the cleanup officially concluded on the basis that there would be no further net environmental benefit from continuing the effort.

Shortly after the Exxon Valdez spill, it was suggested that bioremediation may be able to enhance the rates of oil removal from the contaminated beaches [108]. As a preliminary step, the number of oil degrading microorganisms on oiled beaches in comparison with untreated controls was determined. Pritchard et al. (2005) reported that the hydrocarbon-degrading microorganisms on oiled shorelines had increased by as much as 10,000 times to an average level of 106 cells per g of beach material. Once it was clear that hydrocarbon degraders were present in abundance, it was necessary to establish which factors were likely to limit biodegradation and which specific hydrocarbon components were biodegradable. The research was conducted in the laboratory with Prudhoe Bay crude oil weathered by distillation to remove the volatile fraction. Biodegradation by indigenous microorganisms was monitored by noting changes in the concentration of components of the oil by GC-MS, by monitoring carbon dioxide evolution and oxygen consumption by the microorganisms, and by determining the evolution of radioactive 14CO_2 from specific 14C-labeled oil components such as phenanthrene [109].

The experiments demonstrated unequivocally that the microbial population in Prince William Sound could rapidly biodegrade the aliphatic and aromatic fractions of Prudhoe Bay crude in the presence of suitable nitrogen and phosphorus sources. The microbial community decomposed C1 dibenzothiphene, C_2 fluorenes, C_3 naphthalenes, phenanthrene, and anthracene among others (113). Studies of CO_2 production suggested that the oil was not just being biotransformed but that it was being completely mineralized to CO_2 and H_2O. For example, over 30% of [U-14C]phenanthrene could be mineralized to 14CO_2 within 4.5 days when incubated with oil-contaminated beach material from Prince William Sound [83].

The highest mineralization rates were noted in the test systems treated with the highest concentration of nitrogen. From these results, it is clear that the main factor limiting the biodegradation of oil on the beaches in Prince William Sound was the concentration of nutrients, particularly nitrogen. A substantial microbial biomass had already developed in the contaminated areas of Prince William Sound which was able to decompose many components within the contaminant oil. Hence, addition of nutrients, and not seeding, was thought to be the most appropriate bioremediation strategy [17, 80].

11. Conclusion and future prospects

Despite the growing acceptance of bioremediation as a means to treat spilled oil in marine environments the mechanisms that promote the process under field conditions remain poorly constrained. Although general statements can be made regarding the enhancement of biodegradation by nutrient amendment, there is no consensus on how to best optimize nutrient additions. Subsequently, oil spill treatment strategies are largely developed empirically from previous experience and/or from laboratory feasibility studies. Introduction of a theoretical framework to explain observations from primarily empirical studies of oil-spill bioremediation would be a fundamental step towards the development of more objective spill management practices. Resource ratio theory has recently been put forward as a theoretical basis to explain some of the effects of bioremediation and many of the observations made in bioremediation studies are consistent with the theory's predictions. Although the introduction of this theory may simply augment current empirical approaches, in the longer term it has the potential to form the basis of more predictable bioremediation strategies, and the introduction of theory to the field of bioremediation is an important progression. To further test the applicability of resource-ratio theory it will be necessary to conduct systematic studies on the effect of different nutrient amendments on bacterial populations and concomitant alterations in biodegradation rates, to identify patterns of microbial diversity associated with optimum contaminant removal. Until recently, such an approach would not have been possible due to the limitations of the methods available to characterize the composition of microbial communities. With the introduction of molecular methods to study indigenous microorganisms, this limitation has been alleviated to some extent. Integrated studies combining careful field evaluation of crude oil biodegradation with molecular approaches to study microbial populations involved in degradation of spilled oil have already begun and promise to reveal much regarding the relationship between microbial population structure and the progress of bioremediation. Anaerobic hydrocarbon degradation in marine environments has only recently been widely accepted and there is a need to determine both how widespread an occurrence this is and in what circumstances it will have a significant impact on the dissipation of crude oil contamination. The environmental factors that promote the process must also be identified if it is to be exploited for the treatment of spilled oil.

Author details

Mehdi Hassanshahian[1*] and Simone Cappello[2,3*]

*Address all correspondence to: mshahi@uk.ac.ir

1 Department of Biology - Faculty of Science- Shahid Bahonar University of Kerman - Kerman, Iran

2 Istituto per l'Ambiente Marino Costiero (IAMC) - C.N.R. U.O.S. di Messina, Italy

3 Istituto Sperimentale Talassografico (IST) di Messina, Italy

References

[1] Alloway BJ, Ayres DC. Organic Pollutants. In: Chemical Principles of Environmental Pollution. 1st edition. Chapman and Hall, India; 1993.

[2] Atlas RM. Microbial degradation of petroleum hydrocarbons: an environmental perspective. Microbiology Review 1981; 45 180-209.

[3] Marshall AG, Rogers RP. Petroleomics: the next grand challenge for chemical analysis. Analytical Chemistry Research 2003; 37 53–59.

[4] Head IM, Jones DM Larter S R. Biological activity in the deep subsurface and the origin of heavy oil. Nature 2003; 426344–352.

[5] Texas B. In Proceedings of the Fifteenth Arctic and Marine Oil Spill Program Technical Seminar. Canada Ottawa Canada, 2001.

[6] Cappello S, Denaro R, Genovese M, Giuliano L, Yakimov MM. Predominant growth of Alcanivorax during experiments on oil spill bioremediation in mesocosms. Microbiology Research 2006; 162 185-190.

[7] National Research Council. Oil in the Sea III: Inputs, Fates and Effects. National Academy of Sciences Washington DC; 2002.

[8] Mahon A, Labelle RP. Update of comparative occurrence rates for offshore oil spills. Spill Science Technology Bulletin 2000; 6 303-321.

[9] Cappello S, Caruso G, Zampino D, Monticelli LS, Maimone G, Denaro R, Tripodo B, Troussellier M, Yakimov MM, Giuliano L. Microbial community dynamics during assays of harbour oil spill bioremediation: a microscale simulation study. Journal of Applied Microbiology 2007. 102 (1), 184-194.

[10] Department of the Environment. Transport and the Regions: Digest of Environmental Statistics. London HMSO; 1998.

[11] Prince RC. Bioremediation of oil spills. Trends Biotechnology 1997, 15158-160.

[12] Swannell RPJ, Lee K, McDonagh M. Field evaluations of marine oil spill bioremediation. Microbiology Review 1996; 60 342-365.

[13] Cappello S, Gabriella G, Vivia B. Crude oil-induced structural shift of coastal bacterial communities of rod bay and characterization of cultured cold-adapted hydrocarbonoclastic bacteria. FEMS Microbiology Ecology 2004; 49 419–432.

[14] Cappello S, Crisari A, Hassanshahian M, Genovese M, Santisi S, Yakimov MM. "Effect of a Bioemulsificant Exopolysaccharide (EPS2003) on Abundance and Vitality of Marine Bacteria." Water Air Soil Pollution 2012; DOI 10.1007/s11270-012-1159-8.

[15] U.S. Environmental Protection Agency. Interim report. Oil Spill Bioremediation Project. Office of Research and Development, U.S. Environmental Protection Agency 1990; Gulf Breeze, Fla.

[16] Cappello S, Genovese M, Torre CD, Crisari A, Hassanshahian M, Santisi S, Calogero R, Yakimov MM. Effect of bioemulsificant exopolysaccharide (EPS2003) on microbial community dynamics during assays of oil spill bioremediation: A microcosm study. Marine Pollution Bulletin 2012; http://dx.doi.org/10.1016/j.marpolbul.2012.07.046.

[17] Cappello S, Santisi S, Calogero R, Hassanshahian M, Yakimov MM. Characterization of Oil-Degrading Bacteria Isolated from Bilge Water. Water Air Soil Pollution 2012; 223 3219-3226.

[18] Caruso G, Denaro R, Genovese M, Giuliano L, Mancuso M, Yakimov MM. New methodological strategies for detecting bacterial indicators. Chemistry and Ecology 2004; 20 (3) 167–181.

[19] Kohno T, Sugimoto Y, Sei K, Mori K. Design of PCR Primers and gene probes for general detection alkane-degrading bacteria. Microbes and Environment 2002; 17 (3) 114-212.

[20] Hassanshahian, M., Emtiazi, G., Kermanshahi, R., Cappello, S. 2010. Comparison of oil degrading microbial communities in sediments from the Persian Gulf and Caspian Sea. Soil and Sediment Contamination. 19 (3), 277-291.

[21] Das K, Ashis K, Mukherjee F. Crude petroleum-oil biodegradation efficiency of Bacillus subtilis and Pseudomonas aeruginosa strains isolated from a petroleum-oil contaminated soil from North-East India. Bioresource Technology 2006; 98 1339-1345.

[22] Naughton SJ, Stephen JR, Venosa AD, Davis GA, Chang Y-J, White DC. Microbial population changes during bioremediation of an experimental oil spill. Applied Environmental Microbiology 1999, 65 3566-3574.

[23] Söhngen NL. Benzin Petroleum, Paraffinöl und Paraffin als Kohlenstoff- und Energiequelle für Mikroben. Zentralbl. Bakteriol 1913; 2 37 595–609 (in German).

[24] Prince RC. Petroleum Microbiology. American Society for Microbiology Press Washington DC; 2005.

[25] Emtiazi G, Hassanshahian M, Golbang N. Development of a microtiter plate method for determination of phenol utilization, biofilm formation and respiratory activity by environmental bacterial isolates. International Biodeterioration & Biodegradation 2005; 56 231-235.

[26] Hedlund BP, Geiselbrecht AD, Bair TJ, Staley JT. Polycyclic aromatic hydrocarbon degradation by a new marine bacterium, Neptunomonas naphthovorans gen. nov., sp. nov. Applied Environmental Microbiology 1999; 65 251-259.

[27] Emtiazi G, Saleh T, Hassanshahian M. The effect of bacterial glutathione S-transferase on morpholine degradation. Biotechnology Journal 2009 ; 4, 202–205.

[28] Chaillana F, Flècheb A, Burya E, Phantavonga Y, Saliot A, Oudot J. Identification and biodegradation potential of tropical aerobic hydrocarbon-degrading microorganisms. Research Microbiology 2004; 155(7) 587-595.

[29] Lliros M, Munill X, Sole A, Martinez-Alonso M, Diestra E, Esteve I. Analysis of cyanobacteria biodiversity in pristine and polluted microbial mats in microcosms by confocal laser scanning microscopy (CLSM). Science Technology and Education of Microscopy 2003; 52 483–499.

[30] Barth HJ. The influence of cyanobacteria on oil polluted intertidal soils at the Saudi Persian Gulf shores. Marine Pollution Bulletin 2003; 46 1245-52.

[31] Hassanshahian M, Emtiazi G, Cappello S. Isolation and characterization of crude-oil-degrading bacteria from the Persian Gulf and the Caspian Sea. Marine Pollution Bulletin 2012; 64 7–12.

[32] Ghanavati H, Emtiazi G, Hassanshahian M. Synergism effects of phenol degrading yeast and Ammonia Oxidizing Bacteria for nitrification in coke wastewater of Esfahan Steel Company. Waste Management & Research 2008; 26(2) 203-208.

[33] Yakimov MM, Golyshin PN, Lang S, Moore ER, Abraham WR, Lunsdorf H, Timmis KN. Alcanivorax borkumensis gen. nov., sp. nov., a new hydrocarbon-degrading and surfactant producing marine bacterium. International Journal Systematic Bacteriology 1998; 48 339-348.

[34] Dyksterhouse SE, Gray JP, Herwig RP, Lara JC, Staley JT. Cycloclasticus pugetii gen. nov., sp. nov., an aromatic hydrocarbon-degrading bacterium from marine sediments. International Journal Systematic Bacteriology 1995; 45 116-123.

[35] Gauthier MJ, Lafay B, Christen R, Fernandez L, Acquaviva M, Bonin P, Bertrand JC: Marinobacter hydrocarbonoclasticus gen. nov., sp. nov., a new extremely halotolerant, hydrocarbondegrading marine bacterium. Int J Syst Bacteriol 1992, 42:568-576.

[36] Golyshin PN, Chernikova TN, Abraham WR, Lunsdorf H, Timmis KN, Yakimov MM. Oleiphilaceae fam. nov., to include Oleiphilus messinensis gen. nov., sp. nov., a

novel marine bacterium that obligatory utilizes hydrocarbons. International Journal Systematic Bacteriology 2002; 52 901-911.

[37] Yakimov MM, Giuliano L, Gentile G, Crisafi E, Chernikova TN, Abraham WR, Lunsdorf H, Timmis KN, Golyshin PN. Oleispira antarctica gen. nov., sp. nov., a novel hydrocarbonoclastic marine bacterium isolated from Antarctic coastal sea water. International Journal Systematic Evolutionary Microbiology 2003; 53 779-785.

[38] Engelhardt MA, Daly K, Swannell RP, Head IM. Isolation and characterization of a novel hydrocarbon-degrading, Gram positive bacterium, isolated from intertidal beach sediment, and description of Planococcus alkanoclasticus sp. nov. Journal Applied Microbiology 2001; 90 237-247.

[39] Melcher RJ, Apitz SE, Hemmingsen BB. Impact of irradiation and polycyclic aromatic hydrocarbon spiking on microbial populations in marine sediment for future aging and biodegradability studies. Applied Environment Microbiology 2002; 68 2858-2868.

[40] Zhuang WQ, Tay JH, Maszenan AM, Tay ST. Isolation of naphthalene-degrading bacteria from tropical marine sediments. Water Science Technology 2003; 47 303-308.

[41] Gilewicz M, Nimatuzahroh T, Nadalig H, Budzinski P, Doumenq V, Michotey JC, Bertrand JC. Isolation and characterization of a marine bacterium capable of utilizing 2-methylphenanthrene. Applied Microbiology Biotechnology 1997; 48 528-533.

[42] Button DK, Schut F, Quang P, Martin R, Robertson BR. Viability and isolation of marine bacteria by dilution culture: theory, procedures, and initial results. Applied Environmental Microbiology 1993; 59 881-891.

[43] Atlas RM. Petroleum Microbiology. Macmillan Publishing New York; 1984.

[44] Gibson DT. Microbial Degradation of Organic Compounds. Dekker New York; 1984.

[45] Ensign SA. Microbial metabolism of aliphatic alkenes. Biochemistry 2001; 40 5845–5853.

[46] Van Beilen JB, Wubbolts MG, Witholt B. Genetics of alkane oxidation by Pseudomonas oleovorans. Biodegradation 1994; 5 161– 174.

[47] Atlas RM, Bartha R. Inhibition by fatty acids of the biodegradation of petroleum. Antonie van Leeuwenhoek 1973; 39 257– 271.

[48] Cerniglia CE. Biodegradation of polycyclic aromatic hydrocarbons. Biodegradation 1992; 3 351– 368.

[49] Maneerat S, Kulnaree P. Isolation of biosurfactant-producing marine bacteria and characteristics of selected biosurfactant. Applied Microbiology 2007; 29 783-791.

[50] Mathew M. Obbard JP. Optimization of the dehydrogenase assay for measurement of indigenous microbial activity in beach sediments contaminated with petroleum. Biotechnology Letters 2001; 23 227–230.

[51] Evans WC, Fuchs G. Anaerobic degradation of aromatic compounds. Annual Review Microbiology 1988; 42 289– 317.

[52] Pruthi V, Cameotra SS. Rapid identification of biosurfactant-producing bacterial strains using a cell surface hydrophobicity technique. Biotechnology Techniques 1997; 11 671-674.

[53] Aeckersberg F, Bak F, Widdel F. Anaerobic oxidation of saturated hydrocarbons to carbon dioxide by a new type of sulfate-reducing bacterium. Archive Microbiology 1991; 156 5 – 14.

[54] Spormann AM, Widdel F. Metabolism of alkylbenzenes, alkanes, and other hydrocarbons in anaerobic bacteria. Biodegradation 2000; 11 85– 105.

[55] Anderson RT, Lovley DR. Hexadecane decay by methanogenesis. Nature 2000; 404, 722–723.

[56] So CM, Phelps CD, Young LY. Anaerobic transformation of alkanes to fatty acids by a sulfate-reducing bacterium, strain Hxd3. Applied Environmental Microbiology 2003; 69 3892– 3900.

[57] Edwards EA, Grbic-Galic D. Anaerobic degradation of toluene and o-xylene by a methanogenic consortium. Applied Environmental Microbiology 1994; 60 313– 322.

[58] Wilkes H, Rabus R, Fischer T, Armstroff A, Behrends A, Widdel F. Anaerobic degradation of n-hexane in a denitrifying bacterium: further degradation of the initial intermediate (1-methylpentyl) succinate via C-skeleton rearrangement. Achieve Microbiology 2002; 177 235– 243.

[59] Tannenbaum E, Starinsky A, Aizenshtat Z. Light-oils transformation to heavy oils and asphalts—assessment of the amounts of hydrocarbons removed and the hydrological-geological control of the process. Exploration for Heavy Crude Oil and Natural Bitumen. The American Association of Petroleum Geologists 1987, Tulsa, Oklahoma

[60] Sorkhoh NA, Ghannoum MA, Ibrahim AS, Stretton RJ, Radwan S. Crude oil and hydrocarbon-degrading strains of Rhodococcus rhodochrous isolated from soil and marine environments in Kuwait. Environmental Pollution 1990; 65 1 – 17.

[61] Hfpner T, Felzmann H, Struck H, van Bernem KH. The nature and extent of oil contamination on Saudi Persian Gulf beaches: examinations of beaches of Dawhat ad Dafi and Dawhat ad Musallamiya in summer 1991 and winter 1991/92. Arab Journal Science Engineering 1993; 18 243–255.

[62] Hasanshahian M, Emtiazi G. Investigation of alkane biodegradation using the microtiter plate method and correlation between biofilm formation, biosurfactant production and crude oil biodegradation. International Biodeterioration & Biodegradation 2008; 62 170-178.

[63] Hassanshahian M, Tebyanian H, Cappello S. Isolation and characterization of two crude-oil degrading yeast strains, Yarrowia lipolytica PG-20 and PG-32 from Persian Gulf. Marine Pollution Bulletin 2012; 64 1386-1391.

[64] Kubota M, Nodate M, Yasomoto H, Taku U, Osamu K, Misawa R. Isolation and functional analysis of cytochrom P450 genes from various environment. Biosience Biotechnology and Biochemistry 2005; 69 (12) 2421-2430.

[65] Lebaron P, Servais P, Trousellier M. Changes in bacterial community structure in seawater mesocosms differing in their nutrient status. Acquatic Microbial Ecology 1999; 19 255-267.

[66] Kloos J, Charles M, Schloter M. New method for the detection of alkane-monooxygenase homologous genes (alkB) in soils based on PCR-hybridization. Journal of Microbiological Methods 1999 ; 66 486–496.

[67] Lee M, Hwang G, Hung J, Young K, Kyung H. Physical structure and expression of alkb encoding alkane hydroxylase and rubredoxin reductase from Pseudomonas maltophilia. Biochemical and Biophysical Research Communications 1996; 218 17–21.

[68] Pukall R, Pauker O, Buntefu BD, Ilichs G, Lebaron P, Bernard, L, Guindulain T, Vives-Rego J, Stackebrandt E. High sequences diversity of Alteromonas macleodii-related cloned and cellular 16S rDNAs from Mediterranean seawater mesocosm experiment. FEMS Microbiology Ecology 1999; 28 335-344.

[69] Radwan SS, Al-Hasan RH, Salamah A, Khanafer M. Oil-consuming microbial consortia floating in the Persian Gulf. International Biodeterioration & Biodegradation 2005; 56 28-33.

[70] Li ZY, Kravchenko I, Xu H, Zhang C. Dynamic changes in microbial activity and community structure during biodegradation of petroleum compounds: A laboratory experiment. Journal of Environmental Science 2007; 19 1003–1013.

[71] Liu C, Zongze S. Alcanivorax dieselolei sp. nov., a novel alkane-degrading bacterium isolated from sea water and deep-sea sediment. International Journal of Systematic and Evolutionary Microbiology 2005; 55 1181–1186.

[72] Maa FB, Jing B, Guo L, Zhao C, Chein-chi C, Di C. Application of bioaugmentation to improve the activated sludge system into the contact oxidation system treating petrochemical wastewater. Bioresource Technology 2009; 100 597–602.

[73] Macnaughton S J, Stephen JR, Venosa AD, Davis GA, Chang YJ, White DC. Microbial population changes during bioremediation of an experimental oil spill. Applied and Environmental Microbiology 1999; 65 3566-3574.

[74] Rahman KSM, Thahira-Rahman J, Lakshmanaperumalsamy P, Banat IM. Towards efficient crude oil degradation by a mixed bacterial consortium. Bioresource Technology 2004; 85 257–261.

[75] Makrarn T. Suidan Effects of nitrogen source on crude oil biodegradation. Journal of Industrial Microbiology 1994; 13 279-286.

[76] Malatova A, Sbirova P, Rastosdia R. Isolation and characterization of hydrocarbon degrading bacteria from enviromental habitats in western New York State. Journal of Applied Microbiology 2005; 65 780-790.

[77] Manee P, Prayad P, Edward S, Upatham A, Ladda T. Biodegradation of crude oil by soil microorganisms in the tropic. Biodegradation 1998; 9 83–90.

[78] Margesin R, Labbe D, Schinner FC, Greer W, Whyte LG. Characterization of hydrocarbon-degrading microbial populations in contaminated and pristine alpine soils. Applied and Environmental Microbiology 2003; 69 3085–3092.

[79] Margesin R, Feller G, Hämmerle M, Stegner U, Schinner F. colorimetric method for the determination of lipase activity in soil. Biotechnology Letters 2002; 24 27–33.

[80] Marquez MC, Ventosa A. Marinobacter hydrocarbonoclasticus Gauthier et al. 1992 and Marinobacter aquaeolei Nguyen et al. 1999 are heterotypic synonyms. International Journal of Systematic and Evolutionary Microbiology 2005; 55 1349–1351.

[81] Mckew B, Coulon F, Yakimov MM, Denaro R, Genovese M, Smith J, Osborn M, Timmis KN, Mcgenity TJ. Efficacy of intervention strategies for bioremediation of crude oil in marine systems and effects on indigenous hydrocarbonoclastic bacteria. Environmental Microbiology 2007; 9 (6) 1562–1571.

[82] Muratova AY, Turkovskaya V. Degradation of petroleum oils by a selected microbial association. Applied Biochemistry and Microbiology 2001; 37 155–159.

[83] Muyzer G, Kornelia S. Application of denaturing gradient gel electrophoresis (DGGE) and temperature gradient gel electrophoresis (TGGE) in microbial ecology. Antonie van Leeuwenhoek 1998; 73 127–141.

[84] Muyzer G, Waal EC, Uitterlinden AG. Profiling of complex microbial populations by denaturing gradient gel electrophoresis analysis of polymerase chain reaction-amplified genes encoding for 16S rRNA. Applied and Environmental Microbiology 1993; 59, 695-700.

[85] Nakamuraa S, Sakamotoa Y, Ishiyamaa M, Tanakaa M, Kuniib K, Kubob C, Sato P. Characterization of two oil-degrading bacterial groups in the Nakhodka oil spil. International Biodeterioration & Biodegradation 2007; 60 202–207.

[86] Narhi LO, Wen LP, Fulco AJ. Characterization of the protein expressed in Escherichia coli by a recombinant plasmid containing the Bacillus megaterim cytochrome P-450 BM-3 gene. Molecular Cell Biochemistry 1988; 79 63-71.

[87] Nodate M, Mitsutoshi K, Norihiko M. Functional expression system for cytochrome P450 genes using the reductase domain of self-sufficient P450RhF from Rhodococcus sp. NCIMB 9784. Applied Microbiology Biotechnology 2006; 71 455–462.

[88] Odum EP. The mesocosm. Bioscience 1984; 34 558-562.

[89] Okerentugba PO, Ezeronye OU. Petroleum degrading potentials of single and mixed microbial cultures isolated from rivers and refinery effluent in Nigeria. African Journal of Biotechnology 2003; 2 (9) 288-292.

[90] Hoff R. A summary of bioremediation applications observed at marine oil spills. Report HMRB 91-2. Hazardous Materials Response Branch, National Oceanic and Atmospheric Administration 1991; Washington DC.

[91] Petersen JE, Cornwell JC, Kemp WM. Implicit scaling in the design of experimental aquatic ecosystems. Oikos 1999; 85 3-18.

[92] Powell SM, Ferguson SH, Bowman P, Snape I. Using real-time PCR to assess changes in the hydrocarbon-degrading microbial community in antarctic soil during bioremediation. Microbial Ecology 2006; 52 523–532.

[93] National Academy of Sciences. Oil in the Sea: Inputs, Fates and Effects. Washington DC: National Academy Press; 1985.

[94] Hoff R. Bioremediation: an overview of its development and use for oil spill cleanup. Marine Pollution Bulletin 1993; 26 476–481.

[95] Atlas RM. Microbial hydrocarbon degradation—bioremediation of oil spills. Journal Chemistry Technology Biotechnology 1991; 52 149–156.

[96] Bocard CP, Renault J, Croquette S. Cleaning products used in operations after the Amoco Cadiz disaster. In Proceedings of the International Oil Spill Conference. American Petroleum Institute 1979; Washington D.C.

[97] Pruthi V, Cameotra S S. Rapid identification of biosurfactant-producing bacterial strains using a cell surface hydrophobicity technique. Biotechnology Techniques 1997; 11 671-674.

[98] Mearns A J. 1991. Observations of an oil spill bioremediation activity in Galveston Bay, Texas. NOAA Technical Memorandum NOS OMA 57. National Oceanic and Atmospheric Administration 1997; Washington D.C.

[99] Texas General Land Office. Combating oil spills along the Texas coast: a report on the effect of bioremediation. Texas General Land Office 1989; Austin.

[100] Nadeau R, Singhvi J, Ryabik YH, Lin J, Syslo C. Bioremediation efficacy in Marrow Marsh following the Apex Oil Spill, Galveston 1992.

[101] Greene TC. The Apex Barges spill, Galveston Bay, July 1990, In Proceedings of the 1991 Oil Spill Conference. American Petroleum Institute 1991; Washington, D.C.

[102] Leveille TP. The Mega Borg fire and oil spill: a case study. In Proceedings of the 1991 International Oil Spill Conference. American Petroleum Institute 1991; Washington, D.C.

[103] Texas General Land Office. Mega Borg oil spill off the Texas coast: an open water bioremediation test. Texas General Land Office 1990; Texas Water Commission Austin 273–278.

[104] Lee K, Levy EM. Bioremediation: waxy crude oils stranded on low-energy shorelines, In Proceedings of the 1991 Oil Spill Conference (Prevention, Behaviour, Control, Cleanup). American Petroleum Institute 1991; Washington, D.C 541–547.

[105] Bragg JR, Prince JB, Wilkinson RM, Atlas D. Biore- 362 SWANNELL ET AL. MICROBIOL. REV. mediation for shoreline cleanup following the 1989 Alaskan oil spill. Exxon Co., USA 1992; Houston.

[106] Owens EH. Changes in shoreline oiling conditions 1 1/2 years after the 1989 Prince William Sound spill. Technical Report 1991; Woodward-Clyde Consultants, Coastal Science and Engineering Center, Seattle.

[107] Jahns HO, Bragg LC, Dash EH, Owens. Natural cleaning of shorelines following the Exxon Valdez oil spill. In Proceedings of the 1991 Oil Spill Conference. American Petroleum Institute 1991; Washington, D.C., 167–176.

[108] National Oceanic and Atmospheric Administration. Excavation and rock washing treatment technology—net environmental benefit analysis. Hazardous Materials Response Branch, National Oceanic and Atmospheric Administration 1990; Seattle.

[109] Pritchard PH, Costa C. EPA's Alaska oil spill bioremediation project. Environment. Science. Technology 1991; 25 372–379.

Section 2

Biodegradation and Anaerobic Digestion

Chapter 6

Challenges for Cost-Effective Microalgae Anaerobic Digestion

Álvaro Torres, Fernando G. Fermoso,
Bárbara Rincón, Jan Bartacek, Rafael Borja and
David Jeison

Additional information is available at the end of the chapter

http://dx.doi.org/10.5772/55975

1. Introduction

Microalgae, the common denomination for a broad group of photosynthetic prokaryotes and eukaryotes, are characterized by an efficient conversion of the solar energy into biomass. They are a promising feedstock for the production of third generation biofuels for several reasons:

1. Microalgae photosynthesis allows biological CO_2 fixation, which is expected to mitigate atmospheric CO_2 increase (Amin 2009; Brennan & Owende 2010; Mutanda *et al.* 2011).
2. Microalgae are 10 – 50 times more efficient than plants in terms of CO_2 fixation (Wang *et al.* 2008). Thus, microalgae can fix 1.83 tonnes of CO_2 per 1 tonne of produced microalgae (Chisti 2007).
3. Microalgae can be produced on non-arable areas such as lakes, oceans or deserts, thus reducing competition with food production (Mussgnug *et al.* 2010; Stephens *et al.* 2010). This advantage is a key factor when energy supply is considered in desert zones near oceans.
4. Some microalgae can grow under saline conditions, which strengthen the use of microalgae as feedstock for biofuel production in desert zones near the ocean when freshwater supply is not feasible.

Most of current efforts to take advantage of microalgae as a source of bioenergy are directed to biodiesel production, considering the ability of certain types of microalgae to accumulate lipids under controlled culture conditions. Microalgae biodiesel produced from microalgae lipids also presents technical advantages compared to lignocellulosic biomass based biodiesel.

Biodiesel from microalgae has a higher calorific value (30 and 29 MJ/kg for *C. protothecoides* and *Microcystis aeruginose*, respectively) and lower viscosity and density than plants-based biodiesel (Costa & de Morais 2011). However, the biodiesel yield from algae is rather low compared to biodiesel from lignocellulose energy (Chisti 2007; Sialve *et al.* 2009; Scott *et al.* 2010; Stephens *et al.* 2010). Indeed, with current technology, a negative energy balance was calculated by Lardon *et al.* (2009) when evaluating biodiesel production from *C. vulgaris*, considering biomass drying and further lipid extraction by solvents. During biodiesel production from microalgae, energy consumption associated with culture mixing and pumping, lipid extraction, nutrients addition, drying is of particular importance (Scott *et al.* 2010). Indeed, Lardon *et al.* (2009) estimated that the necessary energy consumption for drying was near 85% of the total energy consumption in a biodiesel production process from microalgae. Another drawback of biodiesel process is associated with the microalgae cultivation step, as nutrient requirements are 55-111 times higher than for e.g. rapeseed cultivation (Halleux *et al.* 2008). Under these conditions, biodiesel production from microalgae may not be energetically and environmentally sustainable (Sialve *et al.* 2009; Ras *et al.* 2011).

2. Microalgae as a source of biogas

Biogas production through anaerobic digestion is an established technology where a wide variety of residues can be used as substrate. In 2011, 8,760 anaerobic digesters were reported in Europe (IEA, 2011). The contribution of this technology to the reduction of carbon emissions, green energy and green gas policies has generated intense interest, especially over the past decade.

When considering biogas production from microalgae two alternatives can be conceived: Microalgae biodiesel production and further anaerobic digestion of microalgae residues for biogas production (Process 1, Figure 1A) and anaerobic digestion of whole microalgae with biogas as sole biofuel (Process 2,Figure 1B).

Process 1: Biodiesel production and subsequent biogas production from spent microalgae. Two principal drawbacks are identified when biodiesel production from microalgae is considered: high nutrients requirements for microalgae growth and low energy efficiency of biodiesel production process. Anaerobic digestion may contribute to overcome such limitations, by enabling nutrients recovery and biogas production when spent microalgae after lipid extraction is used as substrate. This is based on the fact that biogas can be used as source of renewable energy and that during anaerobic digestion process, nitrogen and phosphorus may be recovered, creating opportunities for their reuse as nutrients. Theoretical energy contribution of anaerobic digestion is presented in Figure 1A, assuming microalgae content of lipids, proteins and carbohydrates to be 30, 45 and 25%, respectively.

Figure 1A shows that an energy yield of 11MJ per kilogram of gross microalgae is reached when biodiesel production is considered. If oil extracted microalgae is used as substrate in anaerobic digestion process, methane produced would have a maximum theoretical contribution of 17MJ per kilogram of gross microalgae (thermal). Such value has been computed

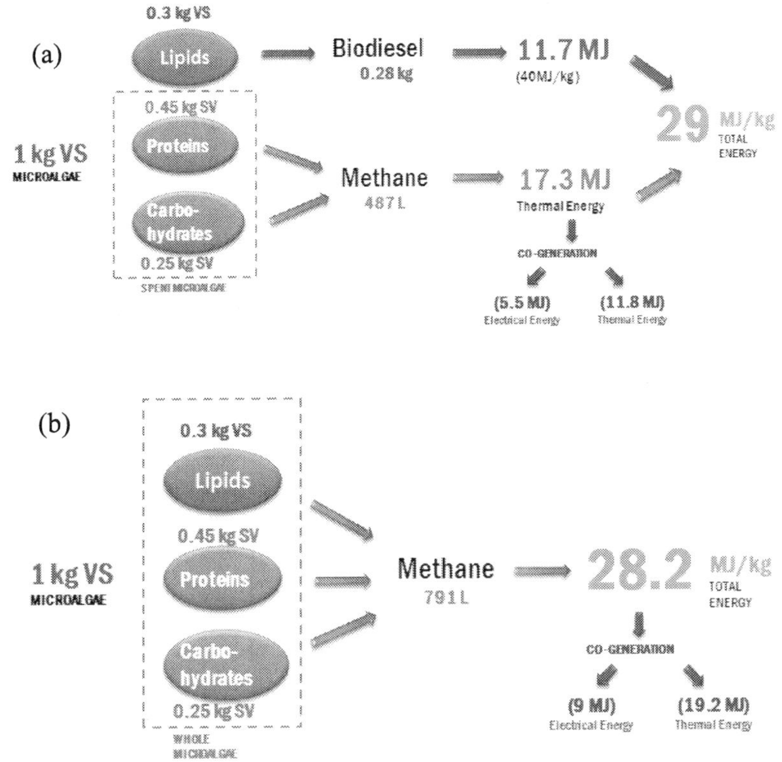

Figure 1. Energy potential of microalgae considering: a) Biodiesel production and further anaerobic digestion of microalgae residues for biogas production or b) Anaerobic digestion of whole microalgae only for biogas production.

assuming carbohydrate and protein methanogenic potentials of 0.415 and 0.851 L CH_4/kg VS, respectively (Angelidaki & Sanders 2004). If the latter thermal energy is transformed into electricity, a maximum energy yield of 5.5 MJ per kilogram of gross microalgae would be achieved (assuming a conversion efficiency of 32%). Thus, a substantial increase in energy yield could be theoretically achieved, representing a considerable contribution to biodiesel sustainability and economic feasibility. Energy contained in biogas can be used for both anaerobic digestion and trans-esterification reactor heating. Electricity obtained via co-generation can be used for different purposes such as photobioreactor mixing, microalgae harvesting and drying (Harun *et al.* 2010; Razon & Tan 2011). Neumann et al. (2011) evaluated energy contribution of biogas production in Process 1 for *Botryococcus braunii* with 30% lipid content. The latter study considered a nutrient recovery step through membrane liquid/solid separation from anaerobic digestion reactor and heptane evaporating step in order to recovery this solvent. Biogas production could theoretically contribute with close to 50% of the overall energy yield of Process 1.

Process 2: Biogas production from whole microalgae. Another alternative to recover energy from microalgae consists of methane production from whole microalgae. In such process, all organic matter (proteins, carbohydrates and lipids) present in microalgae biomass would be converted into methane and carbon dioxide, without considering biodiesel production (De Schamphelaire & Verstraete 2009; Douskova *et al.* 2010; Zamalloa *et al.* 2011). Several advantages are recognized when energy production from whole microalgae through biogas generation is considered: Biogas productions involves high energy yields, biogas production would not require microalgae biomass drying (it involves wet fermentation), biogas can be used to produce heat and electricity through co-generation, microalgae cultures can be used for biogas upgrading (i.e. CO_2 biosequestration), microalgae species not capable of accumulating lipids may be also used as feedstock. Moreover, co-digestion with other types of biomass such as solid or liquid wastes is feasible. Anaerobic digestion of algal and microalgae biomass has been previously studied by some researches (Vergara-Fernańdez *et al.* 2008; De Schamphelaire & Verstraete 2009; Mussgnug *et al.* 2010; Zamalloa *et al.* 2011). Figure 1B shows the energy potential of Process 2, in which whole microalgae is used as substrate in order to produce biogas. In this estimation, all energy is produced as methane, which allows theoretical maximum energy recovery of 27 MJ per kg of volatile solids of microalgae (8.6MJ of electricity and 18.4 MJ of heat, if co-generation is considered). The lower operational energy demands for biogas production, compared with biodiesel together with biogas, makes Process 2 very promising for energy recovery.

3. Anaerobic digestion of microalgae

Reports of the anaerobic digestion of microalgae go back to the fifties when Golueke *et al.* (1957) was one of the first authors studying the feasibility of sunlight energy conversion to methane by algae sunlight fixation followed by biomass anaerobic fermentation. In this early study, 0.5 m^3 of biogas was obtained per volatile kg of algal biomass, with methane content 63%. More than two decades later, Nair *et al.* (1983) reported a lower yield, close to 0.22 m^3/kg VSS, at loading rate 1.7 $kg/(m^3\ d)$. Despite those early reports, biogas production from algae and microalgae has not yet widely researched (Foree & McCarty 1970; Samson & Leduy 1983; Tarwadi & Chauhan 1987; Vergara-Fernańdez *et al.* 2008; De Schamphelaire & Verstraete 2009; Mussgnug *et al.* 2010; Zamalloa *et al.* 2011).

3.1. Choosing microalgal culture for direct biogas production

The ideal microalgae specie for a maximum biogas production would that presenting:

1. thin or no cell wall
2. large cells
3. high growth rate in non-sterile media
4. high resistivity against natural contaminants

5. carbohydrate-based cell wall.

Of the above mentioned factors, the quality of cell wall is crucial for anaerobic digestion of algae. This is because cell walls are hard to degrade biologically and their presence avoids contact of anaerobic bacteria with the readily degradable content of algal cells. Therefore, the influence of cell wall presence is described in detail in the following text.

3.1.1. *Composition of algal cell wall*

Cell wall in microalgae represents 12-36% of total cell mass (cell wall weight/cell weight) in different microalgae (Table 1). Microalgae cell wall is composed mainly of carbohydrates and proteins which represent 30-75% and 1-37% of cell wall, respectively.

Microalgae	**Cell Wall**	**Cell Wall composition (%)**			**References**
	(% w/w)	Carbohydrates	Protein	n.d.*	
Chlorella vulgaris (F)	20.0	30.00	2.46	67.54	(Abo-Shady *et al.* 1993)
Chlorella vulgaris (S)	26.0	35.00	1.73	63.27	(Abo-Shady *et al.* 1993)
Kirchneriella lunaris	23.0	75.00	3.96	21.04	(Abo-Shady *et al.* 1993)
Klebsormidium flaccidum	36.7	38.00	22.60	39.40	(Domozych *et al.* 1980)
Ulothrix belkae	25.0	39.00	24.00	37.00	(Domozych *et al.* 1980)
Pleurastrum terrestre	41.0	31.50	37.30	31.20	(Domozych *et al.* 1980)
Pseudendoclonium basiliense	12.8	30.00	20.00	50.00	(Domozych *et al.* 1980)
Chlorella Saccharophila	-	54.00	1.70	44,30	(Blumreisinger *et al.* 1983)
Chlorella fusca	-	68.00	11.00	20.00	(Blumreisinger *et al.* 1983)
Chlorella fusca	-	80.00	7.00	13.00	(Loos & Meindl 1982)
Monoraphidium braunii	-	47.00	16.00	37.00	(Blumreisinger *et al.* 1983)
Ankistrodesmus densus	-	32.00	14.00	54.00	(Blumreisinger *et al.* 1983)
Scenedesmus obliquos	-	39.00	15.00	46.00	(Blumreisinger *et al.* 1983)

* not determined.

Table 1. Cell wall composition of microalgae.

Other compounds found in microalgal cell wall are uronic acid, glucosamine, hidroxyproline, proline, sporopollenin, carotenoids and another resistant biopolymers (Punnett & Derrenbacker 1966; Domozych *et al.* 1980; Blumreisinger *et al.* 1983; Brown 1991, 1992; Abo-Shady *et al.* 1993).

In relation to carbohydrates in microalgae cell wall, neutral sugars, cellulose and hemicelluloses are the main components. Blumreisinger *et al.* (1983) studied five different microalgae in relation to carbohydrate composition in cell wall, obtaining a prominent neutral sugar component. Composition of cellulose and hemicelluloses has ranged between 6-17% and 18-32% for microalgae studied in other researches carried out by Abo-Shady *et al.* (1993) and Domozych *et al.* (1980), respectively. On the other hand, Northcote *et al.* (1958) reported contents of cellulose near to 45% in cell wall of *Chlorella pirenoidosa.* Unlike these researches, Loos and Meindl (1982) found no presence of cellulose in cell wall of *Clhorella fusca.* In relation to proteins, peptides, proline and hidroxyproline are the main components. According to Punnett and Derrenbacker (1966), the cell wall of six different microalgae consisted of peptides (simple amino acid composition) but it contained no protein. In addition, this research revealed the existence of proline in the cell wall of *Chlorella vulgaris* and hidroxyproline in the cell wall of *Chlorella pyrenoidosa* and *Scenedesmus obliquos.*

3.1.2. Degradability of algal cell wall

Although methane yield is dependent on microalgae composition (Sialve *et al.* 2009), the resistance of cell wall is considered to be the limiting factor for the anaerobic digestion of microalgae (Afi *et al.* 1996; Chen & Oswald 1998). The kinetics of anaerobic digestion is highly dependent on the degradability of the given microalgae species (Sialve *et al.* 2009). Mussgnug *et al.* (2010) studied the methane production from six different microalgae, obtaining from 287 to 587 mL CH_4/ g VS. The low levels of methane yield were related to low cell degradation and high amount of indigestible residues. According to these results, easily degradable microalgae had no cell wall or a protein-based cell wall not containing cellulose/hemicellulose. Batch tests with low methane yields, intact cell walls of microalgae were found with light microscopy in this study. Thus, the intracellular content was not available for efficient digestion. The presence of biopolymers resistant to anaerobic degradation has been reported in the outer cell wall of microalgae species such as *Botryococcus braunii* (Templier *et al.* 1992; Banerjee *et al.* 2002). Moreover, microalgae degradability is related to cell wall structures containing these resistant biopolymers. Some microalgae have a protective tri laminar outer wall called tri laminar sheath (*TLS*), which hinders efficient microalgae degradation (Derenne *et al.* 1992). Thus, higher *TLS* resistance to degradation reported by Derenne *et al.* (1992) for microalgae *B. braunii* has been associated to the presence of sporopollenin-like biopolymers (Kadouri *et al.* 1988; Derenne *et al.* 1992). Other indigestible compound found in microalgae cell wall is algaenan, which has been reported as non-hydrolysable resistant biopolymer composed of polyether linked long-chain (up to C36) n-alkyl units (Gelin *et al.* 1997; Blokker *et al.* 1998; Gelin *et al.* 1999; Simpson *et al.* 2003).

3.1.3. Source of methane in algae

Many authors have related methane yield from microalgae to their composition (Sialve *et al.* 2009; Mairet *et al.* 2011; González-Fernández *et al.* 2012; Mairet *et al.* 2012), especially with the content of lipids, carbohydrates and proteins. However, the experimental data collected from literature do not show strong correlation between lipids, carbohydrates and proteins found in various algal species and the methane yield obtained by various authors (Fig. 2).

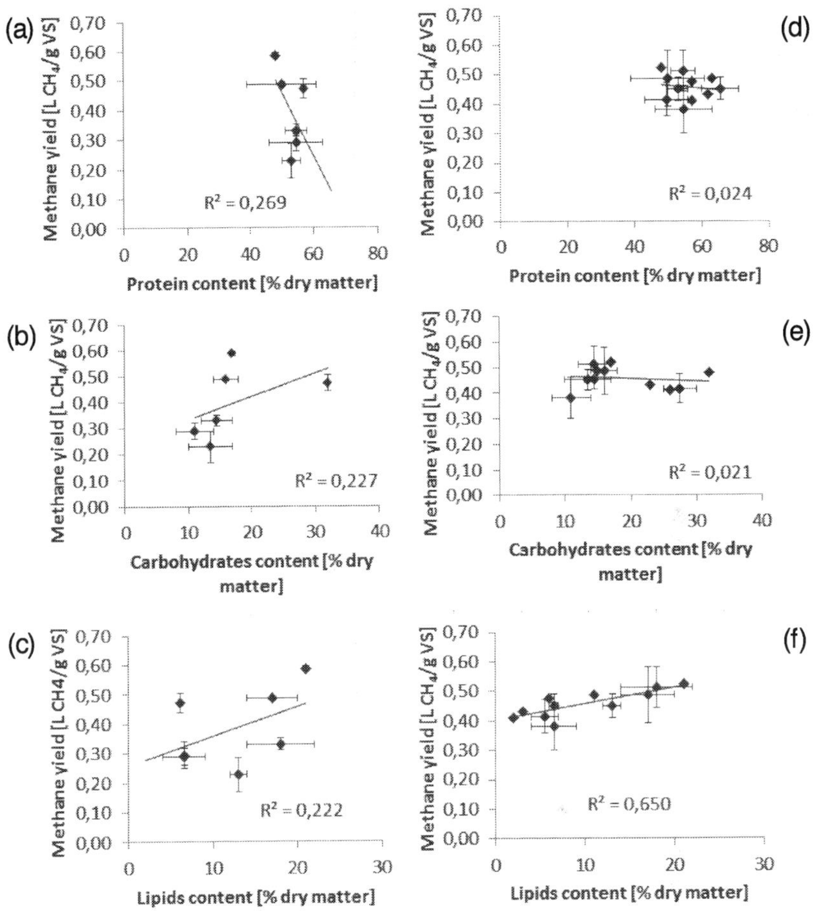

Figure 2. Dependence between methane yield from microalgae and their lipids, carbohydrates and proteins content. Each data point represents one algae species while the error bars show the range found in the literature. Figures (a), (b) and (c) show experimentally obtained methane yields, figures (d), (e) and (f) represent theoretical methane yield for the given algae composition calculated according to Angelidaki and Sanders (2004). Data were extracted from multiple authors (Becker 2007; Griffiths & Harrison 2009; Sialve *et al.* 2009; Mairet *et al.* 2011; González-Fernández *et al.* 2012; Mairet *et al.* 2012).

Angelidaki and Sanders (2004) presented theoretical methane yields from proteins, carbohydrates and lipids of 0.50, 0.42 and 1.01 L/g VS, respectively (Fig. 3). Even when these values are used for calculation of the potential methane yield from various algal species, no strong correlation can be found (Fig. 2d, e and f). Theoretically, lipids content has the biggest influence on methane yield, but as lipids are usually not the mayor source of methane (Fig. 2), the correlation between lipids content and methane yield is still rather vague (Fig. 2).

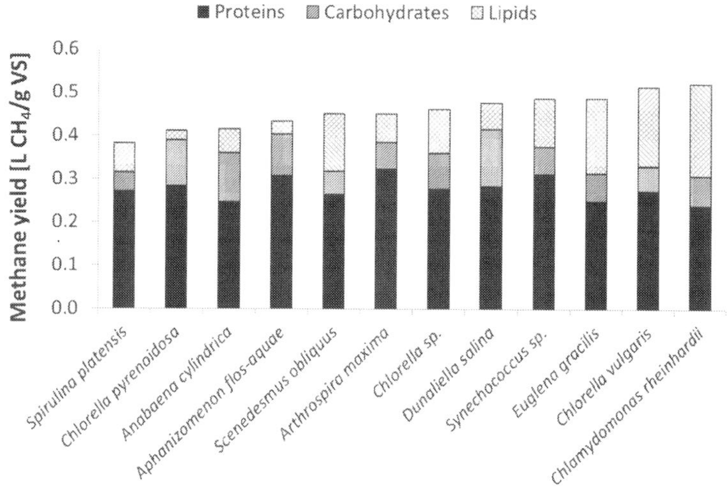

Figure 3. Potential methane yield from proteins, carbohydrates and lipids present in various algae species calculated according to Angelidaki and Sanders (2004). The data on proteins, carbohydrates and lipids content in algae were extracted from Becke (2007), Sialve (2009), Griffiths and Harrison (2009) and González-Fernández et al. (2012).

These facts clearly show that the ration between various macromolecules is not the most important parameter determining the actual methane yield from algae. As it was mentioned before, content of inert organic matter (e.g. cell wall) would play more important role (González-Fernández *et al.* 2012).

These findings show that plain composition of algal biomass indeed cannot be the main factor while choosing the best algal strain for methane production. Biomass production rate and the content of cell-walls will be of higher importance. Moreover, environmental conditions such as the salinity of available water source must be taken into account.

3.2. Pretreatment

In order to overcome limitation caused by cell wall degradability, which is necessary to access the intracellular content, cell disruption (pretreatment) has been pointed out as an important contributor in order to enhance anaerobic digestion efficiency. As mentioned above, cell wall degradability affects both Processes 1 and 2. However, in Process 1, cell wall degradability should not be as critical as in Process 2 since lipid extraction itself may be considered a pretreatment step.

There are different pretreatment techniques applied to microalgae, which can be classified as enzymatic, chemical and mechanical treatments. Mechanical pretreatment include autoclaving, homogenizers, microwaves and sonication, which increases the availability of organic matter (Angelidaki & Ahring 2000). Chemical pretreatment will increase availability of compounds resistant to anaerobic hydrolysis due to the enhanced disintegration (Bonmatí *et al.* 2001).

Chemical pretreatment can be clasified as acid or alkaline treatment. An increase in soluble hemicellulose present in cell wall is expected when alkaline pre-treatment is used (Abo-Shady *et al.* 1993). Thus, chemical pre-treatment is suitable when microalgae cell wall is rich on hemicelluloses. Also, enzymatic pretreatment has been used in order to attack cell wall and improve compounds extraction from microalgae. Enzymatic pretreatment with α-amilase, amylo-glucosidase and cellulase have shown a positive effect on cell wall hydrolysis (Choi *et al.* 2010; Fu *et al.* 2010). Fu *et al.* (2010) reported a 62% increase in cell wall hydrolysis, when *Chlorella sp.* was pretreated by immobilized cellulase.

Few studies report the effect of cell disruption pretreatment in anaerobic digestion (Samson & Leduy 1983; Chen & Oswald 1998). Samson and Leduy (1983) reported an increase of 78% in soluble COD when algae *Spirulina maxima* was mechanically pretreated (sonication and mechanical disintegration). However, no increase in methane yield was observed.

Finally, two considerations should be taken into account when cell disruption pretreatment is evaluated in the context of anaerobic digestion: On one hand, energy consumption associated with pretreatment should be low in order to avoid a negative contribution to the energy balance of anaerobic digestion process. On the other hand, contribution to the biodegradability of the given substrate should be a response variable when the effect of pretreatment on anaerobic digestion is evaluated. In other words, some pretreatment techniques increase solubility of organic matter but do not increase its biodegradability.

3.3. Inhibiting factors related to anaerobic digestion

Figure 1B shows the energy potential when microalgae are used as substrate in order to produce biogas. In this estimation, total energy is produced as methane, which allows a theoretical maximum energy recovery of 27MJ per kg of volatile solids of microalgae. As in Process 1, part of energy produced will be spent for supplying the energy necessary for microalgae harvesting and up-concentration, photobioreactor mixing, photobioreactor and anaerobic reactor heating, etc. The theoretical estimations of energy production from anaerobic digestion presented in this review have been so far computed considering 100% of microalgae biodegradability and high performance of anaerobic digestion. However, an energy production lower than ideal can be expected when limiting factors in anaerobic digestion process are considered. For this reason, this book chapter examines different limiting factors of anaerobic digestion, which are necessary to overcome in order to improve performance of this process.

3.3.1. Ammonium inhibition

Ammonium is presented as protonated form (NH_4^+) and deprotonated form (NH_3, ammonia). The latter is considered to be the specie responsible for the inhibition of anaerobic digestion, due to its permeability through cell membrane (De Baere *et al.* 1984). There are several mechanisms by which ammonia will act as inhibitor of anaerobic bacteria among which are intracellular pH changes, increase in energy requirements for maintenance and inhibition of specific enzymes (Wittmann *et al.* 1995).

Several factors determining ammonia concentration in anaerobic reactor has been reported, but substrate concentration is a major one (Sialve *et al.* 2009). Distribution of total ammonia between protonated and deprotonated forms strongly depends on factors such as pH and temperature. At high pH values ammonium gets deprotonated forming toxic ammonia (NH_3) (Borja *et al.* 1996). Its inhibitory effect can result in volatile fatty acids accumulation due to a decrease in methanogenic activity, which generates a decrease in pH and ammonia concentration (Chen *et al.* 2008). This interaction may generate an inhibited steady-state, in which the process remains stable despite inhibition (Angelidaki & Ahring 1993; Angelidaki *et al.* 1993). Temperature is another variable that determine NH_4^+/NH_3 ratio, which is directly related to the increase of ammonia fraction and thus, inhibition level (Braun *et al.* 1981; Angelidaki & Ahring 1994).

Microalgal biomass can be expected to have low C/N ratio due to the high protein content in microalgae (Becker 2007). Then, anaerobic degradation of these residues is expected to generate a high ammonium concentration that may cause inhibition of anaerobic microbial consortia, especially methanogenic bacteria (Angelidaki & Ahring 1993; Chen *et al.* 2008). In addition, high ammonium concentration may affect biogas quality since ammonia can be stripped into gas phase (Sialve *et al.* 2009).

During anaerobic digestion of oil extracted microalgae (Process 2 on Figure 1), ammonia inhibition is expected to be especially of concern, since oil extraction will decrease C/N ratio. Figure 4 shows an estimation of the effect of substrate concentration and free ammonia levels in a hypothetical anaerobic digestion reactor. Estimation was calculated considering protein content reported by Becker (2007), operation pH value 8, temperature 35º C, ammonia conversion 90% and total lipid extraction efficiency. Figure 4 shows that inhibitory ammonia concentrations will develop whenever solids concentration exceeding 2% are applied during the anaerobic digestion step. This result was evaluated considering free ammonia inhibition at 100 mg/L NH_3 (dotted line in Figure 4).

Results shown in Figure 4 indicate that that either anaerobic digestion has to be performed at very low levels of solids concentration, or mechanisms for ammonia removal must be implemented. It has to be remained that Figure 4 assumes 90% of conversion of proteins. Lower protein conversions will reduce the chances of ammonia inhibition. However it is clear that this phenomena needs to be addressed if high rate digestion of microalgae is of interest.

One way to overcome this drawback is the possibility of co-digestion in order to provide an optimal C/N ratio for anaerobic digestion process (Yen & Brune 2007; Ehimen *et al.* 2011). Thus, a higher C/N ratio co-substrate should be mixed with microalgae in order to increase anaerobic digestion yield. This strategy is more attractive considering the fact that some co-substrate can stimulate enzymatic synthesis and, hence, increase hydrolysis and degradability (Yen & Brune 2007). Also, co-digestion can dilute toxic compounds decreasing their concentration below toxic/inhibition levels (Sialve *et al.* 2009).

http://dx.doi.org/10.5772/55975

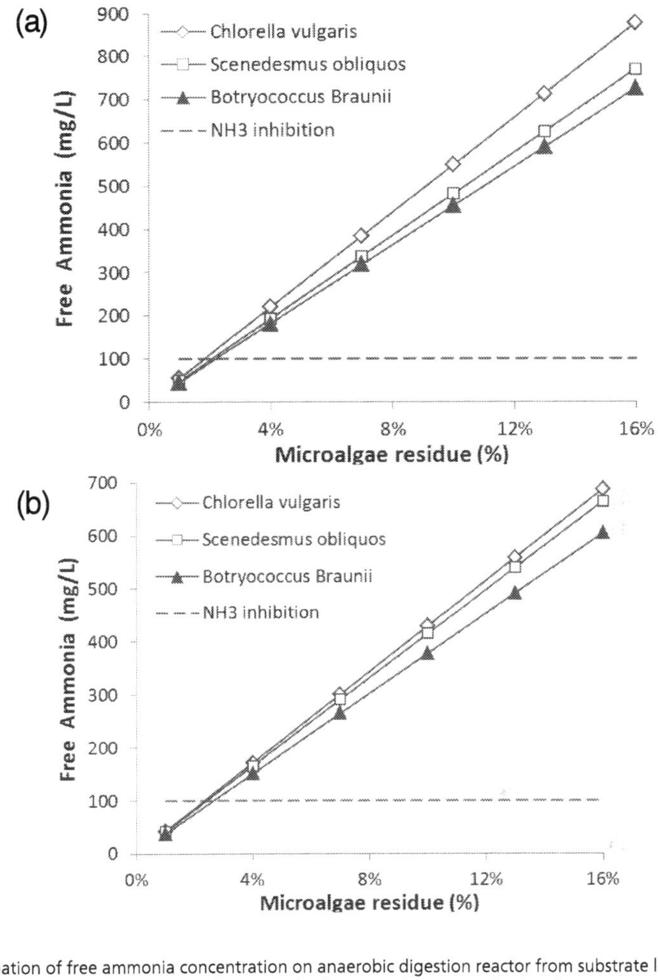

Figure 4. Estimation of free ammonia concentration on anaerobic digestion reactor from substrate level of feedstock, considering (a) processes 1, Biodiesel production and subsequent biogas production from spent microalgae and (b) process 2, Biogas production from whole microalgae.

3.3.2. Salt inhibition

Salt inhibition is expected to be relevant when saline microalgae are used as substrate for biogas production. In those locations where freshwater is not abundant or available, saline microalgae may be of interest, if cultivation takes place close to the sea. In those situations, salinity may even be higher than sea water when open pounds are used, as a result of water evaporation. If biomass is not diluted with fresh water after harvesting, downstream processes such as anaerobic digestion may need to deal with the salinity present in the biomass.

At low concentrations, sodium is essential for methanogenic bacteria. Probably, it is due to its role in ATP formation or NADH oxidation (Dimroth & Thomer 1989). Sodium concentration ranges 100-350mg/L have been reported as beneficial for mesophilic methanogenic growth (McCarty 1965; Patel & Roth 1977). Although moderate concentrations can stimulate bacteria growth, excessive amounts of salt reduce growth rate, and can cause severe inhibition or toxicity (Soto *et al.* 1991). Moreover, high salt levels can cause dehydration in bacteria due to osmotic pressure (De Baere *et al.* 1984; Yerkes *et al.* 1997).

Different levels of saline tolerance in anaerobic bacteria have been reported (Lefebvre & Moletta 2006). Easily degradable substrates seem to increase salt tolerance, most likely as a result of higher energy availability to cope with the energetic requirements of salt tolerance mechanisms (Xiao & Roberts 2010). Rinzema *et al.* (1988) found non acetoclastic methanogenic activity at 16 g/L of sodium concentration. The concentration that generated 50% of activity reduction (IC50) was 10 g/L and no bacteria adaptation after 12 weeks was observed. Similar saline tolerance was observed by Liu and Boone (1991). Feijoo *et al.* (1995) analyzed sodium inhibition for anaerobic bacteria from different reactors. A high tolerance in anaerobic bacteria from reactor treating wastewater under salinity conditions was observed, which was interpreted as consequence of bacteria adaptation. IC50 value for these bacteria was 16.3 g Na^+/L and entire inhibition was observed at 21 g Na^+/L.

Several reports indicate that biomass acclimation may significantly increase the activity under saline conditions (Soto *et al.* 1991; Omil *et al.* 1995; Chen *et al.* 2008; Kimata-Kino *et al.* 2011). However, reports are also available where no or little acclimating was observed (Aspe *et al.* 1997). Then, selection rather than adaptation is likely to be the mechanisms to provide high activity when big changes in salinity are imposed, requiring the presence of salinity-tolerant microorganisms in the inoculum (Gebauer 2004). It is indeed a common practice to use inoculums containing sources of saline resistant microorganisms, such as marine sediments (Xiao & Roberts 2010).

3.4. Biogas upgrading

Many biogas applications such as vehicle use, household distribution and electricity production, require some level of biogas upgrading to remove impurities or to increase methane content.

CO_2 removal is a key factor in order to obtain a higher calorific value of biogas. Processes such as solvent absorption, activated carbon adsorption and membrane filtration have been used for CO_2 removal (Kapdi *et al.* 2005; Makaruk *et al.* 2010; Ryckebosch *et al.* 2011).

Photosynthetic microorganisms such microalgae can also be used to remove CO_2 from biogas. Microalgae cultures are regarded as an interesting tool for carbon dioxide capture from gases such as flue gases from boilers, combustion engines or thermal power plants. This would not only alleviate impact of CO_2 emissions on the environment, but it would also reduce the cost of microalgae production (Doucha *et al.* 2005; Ryu *et al.* 2009). Stabilization ponds have been already recognized as potential CO_2 scrubbers due to their (micro-) algae growth (Shilton *et al.* 2008). Several authors have reported the successful growth of microalgae using flue gases. Negoro *et al.* (1993) reported productivities similar to those using pure CO_2, and showed that growth was barely influenced by the content of SO_X and NO_X contained in flue gases. Similar results were obtained by Hauck *et al.* (1996) who found no inhibition of *Chlorella sp.* by the levels of NO_X typically contained in flue gases. Doucha *et al.* (2005) reported 50% of flue gas decarbonization when working with a photobioreactor. In this study, 4.4 kg of CO_2 was needed for the production of 1 kg of dried algal biomass.

Conde *et al.* (1993) achieved biogas purification in laboratory experiments up to methane content of 97% with algae grown on synthetic nutrient medium. Mandeno *et al.* (2005) achieved CO_2 reduction from 40 to less than 5% using synthetic biogas, observing little transfer of oxygen to the biogas, so explosive methane/oxygen mixtures would not be formed. Similar results in terms of CO_2 reduction were obtained by Travieso *et al.* (1993) working with real biogas. Several microalgae species such as *Chlorococcum littorale, Chlorella sp., Chlorella sp.* UK001, *Chlorella vulgaris, Chlorella kessleri, Scenedesmus obliquus, Spirulina sp., Haematococcus pluvialis* or *Botryococcus braunii* have shown high levels of tolerance to high partial pressures of CO_2 (Wang *et al.* 2008; Brennan & Owende 2010). Mass transfer of carbon dioxide from gas to liquid phase is dependent on several factors highlighting chemical balance in microalgae media, pH and flow pattern of reactor in which culture is growing (Kumar *et al.* 2010). However, no full scale installations are under operation with this concept.

Available publications do not report negative effects of high methane partial pressures over microalgae cultures. Moreover, Meier *et al.* (2011) reported no inhibition effect when exposing a culture of *N. gaditana* to atmospheres containing methane up to 100%.

Hydrogen sulfide is present in biogas at low concentrations although its treatment should be considered. Some studies have reported a hydrogen sulphide decrease after biogas is upgraded in microalgae culture (Conde *et al.* 1993; Heubeck *et al.* 2007; Sialve *et al.* 2009). Most likely, hydrogen sulphide removal should be attributed to relative high solubility in growth medium (Conde *et al.* 1993; Sialve *et al.* 2009). Solubilised hydrogen sulphide can be easily oxidized into sulphate due to oxygen presence in growth medium.

4. Conclusions

Microalgal biomass is a promising substrate for renewable energy production. In this book chapter, direct anaerobic digestion without previous biodiesel extraction was shown to be the most promising method of energy production from microalgae. Lipids used for biodiesel production can also serve as a rich source of biogas with energetic efficiency higher than when microalgae are used for subsequent biodiesel and biogas production. The higher energy efficiency is given mostly by the simple technology with low energy demand used for methane production. These benefits combined with the possibility of CO_2 and nutrients recycling from the anaerobic effluents make anaerobic digestion the best technology for removable energy production from microalgae.

Acknowledgements

The authors would like to thank to Marie Curie's International Research Staff Exchange Scheme (IRSES) "Renewable energy production through microalgae cultivation: closing material cycles" (PIRSES-GA-2011-295165) for providing financial support. Dr. Rincón wishes to express her gratitude to the Ramón y Cajal Program from the Spanish Science and Innovation Ministry for providing financial support. Dr. Jeison wishes to thank CONICYT Chile and FONDECYT Project 1120488.

Author details

Álvaro Torres[1], Fernando G. Fermoso[2], Bárbara Rincón[2], Jan Bartacek[3], Rafael Borja[2] and David Jeison[1]

1 Chemical Engineering Department, Universidad de La Frontera, Temuco, Chile

2 Instituto de la Grasa (CSIC). Avenida Padre García Tejero, Sevilla, Spain

3 Department of Water Technology and Environmental Engineering, Institute of Chemical Technology Prague, Prague, Czech Republic

References

[1] Abo-Shady A. M., Mohamed Y. A. and Lasheen T. (1993). Chemical composition of the cell wall in some green algae species. Biologia Plantarum 35(4), 629-32, ISSN: 1573-8264

[2] Afi L., Metzger P., Largeau C., Connan J., Berkaloff C. and Rousseau B. (1996). Bacterial degradation of green microalgae: Incubation of Chlorella emersonii and Chlorella vulgaris with Pseudomonas oleovorans and Flavobacterium aquatile. Organic Geochemistry 25(1-2), 117-30, ISSN: 0146-6380

[3] Amin S. (2009). Review on biofuel oil and gas production processes from microalgae. Energy Conversion and Management 50(7), 1834-40, ISSN: 0196-8904

[4] Angelidaki I. and Ahring B. K. (1993). Thermophilic anaerobic digestion of livestock waste: The effect of ammonia. Applied Microbiology and Biotechnology 38(4), 560-4, ISSN: 1432-0614

[5] Angelidaki I. and Ahring B. K. (1994). Anaerobic thermophilic digestion of manure at different ammonia loads: Effect of temperature. Water Research 28(3), 727-31, ISSN: 0043-1354

[6] Angelidaki I. and Ahring B. K. (2000). Methods for increasing the biogas potential from the recalcitrant organic matter contained in manure. Water Science and Technology 41 (3), 189-94, ISSN: 0273-1223

[7] Angelidaki I., Ellegaard L. and Ahring B. K. (1993). A mathemathical model for dynamic simulation of anaerobic digestion of complex substrates, focusing on ammonia inhibition. Biotechnology and Bioengineering 42, 159-66, ISSN: 1097-0290

[8] Angelidaki I. and Sanders W. (2004). Assessment of the anaerobic biodegradability of macropollutants. Reviews in Environmental Science and Biotechnology 3(2), 117-29, ISSN: 1569-1705

[9] Aspe E., Marti M. C. and Roeckel M. (1997). Anaerobic treatment of fishery wastewater using a marine sediment inoculum. Water Research 31(9), 2147-60 ISSN: 0043-1354

[10] Banerjee A., Sharma R., Chisti Y. and Banerjee U. C. (2002). Botryococcus braunii: A renewable source of hydrocarbons and other chemicals. Critical Reviews in Biotechnology 22(3), 245-79, ISSN: 1549-7801

[11] Becker E. W. (2007). Micro-algae as a source of protein. Biotechnology Advances 25(2), 207-10, ISSN: 0734-9750

[12] Blokker P., Schouten S., Van den Ende H., De Leeuw J. W., Hatcher P. G. and Sinninghe Damsté J. S. (1998). Chemical structure of algaenans from the fresh water algae Tetraedron minimum, Scenedesmus communis and Pediastrum boryanum. Organic Geochemistry 29(5-7 -7 pt 2), 1453-68, ISSN: 0146-6380

[13] Blumreisinger M., Meindl D. and Loos E. (1983). Cell wall composition of chlorococcal algae. Phytochemistry 22(7), 1603-4, ISSN: 0031-9422

[14] Bonmatí A., Flotats X., Mateu L. and Campos E. (2001). Study of thermal hydrolysis as a pretreatment to mesophilic anaerobic digestion of pig slurry. Water Science and Technology 44 (4), 109-16, ISSN: 0273-1223

[15] Borja R., Sánchez E. and Durán M. M. (1996). Effect of the clay mineral zeolite on ammonia inhibition of anaerobic thermophilic reactors treating cattle manure. Journal of Environmental Science and Health - Part A Toxic/Hazardous Substances and Environmental Engineering 31(2), 479-500, ISSN: 1532-4117

[16] Braun R., Huber P. and Meyrath J. (1981). Ammonia toxicity in liquid piggery manure digestion. Biotechnology Letters 3(4), 159-64, ISSN: 1573-6776

[17] Brennan L. and Owende P. (2010). Biofuels from microalgae-A review of technologies for production, processing, and extractions of biofuels and co-products. Renewable and Sustainable Energy Reviews 14(2), 557-77, ISSN: 1364-0321

[18] Brown M. R. (1991). The amino-acid and sugar composition of 16 species of microalgae used in mariculture. Journal of Experimental Marine Biology and Ecology 145(1), 79-99, ISSN: 0022-0981

[19] Brown M. R. (1992). Biochemical composition of microalgae from the green algal classes Chlorophyceae and Prasinophyceae. 1. Amino acids Sugars and pigments. Journal of Experimental Marine Biology and Ecology 161(1), 91-113, ISSN: 0022-0981

[20] Conde J. L., Moro L. E., Travieso L., Sanchez E. P., Leiva A., Dupeiron R. and Escobedo R. (1993). Biogas purification process using intensive microalgae cultures. Biotechnology Letters 15(3), 317-20, ISSN: 1573-6776

[21] Costa J. A. V. and de Morais M. G. (2011). The role of biochemical engineering in the production of biofuels from microalgae. Bioresource Technology 102(1), 2-9, ISSN: 0960-8524

[22] Chen P. H. and Oswald W. J. (1998). Thermochemical treatment for algal fermentation. Environment International 24(8), 889-97, ISSN: 0160-4120

[23] Chen Y., Cheng J. J. and Creamer K. S. (2008). Inhibition of anaerobic digestion process: A review. Bioresource Technology 99(10), 4044-64, ISSN: 0960-8524

[24] Chisti Y. (2007). Biodiesel from microalgae. Biotechnology Advances 25(3), 294-306, ISSN: 0734-9750

[25] Choi S. P., Nguyen M. T. and Sim S. J. (2010). Enzymatic pretreatment of Chlamydomonas reinhardtii biomass for ethanol production. Bioresource Technology 101(14), 5330-6, ISSN: 0960-8524

[26] De Baere L. A., Devocht M., Van Assche P. and Verstraete W. (1984). Influence of high NaCl and NH4Cl salt levels on methanogenic associations. Water Research 18(5), 543-8, ISSN: 0043-1354

[27] De Schamphelaire L. and Verstraete W. (2009). Revival of the biological sunlight-to-biogas energy conversion system. Biotechnology and Bioengineering 103(2), 296-304, ISSN: 1097-0290

[28] Derenne S., Largeau C., Berkaloff C., Rousseau B., Wilhelm C. and Hatcher P. G. (1992). Non-hydrolysable macromolecular constituents from outer walls of Chlorella fusca and Nanochlorum eucaryotum. Phytochemistry 31(6), 1923-9, ISSN: 0031-9422

[29] Dimroth P. and Thomer A. (1989). A primary respiratory Na+ pump of an anaerobic bacterium: the Na+-dependent NADH:quinone oxidoreductase of Klebsiella pneumoniae. Archives of Microbiology 151(5), 439-44, ISSN: 1432-072X

[30] Domozych D. S., Stewart K. D. and Mattox K. R. (1980). The comparative aspects of cell wall chemistry in the green algae (Chlorophyta). Journal of Molecular Evolution 15(1), 1-12, ISSN: 1432-1432

[31] Doucha J., Straka F. and Lívansk@ K. (2005). Utilization of flue gas for cultivation of microalgae (Chlorella sp.) in an outdoor open thin-layer photobioreactor. Journal of Applied Phycology 17(5), 403-12, ISSN: 1573-5176

[32] Douskova I., Kastanek F., Maleterova Y., Kastanek P., Doucha J. and Zachleder V. (2010). Utilization of distillery stillage for energy generation and concurrent production of valuable microalgal biomass in the sequence: Biogas-cogeneration-microalgae-products. Energy Conversion and Management 51(3), 606-11, ISSN: 0196-8904

[33] Ehimen E. A., Sun Z. F., Carrington C. G., Birch E. J. and Eaton-Rye J. J. (2011). Anaerobic digestion of microalgae residues resulting from the biodiesel production process. Applied Energy 88(10), 3454-63, ISSN: 0306-2619

[34] Feijoo G., Soto M., Méndez R. and Lema J. M. (1995). Sodium inhibition in the anaerobic digestion process: Antagonism and adaptation phenomena. Enzyme and Microbial Technology 17(2), 180-8, ISSN: 0141-0229

[35] Foree E. G. and McCarty P. L. (1970). Anaerobic decomposition of algae. Environmental Science and Technology 4(10), 842-9, ISSN: 1520-5851

[36] Fu C. C., Hung T. C., Chen J. Y., Su C. H. and Wu W. T. (2010). Hydrolysis of microalgae cell walls for production of reducing sugar and lipid extraction. Bioresource Technology 101(22), 8750-4, ISSN: 0960-8524

[37] Gebauer R. (2004). Mesophilic anaerobic treatment of sludge from saline fish farm effluents with biogas production. Bioresource Technology 93(2), 155-67, ISSN: 0960-8524

[38] Gelin F., Boogers I., Noordeloos A. A. M., Sinninghe Damsté J. S., Riegman R. and De Leeuw J. W. (1997). Resistant biomacromolecules in marine microalgae of the classes eustigmatophyceae and chlorophyceae: Geochemical implications. Organic Geochemistry 26(11-12), 659-75, ISSN: 0146-6380

[39] Gelin F., Volkman J. K., Largeau C., Derenne S., Sinninghe Damsté J. S. and De Leeuw J. W. (1999). Distribution of aliphatic, nonhydrolyzable biopolymers in marine microalgae. Organic Geochemistry 30(2-3), 147-59, ISSN: 0146-6380

[40] Golueke C. G., Oswald W. J. and Gotaas H. B. (1957). Anaerobic digestion of Algae. Applied microbiology 5(1), 47-55, ISSN: 1365-2672

[41] González-Fernández C., Sialve B., Bernet N. and Steyer J. P. (2012). Impact of microalgae characteristics on their conversion to biofuel. Part II: Focus on biomethane production. Biofuels, Bioproducts and Biorefining 6(2), 205-18, ISSN: 1932-1031

[42] Griffiths M. and Harrison S. (2009). Lipid productivity as a key characteristic for choosing algal species for biodiesel production. Journal of Applied Phycology 21(5), 493-507, ISSN: 0921-8971

[43] Halleux H., Lassaux S., Renzoni R. and Germain A. (2008). Comparative life cycle assessment of two biofuels: Ethanol from sugar beet and rapeseed methyl ester. International Journal of Life Cycle Assessment 13(3), 184-90, ISSN: 1614-7502

[44] Harun R., Davidson M., Doyle M., Gopiraj R., Danquah M. and Forde G. (2010). Technoeconomic analysis of an integrated microalgae photobioreactor, biodiesel and biogas production facility. Biomass and Bioenergy 35(1), 741-7, ISSN: 0961-9534

[45] Hauck J. T., Olson G. J., Scierka S. J., Perry M. B. and Ataai M. M. (1996). Effects of simulated flue gas on growth of microalgae. ACS Division of Fuel Chemistry, Preprints 41(4), 1391-4

[46] Heubeck S., Craggs R. J. and Shilton A. (2007). Influence of CO2 scrubbing from biogas on the treatment performance of a high rate algal pond. Water Science and Technology 55 (11), 193-200, ISSN: 0273-1223

[47] Kadouri A., Derenne S., Largeau C., Casadevall E. and Berkaloff C. (1988). Resistant biopolymer in the outer walls of Botryococcus braunii, B race. Phytochemistry 27(2), 551-7, ISSN: 0031-9422

[48] Kapdi S. S., Vijay V. K., Rajesh S. K. and Prasad R. (2005). Biogas scrubbing, compression and storage: Perspective and prospectus in Indian context. Renewable Energy 30(8), 1195-202, ISSN: 0960-1481

[49] Kimata-Kino N., Ikeda S., Kurosawa N. and Toda T. (2011). Saline adaptation of granules in mesophilic UASB reactors. International Biodeterioration & Biodegradation 65(1), 65-72, ISSN: 0964-8305

[50] Kumar A., Ergas S., Yuan X., Sahu A., Zhang Q., Dewulf J., Malcata F. X. and van Langenhove H. (2010). Enhanced CO2 fixation and biofuel production via microalgae: Recent developments and future directions. Trends in Biotechnology 28(7), 371-80, ISSN: 0167-7799

[51] Lardon L., Hélias A., Sialve B., Steyer J. P. and Bernard O. (2009). Life-cycle assessment of biodiesel production from microalgae. Environmental Science and Technology 43(17), 6475-81, ISSN: 1520-5851

[52] Lefebvre O. and Moletta R. (2006). Treatment of organic pollution in industrial saline wastewater: A literature review. Water Research 40(20), 3671-82, ISSN: 0043-1354

[53] Liu Y. and Boone D. R. (1991). Effects of salinity on methanogenic decomposition. Bioresource Technology 35(3), 271-3, ISSN: 0960-8524

[54] Loos E. and Meindl D. (1982). Composition of the cell wall of Chlorella fusca. Planta 156(3), 270-3, ISSN: 1432-2048

[55] Mairet F., Bernard O., Cameron E., Ras M., Lardon L., Steyer J. P. and Chachuat B. (2012). Three-reaction model for the anaerobic digestion of microalgae. Biotechnology and Bioengineering 109(2), 415-25, ISSN: 00063592

[56] Mairet F., Bernard O., Ras M., Lardon L. and Steyer J.-P. (2011). Modeling anaerobic digestion of microalgae using ADM1. Bioresource Technology 102(13), 6823-9, ISSN: 0960-8524

[57] Makaruk A., Miltner M. and Harasek M. (2010). Membrane biogas upgrading processes for the production of natural gas substitute. Separation and Purification Technology 74(1), 83-92, ISSN: 1383-5866

[58] Mandeno G., Craggs R., Tanner C., Sukias J. and Webster-Brown J. (2005). Potential biogas scrubbing using a high rate pond. Water Science and Technology 51(12), 253-6, ISSN: 0273-1223

[59] McCarty P. L. (1965). Thermodynamics of biological synthesis and growth. Air and water pollution 9(10), 621-39, ISSN: 0568-3408

[60] Meier L., Torres A., Azocar L., Neumann P., Vergara C., Rivas M. and Jeison D. (2011). Biogas upgrading through microalgae culture: Effect of methane concentration on microalgae activity. In: X Latinamerican and Symposium on Anaerobic Digestion, Ouro Preto, Brazil.

[61] Mussgnug J. H., Klassen V., Schlüter A. and Kruse O. (2010). Microalgae as substrates for fermentative biogas production in a combined biorefinery concept. Journal of Biotechnology 150(1), 51-6, ISSN: 0168-1656

[62] Mutanda T., Ramesh D., Karthikeyan S., Kumari S., Anandraj A. and Bux F. (2011). Bioprospecting for hyper-lipid producing microalgal strains for sustainable biofuel production. Bioresource Technology 102(1), 57-70, ISSN: 0960-8524

[63] Nair K. V. K., Kannan V. and Sebastian S. (1983). Bio-gas generation using microalgae and macrophytes. Indian Journal of Environmental Health 24(4), 277-84, ISSN: 0367-827X

[64] Negoro M., Hamasaki A., Ikuta Y., Makita T., Hirayama K. and Suzuki S. (1993). Carbon dioxide fixation by microalgae photosynthesis using actual flue gas discharged from a boiler. Applied biochemistry and biotechnology 39-40(1), 643-53, ISSN: 1559-0291

[65] Neumann P., Torres A., Azocar L., Meier L., Vergara C. and Jeison D. (2011). Biogas production as a tool for increasing sustainability of biodiesel production from microalgae Botryococcus braunii. In: X Latinamerican and Symposium on Anaerobic Digestion, Ouro Preto, Brazil.

[66] Northcote D. H., Goulding K. J. and Horne R. W. (1958). The chemical composition and structure of the cell wall of Chlorella pyrenoidosa. The Biochemical journal 70(3), 391-7, ISSN: 1470-8728

[67] Omil F., Mendez R. and Lema J. M. (1995). Anaerobic treatment of saline wastewaters under high sulphide and ammonia content. Bioresource Technology 54(3), 269-78, ISSN: 0960-8524

[68] Patel G. B. and Roth L. A. (1977). Effect of sodium chloride on growth and methane production of methanogens. Canadian Journal of Microbiology 23(7), 893-7, ISSN: 1480-3275

[69] Punnett T. and Derrenbacker E. C. (1966). The amino acid composition of algal cell walls. Journal of General Microbiology 44(1), 105-14, ISSN: 0022-1287

[70] Ras M., Lardon L., Bruno S., Bernet N. and Steyer J.-P. (2011). Experimental study on a coupled process of production and anaerobic digestion of Chlorella vulgaris. Bioresource Technology 102(1), 200-6, ISSN: 0960-8524

[71] Razon L. F. and Tan R. R. (2011). Net energy analysis of the production of biodiesel and biogas from the microalgae: Haematococcus pluvialis and Nannochloropsis. Applied Energy 88(10), 3507-14, ISSN: 0306-2619

[72] Rinzema A., Van Lier J. and Lettinga G. (1988). Sodium inhibition of acetoclastic methanogens in granular sludge from a UASB reactor. Enzyme and Microbial Technology 10(1), 24-32, ISSN: 0141-0229

[73] Ryckebosch E., Drouillon M. and Vervaeren H. (2011). Techniques for transformation of biogas to biomethane. Biomass and Bioenergy 35(5), 1633-45, ISSN: 0961-9534

[74] Ryu H. J., Oh K. K. and Kim Y. S. (2009). Optimization of the influential factors for the improvement of CO2 utilization efficiency and CO2 mass transfer rate. Journal of Industrial and Engineering Chemistry 15(4), 471-5, ISSN: 1226-086X

[75] Samson R. and Leduy A. (1983). Influence of mechanical and thermochemical pretreatments on anaerobic digestion of Spirulinamaxima algal biomass. Biotechnology Letters 5(10), 671-6, ISSN: 1573-6776

[76] Scott S. A., Davey M. P., Dennis J. S., Horst I., Howe C. J., Lea-Smith D. J. and Smith A. G. (2010). Biodiesel from algae: Challenges and prospects. Current Opinion in Biotechnology 21(3), 277-86, ISSN: 0958-1669

[77] Shilton A. N., Mara D. D., Craggs R. and Powell N. (2008). Solar-powered aeration and disinfection, anaerobic co-digestion, biological CO2 scrubbing and biofuel production: The energy and carbon management opportunities of waste stabilisation ponds. Water Science and Technology 58 (1), 253-8, ISSN: 0273-1223

[78] Sialve B., Bernet N. and Bernard O. (2009). Anaerobic digestion of microalgae as a necessary step to make microalgal biodiesel sustainable. Biotechnology Advances 27(4), 409-16, ISSN: 0734-9750

[79] Simpson A. J., Zang X., Kramer R. and Hatcher P. G. (2003). New insights on the structure of algaenan from Botryoccocus braunii race A and its hexane insoluble botryals based on multidimensional NMR spectroscopy and electrospray-mass spectrometry techniques. Phytochemistry 62(5), 783-96, ISSN: 0031-9422

[80] Soto M., Mendez R. and Lema J. M. (1991). Biodegradability and toxicity in the anaerobic treatment of fish canning wastewaters. Environmental Technology 12(8), 669-77, ISSN: 0959-3330

[81] Stephens E., Ross I. L., King Z., Mussgnug J. H., Kruse O., Posten C., Borowitzka M. A. and Hankamer B. (2010). An economic and technical evaluation of microalgal biofuels. Nature Biotechnology 28(2), 126-8, ISSN: 1087-0156

[82] Tarwadi S. J. and Chauhan V. D. (1987). Seaweed biomass as a source of energy. Energy 12(5), 375-8, ISSN: 0360-5442

[83] Templier J., Largeau C., Casadevall E. and Berkaloff C. (1992). Chemical inhibition of resistant biopolymers in outer walls of the A and B races of Botryococcus braunii. Phytochemistry 31(12), 4097-104, ISSN: 0031-9422

[84] Travieso L., Sanchez E. P., Benitez F. and Conde J. L. (1993). Arthrospira sp. intensive cultures for food and biogas purification. Biotechnology Letters 15(10), 1091-4, ISSN: 1573-6776

[85] Vergara-Fernańdez A., Vargas G., Alarcoń N. and Velasco A. (2008). Evaluation of marine algae as a source of biogas in a two-stage anaerobic reactor system. Biomass and Bioenergy 32(4), 338-44, ISSN: 0961-9534

[86] Wang B., Li Y., Wu N. and Lan C. Q. (2008). CO2 bio-mitigation using microalgae. Applied Microbiology and Biotechnology 79(5), 707-18, ISSN: 1432-0614

[87] Wittmann C., Zeng A. P. and Deckwer W. D. (1995). Growth inhibition by ammonia and use of a pH-controlled feeding strategy for the effective cultivation of Mycobacterium chloropheolicum. Applied Microbiology and Biotechnology 44(3-4), 519-25, ISSN: 1432-0614

[88] Xiao Y. Y. and Roberts D. J. (2010). A review of anaerobic treatment of saline wastewater. Environmental Technology 31(8-9), 1025-43, ISSN: 0959-3330

[89] Yen H.-W. and Brune D. E. (2007). Anaerobic co-digestion of algal sludge and waste paper to produce methane. Bioresource Technology 98(1), 130-4, ISSN: 0960-8524

[90] Yerkes D. W., Boonyakitsombut S. and Speece R. E. (1997). Antagonism of sodium toxicity by the compatible solute betaine in anaerobic methanogenic systems. Water Science and Technology 36 (6-7), 15-24, ISSN: 0273-1223

[91] Zamalloa C., Vulsteke E., Albrecht J. and Verstraete W. (2011). The techno-economic potential of renewable energy through the anaerobic digestion of microalgae. Bioresource Technology 102(2), 1149-58, ISSN: 0960-8524

Chapter 7

Advanced Monitoring and Control of Anaerobic Digestion in Bioreactor Landfills

Mohamed Abdallah and Kevin Kennedy

Additional information is available at the end of the chapter

http://dx.doi.org/10.5772/55715

1. Introduction

Despite recent increases in recycling, composting, and incineration, the sanitary landfill remains the predominant and most economical municipal solid waste (MSW) management alternative. Modern MSW landfills strive to optimize the design, construction, and operation processes in order to mitigate many of the potentially negative impacts, and improve the profitability. The bioreactor landfill (BL) is considered one of the promising developments that have recently gained significant attention. This waste-to-energy technology requires specific management activities and operational procedures that enhance the microbial decomposition processes inside the landfill resulting in higher production of landfill gas [1]. The recirculation of leachate, which is conducted by recycling the water passing through and collected from the landfill, is considered the main operational characteristic in the BL to increase moisture, and consequently stimulate the biodegradation process (Figure 1). The potential benefits of the BL include increased waste settlement rates and airspace utilization, decreased costs for leachate treatment, more rapid gas production (which improves the economics of gas recovery), and more rapid waste stabilization (which may reduce the post-closure maintenance period). These potential benefits have led to many full-scale BL applications in the last decade, mostly in the United States, resulting in the generation of design and operation data. In 2004, the Solid Waste Association of North America conducted an inventory that identified over 70 BLs in North America [2]. Many of these experiences revealed scale-up issues and technical limitations that merit further research and development [3-5].

One of the most critical, yet little studied, issues in the operation of BLs is process control. In field applications, unsupervised operational procedures can disturb the dynamics of the landfill biological processes causing serious consequences on the overall evolution of the ecosystem, i.e., unstable and sometimes unsuccessful transition from one operational phase to

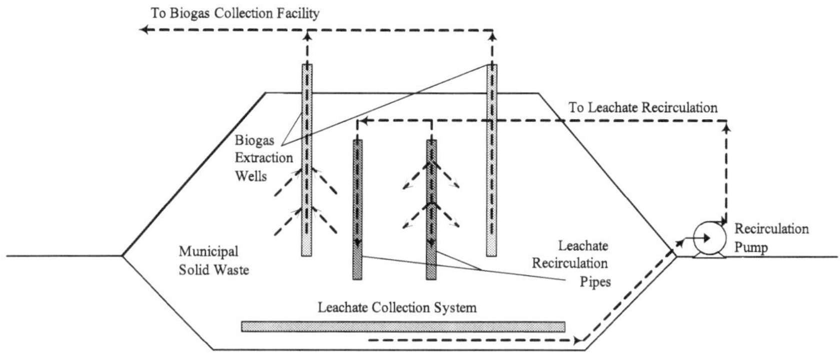

Figure 1. Schematic of an anaerobic bioreactor landfill

another. Dealing with the BL as a dynamic and evolving biological system could solve many of the BL control issues especially those pertinent to daily operation such as leachate recirculation. For example, one of the main operational issues, which are addressed in the present work, is the large variation in the characteristics of the collected leachate, which sometimes makes the leachate (as produced) unsuitable for recirculation. At the same time, the physical, chemical, and biochemical growth requirements of the bacterial consortia inside the BL change significantly during the different operational phases. *It is therefore necessary to manipulate the collected leachate before recirculation in order to suit the prevailing reactions and conditions inside the BL.* Several techniques have been tested in laboratory studies to enhance the performance of BLs either directly or indirectly through the manipulation of the recirculated leachate: pH adjustment, nutrients addition, and biosolids addition [1, 6-8]. However, these techniques are rarely, if ever, used in field applications due to lack of well-defined methodologies and the huge cost if applied excessively in an uncontrolled fashion. Applying advanced process control techniques offers an alternative solution for this problem. *Developing a control system that optimizes the leachate recirculation and manipulation processes based on real-time conditions of the controlled BL can provide a flexible engineered solution that is applicable to any typical landfill site..*

The proposed Sensor-based Monitoring and Remote-control Technology (SMART) features an expert controller that manipulates the controllable variables of the bioreactor process based on online monitoring of key system parameters. The objective of this control framework is to provide the optimal operational conditions for the biodegradation of MSW, and also, to enhance the performance of the BL in terms of biogas production. A comprehensive analysis of the process control of BLs is presented, followed by the conceptual framework of SMART including its structure, components, and instrumentation. In conclusion, a pilot-scale implementation of the control system is discussed.

1.1. Bioreactor landfill ecosystem

Controlling the BL requires a good understanding of the system and its dataflow including inputs, outputs, and interconnecting processes. The basic principles and mechanisms of the BL are well documented in the literature [9-11]. A simplified dataflow diagram for an anaerobic BL is shown in Figure 2. The BL can be considered as an anaerobic fixed-bed reactor in which the biodegradable organic fraction of the solid waste is the substrate. The factors affecting the biological processes in landfills can be grouped to: (1) factors related to the microbial environment (e.g., moisture, temperature, nutrients availability, and toxicity), and (2) factors related to the landfill site including: climate conditions (e.g., air temperature and precipitation), waste characteristics (e.g., particle size and composition), and site-specific settings (e.g., collection and injection systems). The BL concept is based on employing specific operational activities to control the influencing factors in a positive manner, e.g., applying leachate recirculation to optimize waste moisture. From the process control point of view, i.e., based on the feasibility of real-time manipulation, the first group of factors can be considered *controllable* inputs to the BL process, while the second group of factors is *uncontrollable*. The management techniques through which the *controllable* factors can be controlled are discussed below.

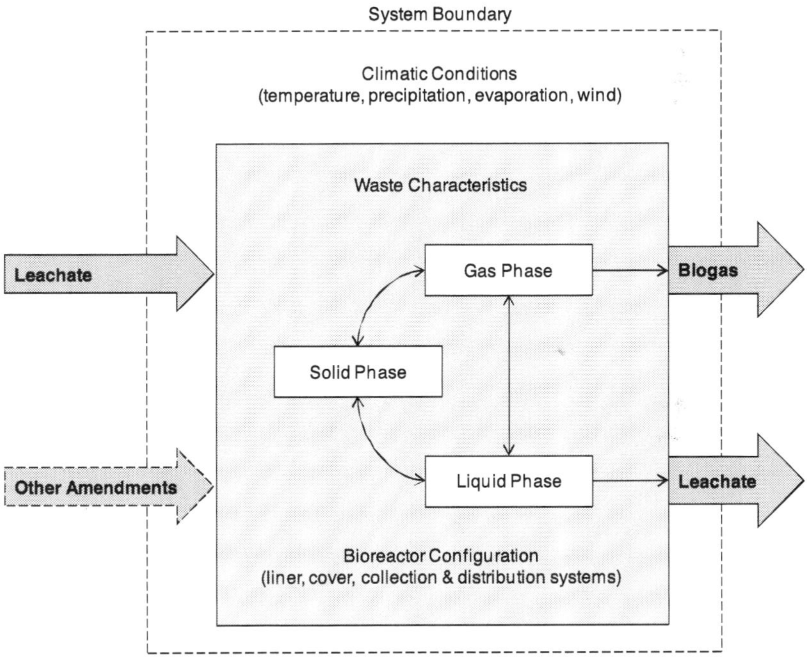

Figure 2. Data flow diagram of the bioreactor landfill ecosystem

1.1.1. Leachate recirculation

Moisture addition has been proved repeatedly to stimulate the methanogenic population in the landfill waste matrix. Leachate recirculation is considered the most effective method to increase moisture content of waste in a controlled fashion, which could reduce the time required for landfill stabilization from several decades to two to three years [12]. Leachate recirculation has been proven to achieve better BL performance in terms of biogas production by several lab-, pilot-, and full-scale studies [13-17]. In full-scale applications, leachate recirculation at Trail Road landfill enhanced waste settlement and resulted in 30% airspace recovery, which was used for landfilling more waste [4]. In another full-scale study by [18], leachate recirculation achieved more rapid biogas production, increased settlement rates, and accelerated decreases in the concentration of certain contaminants in leachate. According to [19], moisture increase alone does not enhance methane production. It is the nutrients, inocula and buffers, which in addition to moisture, enhances biodegradation to the greatest extent. It was shown in [8] that added alkalinity, dissolved oxygen level, and presence of methanogenic bacteria in the recirculated liquid considerably influenced the hydrolysis rate and onset of methanogenesis. *Therefore, it is suggested that, not only moisture addition, but also the quality of the leachate affects the impact/outcome of recirculation significantly.* Hence, there are two main aspects of the recirculation process that can be controlled: the **quantity** and **quality** of the recirculated leachate.

The **quantity** of the leachate generated is site-specific and a function of water availability, weather conditions, characteristics of the waste, as well as the liner and cover design [10]. In order to achieve the benefits of leachate recirculation, leachate has to be recycled at optimal rates that achieve sufficient contact with waste. The effect of varying leachate recirculation rates was studied in lab simulations [13, 16, 17, 20]. These studies demonstrated that higher recirculation rates result in better BL performance in terms of biogas production. It was suggested that leachate recirculation should be adjusted according to the phases of waste stabilization [21]. This practice was applied successfully in [13] as well as [22] who varied the leachate recirculation rates in lab scale BLs based on 7 and 4 operational stages, respectively. Unsupervised high rate of recirculation may result in: (1) washout of large amounts of organic matter before the methanogenic phase, thereby reducing the biological methane potential, (2) production of leachate containing high concentrations of short chain fatty acids which either inhibits methanogenesis directly or by lowering the pH, (3) excessive accumulation of leachate within the landfill, which may breakout from landfill slopes, (4) increase of pore water pressure and decrease of the shear strength of the waste matrix which affect the geotechnical slope stability, (5) increase in the hydrostatic head on the base liner, leading to higher risk for ground water contamination, and (6) drop in the internal temperature of the landfill especially in cold regions. *Therefore, leachate recirculation rate has to be selected such that the desired moisture content levels, moisture movement, and supplements distribution are provided, and at the same time, the pre-mentioned issues are monitored and incorporated in the decision-making process.*

The **quality** of leachate is highly dependent on waste composition and operational phase [10]. Leachate has been reported to contain a wide range of inorganic and organic compounds including toxicants such as aliphatic/aromatic hydrocarbons and halogenated organics [23]. Typically, the concentration of constituents, including pollutants, in leachate decreases with the waste age. The large temporal variation in the biochemical characteristics of leachate - as produced - makes it sometimes unsuitable for recirculation. For example, the concentrations of dissolved organic substances in young leachate are usually much higher than in older leachate. Continuous recirculation of young leachate in early phases of operation will increase the concentration of short chain fatty acids inside the BL which either inhibits methanogenesis directly or indirectly by lowering the pH of the system. Recently, researchers examined the use of different leachate (e.g., mature leachate from older landfill cells) for recirculation [16, 17, 21]. Alternatively, young and mature leachates were used interchangeably over four operational stages along the BL lifespan [22]. They used young leachate in phase I, then mature leachate in phase II and when the characteristics of produced leachate became suitable, they switched back to young leachate in phases III and IV. The same concept was applied by [20] who rotated the recirculated leachate between fresh waste and stabilised waste reactors until a balanced microbial population was established. Other studies combined leachate with water, as simulated rainfall, which simulated field conditions and diluted the leachate [13, 24]. The addition of supplemental water to the recirculated leachate in early operational phases could promote dilution of inhibitory substances and reduce leachate strength resulting in more favourable methanogenic conditions [25]. *Therefore, supplemental water can be used in combination with other leachate manipulation techniques – shown below - to correct certain process deviations, reduce the impact of detrimental substances, and/or enrich the concentration of other beneficial compounds.*

1.1.2. pH Adjustment

Methanogenic bacteria are sensitive to pH, with an optimal range between 6.8 and 7.2, and could be inhibited by acidic conditions at pH less than 6.7. Therefore, pH of recycled leachate can have a significant effect on waste stabilization and methane production. This understanding of microbial ecology has promoted the addition of buffer to adjust the pH of leachate prior to recycling it back to landfill. Buffering as a control option may be best used in response to changes in leachate characteristics (i.e., a drop in pH or increase in volatile acids' concentration). Leachate recirculation with a buffering system to control the pH has been found to result in shorter acidogenic stage leading to earlier initiation of the methanogenic stage, and concomitant higher gas production [7, 8, 25].

1.1.3. Bioaugmentation

Bioaugmentation or inoculation of the landfill has been investigated, usually through the addition of bio-solids from wastewater treatment facilities [1]. The optimal inoculum should provide suitable consortia and concentration of microorganisms, as well as nutrients such as nitrogen and phosphorus. It was stated in [23] that initiating fermentation in BLs can be promoted by addition of large amounts of methanogenic microorganisms in the form of effluent and sludge from an anaerobic sewage digester since the population of such microorganisms in fresh MSW is typically low. In [6], moisture saturation conditions was examined with digested sewage sludge, with fertilizer, and without additives. It was found that moisture and sewage sludge additions resulted in the shortest acidogenic phase and highest gas production. However, it has been suggested that any measured beneficial effects associated with the addition of biosolids may be due to buffering or moisture addition rather than inoculation [26]. Moreover, generic conclusions regarding the effect of sludge addition cannot be drawn, since different types and percentages of sludge might have been used in different experiments.

1.1.4. Nutrients addition

Nutrients required for anaerobic degradation of waste are generally low, and therefore, nutrients are expected to be available especially during early phases of biodegradation [7]. It was found that all the necessary nutrients and trace heavy metals are available in most landfills, but insufficient mixing and heterogeneity of the wastes may result in nutrient-limited zones [23, 27]. Experimentally, it was proven that the addition of nitrogen and phosphorous stimulated methane production, rapidly decreased organic concentration in leachate, and shortened the initial phase before methane generation commenced [1, 28].

1.2. Identification of control problem

While most studies reported process improvements associated with leachate recirculation and manipulation processes, other studies found the contrary, such as toxicity and souring conditions. The results reported in many studies are different, and sometimes contradicting, since the same substance can be useful or harmful depending on its dose. This can be explained by the general effect of increasing salt concentration in anaerobic systems shown in Figure 3. A substance which is essential to a biological process can stimulate the bacterial growth at low concentrations. However, as concentrations increase above optimal, the rate of microbial activity decreases until the process is inhibited. Similarly, this trend can describe the effects of adding leachate and other amendments on the BL performance. In addition to the dose, other factors may affect the results: (1) operational factors, such as the type and characteristics of amendments, and (2) operational phase and progressive evolution of the BL.

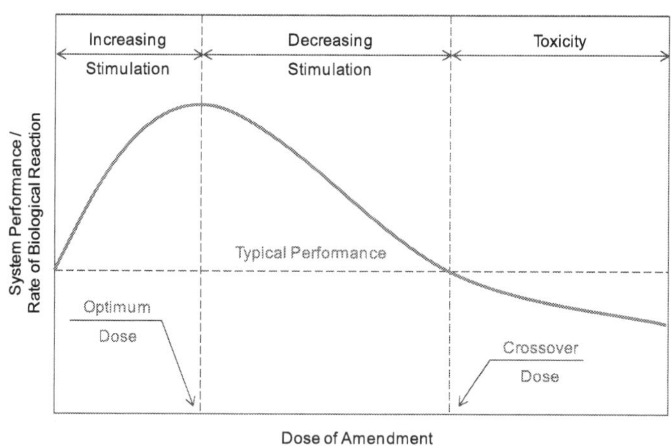

Figure 3. Effect of adding amendments on BL performance (modified from [29])

http://dx.doi.org/10.5772/55715

In conclusion, specific growth needs of the BL bacterial consortia changes with time, concomitantly the required leachate characteristics are continuously changing such that leachate as produced in its original form may not always be ideal for recirculation. The goal of the present research is the development of a real-time monitoring and expert decision making system that can adjust both, leachate characteristics and rates of recirculation according to the ecological requirements of each operational phase to provide the optimum conditions for waste biodegradation in BLs.

2. The proposed control system

The main real-time control tool in an anaerobic BL is leachate recirculation combined with amendment addition to provide both optimal moisture content and distribution of essential additives. The pH of the recirculated leachate can be adjusted by adding buffer, while inoculum in the form of anaerobic digested sludge, can be used both as a buffer and a rich source of methanogenic bacteria. At later BL operational phases, nutrients can be added as needed to supply the nutritional needs of the bacterial consortia. Supplemental water can be added to dilute concentrated leachate (as a remedy for toxicity) and to account for any shortage in available recyclable leachate for moisture control. The rate of application of any of these amendments can be decided based on measurable parameters in the leachate as well as the specific requirements of each BL operational phase. In conjunction with recirculation, certain parameters such as pore water pressure, landfill internal temperature, and hydrostatic head on the liner must be monitored and considered as they are influenced by recirculated leachate, and can affect BL operation.

2.1. Control scheme

The biological processes occurring in the landfill are largely anaerobic. Similar to anaerobic digesters, the landfill ecosystem is sensitive to environmental conditions such as pH, temperature, moisture, toxic compounds, and presence of oxygen. In fact, much of what is known or assumed concerning processes in landfills has primarily come from experiences with anaerobic digesters [10]. For this reason, the required control for an anaerobic BL is analogous to that of an anaerobic digester, with the latter more easily to control being a well-mixed reactor [7]. There are various control schemes that can be applied in managing biochemical systems. The most widespread control schemes are: feedback, feed-forward, and open-loop. Feedback control is a control mechanism that uses information from measurements to manipulate a variable so that the desired result is achieved. Alternatively, feed-forward control mechanism predicts the effects of measured disturbances and takes corrective action to achieve the desired result. On the other hand, the open-loop controller does not utilize feedback to determine whether the input achieved the desired goal or not, and can neither engage in machine learning nor correct any errors that it could make. Thus far in landfill sites, process control is accomplished, if ever, based on a non-feedback scheme. Therefore, the present study aims at applying feedback control in the management of BLs.

In feedback control, the variable being controlled is measured and compared with a target value. The difference between the measured and desired value is called the *error*. Feedback control manipulates inputs of the system to minimize this error. Figure 4 shows a generic component block diagram of an elementary feedback controller. The output of the system is measured by a *sensor* and the *control element* represents an actuator or control device. The *error* in this system would be the *Measured Output - Desired Output*.

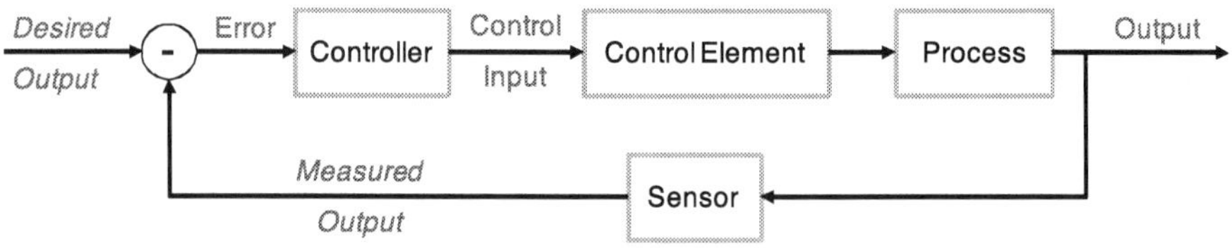

Figure 4. Block diagram of a basic feedback control loop

The potential advantages of feedback control lie in the fact that it obtains and utilizes data at the process output [30]. Therefore, the controller takes into account unforeseen disturbances in the process. Feedback control architecture ensures the desired performance by altering the inputs immediately once deviations are observed regardless of their reason. Thus, it reduces operator workload by eliminating the need for human adjustment of the control variable. An additional advantage is that by analyzing the output of a system, unstable processes may be stabilized. Feedback controls do not require detailed knowledge of the system and, in particular, do not require a mathematical model of the process. The controller can be easily duplicated from one system to another.

On the other hand, the time lag in the system is potentially the main disadvantage of feedback control. A process deviation occurring near the beginning of the process will not be recognized until the process output. The feedback control will then have to adjust the process inputs in order to correct this deviation. This results in the possibility of substantial deviation throughout the entire process [30]. The system could possibly miss process output disturbances and the error could continue without adjustment resulting in a steady state error. When the feedback controller proves unable to maintain stable closed-loop control, operator intervention is then required. Finally, feedback control does not take predictive control action towards the effects of known disturbances, and depends entirely on the accuracy with which the controlled output is measured.

2.2. Control framework

The proposed Sensor-based Monitoring and Remote-control Technology (SMART) system features software and hardware interacting components that provide real-time monitoring and expert control of BLs. Figure 5 shows a general diagram of the control system. The dashed lines indicate the sensory data, while the dot-dashed lines represent the commands.

http://dx.doi.org/10.5772/55715

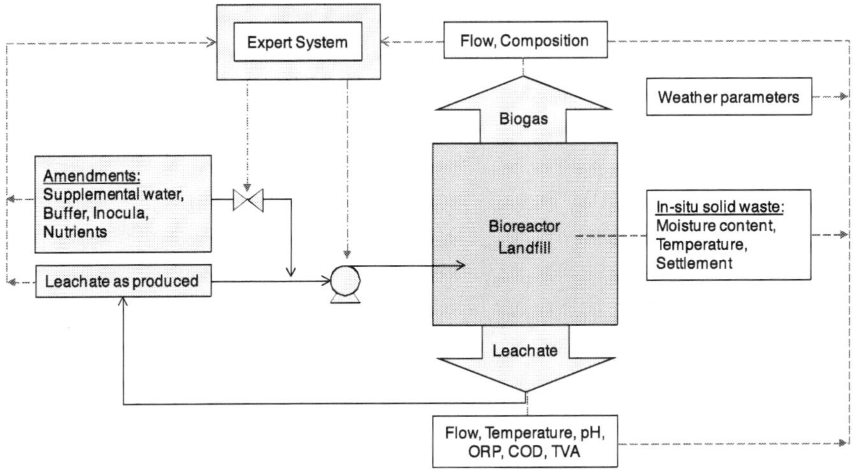

Figure 5. Schematic of the SMART control system

The control system has a geographically and functionally distributed architecture in which the BL is divided into basic blocks. Each block has its own local sensory data acquisition and control units. In addition, global sensory units are to provide measurements for the landfill body altogether as one block. All these local and global components are connected and remotely controlled by a global data processing and decision making unit. The controller continuously monitors two types of sensory data: process parameters (such as moisture and temperature), and returned feedback from performance indicators (such as biogas production and settlement). The decision made by the control algorithm is transmitted to the actuators, after authorization from the site operator, to inject the computed volumes of the selected amendments in order to manipulate the characteristics of the recirculated leachate. This batch control process runs continuously along the lifetime of the BL cell.

2.3. System components

The SMART system incorporates six interacting components: (1) Local Sensory Unit, (2) Global Sensory Unit, (3) Primary Sensory Data Processor, (4) Main Controller Unit, (5) Primary Driving Controller, and (6) Local Driving Unit. The main components of the system are shown in Figure 6, and described in detail below.

Local Sensory Unit (LSU)

The LSU is placed in each block, i.e., *n* sensory units for the *n* blocks. Each unit includes a set of analog sensors which quantify the values of different system parameters, such as temperature and moisture content, in the corresponding block. The installed units form a three dimensional grid in order to show the spatial dynamic status of the main parameters within the BL. All LSUs are designed to send the measured data to the Primary Sensory Data Processor.

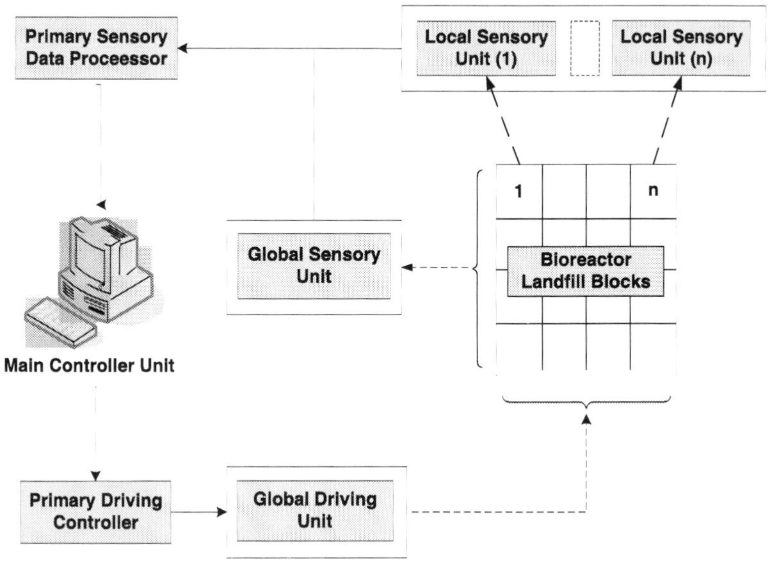

Figure 6. Main components of the SMART control system

Global Sensory Unit (GSU)

The GSU provides global measurements for the landfill body altogether as one block. These measurements include the parameters that are impractical to be determined for each block individually such as leachate characteristics, settlement, hydrostatic head on the liner, as well as biogas quantity and quality. Other examples of global measurements are the weather condition parameters such as air temperature, wind speed and direction, humidity, solar radiation, precipitation, and evaporation. All GSUs are connected to the Main Controller Unit through the Primary Sensory Data Processor.

Primary Sensory Data Processor (PSDP)

The PSDP is responsible for analyzing the acquired data from the Local and Global Sensory Units, and arranging them in a new frame to be delivered to the Main Controller Unit. Although this work could be done by the Main Controller Unit, employing an intermediate device here provides more modularity and flexibility to the system by providing an interface between the software of the Main Controller Unit from one side, and the LSUs from the other side.

Main Controller Unit (MCU)

The MCU is considered the driving brain of the control system. It receives the measured data (inputs), processes them within the developed expert system, and makes the control decision. The operator is prompted with the decision made by the MCU in order to evaluate it, and then approves it to be sent to the Primary Driving Controller in the form of quantified commands.

http://dx.doi.org/10.5772/55715

The operator can overwrite the decision to deal with any unexpected problem or unconsidered scenario in the expert system. The control program was programmed on the LabVIEW™ graphical programming platform (National Instruments, USA). The control program and expert system of MCU are discussed in *Section 2.5.*

Primary Driving Controller (PDC)

The PDC receives the commands from the MCU and distributes it to the different Local Driving Units. Basically, it is a device that de-multiplexes the received data set which holds the commands for all the driving units, and then delivers the commands to each unit separately. This unit combines analog/digital conversion, signal conditioning, and signal connectivity.

Local Driving Unit (LDU)

The LDU receives the commands and performs the required action by driving the corresponding actuator (motorized valves and/or pumps). Similar to the LSU, each of these units is responsible for controlling a single block, i.e., n driving units for the n blocks. Each actuator receives from the PDC the exact quantity required of the amendment it controls.

2.4. Instrumentation

In order to build the on-line monitoring and real-time control system of SMART, all sensors and control elements must be adaptable to automatic operation and because of the aggressive environment of landfills, instruments have to be durable, chemical and corrosion resistant, and robust (especially against overburden pressure). Typical sensor requirements to monitor in-place waste, leachate, and biogas for a generic block in the SMART system are shown in Figure 7. In this instrumentation system, sensors are controlled remotely by the PSDP, whereas the final control elements are controlled by the PDC. The PSDP/PDC unit transmits/receives the input/output signals via standard communication protocols (such as RS-232 or RS-485) to/ from the MCU.

In-place waste is monitored by LSU bundles which are evenly distributed in the BL body forming a three-dimensional grid. Each bundle measures moisture content, temperature, and water pressure (Figure 7, objects 10-12, respectively). Electrical resistivity and capacitance (frequency domain) technologies are suitable technologies for moisture measurements, and are compatible with automated monitoring systems. Waste temperature can be measured using thermocouples or thermistors, with the latter built into most commercial moisture and pressure sensors. However, thermocouples are still the preferred stand-alone temperature monitoring devices because they are reliable, inexpensive and the higher accuracy of thermistors is not needed in landfill applications. Thermocouples of types T (-250 to 350°C) or K (-200 to +1350°C) or J (-40 to +750°C) are widely used in landfill applications. Pore water pressure is measured using vibrating wire or solid state piezometers. Settlement is measured using settlement plates, whereas hydrostatic head on the liners is monitored by differential pressure transducers (Figure 7, objects 13 and 14, respectively). Landfill biogas flow is metered and totalized onsite using turbine or thermal dispersion flow meters (Figure 7, object 15). Biogas is analyzed for carbon dioxide and methane with dual wavelength infrared gas analyzers, whereas, oxygen is monitored via a zirconium dioxide sensor (Figure 7, object 16).

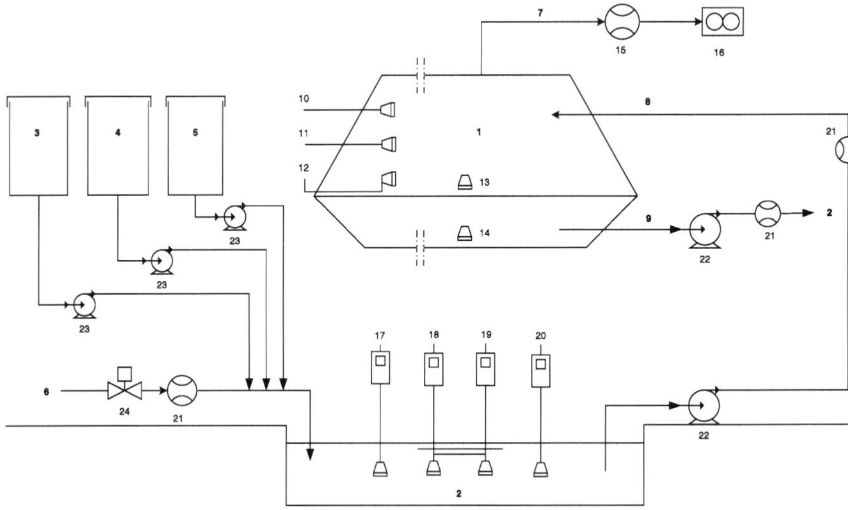

Figure 7. Schematic instrumentation diagram of the SMART system: (1) bioreactor landfill cell; (2) leachate storage facilities; (3) buffer tank; (4) inocula tank; (5) nutrient tank; (6) water supply; (7) collected biogas flow line; (8) recirculated leachate flow line; (9) collected leachate flow line; (10) moisture sensors; (11) thermocouples; (12) piezometers; (13) settlement plates; (14) pressure transducers; (15) gas flow meter; (16) inline gas analyzer; (17) liquid thermistor; (18) pH probe; (19) ORP electrode; (20) ammonia-selective electrode; (21) liquid flow meters; (22) pumps; (23) dosing pumps; and (24) electrically actuated valves.

Collected leachate is analyzed for major parameters such as chemical oxygen demand (COD), volatile fatty acids (VFA), oxygen reduction potential (ORP), and pH. The pH, ORP, and ammonia are measured by inline double-junction temperature-compensated pH, ORP, and ion-selective electrodes, respectively connected to a transmitter (Figure 7, objects 18-20, respectively). Online analyzers for COD and VFA are commercially available, however due to their high capital and maintenance costs as well as the slow reaction time in landfill processes, determination of these parameters by standard offline analytical methods is still the most economic and practical approach, and therefore is used in SMART. Leachate flow rate and cumulative flow are measured via Coriolis mass flow sensors equipped with totalizers (Figure 7, object 21). On the control side, GDU units include electrically actuated double-diaphragm or peristaltic pumps, and diaphragm valves that can safely handle particulate-laden and corrosive liquids (Figure 7, objects 22-24, respectively).

2.5. Expert system

The control program receives the measured data (inputs), processes them within the MCU expert system, makes the control decision, and sends it to the LDUs in the form of quantified commands. The expert system is designed to determine the required volumes of leachate, make-up water as well as bioaugmentation and nutritional amendments necessary to provide the BL microbial consortia with their optimum growth requirements. It was assumed that the

http://dx.doi.org/10.5772/55715

chemical/biochemical characteristics of the effluent leachate are representative of the conditions within the whole BL waste matrix. Regulating the characteristics of the recirculated leachate alters the characteristics of the waste matrix through which it percolates, in a gradual stepwise manner, over a number of cycle times. It is the premise of the system to identify the current operational phase of the controlled bioreactor, and accordingly determines quantities of leachate, buffer, supplemental water, and inoculum/nutrition amendments required to provide the landfill microbial consortia with their growth needs.

The data flow diagram and hierarchy of the developed control program are shown in Figure 8. The structure of the program is composed of multiple cascading mathematical calculations (MCs 1-5) based on a main logic controller (LC). The control sequence in Figure 8 is repeated every operational cycle. The LC is discussed below (why a logic controller is needed? which method should be used? how the model is developed?), and then the mathematical calculations are presented.

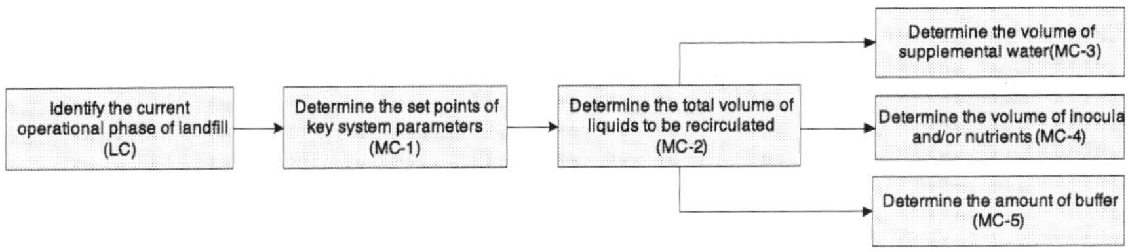

Figure 8. Dataflow diagram of the control program

2.5.1. Logic controller

Bioreactor landfills undergo the typical waste decomposition phases of sanitary landfills (in the order of: *initial/aerobic, transition, acid formation, methane generation,* and *final maturation* phases) but in a shorter time frame [7, 9, 31]. The determination of the current operational phase of the BL is vital because the bacterial consortia change significantly throughout the BL lifetime, and accordingly so do the conditions for their optimal growth. In order to stimulate the decomposition process and consequently biogas generation, those requirements have to be adequately provided. Practically, the identification of the dominant operational phase of the BL at a given time is challenging especially because of factors such as the heterogeneity of the waste which may cause system parameters not to follow their normal expected trends. Moreover, since landfills receive waste continually over several years, these progressive phases occur simultaneously, but in different neighbouring locales. The temporal and spatial dimensions of each phase depends on many factors such as waste characteristics, landfill design, operational strategy, and environmental conditions, that can be characterized by changes in various physical and biochemical indicator parameters.

In recent years, intelligent control of large-scale industrial processes has brought about a revolution in the field of advanced process control [32]. Knowledge-based techniques, such as fuzzy logic which uses linguistic control rules capturing the know-how of the experienced human operators, proved to be robust and reliable solutions for dealing with complex and ill-defined processes, such as those encountered in the operation of a BL. In fact, no conventional controller could efficiently operate such a complex process because it is practically impossible to predict its behaviour especially with the heterogeneity of waste. Fuzzy logic has been applied successfully to control various biological treatment systems such as anaerobic digesters [33], biological reactors [34], and wastewater treatment plants [35].

Therefore, the objective was to employ the modeling capabilities of fuzzy logic in developing a knowledge-based controller that determines the operational phase based on quantifiable input parameters of leachate and biogas, while taking uncertainty issues into consideration. The selected input variables include the leachate's COD, total volatile acids (TVA), pH, ORP, and methane content (%CH_4) in biogas, whereas, the single output variable is an index that defines the current operational phase of the BL, hereafter named the *Phase Index*.

Model development

The **first step** in the design of a Fuzzy Logic Controller (FLC) is to build the *data base* which contains the membership functions defined for each input and output variable. Each variable is expressed by linguistic terms (fuzzy sets) within its predefined range (universe of discourse). The degree of truth of a fuzzy set A is defined by a membership function μ_A, which is represented by a real number in the interval [0, 1] depending on the degree at which it belongs to the set. This is different from conventional numerical sets where an element either belongs or does not belong to a particular set (membership = 0 or 1). This distinctive feature is advantageous for controlling biological ecosystems, like the BL, where the change in input variable does not cause the controlled process to shift abruptly from one state to another. Instead, as the variable changes, it loses its membership in one fuzzy set while gaining membership in the next. This is a logical approach to account for the fact that a part of the BL may be in a particular operational phase, while adjacent parts may be in other phases.

Membership functions (MFs) can have different shapes such as triangular, trapezoidal, bell-shaped (Gaussian), or wave-shaped (Sigmoid). In the present FLC, fuzzy sets were defined by trapezoidal and/or triangular (special case of the trapezoidal shape) MFs where the uncertainty in each variable is represented by the most likely interval (i.e., the range at membership degree = 1.0) and the largest likely interval (i.e., the range at membership degree = 0.0) as shown in Figure 9. These intervals facilitate the interpretation of overlapping and disagreement in the compiled data ranges. The membership value is constant in the most likely interval [b, c], and increasing linearly from 0 to 1 between (a & b) and decreasing linearly from 1 to 0 between (c & d), thus providing the trapezoidal shape. For the special case of the triangular MF, the only difference to the trapezoidal MF is that the most likely interval [b, c] is a single point.

Figure 10 shows the MFs defined for a sample input (ORP) and the single output (*Phase Index*). The linguistic labels (fuzzy sets) used to describe the ORP values are *positive* (P), *zero* (Z), *negative* (N), and *very negative* (VN). The '*Phase Index*' variable was defined by the basic phases that typically characterize the BL lifespan; *aerobic* (A), *transition* (T), *acid formation* (AF), and *methane generation* (MG).

http://dx.doi.org/10.5772/55715

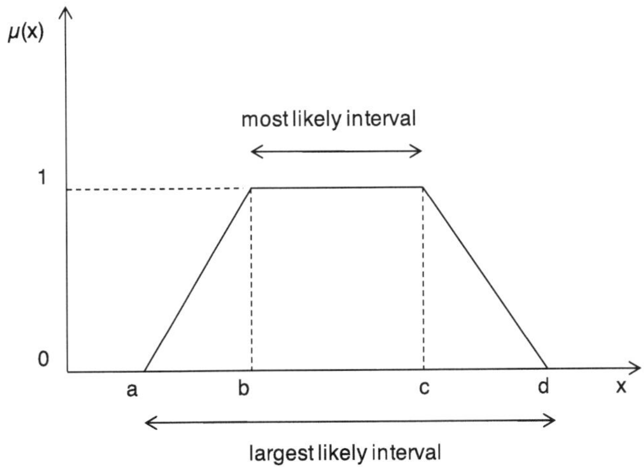

Figure 9. Typical trapezoidal membership function

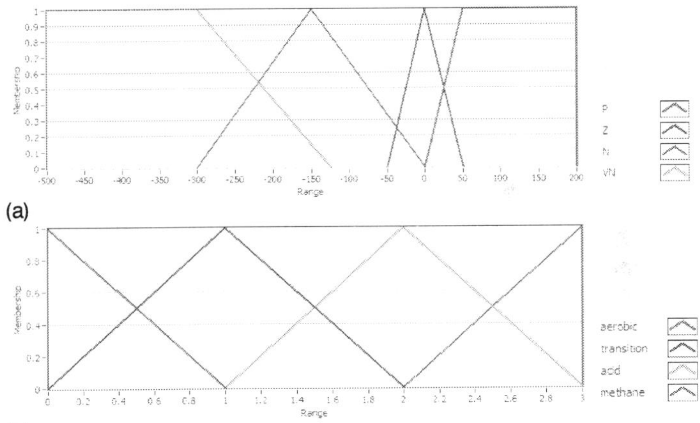

Figure 10. Membership functions for: a) ORP, and b) Phase Index

The **second step** in the design of FLC is developing the *rule base* for the controlled process. The *rule base* consists of fuzzy rules which are stated as IF–THEN statements that define the system behavior and predict the output variable. A typical fuzzy rule can include several variables in the antecedent (IF part) and consequent (THEN part) of the rule. If a rule has more than one antecedent, a fuzzy operator such as AND, OR, or NOT, is used to connect them, and to determine how to calculate the truth value of the aggregated rule antecedent. In the present

FLC, five basic statements (rules) were created to define the expected operational phase based on different quantifiable parameters. The probabilistic-type of the OR operator, which uses the probabilistic sum of the degrees of membership of the antecedents, was applied in the formulated rules. The following is an example of the developed rule base statements:

IF 'ORP' is 'VN' OR 'pH' is 'HN' OR 'COD' is 'H' OR 'TVA' is 'I' OR '%CH_4' is 'H'

THEN 'Phase Index' is 'MG'

In the above rule, VN, HN, H, I, H, and MG are fuzzy sets that denote *very negative, high neutral, high, intermediate, high,* and *methane generation,* respectively. The complete fuzzy rules as well as parameters of membership functions defined in the FLC are presented in [36].

Example: Based on the compiled knowledge base, when the ORP of the leachate is -250 mV, it has a 0.3 membership in the "*negative*" fuzzy set, and a 0.7 membership in the "*very negative*" fuzzy set (see Figure 10a). This allows the single input (-250 mV) to be processed with multiple rules, i.e., the fuzzy rules that include "*negative*" and "*very negative*" ORP in their antecedents. Although all the invoked rules influence the output, the rules with higher truth values ("*very negative*" in this case) have the greatest effect. This weighing system helps in dealing with the uncertainties in the landfill ecosystem, as well as simplifying the complexity of the controlled process.

The data base and rule base represent the knowledge components based on which the FLC makes the decision. The knowledge was compiled from information presented in [7, 37-39]. Table 1 shows the reported ranges of the input system parameters in the compiled studies.

Parameter	Study	Phase II Transition	Phase III Acid Formation	Phase IV Methane Generation	Phase V Maturation
COD, mg/l	[7]	20 - 20,000	11,600 - 34,550	1,800 - 17,000	770 - 1,000
	[38]	-	15,000 - 41,000	1,000 - 41,000	-
TVA, mg/l	[7]	200 - 2,700	1 - 30,730	0 - 3,900	0
	[38]	-	7,000 - 15,000	10,000	0
pH	[7]	5.4 - 8.1	5.7 - 7.4	5.9 - 8.6	7.4 - 8.3
	[38]	-	5 - 6	5.6 - 7.1	-
	[37]	-	5.8 - 6	6 - 7.8	7.1
%CH_4	[38]	-	0	0 - 50	40
	[37]	-	-	23 - 62	-
ORP, mV	[38]	50 - (-50)	50 - 0	0 - (-125)	-
	[39]		(-100)	(-300)	-

Table 1. Ranges of selected system parameters at the main operational phases

The data base and rule base are incorporated in the typical FLC components, shown in Figure 11, which includes: (1) *fuzzification unit,* (2) *inference engine,* and (3) *defuzzification unit.* The

http://dx.doi.org/10.5772/55715

fuzzification unit converts the input variables into fuzzy sets using the predefined membership functions. The *inference engine* then processes the fuzzy inputs based on their relevant fuzzy rules, and determines the fuzzy output(s). As mentioned above, the *inference engine* invokes more than one rule, which results in having different memberships in multiple output fuzzy sets. In the present LC, the *inference engine* uses the product implication method in which each output MF is scaled down at the truth value of the corresponding aggregated rule antecedent. The output from this step is an irregular area under the scaled-down membership functions. Finally, the *defuzzification unit* incorporates a number of fuzzy sets in a calculation that gives a single numeric value for each output.

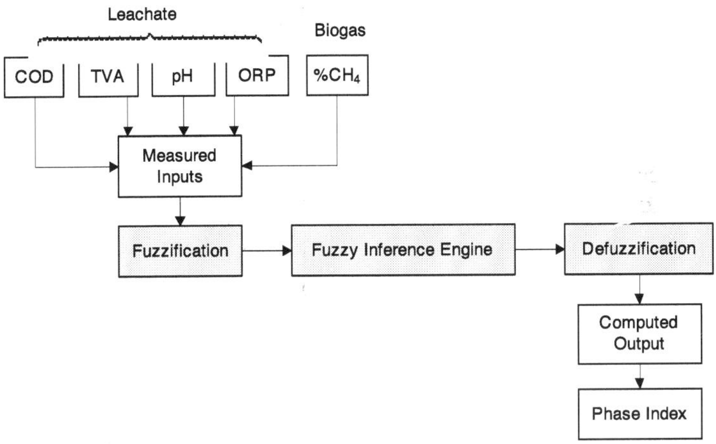

Figure 11. Typical structure of a fuzzy logic controller

In order to help visualize the non-linear characteristics of the *Phase Index*, surface plots were generated by varying two variables while the other variables remained constant. This can generate an infinite number of response surface, however if grouped for each pair of inputs, the number of possible groups of response surfaces becomes equal to the combination $C(n, 2) = n! / 2! (n - 2)!$ where n is the number of input variables. In the present FLC, 10 groups of response surfaces can be established for the 10 possible pairs of input variables. Figure 12 shows the response of the output variable '*Phase Index*' to changes in two pairs of the input variables, namely ORP and COD as well as TVA and pH, at the average defined value for the other input variables. The non-linear variation of the response intensity for the different values of input variables is considered one of the main advantages of the fuzzy logic system. Moreover, SMART's numeric representation for the operational phase offers a unique feature being able to obtain the transitional stage of the controlled BL. For example, when the '*Phase Index*' is equal to 2.7, this means that the bioreactor is transitioning from the acid formation phase (2.0) to the methane generation phase (3.0). The value (2.7) indicates also that the BL microbial ecosystem is closer to the methanogenic stage.

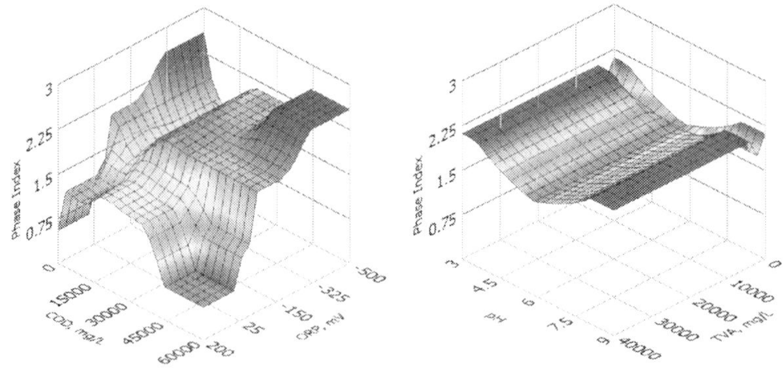

Figure 12. Response surfaces for two pairs of inputs: 1. COD/ORP (left), and 2. TVA/pH (right)

2.5.2. Mathematical calculations

As shown in Figure 8, the program sequence starts with the logic controller (LC) which identifies the current operational phase of the BL based on quantifiable characteristics of the generated leachate and biogas. The output of LC is a real number in the interval [0, 3] that expresses the BL operational phase, where 0 is the aerobic phase and 3 is the methanogenic phase. The output from LC is the input to the first mathematical step (MC-1).

Target set points

In MC-1, set points of pH (leachate), Carbon/Nitrogen (C/N) ratio (leachate), and moisture content (solid waste matrix) are computed based on the BL operational phase determined from LC. Table 2 shows default set points used in the present study for the two main BL operational phases. It should be noted that these set points may vary depending on several site-specific factors such as holding capacity of waste matrix, degree of compaction, and waste composition.

Parameter	Medium	Phase III Acid Formation	Phase IV Methane Generation
pH	Leachate	5.5-6.5	6.8-7.2
C/N ratio	Leachate	10	15
Moisture content, %	Waste Matrix	50	60

Table 2. Set points of process parameters at the Acid Formation and Methane Generation phases

MC-1 applies linear interpolation between the predefined parameter values (shown in Table 2). The parameter setpoint (*S*) at a given phase (*P*) can be calculated as follows:

$$S_P = S_i + [(S_{i+1} - S_i) \times (P - i)] \quad (1)$$

http://dx.doi.org/10.5772/55715

Where P is the computed phase from LC, i is the integer part from the computed *Phase Index* P, S_i is the setpoint at phase i, and S_{i+1} is the setpoint at phase $i+1$.

Recirculation volume

MC-2 computes the total required volume of recirculated liquids to raise the water content of the waste matrix from its current level to the desired setpoint. The liquid volume is calculated as follows:

$$V_{liquid} = \left(\frac{S_{mc} \times w}{\rho_{water}} \right) - \left(\theta \times V_{waste} \right) \tag{2}$$

Where V liquid is the total required volume of liquids to be added in a cycle (m^3), S_{mc} is the setpoint for the gravimetric water content (calculated in MC-1), θ is the measured volumetric water content, ρ *water* is the water density (t/m^3), and w is the bulk weight of the waste (t).

Supplemental water addition

One of the main benefits of supplemental water addition is to dilute elevated concentrations of pollutants in leachate which may inhibit the microbial consortia in the waste matrix. The primary inhibitors in MC-3 can include, but are not limited to, ammonia-nitrogen, VFA, and their free unionized fractions, as well as alkali cations. The concentrations of selected inhibitors are used to compute a factor (D) for the required dilution (i.e., dilution water as a fraction of the liquid recirculated). D is calculated as the greatest of individually calculated dilution indices required to bring each of the potential inhibitors, if any, to its nontoxic range, as follows:

$$D = Max\left(1 - \frac{C_{target}}{C_{inhibitor}} \right) \tag{3}$$

Where $C_{inhibitor}$ is the concentration of an inhibitor in leachate (g/m^3), and C_{target} is the nontoxic concentration of that inhibitor (g/m^3). The required supplemental water volume can then be calculated by multiplying the volume of leachate produced in previous operational cycle by the dilution factor.

Nutritional requirements

Next, MC-4 determines additional nutrient requirements using the set point for C/N ratio as well as the concentrations of TOC and TN of the generated leachate. The addition of a nitrogen source to the BL is controlled according to the C/N ratio. The volume of nutritional source is calculated as:

$$V_{nutrients} = \frac{\left(\frac{TOC}{S_{C/N}} \right) - TN}{TN_{nutrients}} \times V_{liquid} \tag{4}$$

Where $V_{nutrients}$ is the required volume of the nutritional source (m^3), $S_{C/N}$ is the setpoint calculated for the C/N ratio, V_{liquid} is the volume of liquid calculated in MC-2 (m^3), and TN, TOC and TN nutrients are the concentrations of total nitrogen of diluted leachate, total organic carbon of diluted leachate, and total nitrogen of the nutritional source to be used, respectively (g/m^3).

Buffering requirements

Next, the required amount of buffer is calculated in MC-5. The buffer salt is used to adjust the pH and provide external source of alkalinity to the system. MC-5 calculates the required bicarbonate alkalinity to be added to the leachate regardless of the resultant pH. The buffer is added to provide the difference between the required alkalinity (CO_2/water buffering system) and the available alkalinity in the system. The available bicarbonate alkalinity can be calculated as:

$$BA = ALK - 0.83 \times f \times VFA \tag{5}$$

Where *BA* is the bicarbonate alkalinity (mg $CaCO_3$/L), *ALK* is the total alkalinity (mg $CaCO_3$/L), *VFA* is the concentration of the volatile fatty acids (mg/L), 0.83 is a unit conversion factor (Equivalent weight of $CaCO_3$/Equivalent weight of VFA), and *f* is a factor for the percentage of VFA titrated at the pH endpoint of the alkalinity test. On the other hand, the required alkalinity (*RA*) for the CO_2/water buffering system can be calculated as:

$$RA = K_1 \times K_H \times P_{CO_2} \times E_{CaCO_3} \times 10^{S_{pH}} \tag{6}$$

Where *RA* is the required concentration of bicarbonate ion for CO_2 neutralization (g $CaCO_3$/L), K_1 is the ionization constant for carbonic acid, K_H is the hydration equilibrium constant, P_{CO2} is the partial pressure of CO_2 in the system (fraction of CO_2 in the composition of air), S_{pH} is the target pH as computed in MC-1, and E_{CaCO3} is the equivalent weight of $CaCO_3$. The added alkalinity is the difference between the required and available alkalinity in the system. The volume of buffer to provide the required alkalinity can be calculated as:

$$V_{buffer} = \frac{\left[RA - BA\right] \times E_{buffer} \times V_{liquid}}{C_{buffer}} \tag{7}$$

Where V_{buffer} is the required volume of buffer, E_{buffer} is the equivalent weight of buffer salt, C_{buffer} is the concentration of buffer salt in solution, and V_{liquid} is the volume of recirculated liquid. The amount of buffer to be added should be equal or greater than the amount required to bring the pH up to the setpoint calculated from MC-1.

http://dx.doi.org/10.5772/55715

3. Application and evaluation of SMART

The new concepts proposed and incorporated in SMART were demonstrated in a real operational prototype. Specifically, the concept of temporal determination of the BL operational phase as the starting step for initiating the other subsequent computations to determine the various amendments to be added to manipulate the leachate recirculated. Concomitantly, the main objectives of this research phase were to: (1) implement the software and hardware components of SMART on a pilot-scale prototype, and (2) evaluate the system viability to control the BL versus a conventional open-loop leachate control (OLLC) scheme, in which recirculation rate is fixed and the leachate quality is not changed.

3.1. Experimental setup

Experimental work was conducted on two bioreactor setups; Cell-1 and Cell-2. Figure 13 shows the configuration of a single bioreactor cell (675 litres volume) with its leachate collection and recycling tanks. An equal mixture of residential and food wastes were thoroughly mixed while loaded to the bioreactor cells. The average total organic fraction and water content of the mixed waste was 73%, and 48%, respectively. Each bioreactor cell was equipped with three type-T thermocouples measuring temperature in different radial positions at three equidistant vertical levels in the waste matrix. In addition, three moisture sensors were placed at the same monitoring spots in order to measure the volumetric water content using frequency domain technology. The biogas generated went through a micro-turbine wheel flow meter, followed by an inline infrared methane analyzer. Leachate was collected by gravity from a lower outlet port connected to a collection tank with a mechanical mixer. This tank also received the flow from the amendments' tanks through tube lines with actuated solenoid valves (SMART-controlled). The recirculated leachate was manipulated by adding amendments such as inoculum (anaerobic digester sludge), nutritional source (plant fertilizer), buffer (sodium bicarbonate), and supplemental water. After mixing with amendments, leachate was recycled in a cyclic batch mode using a submersible pump (SMART-controlled).

After loading the bioreactor cells, the first nine months were used to examine the communication and synchronization between system components, as well as test run of the system. By the end of this period, Cell-2 has already started producing methane and surpassed Cell-1 in terms of all performance and evolution parameters. In order to effectively assess the system, SMART was applied on Cell-1 (the inadequately performing cell) for four months so as to evaluate the performance. In parallel, Cell-2 was running according to an OLLC scheme, at a constant rate of leachate recirculation equal to a predetermined percentage (8%) of the initial volume of waste matrix. The discussion is presented in two main sections: (1) assessment of the control actions made by SMART, and (2) evaluation of the system performance through its effect on leachate and biogas.

3.2. Evaluation of SMART control decisions

There has been no consensus in the literature on the optimal leachate recirculation rates in BLs, and the reported rates are extremely diverse to over 400 fold [17]. It was also found

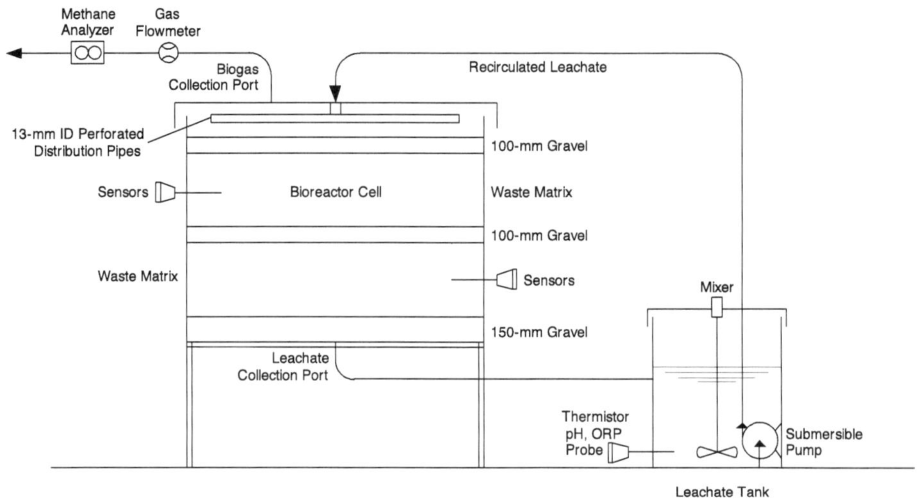

Figure 13. Configuration and instrumentation of the prototype bioreactor cell

that higher recirculation rates do not necessarily achieve better performance of the BL [1, 24]. Alternatively in SMART, recirculation rates vary based on the site-specific and real-time conditions, and so every BL is controlled according to its own evolution. Figure 14 shows the different recirculated volumes of leachate as determined by SMART for Cell-1, as well as the various fractions of leachate, water, buffer, and sludge in the recirculated liquid in each cycle. It can be observed that the calculated volumes of leachate and other amendments did not follow a predictable trend, and they varied significantly over time (34±7 L/cycle). However, the volumes of amendments followed a decreasing trend that seemed to restart every four operational cycles (1-4 & 5-8).

3.2.1. System evolution

The *Phase Index,* determined by the logic controller, for the two cells is shown in Figure 15. The progress of Cell-1 surpassed that of Cell-2 which was also evolving but at slower rate. It can be seen that, while at the beginning of this test, Cell-2 was ahead of Cell-1 with a PI of 1.6 (Cell-1) versus 2.0 (Cell-2), the SMART-controlled Cell-1 was able to catch up and actually surpassed Cell-2 in four operational cycles. It is clear that since Cell-2 was running with an open-loop control scheme, the improvement in the evolution pattern of Cell-1 can be mostly attributed to the implementation of SMART. The fuzzy logic controller was able to track the BL evolution by identifying the operational phase at any time based on multiple parameters of leachate and biogas. The computed *Phase Index* described the transitioning progress between the main phases of BL, which enabled the interpolation of the evolving growth requirements for the bacterial population inside the BL, and led to successful transition from one phase to another.

http://dx.doi.org/10.5772/55715

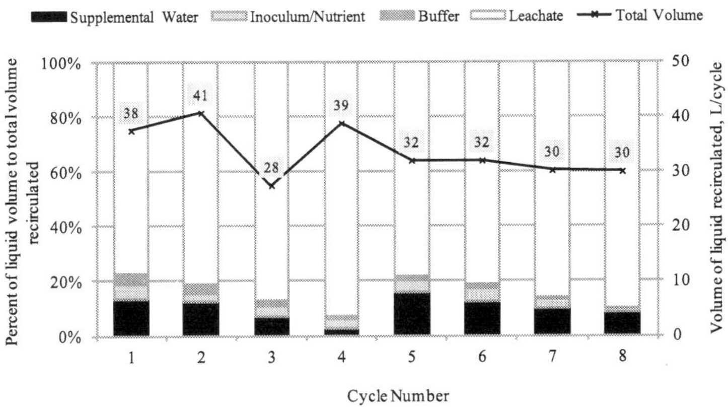

Figure 14. Cyclic recirculated liquid volumes and amendment fractions in Cell-1

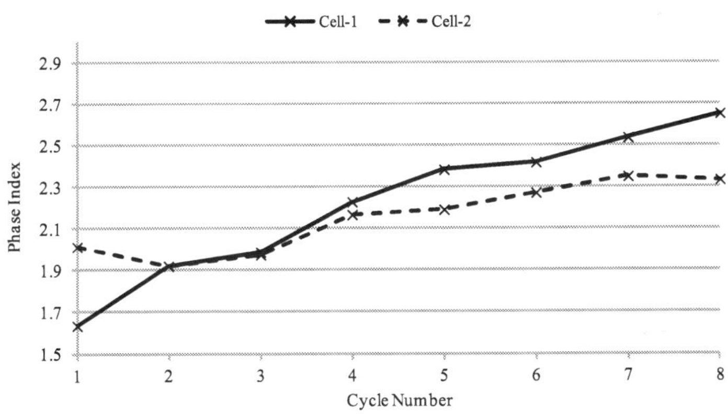

Figure 15. Progress of the Phase Index of Cell-1 and Cell-2

3.2.2. Control strategy

During the operation period, the operator had to interfere occasionally so as to insure the control actions address all potential problems. This man-computer interaction was crucial due to: (1) the instability and unexpected behavior of the BL system, in part due to its complexity and nonlinear responses, and (2) the fact that the reasoning of the fuzzy logic is limited to its knowledge base. Therefore, applying a semi-automated control strategy, rather than a fully automated one, was found to achieve more stable performance of the system. In this control strategy, SMART collects and analyzes the data, performs the computational effort to determine the optimum operational strategy, and then aids the site operator to apply the final operational decision through the computer interface.

3.2.3. Feedback control scheme

The control actions determined by SMART were based on multiple leachate and biogas parameters acquired from previous cycles. The response time of the BL ecosystem, i.e., time from changing a system parameter to when its effect (feedback) on system performance is detected, was found to be sufficient to facilitate the application of the feedback control scheme. The BL performance was significantly improved with the application of closed-loop control (in Cell-1) as opposed to an open-loop strategy (in Cell-2).

3.3. Evaluation of process parameters

3.3.1. Organic matter

The development of oxidizable organic concentration in the leachate produced is plotted in terms of COD and VFAs in Figure 16. The average degradation rate of COD in Cell-1 (controlled by SMART) was 330 mg/L.d compared to 110 mg/L.d in Cell-2. The COD concentrations in leachate from Cell-2 were fluctuating and the final COD was about 10% less than the initial concentration (from 116 to 105 g/L). After 40 days, COD concentration in leachate from Cell-1 was consistently less than that of Cell-2 which shows that the implementation of SMART had a positive effect on the degradation of organic matter.

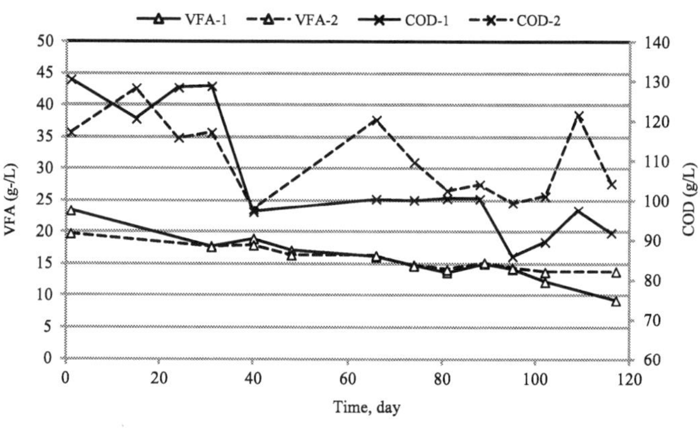

Figure 16. Evolution of organic concentration of leachate from Cell-1 and Cell-2

As shown in Figure 16, the conversion of VFAs to methane was increasing slowly resulting in lower and mostly similar concentrations of VFA in leachate from both cells. However at day 95, the VFA concentration in Cell-1 started to drop, leading to an overall conversion rate of 120 mg/L.d compared to 50 mg/L.d in Cell-2. The last recorded VFA concentration was less than

10 g/L in leachate from Cell-1 compared to 14 g/L in Cell-2. It is therefore clear that the SMART control system stimulated the methanogenic activity which gradually consumed the produced VFAs, until the conversion rate of VFA became greater than the production rate (starting from the 95th day).

3.3.2. *Biogas production*

The CH_4 fractions of the biogas produced from both cells are shown in Figure 17. The performance of Cell-1 in terms of the rate of increase of the CH_4 fraction was improved. SMART was successful in leading the cell through the transitional stage from acid formation to methane generation. The CH_4 content increased from 10 to 62% in Cell-1 in a four-month period, while Cell-2 continued to increase but at slower rate going from 40 to 58%. Based on the equations of the trend lines fitted to the actual data of cumulative production in Figure 17, the rate of increase in methane production in Cell-1 was 1.7 fold higher than that of Cell-2. By the end of operation, the cumulative biogas production reached 23 and 14 m³ which corresponds to a specific production of 61 and 35 L/kg of waste in Cell-1 and Cell-2, respectively.

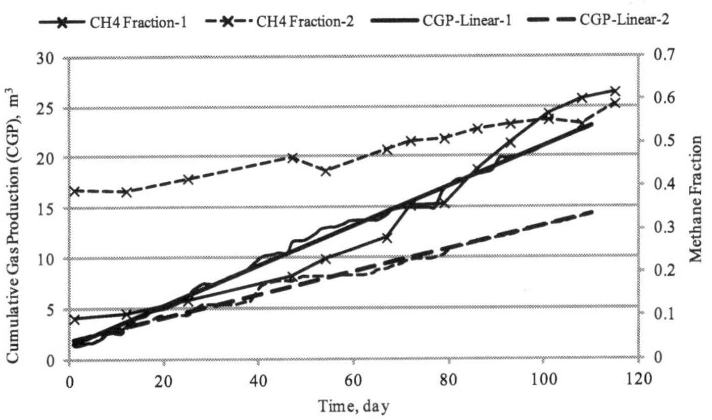

Figure 17. Development of methane production and methane content in the biogas produced

3.4. Future aspects and potential implications

The implementation of a sensor-based control strategy in full-scale BL faces two main issues: (1) instrumentation of the system, and (2) the heterogeneity of the waste matrix which affects the degree to which the measurements are representative. While in-situ measurements of leachate and biogas are well established, the main instrumentation problem is the subsurface monitoring for in-place waste, such as: moisture content and temperature. The difficulty arises from the following issues: (1) instrument failure is most likely to occur since no specialized sensing technologies and installation procedures exist for landfill application, (2) installation techniques are very challenging and obstruct daily site operations, (3) cables are subject to physical damage due to heavy equipments, differential settlement, and aggressive environment, and (4) cable conduits create pathways for lateral breakout of leachate and gas. It is clear that, with all these operational issues, current monitoring techniques are neither robust nor efficient. The solution for these issues can be realized via two approaches: (1) using non-intrusive surface methods for subsurface monitoring; e.g., for moisture measurements: seismic waves [40], ground penetrating radar [41], and fiber optics [42], or (2) using wireless communication techniques to eliminate the huge capital cost and operational problems associated with conventional wired techniques [43]. Both approaches can also solve the heterogeneity problem in a way that: in the first approach, a three-dimensional image of moisture distribution can be produced, and in the second approach, more wireless sensors can be used to give higher resolution data. In addition, soft computing methods can be used to deal with the uncertainty in measurements, and by using adaptive systems, monitoring and control programs can learn and adapt to the controlled BL.

Given the rapid development in both instrumentation and full-scale applications of BLs, it is expected that robust subsurface monitoring techniques will appear in the near future. However, research in the area of advanced BL process control like the present research, has to move in-parallel and not to wait until a flawless method to measure subsurface parameters is ready. In fact, process control research can motivate the search for robust and reliable sensory equipment. Therefore, SMART can be currently applied in full-scale BLs if some technical modifications of in-situ monitoring are considered, e.g., monitoring in/out liquid to/from the BL can effectively replace the in-situ measurements of moisture content by means of continuously conducting a real-time water balance.

4. Conclusions

The present work developed a control framework in which an expert system is responsible for the operation of BLs. The main control objective of the system was to optimize the performance of the BL by manipulating the quantity and quality of leachate recirculated so as to supply the microbial consortia inside the BL with their optimal growth requirements. The proposed control framework and guidelines were described, and an assessment was conducted for a SMART-controlled pilot-scale BL in order to examine the applicability, feasibility, and effectiveness of the technology. The following conclusions were drawn:

1. The control system successfully determined the quantity and quality of recirculated liquid based on the BL operational stage and multiple process leachate and biogas parameters.

2. The performance of the BL was significantly improved with the application of closed-loop control as opposed to an open-loop strategy.

3. Leachate manipulation techniques, such as buffering, bioaugmentation, and supplemental water addition, were proven to be potentially effective control tools that are able to adjust/optimize the leachate characteristics.

4. Recirculating variable calculation-based amounts of leachate and other amendments resulted in a positive influence on the overall performance of the BL system.

5. The pilot-scale implementation of SMART demonstrated the feasibility of the system. Since all the incorporated hardware components are commercially available, the system can be readily scaled-up to a larger scale application.

Author details

Mohamed Abdallah and Kevin Kennedy

Department of Civil Engineering, University of Ottawa, Ottawa, Canada

References

[1] Warith M. Bioreactor Landfills: Experimental and Field Results. Waste Management. 2002;22(1):7-17.

[2] SWANA. Summary List of North American Bioreactor Landfill Projects 2004 [cited 2012 December 1st]. Available from: www.swana.org/pdf/swana_pdf_295.pdf.

[3] Reinhart DR, McCreanor PT, Townsend T. The Bioreactor Landfill: Its Status and Future. Waste Management & Research. 2002;20(2):172-86.

[4] Warith M, Smolkin P, Caldwell J. Effect of Leachate Recirculation on Enhancement of Biological Degradation of Solid Waste: Case Study. Practice Periodical of Hazardous, Toxic, and Radioactive Waste Management. 2001;5(1):40-6.

[5] Zhao XD, Musleh R, Maher S, Khire MV, Voice TC, Hashsham SA. Start-up Performance of a Full-scale Bioreactor Landfill Cell under Cold-Climate Conditions. Waste Management. 2008;28(12):2623-34.

[6] Cossu R, Blakey N, Trapani P, editors. Degradation of mixed solid waste in conditions of moisture saturation. International Sanitary Landfill Symposium; 1987; Cagliari, Italy.

[7] Reinhart DR, Townsend TG. Landfill Bioreactor Design and Operation. Boca Raton, FL: Lewis Publishers; 1998.

[8] He PJ, Shao LM, Qu X, Li GJ, Lee DJ. Effects of Feed Solutions on Refuse Hydrolysis and Landfill Leachate Characteristics. Chemosphere. 2005;59(6):837-44.

[9] Pohland FG, Alyousfi B. Design and Operation of Landfills for Optimum Stabilization and Biogas Production. Water Science and Technology. 1994;30(12):117-24.

[10] ElFadel M, Findikakis AN, Leckie JO. Environmental Impacts of Solid Waste Landfilling. Journal of Environmental Management. 1997;50(1):1-25.

[11] Hilger HH, Barlaz MA. Anaerobic decomposition of refuse in landfills and methane oxidation in landfill covers. 2007 (Ed.3):818-42.

[12] Pohland FG, Behavior and Assimilation of Organic and Inorganic Priority Pollutants Codisposed with Municipal Refuse: Project Summary: U.S. Environmental Protection Agency, Center for Environmental Research Information; 1993.

[13] San I, Onay TT. Impact of Various Leachate Recirculation Regimes on Municipal Solid Waste Degradation. Journal of Hazardous Materials. 2001;87(1-3):259-71.

[14] Mehta R, Barlaz MA, Yazdani R, Augenstein D, Bryars M, Sinderson L. Refuse Decomposition in the Presence and Absence of Leachate Recirculation. Journal of Environmental Engineering-ASCE. 2002;128(3):228-36.

[15] Bilgili MS, Demir A, Ozkaya B. Influence of leachate recirculation on aerobic and anaerobic decomposition of solid wastes. Journal of Hazardous Materials. 2007;143(1-2):177-83.

[16] Filipkowska U. Effect of Recirculation Method on Quality of Landfill Leachate and Effectiveness of Biogas Production. Polish Journal of Environmental Studies. 2008;17(2):199-207.

[17] Benbelkacem H, Bayard R, Abdelhay A, Zhang Y, Gourdon R. Effect of Leachate Injection Modes on Municipal Solid Waste Degradation in Anaerobic Bioreactor. Bioresource Technology. 2010;101(14):5206-12.

[18] Morris JWF, Vasuki NC, Baker JA, Pendleton CH. Findings from Long-Term Monitoring Studies at MSW Landfill Facilities with Leachate Recirculation. Waste Management. 2003;23(7):653-66.

[19] Buivid MG, Wise DL, Blanchet MJ, Remedios EC, Jenkins BM, Boyd WF, et al. Fuel Gas Enhancement by Controlled Landfilling of Municipal Solid-Waste. Resources and Conservation. 1981;6(1):3-20.

[20] Chugh S, Clarke W, Pullammanappallil P, Rudolph V. Effect of Recirculated Leachate Volume on MSW Degradation. Waste Management & Research. 1998;16(6): 564-73.

[21] Jiang JG, Yang GD, Deng Z, Huang YF, Huang ZL, Feng XM, et al. Pilot-scale Experiment on Anaerobic Bioreactor Landfills in China. Waste Management. 2007;27(7): 893-901.

[22] Suna Erses A, Onay TT. Accelerated Landfill Waste Decomposition by External Leachate Recirculation from an Old Landfill Cell. Water science and technology. 2003;47(12):215-22.

[23] Lisk DJ. Environmental Effects of Landfills. Science of the Total Environment. 1991;100:415-68.

[24] Sponza DT, Agdag ON. Impact of Leachate Recirculation and Recirculation Volume on Stabilization of Municipal Solid Wastes in Simulated Anaerobic Bioreactors. Process Biochemistry. 2004;39(12):2157-65.

[25] Sanphoti N, Towprayoon S, Chaiprasert P, Nopharatana A. The Effects of Leachate Recirculation with Supplemental Water Addition on Methane Production and Waste Decomposition in a Simulated Tropical Landfill. Journal of Environmental Management. 2006;81(1):27-35.

[26] Stegmann R, Spendlin HH. Enhancement of biochemical processes in sanitary landfills. International Sanitary Landfill Symposium; Cagliari, Italy1987. p. 1-16.

[27] Rees JF. Optimization of Methane Production and Refuse Decomposition in Landfills by Temperature Control. Journal of Chemical Technology and Biotechnology. 1980;30(8):458-65.

[28] Pohland F. Anaerobic treatment: fundamental concepts, applications, and new horizons. In: Malina J, Pohland F, editors. Design of Anaerobic Processes for the Treatment of Industrial and Municipal Wastes. Water Quality Management Library. Lancaster, USA: Technomic; 1992.

[29] McCarty P. Anaerobic Waste Treatment Fundamentals, Part Three: Toxic Materials and their Control. Public Works. 1964;92(11).

[30] Kim E, Liu B, Roehm T, Tout S. Feedback Control 2007 [cited 2012 December 1st]. Available from: https://controls.engin.umich.edu/wiki/index.php/Feedback_control.

[31] Kim J, Pohland FG. Process Enhancement in Anaerobic Bioreactor Landfills. Water Science and Technology. 2003;48(4):29-36.

[32] Manesis SA, Sapidis DJ, King RE. Intelligent Control of Wastewater Treatment Plants. Artificial Intelligence in Engineering. 1998;12(3):275-81.

[33] Estaben M, Polit M, Steyer JP. Fuzzy Control for an Anaerobic Digester. Control Engineering Practice. 1997;5(9):1303-10.

[34] Bae H, Seo H, Kim S. Knowledge-based Control and Case-based Diagnosis based upon Empirical Knowledge and Fuzzy Logic for the SBR Plant. Water Science and Technology. 2006 2006;53(1):217-24.

[35] Garcia C, Molina F, Roca E, Lema JM. Fuzzy-based Control of an Anaerobic Reactor Treating Wastewaters Containing Ethanol and Carbohydrates. Industrial & Engineering Chemistry Research. 2007;46(21):6707-15.

[36] Abdallah M. A novel computational approach for the management of bioreactor landfills. Ottawa, Canada: University of Ottawa; 2011.

[37] Barlaz MA, Schaefer DM, Ham RK. Bacterial Population Development and Chemical Characteristics of Refuse Decomposition in a Simulated Sanitary Landfill. Applied and Environmental Microbiology. 1989;55(1):55-65.

[38] Pohland FG, Kim J. Insitu Anaerobic Treatment of Landfills for Optimum Stabilization and Biogas Production. Water Science and Technology. 1999;40(8).

[39] Gerardi MH. The Microbiology of Anaerobic Digesters. Hoboken, N.J.: Wiley-Interscience; 2003.

[40] Catley AJ, Seismic Velocity analysis to determine moisture distribution in a bioreactor landfill: Carleton University (Canada); 2007.

[41] Cassiani G, Fusi N, Susanni D, Deiana R. Vertical Radar Profiling for the Assessment of Landfill Capping Effectiveness. Near Surface Geophysics. 2008;6(2):133-42.

[42] Sayde C, Gregory C, Gil-Rodriguez M, Tufillaro N, Tyler S, van de Giesen N, et al. Feasibility of Soil Moisture Monitoring with Heated Fiber Optics. Water Resources Research. 2010;46:8.

[43] Nasipuri A, Subramanian KR, Ogunro V, Daniels JL, Hilger HA. Development of a wireless sensor network for monitoring a bioreactor landfill. GeoCongress 2006: Geotechnical Engineering in the Information Technology Age; 2006.

Chapter 8

Sustainable Post Treatment Options of Anaerobic Effluent

Abid Ali Khan, Rubia Zahid Gaur,
Absar Ahmad Kazmi and Beni Lew

Additional information is available at the end of the chapter

http://dx.doi.org/10.5772/56097

1. Introduction

The strategy of treating sewage by the common and known aerobic process has been shifted back to anaerobic processes in the recent years with the advent of high rate anaerobic systems such as up-flow anaerobic sludge blanket reactor (UASB), anaerobic contact process, anaerobic filter (AF) or fixed film reactors and fluidized bed reactors. The high rate anaerobic processes, like UASB have several advantages such as low capital, operation and maintenance costs, energy recovery in the form of biogas, operational simplicity, low energy consumption, and low production of digested sludge (van Haandel & Lettinga, 1994; Gomec, 2010; Khan et al., 2011a).

During early 1970s, due to the energy crisis and the above advantages, the UASB process was recognized as one of the most feasible method for the treatment of sewage in developing tropical and sub-tropical countries like India, Brazil and Colombia; where financial resources are generally scarce. However, the quality of UASB effluent rarely meet the discharge standards despite several modifications; such as settlers at the top of gas-liquid-solid-separator, addition of AF, two UASB reactors placed in series and even the incorporation of an external sludge digester (Lew et al., 2003; Khan et al., 2011a).

Since early 1980, the discussion on the applicability of UASB process for the treatment of sewage has been presented by Lettinga and co-researchers (Lettinga et al., 1980; Lettinga et al., 1993; Lettinga, 2008; Seghezzo et al., 2002; von Sperling and Chernicharo, 2005) and the results indicated that about 70% chemical oxygen demand (COD) removal can be achieved in warm climates countries (Schellinkhout and Collazos, 1992; Souza and Foresti, 1996; Khan et al., 2011a). Since its inception a lot of research has been done on this process and technology has been given wider publicity. Presently about 30 UASB based sewage treatment plants (STPs) are in operation in India and more than 20 are under construction (MoEF, 2005 and 2006). In total, about 200 UASB reactors are used for municipal and industrial applications (Khan, 2012).

The UASB reactor treating domestic wastewater can produce two main valuable resources, which can be recovered and utilized: methane and the effluent. The methane gas, which is produced during the COD removal can be recovered (from 28% to 75%) and transformed into energy (Mendoza et al., 2009). In energy terms, 1m^3 of biogas with 75% methane content is equivalent to 1.4 kWh electricity. The biogas can be used to run dual fuel generators or street lighting (Arceivala and Asolkar, 2007). According to Arceivala and Asolkar (2007) approximately 23% methane gas was observed dissolved in UASB effluents, therefore, the recovery of dissolved methane gas is discretionary and may not be acceptable in case of sewage treatment due to high expenditure costs and complexity. However, the methane gas evolved to the headspace (gas phase) can be of much importance and easily collected. For high strength industrial wastewaters the recovery of dissolved methane gas is favoured in view of the global warming and its fuel value. Moreover, at high temperature the solubility of gaseous compounds decreases. Therefore, the issue of gas recovery especially dissolved methane gas must be carefully reviewed for each individual case in terms of economics and desirability.

The produced effluent can be used in agriculture irrigation or disposed. However, the inability of UASB process to meet international disposal standards, owing to its anaerobic nature has given enough impetus for the subsequent post treatment. Furthermore, the growing concern over the impact of sewage contamination on rivers and lakes and the increasing scarcity of water in the world along with rapid population increase in urban areas give reasons to consider appropriate technologies for the post treatment of anaerobic effluent in order to achieve the desired effluent quality and save receiving water bodies.

A variety of post treatment configurations based on various combinations with UASB have been studied, such as aerobic suspended growth, aerobic attached growth, combined suspended and attached growth aerobic processes, anaerobic processes, natural treatment processes, physical processes and physico - chemical processes. UASB followed by final polishing units (FPU) or polishing pond (PP) is a common process used at several STPs in India, Colombia and Brazil, since the technology is simple in operation (von Sperling and Mascarenhas, 2005; von Sperling et al., 2005; Chernicharo, 2006; Khan et al., 2011a). However, still the final effluent is generally devoid of dissolved oxygen (DO) and rich in nutrient. Moreover, polishing ponds operate at long hydraulic retention time (HRT), around 1 day, leading to a high land requirement (Khan, 2012).

Other post treatment options widely used in India are activated sludge process (ASP) and aeration-polishing pond. A demonstration scale Down-flow hanging sponge reactor is also in operation (Tandukar et al., 2005 and 2006). Several other options such as plain aeration i.e. without using biomass, are the next technology option for the post treatment of anaerobic effluents but, limited studies have been performed. A bench scale batch aeration investigated by Khan (2012) has demonstrated that aeration systems operating at 1 to 2 h HRT are able to reduce the BOD of UASB effluent to discharge standards but, unable to remove nutrient. In the same study, continuous aeration of UASB effluent with and without activated sludge could remove nutrient.

Similarly, sequencing batch reactor (SBR), moving bed bio-film reactor, sand filtration, dissolved air flotation, rotating biological contactors (RBC), wetlands and others are still under investigations at bench and pilot scale. Results are promising; however, more studies are needed at pilot or demonstration level with actual environmental conditions in order to scale-up these technologies for best treatment concept. If stringent disposal standards need to follow, aeration with biomass can effectively reduce the organics, nutrients and odour causing substances like sulfides. Some of these processes are exclusively discussed in subsequent section.

Recently two different aerobic biomass based processes viz. continuous fill intermittent decant (CFID) type SBR and intermittent fill and intermittent decant type SBR were investigated by Khan (2012). Several researchers investigated the CFID and SBR and results revealed that the CFID can reduce the nitrogen to less than 10 mg/L as nitrogen. SBR is highly efficient to remove the nitrogen and phosphorous. Detailed studies were carried out on different aerobic treatment processes by Khan (2012).

Another latest concept of treatment is the 'Natural Biological Mineralization Route' (NBMR), which can be applied for the treatment of anaerobic effluents as suggested by Lettinga (2008) and elucidated in detail, by Khan (2012). This treatment concept enables conserving or recovery of useful by-products in the form of fertilizers, soil conditioners and renewable energy. The whole concept consists of treatment units of several micro aerobic and aerobic systems and dealt in subsequent section.

The objective of this chapter is to summarize different post treatment options for anaerobic effluent in general and specifically effluent of UASB reactor treating sewage. Natural biological mineralization route (NBMR) concept is also explained for an economical and efficient treatment.

2. Anaerobic effluent/ UASB effluent characteristics

The effluent characteristics in terms of biological oxygen demand (BOD), COD, suspended solids (SS), nutrients (N & P), microbial pathogens and reduced species such as sulfides explained as follows:

2.1. Organics and suspended solids

The BOD, COD and SS of various anaerobic treatment systems anaerobic ponds, UASB reactors, Imhoff tank and septic tanks treating sewage without any post treatment system has been reported to vary from 60 to 150; 100 to 200 and 50 to 100 mg/L, respectively (Chernicharo, 2006; Foresti et al., 2006). The process efficiency depends on different factors like strength and composition e.g. fraction of industrial wastewater infiltrated, temperature and diurnal fluctuations. The dissolved mineralized compounds such as ammonia, phosphate and sulfides in the effluent also varied with these factors. The performance of these treatment systems highly depends on temperature and decreases with a decrease in temperature (Lew et al., 2003 and 2004; Elmitwalli et al., 2001). The performance of UASB reactors (COD, BOD and TSS influent, effluent and removal) treating sewage at different temperatures is summarized in Table 1.1.

2.2. Nutrients (N and P)

Insignificant or negligible removal of nutrient may be expected in anaerobic systems treating sewage (Foresti et al., 2006; Moawad et al., 2009). The primary reason of poor removal of nutrients in anaerobic process is organic nitrogen and phosphorous hydrolyzed to ammonia and phosphate, respectively, which are not removed by anaerobic processes and in consequence, their concentration increases in the liquid phase. The concentration of ammonia nitrogen and phosphorous in anaerobically treated sewage has been reported to be from 30-50 and 10-17 mg/L, respectively (Foresti et al., 2006).

2.3. Highly mineralized or reduced compounds

Sulfur compounds exist as sulfides in anaerobic systems effluent treating sewage. The effluent total sulfides concentration to the highest degree depends on concentration of sulfates in influent and sulfate reducing bacterial activity present in the reactor. Generally, sulfide concentrations around 7-20 mg/L have been observed in the UASB effluent treating sewage, which increases the effluent oxygen demand (Khan, 2012). Moreover, the chemical and biochemical oxidation also depends on sulfides concentration along with other reduced species such as Fe^{2+}, mercaptans etc. although low ferrous ion concentration has been observed in the anaerobic effluent of systems treating sewage. However, Vlyssides et al. (2007) investigated the effect of ferrous ions addition to influent to enhance COD removal. The addition of ferrous ion induces a stable and outstanding conversion rate of COD and was proved to enhance the biological activity of UASB reactor; otherwise ferrous ions results by reduced environment if sewage is treated by UASB reactor.

2.4. Indicators of microbial pathogens

The reduction of fecal coliforms is around one order of magnitude (from around 10^8 to 10^7) in UASB systems although they are not designed for pathogenic removal, while helminth eggs removal efficiency has been reported to be 60–90% (Chernicharo et al., 2001; von Sperling et al., 2002; Chernicharo, 2006; von Sperling and Mascarenhas, 2005).

Hence, for ideal and sustainable treatment the high rate anaerobic treatment systems especially UASB rector must be integrated with novel and innovative post treatment systems based on NMBR sequence. Numerous post treatment system or combination of anaerobic pre-treatment (i.e. UASB reactor) followed by aerobic systems were investigated at laboratory and pilot scale levels for the treatment of sewage. Most of these combinations were found viable option for the treatment of effluent of UASB reactor.

Country	Capacity	Temp. (⁰C)	HRT (h)	Influent (mg/L)			Effluent (mg/L)			Removal Efficiency (%)		
				COD	BOD	TSS	COD	BOD	TSS	COD	BOD	TSS
Japan	-	-	6	600	291	333	222	153	-	63	53	-
Japan	1148 L	-	6	532	240	-	197	79	-	63	67	-
India	5 MLD	25	10	590	167	-	201	60	-	66	64	-
-	-	-	8	463	214	174	125	39	47	73	82	73
India	5 MLD	20-31	6	560	210	420	140	53	105	74-78	75-85	75-89
Brazil	106 L	21-25	4.7	265	150	123	133	59	33	50	61	73
-	110 L	12-18	18	465	-	154	163	-	42	65	-	73
Colombia	35m³	23-24	5.2	430-520	-	200-250	170	-	65	66	80	69
-	3.7m³	24-26	10-18	660	300	-	178	66	-	73	78	-
Brazil	106 L	20	4	424	195	188	170	61	59	60	69	69
Netherlands	120 L	8-20	12	500	-	-	225	-	-	60-90	-	65-90
Netherlands	6 m³	20	18	550	-	-	165	-	-	70	-	-

Table 1. Treatment Performances of Lab and Full Scale UASB Reactors Treating Sewage *(adopted from Khan et al., 2011a)*

The discussion for the selection of the sustainable technology for the policymakers, engineers, contractors, consultants and authorities of the public sanitation (PuSan sector) has been presented in discussion/ summary part of this chapter.

In addition, more sustainability and treatment performance of these treatment system can be improved if these systems/ combinations were categorized based on their application to remove the suspended solids with or without chemical coagulants, soluble organic and inorganic matter, and removal of reduced compounds such as ferrous ions, sulfides etc. and recovery of methane.

The foremost categories are:

i. Conventional settling systems and flotation methods with or without chemical coagulants for the removal of suspended solids and soluble organic and inorganic compounds like phosphate or termed as primary post treatment options;

ii. Application of physical chemical methods to remove and recover dissolved methane from the effluent, which is very important issue for the researchers, engineers and scientists;

iii. Biological micro-aerobic methods for the removal of highly reduced (malodours) compounds like sulfides and volatile organic S^- compounds, Fe^{2+} and colloidal matter;

iv. High rate aerobic systems for nitrification, when combined with denitrification step;

v. Polishing methods for high rate removals of pathogens and further polishing of the secondary treated effluent. The post treatment systems thus, categorized can either be used singly or sequentially.

3. Post treatment systems

3.1. Low rate natural settling systems

The highly stabilized suspended matter present in the UASB effluent can be removed by micro-aeration and settling process. Therefore, proper methods of removal of suspended solids are needed. Currently, natural settling processes are widely used at full scale STPs. The natural settling method is often slow and inefficient and sometimes enhanced by addition of chemical which could easily remove the colloidal and finely dissolved solids, which are separated by physical aeration. Further, the recovery of resources in terms of phosphates and treated effluent, if used for irrigation purposes makes it ideal as a sustainable option.

3.1.1. Overland Flow System (OFS)

Chernicharo et al. (2001) investigated extensively OFS operated in two phases in Brazil. This system is a classical example of a full scale natural system in use for UASB effluent post treatment and characterized by constant and transient hydraulic regime respectively. Three slopes (physically identical) for wastewater overland flow constituted the post-treatment system. A very common weed species named *Brachiaria humidicola* was used as vegetative cover on the slopes. This weed is known for its high rate of nutrient absorption and high resistance against flooding.

The good performance of OFS can be achieved at low flow rate application ranging from 0.4 - 0.5 $m^3/m.h$. The final effluent concentration of the combined system (UASB followed by OFS) showed average values of BOD from 48 to 62 mg/L; COD from 98 to 119 mg/L and SS from 17 to 57 mg/L. The combined system removed 2 to 3 log-units of FC thereby reducing the residual FC of effluent to around 8.4×10^4 to 2.4×10^5 MPN/100mL. In addition, a significant removal of helminth eggs was observed with an average effluent concentration of 0.2 Egg/L. However, the final effluent quality of the overland flow system was interfered by the transient flow regime and the high concentrations of solids and organic matter in the UASB reactor effluent. For these situations, the length of the slope was suggested to be kept above 35 meters.

3.1.2. Polishing Ponds (PP)

Cavalcanti et al. (2001) investigated the feasibility of a single flow-through PP for the post-treatment of effluent of UASB reactor in Brazil. The plug flow regime was maintained in pond in order to elevate the fecal coliform removal efficiency of the system. Two distinct HRT of 5d and 15d were maintained in the pond. At 5d HRT, the average BOD, COD and TSS values were reduced to 68, 188 and 68 mg/L, respectively. At HRT of 15d these concentrations lowered down to 24, 108 and 18 mg/L, respectively. Removal of pathogenic microbial indicators was also encouraging, with the complete removal of helminth eggs at 5d HRT. Moreover, at 15d HRT the effluent FC concentration was very close to 1000 MPN/100mL, with conformity to the WHO guideline for unrestricted irrigation.

Again in Brazil, von Sperling and Mascarenhas (2005) investigated the performance of four shallow (0.40 m depth) PP in series for the treatment of UASB effluent at a total HRT of 7.4 d (1.4–2.5 days in each pond). Based on the results, the final effluent average concentration of BOD and COD were 44 and 170 mg/L, respectively. The mean overall FC removal efficiency was remarkably high, 6.42 log units, or 99.99996%. The high FC removal together with total nitrogen concentration of 10 mg/L in the effluent were found compatible with the discharge standards for urban wastewaters from the European Community, 15 mg/L (70% removal). The ammonia nitrogen concentration in effluent from combined system was 7.3 mg/L (67% removal). However, phosphorus removal was only 28% (effluent total phosphorus concentration of 2.8 mg/L). Others studies on integrated anaerobic-aerobic systems carried out in Brazil showed that shallow ponds in series, even at short HRT, are able to produce effluents complying with the WHO guidelines for unrestricted irrigation in respect to coliforms concentration (lower than 1000 MPN/100 mL). All polishing pond systems were able to produce effluents without helminth eggs, what is in compliance with the WHO guidelines for unrestricted and restricted irrigation (≤1 egg/L, arithmetic mean).

Many UASB reactors combined with PP are located in India. Khan (2012) studied short HRT PP, 1d. The treatment performance was insignificant and merely running as settling tanks with a very limited algal activity. The BOD and TSS removal was generally found less than 50%. Due to very limited algal activity, coliform removal was also restricted to generally 1-2 log unit, however, helminth eggs were removed completely.

3.1.3. Constructed Wetland (CW)

The CW system for wastewater treatment is accepted as a technically and economically feasible alternative for small communities (Okurut et al., 1999). The systems used solid medium (sand, soil or gravel) to develop a natural processes under suitable environmental conditions that lead to the treatment of wastewaters. The plants are densely spaced and, together with the shallow water, provide habitats for animal, bird and insect communities. Vegetation in a wetland provides a substrate (roots, stems, and leaves) upon which microorganisms can grow as they break down organic materials. The most important functions of the plants are: (a) utilization of the nutrients and other constituents; (b) oxygen transfer to the solid medium; (c) support medium for bio-films on the roots and rhizomes (Sousa et al., 2001).

Sousa et al. (2001) investigated the demonstration scale wetland system for the treatment of effluent of UASB reactor for the removal of residual organic matter, suspended solids, nutrients (nitrogen and phosphorus) and fecal coliforms. The 1500 liter UASB reactor was operated at varied HRT (3h and 6h) while the effluent of the UASB reactor was treated in four units of CW, each 10 m long and 1.0 m wide, with coarse sand and operated in parallel under different hydraulic and organic loads. Macrophytes (*Juncus sp)* were planted in three CWs, whereas one CW was operated as a control unit without plants. The results revealed that the effluent COD from the four CW units had substantially constant concentration values, indicating that there was no influence of varied hydraulic load applied and presence of plant in CWs on its removal efficiency.

The phosphorous removal was very efficient during entire period of study. The phosphorous removal was mainly due to the utilization by plants and microorganisms as well as adsorption and precipitation. In the CW without plants, the removal was due to precipitation and adsorption as well as assimilation by the bio-film developed on sand grains. The results indicated that there was no adverse affect of varying hydraulic load or retention time on phosphorous removal efficiency.

The nitrogen removal in the four CW units was satisfactory under variable operation conditions. The total nitrogen removal efficiency varied from 59% to 87% in wetlands containing microphytes. The two basic factors for the removal of nitrogen in wetlands containing microphytes were observed to be assimilation by plants and microorganisms present in wetlands and; probably nitrification due to transport of oxygen from atmosphere by plants. The results indicated that the presence of microphytes enhance the nitrogen removal efficiency significantly. The highest removal efficiency occurred in the unit with lowest hydraulic load corresponding to HRT of 10 d. The removal efficiency of fecal coliforms was observed to be very high in wetlands with microphytes. The increase in hydraulic load reduced the removal efficiency.

3.1.4. Duckweed Pond (DP)

The aquatic macrophyte based treatment systems such as DP can be used to recover the nutrient and transformed them into easily harvested protein-rich by-products. The UASB effluents are highly rich in nutrient which should not be removed but, recovered. DP are covered by floating mat of macrophytes, which prevents light penetration into the pond resulting in shading. The high growth rates of the macrophyte permits regular harvesting of the biomass and hence nutrients are removed from the system. The produced biomass has economic value, because it can be applied as fodder for poultry and fish.

El-Shafai et al. (2007) evaluated the performance of a combined UASB and DP system (3 ponds in series). The UASB reactor had a volume of 40 liter and run at 6 h HRT while each pond had 1 m^2 surface area and 0.48 m deapth. The HRT in each pond was 5 d providing total HRT of 15 d in all ponds. The DP were inoculated with *L. gibba*, obtained from a local drain, containing 600 g fresh duckweed per m^2. The system removed 93% COD, 96% BOD and 91% TSS during warm season. Residual values of ammonia, total nitrogen and total phosphorus were 0.41 mg N/L, 4.4 mg N/L and 1.1 mg P/L, with removal efficiencies of 98%, 85% and 78%, respectively. The system achieved 99.998% FC removal during the warm season with final effluent containing 4×10^3 cfu/100 mL. During the winters, the system efficiently removal for COD, BOD and TSS was the same, but not nutrients and fecal coliforms. The coliform count in the effluent was 4.7×10^5 cfu/100 mL. The authors reported that the FC removal in DP was affected by the decline in temperature, nutrient availabilities and duckweed harvesting rate.

3.2. High rate physical chemical methods

3.2.1. Chemically Enhanced Primary Treatment (CEPT) & zeolite column (UASB post treatment)

Aiyuk et al. (2004) proposed an integrated Coagulation and Flocculation- UASB- Zeolite column concept for the low-cost treatment of domestic wastewater. In this integrated treatment system, domestic wastewater is initially treated with CEPT using $FeCl_3$ as a coagulant and polymer to remove suspended material and phosphorus, followed by UASB treatment to remove soluble organics. The effluent of UASB reactor was treated by regenerable zeolites to remove total ammonia nitrogen. The CEPT pre-treatment on average removed 73% COD, 85% SS and 80% PO_4^{3-}. The coagulation/flocculation step of this integrated system produced a concentrated sludge (8.4% solids), which can be stabilized in a conventional anaerobic sludge digester and used as fertilizer for agricultural purposes. After coagulation/ flocculation step, UASB reactor consequently received an wastewater with low total COD, approximately 140 mg/L and it was operated with volumetric loading rate of 0.4 g COD/L.day (HRT of 10 h) and 0.7 g COD/L.day (HRT of 5h). For these conditions, the system removed about 55% COD, thus producing an effluent with a low COD of approximately 50 mg/L (53±28 mg/L). The zeolite removed almost 100% NH_4^+. The integrated coagulation / flocculation–UASB-Zeolite system effectively decreased the TSS and COD upto 88% and more than 90%, respectively. The nitrogen and phosphorus were decreased by 99% and 94%, respectively. The column of zeolite proved most beneficial due to very high removal efficiency of ammonia and the oxidation of residual organic matter. Pathogenic indicators (FC) levels were reduced from 10^7 cfu/L to 10^5 cfu/L, indicating a removal of 99%. The final effluent from the system can be used for crop irrigation or be discharged in surface waters.

Percolation of the UASB effluent through the zeolite ion exchange column resulted in an improved effluent quality (average final effluent total COD of 45±6 mg/L). Still it is possible that the overall integrated system effluent characteristics do not meet desired standards. But, the system operates at low costs, making it suitable for developing countries and rural areas. The final effluent can be used at least for crop irrigation. The recycling/ reuse or disposal of the side streams generated should be explored further and evaluated in future research, together with the energy recovering potential of the CEPT sludge.

3.2.2. Dissolved Air Flotation (DAF)

Based on the results observed from the use of physico-chemical processes for sewage treatment DAF stood up to be an attractive alternative for the post treatment of UASB effluent. DAF system clarifies wastewater by removing floating suspended matter such as oil, fats or solids. The removal is achieved by dissolving air in wastewater under pressure and then releasing the air at atmospheric pressure in a flotation tank. The released air forms tiny bubbles which adhere to the suspended matter causing the suspended matter to float to the surface of the wastewater and form a froth layer where it may then be removed by a skimming device. The feed water to the DAF float tank is often (but not always) dosed with a coagulant (such as ferric chloride or aluminum sulfate) to flocculate the suspended matter. Penetra et. al. (1999) studied a lab scale DAF with previously coagulated effluent from a pilot scale

UASB reactor. Ferric chloride ($FeCl_3$) was used as coagulant and dosages ranged from 30 to 110 mg/L with pH in the range of 5.3 to 6.1, varified with addition of lime. Best results were achieved at a $FeCl_3$ dose of 65 mg/L. The DAF system was found efficient to reduce COD up to 91%, total phosphate up to 96% TSS up to 94%, turbidity up to 97% and sulfides more than 96%. The combined UASB-DAF system was observed to reduce 98% COD, 98% TP, 98.4% TSS and 94% Turbidity.

3.2.3. *Two Stage Flotation and UV disinfection (TSF-UV)*

The $FeCl_3$ coagulant and cationic polymer used in DAF systems presents a faily good removal efficiency of the UASB effluent, but these processes resulted in a significant volume of sludge. Tessele et al. (2005) investigted a pilot scale UASB ($50m^3/d$ flow) reactor followed by conventional two stage flotation and UV disinfection system for nutrient recovery. The proposed two stage flotation unit brings the advantage of separating the biomass and sludge that contain the phosphate and hydoxide. The suspended solids were removed by first stage flotation-flocculation (FF) process referred as F1 followed by second stage DAF referred as F2. Phosphate ions were removed by precipitation and coagulation. The removal mechanism in FF was the formation of small bubble and entrapped in flocs and these flocs floats over the water surface. In second flotation stage, both flocs and fine solids were aimed to removed. The concentration of Fe^{+3} and flocculant varied from 0 to 25 mg/L and 0 to 15 mg/L, respectively. The air flow in FF process was 3.0 L/minutes while DAF air flow rate 0.9 to 1.2 L/minute. The hydraulic loading rate was kept around 49 m/h at an HRT of 2 minutes in DAF, which is higher than in conventional DAF (6-10 m/h). After F2, the effluent was disinfected with low pressure UV lamp operated at a theoretical value of 25 mJ/cm^2. The results present that the combined UASB-TSF-UV process is more efficient than UASB-PP system. The final effluent contained low COD, phosphate ion concentration, turbidity and air/ water surface tension is as high as that of tap water while the ammonia removal was insignificant.

3.2.4. *Coagulation-flocculation*

Feasibility of coagulation and flocculation as a post treatment process for the effluent of UASB reactor treating domestic sewage were studied by Jaya Prakash et al. (2007). Commonly used coagulants (alum, polyaluminium chloride (PAC), ferric chloride, and ferric sulphate) were used in a series of jar tests to determine the optimum coagulant dose. The optimum chemical dosage was 20 mg/L (as Al) for alum, 24 mg/L (as Al) for PAC, 39.6 mg/L (as Fe) for $FeCl_3$ and 17.6 mg/L (as Fe) for $FeSO_4$. All the tested coagulants were found to be effective in reducing the effluent BOD and SS to less than 20 mg/L and 50 mg/L, respectively. However, coagulation–flocculation alone was not found sufficient to reduce the FC concentration to a permissible limit (1000 MPN/100 mL) for unrestricted irrigation. The final concentration of fecal coliform of UASB reactor effluent was 2300 MPN/100 mL using alum and PAC optimum doses. Moreover, the investigators suggested that disinfection by a chlorine dose of 1-2 mg/L with contact time of 30 minutes could reduce the FC concentration to below 1000MPN/100 mL after treating UASB effluent by coagulation-flocculation process. Further, higher doses of chlorine i.e. 3 mg/L removed all the FC from the sample after coagulation together with the above mentioned optimum alum and PAC doses. However, 4 mg/L of chlorine dose was needed after coagulation with iron coagulants to remove all the FC.

3.3. Micro-aerobic methods (Including removal/ or recovery of dissolved gases)

The UASB effluent contains reduced organic and inorganic species and dissolved methane gas which can be removed by micro-aeration. Micro-aeration implies aeration of treated effluent for about 30 min. The role of micro-aeration is to strip off and to oxidize the reduced species such as sulfides, ferrous ions etc. which exert immediate oxygen demand and remaining easily biodegradable organic pollutants and to remove the dissolved methane gas. Generally, these systems have very short HRT and the amount of excess sludge generated is negligible. The simple physical micro-aeration can be sufficiently remove or strip off the dissolved sulfides or methane from the UASB effluent. However the removal of suspended solids is insignificant from this process.

3.3.1. Down-flow Hanging Sponge (DHS)

DHS reactor was developed by Harada and his research group at Nagaoka University of Technology, Japan, for the aerobic post-treatment of the UASB effluent. In DHS, sponge cubes diagonally linked through nylon string have been used to provide a large surface area to accommodate microbial growth under non-submerged conditions. The wastewater trickled through the sponge cubes supplies nutrients to resident microorganisms. Oxygen is supplied through natural draught of air in the downstream without equipment. The system provides for dissolved methane gas to be recovered. Matsuura et al. (2010) investigated a two stage DHS system for the post treatment of UASB effluent in Nagaoka, Japan. Most of the dissolved methane (99%) was recovered by the two stage system, whereas about 76.8% of influent dissolved methane was recovered by the first stage operated at 2h HRT. The second DHS reactor was mainly used for oxidation of the residual methane and polishing of the remaining organic carbons. The removal of COD and BOD in the first stage was insignificant as there was no air supply; however, high removals were expected in the second stage due to sufficient supply of air, which is quickly oxidize the residual dissolve methane in the upper reactor portion before being emitted to the atmosphere as off-gas.

Agrawal et al. (1997) evaluated for the first time the performance of combined UASB reactor and DHS cube process. With post-denitrification and an external carbon source, 84% in average N (NO_3 + NO_2) was removed with an HRT of less than 1 hour, for temperature range of 13 to 30 ^{0}C. The effluent contained a negligible amount of SS and total COD was only in the range of 10 to 25 mg/L. The DHS reactor was capable of stabilizing total nitrogen through nitrification, which ranged from 73-78%. In another study Machdar et al. (2000a & b) observed that the combined UASB+DHS system successfully achieved 96–98% of BOD removal, 91–98% of COD removal, and 93–96% of TSS removal, at an overall HRT of 8 h (6 h for UASB and 2 h for DHS unit). The complete system neither requires external aeration input nor withdrawal of excess sludge. The final BOD effluent concentration was 6- 9 mg/L. Similarly, FC removal was 3.5 log with a final count of 10^3 to 10^4 MPN/100mL in the effluent. Nitrification and denitrification in DHS accounted for 72% removal of total nitrogen (effluent concentration of 11 mg N/L) and

60% removal of ammonium nitrogen (effluent NH_4-N of 9 mg N/L) over the total operational period.

3.3.2. *Trickling Filter (TF)*

The TF consists of a fixed bed of rocks, gravel, slag, polyurethane, foam, sphagnum peat moss, or plastic media over which sewage or other wastewater flows downward promoted a layer or film of microbial slime to grow. Aerobic conditions are maintained by splashing, diffusion, and either by forced air flowing through the bed or natural convection of air if the filter medium is porous. The process mechanism involves sorption of organic pollutants by the layer of microbial slime. Diffusion of the wastewater over the media furnishes dissolved oxygen which the slime layer requires for the biochemical oxidation of the organic compounds and releases CO_2 gas, water and other oxidized end products. Chernicharo and Nascimento (2001) studied the applicability of pilot level TF for polishing the effluent of UASB reactor. The volume of UASB reactor was 416 liter operated at an average HRT of 4h and the TF volume was 60 liters with blast furnace slag of 4 to 6 cm in size used as media. The operational conditions in the UASB reactor was kept constant throughout the study period while the TF was operated at three different phases, low, intermediate and high rate. The performance of UASB reactor was consistent, with removals above 70% in terms of BOD and COD. The final effluent quality was produced when the TF was operated as low and/or intermediate rate. Under these operational conditions the average COD, BOD and SS concentrations were 90, 30 and 30 mg/L, respectively and; hence, complying with the discharging standards. The system proved very efficient under low loading conditions. At high rate conditions the system was not efficient to remove the BOD, COD and SS. The results of this study showed that the TF can be used as the post treatment option for the treatment of UASB effluent for low organic and hydraulic rates in tropical countries.

3.3.3. *Micro aeration methods i.e. flash aeration*

For the last decade progress has been made on the use of high rate micro-aerobic methods for the removal or recovery of dissolved sulfides contained in anaerobic effluents. Besides, sulfide purging into the atmosphere, micro-aeration can also be utilized for biological oxidation of sulfides into elemental sulfur, which offers an excellent potential for reuse and it has been shown to be a cost effective alternative (Vallero et al., 2003; Chuang et al., 2005; Chen et al., 2010; Khan et al., 2011a and 2011c). The process is generally focused on the treatment of biogas, off-gas, natural gas or low strength wastewaters, like in the case of anaerobic effluents. In addition, micro-aeration of anaerobic system may be an option for increase hydrogen sulfide stripping and methane production (van der Zee et al., 2007). Buisman et al. (1990) developed a low-cost, high-rate biotechnological aerobic process for the oxidation of sulfide into elemental sulfur by a group of colorless sulfur bacteria, where the sulfide oxidation rate was dependent on the oxygen level. The biofilm on a reticulated polyurethane was more suitable to produce sulfate than a free cell suspension of biomass, for the same given oxygen and sulfide concentrations. For efficient achievement of elemental sulfur, high sulfide loads or low oxygen concentrations must be applied (Stefess et al., 1996). Vallero et al. (2003) utilized the micro-aerated reactors for the oxidation of sulfides to elemental sulphur from the liquid phase of anaerobically treated sewage. The results were encouraging and partial conversion of soluble sulfides (HS^-) into colloidal elemental sulphur was observed.

The produced element sulfur forms transparent globules of up to 1 micro-meter in diameter, which is deposited inside or outside the bacteria. An important issue is the recovery of the colloidal sulfur particles. Janssen et al. (1999) studied the properties of the colloidal sulfur particles and developed an up-side down cone expanded-bed bioreactor for spatially separation of the aeration and oxidation phases. After 50 days of operation 90% of the sludge settled at a velocity greater than 25 m/h and could be easily removed. Although the results are very encouraging, more studies on the application of high micro-aerobic systems for colloidal matter removal are necessary. One of the most promising technologies for sulfide removal from biogases is a two-step process where gaseous sulfide is dissolved into the liquid in the first step, followed by sulfide oxidation to elemental sulfur. Although little research has been conducted on the subject Chuang et al. (2005) treated a sulfate-rich wastewater in a UASB followed by a floated bed micro-aerated reactor. The floated bed was operated at short HRT (2.8 hours) and during long-term steady state operation results showed that almost all sulfides (>96%) was oxidized to elemental sulfur and sulfate. Annachhatre and Suktrakoolvait (2001) observed a sulfide conversion higher than 90% at sulfide loading rates of 0.13-1.6 kg $S/m^3/d$ and at DOs lower than 0.1 mg/L sulfur was the major end product.

The simplest method of sulfide oxidation is the introduction of micro-aerobic conditions in the anaerobic reactor. Despite the toxicity exerted by oxygen against obligatory anaerobes, its moderate introduction is not expected to have a harmful impact to the biomass, mainly to the limited penetration depth of oxygen in biofilm. Van der Zee et al. (2007) determined the air injected to sulfide ratio to be 8-10:1 (O_2: S in mol units), which was sufficient to reduce the biogas H_2S content to undetectable levels. Element sulfur and sulfate were the main products.

3.3.4. Continuous Diffused Aeration (CDA)

CDA system was investigated to treat the effluent of UASB reactor in India by several authors (Walia, 2007; Khan et al., 2011b; Khan, 2012). The treatment of sewage in a 60 L pilot scale UASB reactor followed by a CDA system and a full scale plant (111MLD capacity; UASB +Aeration+FPU) were investigated by Khan et al. (2011a). The HRT of CDA system was maintained at 15, 30 and 60 min HRT. During aeration at each HRT bulk liquid DO of 5-6 (high) and 1-2 (low) mg/L were maintained. The final COD, BOD and TSS effluent concentrations were 40-60, 25-35, 30- 40 mg/L, respectively, for operating under high DO (5-6 mg/L) and 30 minutes HRT and 30- 50, 18-30, 20-30 mg/L, respectively, at 60 minutes HRT. The combined reduction efficiency of the integrated UASB-CDA system at HRT of 30 and 60 min ranged from 80 to 85% COD, 85 to 90% BOD, 65-75% TSS. A conceptual model was developed wherein it demonstrated that the aerobic nature of the effluent depends on dissolved oxygen (DO), ORP and BOD. Anaerobic UASB effluent becomes aerobic if its BOD is reduced to less than 30 mg/L and minimum values of DO and ORP are observed, 4-5 mg/L and 120-135mV, respectively. Based on experimental results empirical correlations between BOD, ORP and DO have been developed and the results indicated a 50% reduction in BOD of the UASB effluent at HRT of ~100 min. The removal of NH_4-N and total-P was insignificant at any of the maintained HRT. The Integrated UASB-CDA for sewage treatment could be recognized as a sustainable and cost effective option as the combined HRT of the system is still short (8 h for UASB + 0.25-1.0h for aeration, with a total HRT of 8.25-9.0 h). Existing UASB based STPs can be upgraded by installing continuous aeration system through fine pore diffuser and the energy produced by UASB reactor in terms of biogas could be used to operate the aeration system.

3.4. High rate aerobic methods (Including nitrification-denitrification steps)

The poorly biodegradable soluble matter, hazardous compounds or micro pollutants including ammonia-nitrogen and phosphorous present in the UASB effluent sometimes are difficult to be remove by micro-aerobic or simple settling. Therefore, secondary post treatment is required, following the micro-aerobic or settling treatment methods. A number of secondary post treatment processes have been categorized into methods responsible for the removal of (i) poorly biodegradable soluble matter including micro pollutant and hazardous material, (ii) finely dispersed organic matter i.e. colloidal and pathogens removal and (iii) ammonia-N and phosphorous. The removal of residual biodegradable carbon, ammonia nitrogen and phosphorous can also be achieved if the effluent of UASB is treated by high rate aerobic biological treatment methods.

3.4.1. Sequential Batch Reactor (SBR)

The SBR is a fill and draw type modified activated sludge process, where four basic steps of fill, aeration, settle and decant take place sequentially in a single batch reactor. The operation of SBR can be adjusted to obtain aerobic, anoxic and anaerobic phases inside the standard cycles (Droste and Masse, 1995; Surampalli et al., 1997). Sousa and Foresti (1996) proposed a combined system composed of anaerobic-aerobic processes consisting a UASB reactor followed by a SBR. The system performance was evaluated through a bench scale set-up comprising of a 4 litre volume UASB reactor followed by two SBRs of 3.6 litres each. The UASB reactor was fed with partially mixed synthetic substrate in sewage while the SBR received effluent of UASB reactor. The HRT of 4h in UASB was maintained constant throughout the study while the 4 h cycles in the following sequence of fill (0.10h), reaction (1.9h), sedimentation (1.6h), discharge (0.25h); idle (0.15h) were maintained in SBR. The combined system removed ~85% total nitrogen through nitrification. The COD removal in UASB reactor was around 86% while in SBR around 65% of the remaining, thus, combined systems removed 95% (residual effluent COD of 20 mg/L). The performance of combined system was 96% in terms of TSS removal (residual effluent TSS of 9 mg/L) and 98% in terms of BOD removal (residual effluent BOD of 6 mg/L).

Torres and Foresti (2001) studied the effect of aeration on the performance of SBR treating UASB effluent. The UASB reactor was operated at a constant HRT of 6 h while the SBR performance was monitored at four different duration cycles (24, 12, 6 and 4 h) corresponding to aeration times (AT) of 22, 10, 4 and 2 h, respectively. The overall removal efficiencies of COD and TSS were 91% and 84%, respectively and observed independent of aeration time given in the SBR. However, the nutrients removal was found to be dependent on aeration time. Total nitrogen removal of approximately 90% was achieved for AT longer than 10 h; complete nitrification occurred for AT longer than 4 h; significant phosphate removal (72%) occurred only at the AT of 2 h. Moawad et al. (2009) also investigated the performance of the combined UASB-SBR system under different operating conditions for the treatment of domestic wastewater. The retention time in the UASB was changed from 4 h to 3 h and the aeration time in the SBR cycle varied from 2 to 5h, and then to 9 h. The observed average percentage removal for the three runs for COD, BOD and TSS was 94%, 97% and 98%, respectively. The residual COD, BOD, and TSS were 26, 5.8 and 5.0 mg/L, respectively. Complete nitrification of ammonia was achieved after 5 h aeration in the SBR. The average percentage removal of phosphorus reached up to 65%. Increasing the HRT in the SBR from 2 to 9 h caused a significant improvement in FC removal as the geometric count of FC was reduced to 7.5×10^2 MPN/100mL in the effluent of the 3rd run (HRT 9 h).

Khan et al. (2011a) investigated the performance of a pilot scale integrated UASB-SBR system for treatment of sewage. Two different variant of SBR Process were investigated: a Continuous Flow-Intermittent Decant Sequencing Batch Reactor (CFID) and Intermittent Fill- Intermittent Decant Sequencing Batch Rector (IFID) for about 18 months in conjunction with UASB reactor at ambient environment. Initially, the UASB-CFID system was operated at an HRT of 8h in the UASB reactor while it varied in CFID (20, 8 and 6 h),which also had different DO regimes, 4.0 to 5.0 and < 0.5 mg/L, 2.5-3.5 and < 0.5 mg/L and 2.5 to 3.5 and <0.5 mg/L, for the respectively HRT. The BOD and TSS removal efficiency of combined UASB-CFID system was up to 90%. The FC reduction was more than 99%. It was observed that average reactor MLVSS concentration reduced to around 30% at DO of 2.5-3.5 mg/L showing high degree of mineralization. Later, an integrated UASB followed by IFID system for the treatment of sewage was evaluated for the removal of organics and nutrient for more than six months at ambient conditions. The HRT in UASB reactor was maintained constant at 8 h. The IFID was operated at 6h HRT at DO concentration ranged between 2.5 to 3.5 mg/L. Results revealed that the removal of BOD, COD and TSS were 90, 95 and 90%, respectively in IFID. During higher organic loading conditions and low SRT, the removal of phosphorous was significantly higher than that of lower organic loadings and higher SRT. The suitable COD: P ratio of 105~160 helped for the effectively removal of phosphate. The total nitrogen removal was sufficiently good ranged from 80 to 95%.

3.4.2. Activated Sludge Process (ASP)

Activated sludge process is the most widely used process for the treatment of sewage and industrial wastewaters. Atmospheric air is bubbled through wastewater combined with organisms to develop biological flocs which reduce the organic content of the sewage. The combination of wastewater and biological mass is commonly known as Mixed Liquor. von Sperling et al. (2001) monitored a pilot-scale plant comprising of an UASB reactor followed by an activated sludge system treating actual municipal wastewater from a large city in Brazil. The UASB reactor removed 69-84% COD, while ASP only removed remaining COD ranging from 43% to 56%, resulting in 85% to 93% removal achieved through the overall system (residual effluent COD of 50 mg/L avg.). The final effluent SS concentration was 13 - 18 mg/L. Therefore, UASB and ASP configuration was suggested to be a better alternative for warm-climate countries than the conventional activated sludge system alone, considering the total low hydraulic detention time of 7.9h (4.0 h UASB; 2.8 h aerobic reactor; 1.1 h final clarifier), offering the advantages in terms of savings in energy consumption, absence of primary sludge and possibility of thickening and digesting the aerobic excess sludge in the UASB reactor itself.

3.4.3. *Rotating Biological Contactors (RBC)*

A RBC consists of a series of closely spaced circular disks of plastic material such as polystyrene mounted on a shaft that are partially submerged (typically 40%) in wastewater. The microorganisms grow on the surface of circular disks which breakdown and stabilize organic pollutants in presence of oxygen obtained from the atmosphere as the disks rotate. The development of excessive biofilm growth and sloughing problems besides odor and poor performance occurs when oxygen demand has exceeded the oxygen transfer capabilities and is the major drawback of this technology. These rotating biological contactors offer many advantages like the capability of handling a wide range of flows, low power requirements, low sludge production and excellent process control.

Tawfik et al. (2003) examined the removal of organic matter, nitrification and *E. coli* by UASB-RBC system at different operational temperature (11, 20 and 30°C) and at different organic loading rates with constant HRT of 2.5 h in the RBC. The results showed good performance of the system when operated at lower OLRs of 27, 20 and 14.5 g COD/m^2/day at 11, 20 and 30°C, respectively. The residual COD values were 100, 85 and 72 mg/L for the respectively temperatures. Moreover, a high ammonia removal and low residual values of *E. coli* were found for the RBC at operational temperature of 30°C as compared to the situation for treatment of domestic wastewater and UASB effluent at lower temperatures of 11°C and 20°C. The effluent however, did not comply with WHO guidelines for unrestricted irrigation.

Tawfik et al. (2005) investigated the performance of a combined single stage RBC, two stage RBC and an anoxic up-flow submerged bio-filter followed by a segmental two stage aerobic RBC system. This study was carried out in order to assess the impact of biodegradable COD in an UASB effluent applied to the systems on the removal efficiency of different COD fractions, *E. coli,* ammonia and partial nitrate removal. The two (single stage) RBCs were operated at a constant HRT of 2.5 h and temperature of 21 °C but at different OLRs, 10 and 14 g biodegradable COD/m^2/day due to varied UASB effluent qualities. The results clearly show that the residual values of COD, ammonia and *E. coli* in the final effluent are significantly lower at the lower OLR of 10 g biodegradable COD/m^2/day. In view of the results it is recommend to use a single stage RBC system at OLR of 10 g biodegradable COD/m^2/day and at HRT of 2.5 h for post-treatment of the effluent of UASB reactor operated at high temperature of 30 °C, as it generally prevails in tropical countries.

The performance of a single stage versus two stage RBC system for post-treatment of the effluent of an UASB reactor operated at a low temperature of 12 °C was also evaluated. Both systems were operated at the same OLR of 18 g biodegradable COD/m^2/day and at HRT of 2.5 h. The results demonstrated that the COD fractions, ammonia and *E. coli* content in the final effluent of a two stage RBC system were significantly lower than the effluent of the single stage RBC system. Accordingly, results envisaged a two stage RBC system at an HRT of 2.5 h and

OLR of 18 g biodegradable COD/m^2/day for post-treatment of the effluent of a conventional UASB reactor operating at a low temperature of 12 ^{0}C.

The nitrogen removal from the nitrified effluent was investigated using a biofilm system consisting of three stages, namely an anoxic up-flow submerged bio-filter followed by a segmental two stage aerobic RBC. The nitrified effluent of the second stage RBC was recycled to the anoxic up-flow submerged bio-filter reactor. The results obtained reveal that the introduction of an anoxic reactor as a first stage combined with recirculation of the nitrified effluent of the second stage RBC was accompanied with a conversion of nitrate into ammonia, at least in case the content of biodegradable COD in the UASB effluent was low. In such a situation the ammonia needs to be nitrified two times, which obviously should be avoided. Therefore in such situations of a too high quality anaerobic effluent in terms of biodegradable COD content, the introduction of a separate anoxic reactor for denitrification as final post-treatment step cannot be recommended.

3.4.4. Aerated Fixed Bed Reactor (AFBR)

A sequence of denitrification reactor (DN), UASB, AFB and settling units treating sewage was evaluated by Sumino et al. (2007). The DN and AFB reactors contained sponge sheets media fixed to both the surfaces of the boards oriented vertically. The air was supplied to the AFB reactor from the bottom of the tank. Granular sludge obtained from food waste treatment plant was used as the inoculum sludge in the UASB reactor and activated sludge from a sewage treatment plant was used as the inoculum sludge in the AFB reactor. The SS recirculation from settling tank was made to the denitrification tank and the poly aluminium chloride PAC was injected to ABF for phosphorous removal. The whole system was studied for more than 300 days under constant HRT of 24 h in three different seasons, summer, autumn and winter. The performance of the combined system was satisfactory with final mean effluent values of soluble COD of 54, 66 and 65 mg/L in the summer, autumn and winter, respectively, while the mean total soluble BOD were 11, 18 and 25 mg/L for the corresponding periods. The information on nitrogen and phosphorous removal and indicators of pathogens was not discussed in this study.

3.4.5. Submerged Aerated Bio-Filter (SABF)

The SABF system is composed of floating porous media through which wastewater and air flows from the bottom of the reactor. The airflow in the SABF system is always in upflow mode, while the liquid flow can be in upflow or downflow mode. These biofilters backwashed routinely at least once in 3 days. The development of thin, homogeneous and active biofilm layer is the main mechanism of biofilters to remove the soluble organic compound and suspended solids from the wastewater. Besides serving as support medium for microorganisms, the granular material also works as an effective filter (von Sperling and Chernicharo, 2005). Gonclaves et al. (1998) investigated an UASB reactor (46 L) followed by a SABF (6.3 L) for domestic sewage treatment. The floating and totally submerged granular medium in the SABF was made of S5 type polystyrene spheres with 3 mm diameter, 1200 m^2/m^3 specific surface area, 0.04 density and 0.50 m height. The air was injected in the SABF bottom, wastewater co-current through an air compressor.

In the study, the UASB reactor was initially operated at 8h hydraulic retention time and subsequently reduced to 6h and 4h. The 4h HRT in UASB reactor was maintained to investigate the performance of reactor under breakdown situation. Several authors recommended that the HRT in the UASB to be shorter than 5h in order to keep an adequate mechanization activity in UASB reator (Vieira and Garcia Jr., 1992; van Haandel and Lettinga, 1994). However, the performance of the UASB reactor was stable and similar at all HRTs studied. The final mean removal efficiency of the combined system in terms of SS, BOD and COD were 94%, 96% and 91% respectively, which amounts to the final effluent concentration of 10 mg/L, 49 mg/L and 10 mg/L respectively.

Goncalves et al. (1999) studied the combined UASB-SABF system and observed similar results. The experiments were conducted with UASB reactor operated at HRT of 6 h without sludge recirculation and the bio-filter at HRT of 0.5 h. The average removal efficiencies of SS, BOD and COD were 95%, 95% and 88%, respectively, with final effluent quality of 10, 10 and 50 mg/L, respectively. Although the efficiency of the UASB-SABF system was satisfactory in terms of organic matter removal, the removal of the pathogenic microorganisms was very low.

Keller et al. (2004) investigated the combined UASB-SABF system followed by conventional and UV system to enhance the efficiency of the system to remove the pathogenic microorganisms. The results revealed that the 84% of COD (residual effluent COD of 78 mg/L), 86 % of BOD (residual effluent BOD of 26 mg/L) and 86% of TSS (residual effluent TSS of 23 mg/L) removal was achieved. The concentration of *E.coli, salmonellae* and *colliphases* were reduced to very low in the final effluent of the system. The association of UASB-SABF confirms the viability of the system with excellent final effluent quality of the system.

3.4.6. Moving Bed Bio-film Reactor (MBBR)

Tawfik et al. (2010) investigated a laboratory-scale integrated UASB reactor followed by a MBBR for sewage treatment at three different combined HRTs, 13.3 (8+5.3), 10 (6+4) and 5.0 h (3+2) under temperature range of 22–35 °C for a period of 290d in Egypt. The working volumes of UASB reactor and MBBR were 10 and 8.0 L respectively. A cylindrical carrier media of 1.85 cm diameter and 1.8 cm long made of polyethylene was used in MBBR. Its specific gravity and effective specific surface area were 0.95 and 363 m^2/m^3 respectively. The dissolved oxygen was maintained at 2.0 mg/L throughout the experiment. The performance of the integrated UASB-MBBR system was monitored in terms of COD fractions and FC. At the HRT of 5-10 h an overall reduction of 80–86% for total COD; 51–73% for colloidal COD and 20–55% for soluble COD was achieved. The removal efficiencies were increased up to 92, 89 and 80%, for total, colloidal and soluble COD respectively by increasing the HRT to 13.3 h. However, the removal efficiency of suspended COD in the combined system remained unaffected when increasing the total HRT from 5 to 10 h and from 10 to 13.3 h. This indicated that the removal of suspended COD was independent of the HRT. Final effluent total COD at three different HRTs were 54, 95 and 142 mg/L respectively. The final average FC counts were 8.9×10^4, 4.9×10^5 and 9.4×10^5 MPN/100 mL, corresponding to overall log reduction of 2.3, 1.4 and 0.7 respectively. The main mechanisms observed for the removal of FC were adsorption into the media and predation by higher microbes such as protozoa and metazoa.

The removal of ammonia nitrogen was also investigated in MBBR. The results revealed that the removal of ammonia nitrogen greatly depends on organic loading rate. About 62% of ammonia nitrogen was removed at OLR of 4.6g COD/m^2/day but the removal efficiency decreased by 34 and 43% at the higher OLRs of 7.4 and 17.8g COD/m^2/day, respectively. Nitrogen was mainly reduced by assimilation into biomass and denitrification in anoxic zone in the biofilm. The sludge produced by MBBR showed poor settleability, however, the combined system still produced less sludge compared to conventional ASP. The authors reported that the integrated UASB-MBBR system at an HRT of 8 and 5.3 h are technically feasible for sewage treatment.

3.5. Final polishing techniques

To achieve nearly complete removal of pathogens, color and hazardous compounds the UASB effluent needs to be polished after the micro aeration first step or secondary post treatment such as high rate aerobic treatment before reusing for intended purpose or discharging it into receiving water bodies.

3.5.1. Membrane technology

Recently large number of membrane technologies was investigated for secondary and tertiary treatment of sewage. Therefore, in order to achieve the quality of treated effluent up to reuse standard from UASB reactor, YingYu et al. (2009) evaluated the pilot scale cross flow membrane filtration system for polishing the UASB effluent treating low strength sewage in Singapore. A pilot scale UASB reactor (34 litres) was coupled with a side stream membrane module having a centrifugal pump to feed the effluent of UASB reactor into the membrane filtration unit. The HRT of UASB reactor was reduced from initial 10h to 5.5h after 119 days of operation and kept constant throughout the study period. The precise and constant holding tank was used prior to membrane filtration module unit in order to feed constant permeate flow rate. Results clearly showed high performance of UASB reactor for total solids removal at HRT of 10h which, however, significantly were reduced from 91.1 to 83.6% at HRT of 5.5h. At steady state conditions in the UASB reactor, the average TOC removal efficiency was 65% (10 h HRT), which increased to 81% by treating the effluent of UASB reactor through membrane filtration. But, the performance of this system in terms of TOC removal further increased to 73 and 85%, respectively at the HRT of 5.5h. This might be due to the increased up-flow velocity which provides better contact and distribution of wastewater with membrane. But fouling of membrane limits its use for the stated purpose. Therefore, extensive studies were required regarding it controlling factors such as membrane tube diameter and cross flow velocity etc.

YingYu et al. (2010) also proposed membrane filtration for the post-treatment of the effluent of UASB reactor in Singapore. The system comprised of UASB reactor and membrane filtration. The UASB reactor with working volume of 30 liter divided into two parts i.e. a sludge zone and a membrane zone. A gas/liquid separator was installed at the top of the sludge zone to separate the biogas from the liquid suspension. Two flat-sheet membrane modules (0.22 μm, PVDF, 0.1 m^2) were directly submerged into the upper membrane zone of the reactor above gas/liquid separator. The modules of flat sheet membrane were submerged into the UASB reactor to as a barrier to retain the suspended solids present in the effluent of UASB reactor at intermittent permeation and air sparging operating conditions. The whole system was operated at a constant HRT of 12 h at a temperature of 35 °C and no sludge was removed from the reactor, except for sampling. The experimental study was conducted in two phases with varied flux. In phase I, Intermittent permeation was studied at three different flux of 15, 20 and 25 L/m^2/h with varied suction pressure while in phase II, air sparging was investigated at four different air flow rates of 0, 1, 2 and 4 L/h with constant flux of 25 L/m^2/h.

The average supernatant TOC was 10.88 mg/L with fairly stable TOC removal efficiencies of over 89% during the whole operation. Finally this study influence that intermittent permeation was more effective for membrane fouling control compared with air sparging.

The coupling of membrane filtration with UASB represented as an efficient treatment technology for raw municipal wastewater at the ambient temperature. But limited studied are available on this system therefore, detailed investigations on demonstration scale.

3.5.2. Slow Sand Filtration (SSF) system

Various researchers investigated effect of hydraulic loading and sand size on the effectiveness of SSF for tertiary treatment of sewage at laboratory and pilot scale level and found that the SSF was capable of removing BOD, SS, turbidity and total coliforms up to 86%, 68%, 88% and over 99%, respectively (Ellis, 1987; Suhail, 1987; Sawaf, 1986, Adham, 1989; Gersberg, et al., 1989). However, limited data is available on the applicability of SSF on UASB effluent. Recently, Tyagi et al. (2009) studied the applicability of slow sand filter at lab scale as a post treatment option for the treatment of effluent of UASB reactor. The sand filter column operated at hydraulic loading rate of 0.14 m/h was found to be most effective in removing turbidity (91.6%), TSS (89.1%), COD (77 %), BOD (85%), TC (99.95%) and FC (99.99%). The average values of COD, BOD and SS in SSF effluent were 27 mg/L, 12 mg/L and 20 mg/L, respectively. The FC concentration was found below the standards set by WHO 1989 (1000 MPN/100 mL). It was concluded that slow sand filters can be effectively runs upto 7 days at a hydraulic load of 0.14 m/h as compared to the common hydraulic load of 0.19 m/h and 0.26 m/h. Hence, slow sand filtration could also be an effective technology for the post treatment of UASB reactor effluent, where treated effluent can be reuse safely for irrigation and other non-potable reuse purposes. However, the major drawback of SSF system was the frequent cleaning and maintenance requirement.

4. Discussion

The installation of post-treatment system to treat the effluent of UASB reactor treating sewage is a challenging task as to find a proper, reliable and efficient system, that is easy in operation and maintenance; technically feasible, and economically viable (Chernicharo, 2006). Amongst all post treatment systems, four natural wastewater treatment systems were extensively investigated as the post treatment units. The effluent quality of the polishing ponds in series satisfies the effluent pathogen disposal standards, but it has few disadvantages such as large land requirement, poor nutrient removal, odor related problems and occasionally high BOD and TSS concentrations in the effluent. The combination of polishing pond and duckweed pond, duckweed and algae pond system was reported to be very efficient but, large area requirement, low pathogens removal and high TSS concentration in the effluent were the main drawbacks of this system. The combination of polishing pond and coarse rock filter system give an effluent with high FC and occasionally high in BOD. In overland flow system for the treatment of effluent of UASB reactor under low organic loading rate, the performance was observed to be satisfactory, with low solids and organic matter concentration in the final effluent. However, helminthes eggs removal was insignificant.

The duckweed pond and constructed wetland system are also observed to be satisfactory in their respective performances but these systems are dependent on the temperature, hydraulic load, harvesting of plants, etc. Despite their good nutrient removal efficiencies these systems thought to be unable to bring down the effluent quality below discharging standards.

Four high rate physico-chemical processes were presented including CEPT- Zeolite Column system, DAF, TSF-UV and chemical coagulation-flocculation. These processes are capable to reduce organic pollutants and turbidity of UASB reactor effluent up to the level required to meet the reuse standards, but not the fecal coliforms. The other major drawbacks of these processes are high dose and cost of chemicals used, and large sludge volume generation. Further, these systems have only been evaluated on lab-scale models and no scaling up has been investigated so far.

The post-treatment of anaerobic effluents can be carried out by micro-aerobic processes such as flash aeration, trickling filters and DHS, where sulfides are oxidize back to sulfate, specially at low sulfide concentrations. The partial sulfides oxidation to elemental sulfur was observed from the application of these technologies for the anaerobic effluents containing low sulfides. However, the aeration has not been optimized.

Two broad categories of biological wastewater treatment systems were categorised under a high rate aerobic systems and extensive discussed, suspended and attached growth systems. Almost all suspended growth processes were found to be very promising. The SBR was found as one of the most suitable technology for the treatment of UASB effluent due to its high effluent quality with effluent BOD and SS concentrations lower than 10 mg/L. The nutrient removal was also efficient; besides the low energy consumption for aeration and low excess sludge production are other major advantages as compared to other aerobic suspended growth system. In the activated sludge process the final effluent quality follows the discharging standards but, the system requires relatively high energy and land area and, with no nutrient removal capabilities. The continuous aeration system for the treatment of UASB effluent would be able to reduce the BOD of UASB effluent to 50%, but rarely satisfies the effluent discharge standards.

Few attached growth biological treatment processes were also summarized. Among them DHS was reported as a promising technology which reduces the BOD and coliforms well below the effluent discharging standards. However, this process requires high initial investment (sponge cost), it clogs often and no nitrogen and phosphorous removal are observed. Another important attached growth process, RBC was extensively investigated at pilot scale level. The RBC was studied under different combinations, such as one, two, three stage RBC and combination of one, two stage RBC and anoxic biofilter followed by two stage RBC. The best performance was achieved by the post treatment of UASB effluent by a combined one stage RBC, two stage RBC and anoxic biofilter followed by two stage RBC system. The RBC is not very commonly used due to its wear and tear of mechanical moving parts. Additional pre-anoxic unit is required for nitrogen removal. Similarly submerged aerated biofilter systems were evaluated for the post treatment of UASB effluent resulting in high BOD and SS removal but, with no nutrient removal capabilities. Another attached growth process, trickling filter was also evaluated for the UASB reactor effluent. This system was able to maintain the effluent BOD, COD and TSS concentration in the permissible range, however, only under low loading conditions.

The most common physical process, slow sand filtration and membrane filtration as a post treatment unit were also discussed. The systems are able to reduce the physical, chemical and microbiological pollutants not only to the desired standards but, also to satisfy wastewater reuse criteria. However, there are few drawbacks, such as frequent clogging of the filter and membranes.

The performance and effluent concentration of different parameters of various combinations are summarized in Table 2.

Among all discussed post treatment systems few of the alternatives produce final effluent with low COD, BOD and SS concentrations. Between all aerobic post treatment systems presented the SBR was found to be the most compact method and it allows for the removal of nutrient along with residual COD. Scantly information is available in literature on coupling of the SBR with UASB. The major advantage of SBR over other aerobic systems is the system flexibility for BOD and nutrient removal.

Low cost sewage treatment technologies are generally preferred for developing countries. Therefore, it is most important to evaluate the treatment sequence keeping in view of total investment including capital cost, operation and maintenance cost and land requirement. A comparison has been made among UASB reactors and its few post treatment systems with conventional ASP system based on energy requirement and generation from UASB reactor i.e. energy audit of UASB reactor per MLD:

The basis of energy audit of a MLD UASB:

- Negligible energy requirement ~6 kW-h/MLD (only for initial pumping) (Tassou, 1988).
- Energy production in the form of Biogas (60-70% methane) - 50 m^3 biogas/MLD sewage treated (Arceivala, 1998).

Integrated systems	Effluent Concentration*						
	BOD (mg/L)	COD (mg/L)	TSS (mg/L)	NH_4-N (mg/L)	TN (mg/L)	TP (mg/L)	FC (MPN/ 100mL)
CEPT+UASB+Zeolite	32 (85)	45 (91)	24 88)	0.3 (99)	0.5 (99)	0.5 (94)	1.0×10^5 (99)
UASB+DAF	-	17 (98)	4 (98.4)	-	-	0.6 (98)	-
UASB+ Coagulation-flocculation	>20 (91)	>50 (87)	>30 (82)	-	-	-	4.3×10^3 (99.9)
UASB+SSF	12 (92.6)	27 (91)	20 (91)	-	-	-	1.0×10^3 (99.995)
UASB+ Polishing Ponds	24 (92)	108 (79)	18 (96)	20 (50)	25 (55)	-	5.8×10^2 (99.999)
UASB+Constructed Wetlands	-	52 (82)	174 (65)	14 (70)	17.5 (70)	0.74 (89)	1.0×10^3 (99.99)
UASB+ Duckweed ponds	14 (96)	49 (93)	32 (91)	0.41 (98)	4.4 (85)	1.1 (78)	4.0×10^3 (99.998)
UASB+DHS	2 (99)	40 (94)	0 (100)	6 (80)	6 (89)	-	-
	9 (96)	46 (91)	17 (93)	18 (28)	28 (40)	-	3.4×10^4 (99.95)
UASB+SBR	5.8 (97)	26 (94)	5.0 (98)	0 (100)	12.6 (77)	1.2 (65)	7.5×10^2
UASB+ RBC	-	43	-	2.2 (92)	-	-	9.8×10^2 (99.9)
UASB+ Aerated fixed bed reactor	11 (93)	54 (83)	10 (94)	-	30 (21)	3 (40)	-
UASB+ Submerged aerated bio-filter	9.4 (96)	37.8 (92)	9.8 (94)	-	27 (36)	-	
	26 (86)	78 (84)	23 (86)	-	-	-	4.1×10^5 (99)
UASB+ Trickling Filter	17-57 (80-94)	60-120 (74-88)	<30 (73-89)	-	-	-	-
UASB+ Anaerobic Filters	<40 (85-95)	60-90 (85-95)	<25 (77-94)	-	-	-	-
UASB+ Overland Flow System	48-62 (53-83)	98-119 (77-83)	17-57	14-18	-	-	8.4×10^4- 2.4×10^5 (99-99.9)
UASB+ ASP	-	50 (85-93)	13-18 (82)	-	-	-	-
UASB+ Flash Aeration System	22 (89)	57 (86)	47 (83)	-	-	-	5.0×10^3 (99)

*% removal efficiency in parentheses.

Table 2. Treatment Performance of various Integrated UASB Post treatment systems Treating Sewage *(adopted from Khan et al. 2011a)*

- The electricity produced from 1.0 m^3 of methane gas generated by UASB is 36,846 kJ at standard condition and approx.7.0 kW-h under field conditions, since 3600kJ is approximately 1 kW-h (Arceivala, 1998; Metcalf and Eddy, 2003).

- Energy saving through reduced diesel consumption by more than 70% by feeding methane gas into the Dual-Fuel Mode Diesel Engine (Arceivala, 1998).

The basis of energy audit of a MLD aerobic post treatment system:

- Energy requirement of Aerobic Process as the sole wastewater treatment process, including initial pumping is approximately 195 kW-h/MLD (Tassou, 1988).
- *Salient features of comparative energy consumption:*
- Energy requirement of post treatment aerobic system treating only 35% BOD (as 65% BOD removal takes place in anaerobic system) is 195 kW-h/MLD x 0.35 = 68.25 kW-h/MLD
- Hence Total Energy Consumption of integrated UASB-Aerobic Process is (6 + 68.25) kW-h/MLD = 74.25 kW-h/MLD compared to 195 kW-h/MLD for the aerobic process only.

Based on existing waste and wastewater treatment technologies Lettinga (2008) suggested (i) a Natural Biological Mineralization Route followed by physico-chemical methods for achieving the quality of treated wastewater for reuse/ or intended purpose such as for irrigation, industrial reuse etc. and, (ii) decentralization of the sanitation and resource recovery and reuse, that is, a concept which incorporates environmental protection where the waste and wastewater transportation is kept at minimum level and where pollutants are brought to an acceptable value at the location.

4.1. Solutions for sustainability treatment options

Sustainable technologies must be needed in order to make sustainable lifestyle of the society and to protect environment. It is difficult to understand and to implement it due to lack of proper parameters which leads to ambiguously the targets or proposed actions taken by politicians and/ or policy makers. Moreover, the quantification of sustainability is vague. For instance, if government implementing extremely stringent standards for protecting the aquatic environment from pollution many question arises, like why a single country or region pursuing a paradisiacal natural environment while at the same time little if any money or technology is made available to contribute to the highly needed environmental improvement in less prosperous countries. These potential combinations can be considered as sustainable solutions if adopted based on NBMR (Khan et al., 2011a).

4.2. Sustainable technology concept

The superiority of sequential anaerobic – aerobic treatment systems over conventional aerobic is more profound with increase in sewage concentration. In countries of limited per capita share of water, like in Africa, Middle East and India the treatment of concentrated sewage via conventional aerobic system is highly expensive, especially with respect to operational costs (Khan et al., 2011a).

The advantages of introducing UASB reactor ahead of aerobic system is obvious, mainly in terms of sludge production and energy consumption. In view of the fact that aeration costs increase linearly with increasing organic loads, adopting the activated sludge system for polishing of anaerobic effluents may not be the most sustainable option for concentrated sewage. Other aerobic systems, such as DHS, SBR and CFID type SBR for UASB effluents post treatment reviewed in this paper are promising options for sewage management at low cost, low land requirement and low sludge production. Moreover, the potential of nutrients recovery and pathogens removal in an aerobic post-treatment for UASB effluents is considerable and the effluent discharge standards established by various national and international environmental agencies can be achieved.

5. Conclusions

Numerous anaerobic/ aerobic treatment concepts were evaluated in this chapter. The best option observed for the sewage treatment was integrated UASB-SBR system. The organics, nutrients and pathogenic pollutant removal efficiency of the integrated treatment approach was capable to achieve the effluent with low BOD (≈5mg/L; 98 % removal), COD (<25 mg/L; up to 95% removal) and TSS (<10 mg/L; up to 98% removal) and nutrients (TN=4 mg/L; NH_4-N=Nil; P=1 mg/L). Ammonium nitrogen and phosphorus levels were decreased up to 98% and 90%, respectively. Fecal coliforms levels fell to <1000 MPN/100 mL, indicating a significant removal of pathogenic indicators. Thus the final effluent from the integrated UASB-SBR system can be reused for unrestricted irrigation or be discharged safely into the surface waters. However, no information is available regarding the efficacy of integrated UASB-SBR system at full scale level for sewage treatment. The performance of existing UASB based STPs can be improved by installing any of the post treatment system demonstrated in this chapter. The energy conservation, resources recovery and carbon credit were the gaps that still need to be explored for the above suggested post treatment options so that a natural biological mineralization route or sequence can be utilized to make the integrated system a viable sustainable option for treatment of sewage and anaerobically treated effluents.

Author details

Abid Ali Khan[1,2], Rubia Zahid Gaur[3], Absar Ahmad Kazmi[1] and Beni Lew[4,5]

1 Department of Civil Engineering, IIT Roorkee, India

2 Royal HaskoningDHV, India

3 Water & Sanitation Specialist, Plan Environ, H-273, GK1, New Delhi, India

4 The Volcani Center, Institute of Agriculture Engineering, Bet Dagan, Israel

5 Department of Civil Engineering, Ariel University Center of Judea and Samaria, Ariel, Israel

References

[1] Agrawal, L.K., Okui, H., Ueki, Y., Harada, H., Ohashi, A. (1997) Treatment of Raw Sewage in a Temperate Climate using a UASB Reactor and the Hanging Sponge Cubes Process. Wat. Sci. Technol., 36 (6-7), 433-440.

[2] Annachhatre AP, Suktrakoolvait S. (2001). Biological sulfide oxidation in a fluidized bed reactor. Environ Technol; 22:661–72.

[3] Aiyuk, S., Amoako, J., Raskin, L., van Haandel, A., Verstraete, W. (2004) Removal of Carbon and Nutrients from Domestic Wastewater using a Low Investment, Integrated Treatment Concept. Wat. Res., 38, 3031–3042.

[4] Arceivala SJ, Asolkar SR, (2007). Wastewater treatment for pollution control and reuse. 3rd ed. Tata McGraw-Hill Publishing Co. Ltd. New Delhi, India.

[5] Arceivala SJ. (1998). Wastewater treatment for pollution control. 2nd ed. Tata McGraw Hill, New Delhi, India.

[6] A1-Adham, S. S. (1989). Tertiary Treatment of Municipal Sewage via Slow Sand Filtration. MS Thesis, King Fahd University of Petroleum & Minerals, Dhahran, Saudi Arabia.

[7] A1-Sawaf, M. F. (1986). Tertiary Wastewater Treatment by Direct Filtration. MS Thesis, King Fahd University of Petroleum & Minerals, Dhahran, Saudi Arabia.

[8] Arceivala, S.J. (2001). Wastewater Treatment for Pollution Control, New Delhi, Tata McGraw Hill.

[9] Buisman CJN, Geraats BG, Ijspeert P, Lettinga. (1990). Optimization of sulfur production in a biotechnological sulfide removing reactor. Biotech Bioeng; 35: 50–6.

[10] Chuang SH, Pai TY, Horng RY. (2005). Biotreatment of sulfate-rich wastewater in an anaerobic/micro-aerobic bioreactor system. Environ Technol. 26(9):993–1001.

[11] Chen C, Ren N, Wang A, Liu L, Lee DJ. (2010). Enhanced performance of denitrifying sulfide removal process under micro-aerobic condition. J Hazard Mater; 179:1147–51.

[12] Cavalcanti, P.F.F., van Haandel, A., Lettinga, G. (2001). Polishing Ponds for Post-treatment of Digested Sewage Part 1: Flow-through Ponds. Wat. Sci. Technol., 44 (4), 237–245.

[13] Chernicharo, C.A.L., Nascimento, M.C.P. (2001). Feasibility of a Pilot- Scale UASB/ Trickling Filter System for Domestic Sewage Treatment, Wat.Sci.Technol., 44 (4), 221-228.

[14] Chernicharo, C.A.L., Cota, R.S., Zerbini, A.M., von Sperling, M., Brito, L.H.N.C. (2001). Post-treatment of Anaerobic Effluents in an Overland Flow System. Wat.Sci.Technol. 44 (4), 229–236.

[15] Chernicharo, C.A.L. (2006). Post Treatment Options for the Anaerobic Treatment of Domestic Wastewater. Reviews in Environmental Sciences and Bio/Technology, 5, 73-92.

[16] Droste, R.L., Masse, D.I. (1995). Anaerobic Treatment in Sequencing Batch Reactors. International Symposium on Technology Transfer. Pre-prints. Salvador, Bahia, Brazil, pp. 353–363.

[17] Ellis, K.V. (1987). Slow Sand Filtration as Technique for the Tertiary Treatment of Municipal Sewages. Wat. Res., 21 (4), 403- 410.

[18] El-Shafai, S.A., El-Gohary, F.A., Nasr, F.A., van der Steen, P., Gijzen, H.J. (2007). Nutrient Recovery from Domestic Wastewater using a UASB-Duckweed Ponds System. Biores. Technol., 98, 798–807.

[19] Elmitwalli, T., Zeeman, G., Lettinga, G. (2001). Anaerobic Treatment of Domestic Sewage at Low Temperature. Wat.Sci.Technol. 44 (4), 33–40.

[20] Foresti, E., Zaiat, M., Vallero, M. (2006). Anaerobic processes as the core technology for sustainable domestic wastewater treatment: Consolidated applications, new trends, perspectives, and challenges. Reviews in Environmental Science and Bio/ Technology, 5, 3–19.

[21] Gomec, C.Y. (2010). High-rate anaerobic treatment of domestic wastewater at ambient operating temperatures: A review on benefits and drawbacks. J. Environmental Sci Health - A Tox Hazard Subst Environ Eng. 45 (10): 1169 – 84. DOI: 10.1080/10934529.2010.493774.

[22] Gersberg, R.M., Gearheart, R.A., Ives, M. (1989). Pathogen Removal in Constructed Wetlands. In: DA. Hammer, Editor, Constructed Wetlands for Wastewater Treatment: Municipal, Industrial and Agricultural, Lewis Publishers, Inc., Chelsea, MI (1989), Pp. 431–445.

[23] Gnanadipathy, A., Polprasert, C. (1993). Treatment of Domestic Wastewater with UASB Reactor. Wat. Sci. Technol., 27, 195–203.

[24] Goncalves, R.F., Araujo, V.L., Chernicharo, C.A.L. (1998). Association of a UASB Reactor and a Submerged Aerated Biofilter for Domestic Sewage Treatment, Wat. Sci. Technol., 38 (8-9), 189-195.

[25] Goncalves, R.F., de Araujo, V.L., Bof, V.S. (1999). Combining Upflow Anaerobic Sludge Blanket (UASB) Reactors and Submerged Aerated Biofilters for Secondary Domestic Wastewater Treatment. Wat.Sci.Technol. 40 (8), 71-79.

[26] Jaya Prakash K., V.K.Tyagi, A.A.Kazmi & Arwind Kumar, 2007, Post- Treatment of UASB Reactor Effluent by Coagulation and Flocculation Process, AIChE, Environmental Progress, Vol. 26, No.2 pp 164-168.

[27] Janssen AJH, Lettinga G, de Keizer A. (1999). Removal of hydrogen sulphide from wastewater and waste gases by biological conversion to elemental sulphur: colloidal

and interfacial aspects of biologically produced sulphur particles. Colloids Surf A: Physicochem Eng Aspects; 151:389–97.

[28] Khan, AA, Gaur, RZ, Tyagi, VK, Khursheed, A, Lew, B, Kazmi, AA, Mehrotra I (2011a). Sustainable Options of Post Treatment of UASB Effluent Treating Sewage: A Review. Resource, Conservation and Recycling; Vol. 55 (12); 1232-1251.

[29] Khan, AA, Gaur, RZ, Lew, B, Diamantis, V, Mehrotra, I, Kazmi, AA (2011b). UASB/ Flash aeration enable complete treatment of municipal wastewater for reuse. Bioprocess and Biosystem Engineering. Vol. 35(6):907-13.

[30] Khan, AA, Gaur, RZ, Lew, B, Mehrotra, I, Kazmi, AA (2011c). Effect of Aeration on The Quality of Effluent of UASB Reactor Treating Sewage. Journal of Environmental Engineering- ASCE, Vol. 137 (6); 464-472.

[31] Khan, AA (2012). Post treatment of UASB effluent: Aeration and Variant of ASP. PhD Thesis. IIT Roorkee India.

[32] Keller, R., Passamani- Franca, R.F., Passamani, F., Vaz, L., Cassini, S.T., Sherrer, N., Rubim, K., Santa' Ana, T.D. & Goncalves, R.F. (2004). Pathogen Removal Efficiency from UASB+BF Effluent using Conventional and UV Post- Treatment Systems, Wat. Sci. Technol., 50 (1), 1-6.

[33] Khan, Abid A., (2012). Post Treatment of UASB Effluent: Aeration and Variant of ASP. PhD Thesis. IIT Roorkee, India.

[34] Lettinga, G., van Velsen, A. F. M., Hobma S. W., De Zecuw, W., Klapwijk, A. (1980). Use of the Upflow Sludge Blanket (USB) Reactor Concept for Biological Wastewater Treatment, Especially for Anaerobic Treatment. Biotechnol. Bioengg., 22, 699-734.

[35] Lettinga G., deMan A., van der Last, A. R. M., Wiegant, W., van Knippenberg, K., Frijns, J., van Buuren, J. C. L. (1993). Anaerobic Treatment of Domestic Sewage and Wastewater. Wat. Sci. Technol., 27(9), 67-73.

[36] Lettinga G. (2008). Towards feasible and sustainable environmental protection for all. Aquat Ecosyst Health Manage; 11(1):116–24.

[37] Lew B, Belavski M, Admon S, Tarre S, Green M. (2003). Temperature effect on UASB reactor operation for domestic wastewater treatment in temperate climate regions. Water Sci Technol.48 (3):25–30.

[38] Lew B, Tarre S, Belavski M, Green M. (2004). UASB reactor for domestic wastewater treatment at low temperatures: a comparison between a classical UASB and hybrid UASB-filter reactor. Water Sci Technol.49 (11–12):295–301.

[39] Mendoza L, Carballa M, Sitorus B, Pieters J, Verstraete W. (2009). Technical and economical feasibility of gradual concentric chambers reactor for sewage treatment in developing countries. Electron J Biotechnol 2009; 12(2):1–13.

[40] MoEF (2005 and 2006). Management Information System, Technical Report, National River Conservation Directorate, Ministry of Environment and Forests, New Delhi, India.

[41] Metcalf and Eddy. (2003). Wastewater engineering treatment and reuse. 3rd ed. Tata McGraw Hill Co. New Delhi, India.

[42] Machdar, I., Sekiguchi, Y., Sumino, H., Ohashi, A., Harada, H. (2000b). Combination of a UASB Reactor and a Curtain type DHS (Downflow Hanging Sponge) Reactor as a Cost-Effective Sewage Treatment System for Developing Countries. Wat. Sci. Technol., 42 (3–4), 83–88.

[43] Moawad,A., Mahmoud,U.F., El-Khateeb, M.A., El-Molla, E. (2009). Coupling of Sequencing Batch Reactor and UASB Reactor for Domestic Wastewater Treatment. Desalination, 242, 325–335.

[44] Mbuligwe, S.E. (2004). Comparative Effectiveness of Engineered Wetland Systems in the Treatment of Anaerobically Pre-treated Domestic Wastewater, Ecol. Eng., Vol. 23, 269–284.

[45] Okurut, T. O., Rijs, G. B. J., van Bruggen, J. J. A. (1999). Design and Performance of Experimental Constructed Wetlands in Uganda, Planted with Cyperus Papyrus and Phragmites Mauritianus. Wat.Sci.Technol. 40 (3), 265–263.

[46] Penetra, R.G., Reali, M.A.P., Foresti, E., Campos, J.R. (1999). Post-treatment of Effluents from Anaerobic Reactor Treating Domestic Sewage by Dissolved-Air Flotation. Wat.Sci.Technol. 40 (8), 137-143.

[47] Schellinkhout, A., Collazos, C.J. (1992). Full-scale Application of the UASB Technology for Sewage Treatment. Wat.Sci.Technol. 25 (7), 159-166.

[48] Seghezzo, L., Guerra, R.G., Gonza'lez, S.M., Trupiano, A.P., Figueroa, M.E., Cuevas, C.M., Zeeman, G., Lettinga, G. (2002). Removal Efficiency and Methanogenic Activity Profiles in a Pilot-scale UASB Reactor Treating Settled Sewage at Moderate Temperatures. Wat.Sci.Technol. 45 (10), 243–248.

[49] Sousa, J.T., Foresti, E. (1996). Domestic Sewage Treatment in an Up-flow Anaerobic Sludge Blanket – Sequential Batch System. Wat. Sci.Technol. 33 (3), 73-84.

[50] Sousa, J.T., van Haandel, A.C., Guimarães, A.A.V. (2001). Post-treatment of Anaerobic Effluents in Constructed Wetland Systems. Wat.Sci.Technol. 44 (4), 213–219.

[51] Suhail, A. (1987). Tertiary Wastewater Treatment by Sedimentation and Sand Filtration, MS Thesis, King Fahd University of Petroleum & Minerals, Dhahran, Saudi Arabia.

[52] Sumino, H., Takahashi, M., Yamaguchi, T., Abe, K., Araki, N., Yamazaki,S., Shimozaki, S., Nagano, A., Nishio, N. (2007). Feasibility Study of a Pilot-scale Sewage Treatment System Combining an Up-flow Anaerobic Sludge Blanket (UASB) and an

Aerated Fixed Bed (AFB) Reactor at Ambient Temperature. Biores. Technol., 98, 177–182.

[53] Surampalli, R.Y., Tyagi, R.D., Scheible, O.K., Heidman, J.A. (1997). Nitrification, Denitrification and Phosphorus Removal in Sequential Batch Reactors. Biores. Technol., 61, (151–157).

[54] Stefess GC, Torremans RAM, de Schrijver R, Robertson LA, Kuenen JG. (1996). Quantitative measurement of sulfur formation by steady state and transient-state continuous cultures of autotrophic thiobacillus species. Appl Microbiol Biotechnol; 45:169–75.

[55] Tawfik, A. Zeeman, G., Klapwijk, A., Sanders, W., El-Gohary, F., Lettinga, G. (2003). Treatment of Domestic Sewage in a Combined UASB/RBC system. Process optimization for Irrigation purposes. Wat. Sci.Technol. 48 (1), 131-138.

[56] Tawfik, A., Klapwijk, B., El-Gohary, F., Lettinga,G. (2005). Potentials of Using a Rotating Biological Contactors for Post Treatment of Anaerobically Pre- Treated Domestic Wastewater, Biochem. Engg. J., Vol. 25, 89-98.

[57] Tawfik, A., El-Gohary, F., Ohashi, A., Harada, H. (2010). Optimization of the Performance of an Integrated Anaerobic–Aerobic System for Domestic Wastewater Treatment. Wat.Sci.Technol. 58 (1), 185-194.

[58] Torres P., Foresti, E. (2001). Domestic Sewage Treatment in a Pilot System Composed of UASB and SBR Reactors, Wat.Sci.Technol. 44(4), 247-53.

[59] Tyagi VK, Khan AA, Kazmi AA, Mehrotra I, Chopra AK. (2009). Slow sand filtration of UASB reactor effluent: a promising post treatment technique. Desalination; 249:571–6.

[60] Tessele F, Monteggia LO, Rubio J. (2005). Treatment of municipal wastewater UASB reactor effluent by unconventional flotation and UV disinfection. Water Sci Technol. 52(1–2):315–22.

[61] Tassou SA. (1988). Energy conservation and resource utilization in wastewater treatment plants. Appl Energy. 30:2–8.

[62] van Haandel, A.C., Lettinga, G. (1994). Anaerobic Sewage Treatment: a Practical Guide for Regions with a Hot Climate. John Wiley & Sons, Chichester, UK, 226.

[63] von Sperling, M., Freire, V.H., Chernicharo, C.A.L. (2001). Performance Evaluation of a UASB–Activated Sludge System Treating Municipal Wastewater. Wat.Sci.Technol. 43 (11), 323–328.

[64] von Sperling, M., Chernicharo, C.A.L., Soares, A.M.E. and Zerbini, A.M. (2002). Coliform and Helminth Eggs Removal in a Combined UASB Reactor–Baffled Pond System in Brazil: Performance Evaluation and Mathematical Modeling. Wat.Sci. Technol, 45 (10), 237–242

[65] von Sperling, M., Chernicharo, C.A.L. (2005). Biological Wastewater Treatment in Warm Climate Regions. IWA Publishing, London, 1452.

[66] von Sperling,M., Mascarenhas, L.C.A.M. (2005). Performance of Very Shallow Ponds Treating Effluents from UASB Reactors. Wat.Sci. Technol., 51 (12), 83-90.

[67] von Sperling, M., Bastos, R.K.X., Kato, M.T. (2005). Removal of E. coli and Helminthes Eggs in UASB: Polishing Pond Systems in Brazil. Wat.Sci. Technol., 51 (12), 91-97.

[68] Vlyssides A, Barampouti EM, Mai S. (2007). Effect of ferrous ion on the biological activity in a UASB reactor: Mathematical modeling and verification. Biotechnol Bioeng; 96(5):853–61.

[69] Vallero MVG, Sipma J, Annachhatre A, Lens PNL, Hulshoff, Pol LW. (2003). Biotechnological treatment of sulfur-containing wastewaters. In: Fingerman M, Nagabhushanam R, editors. Recent advances in marine biotechnology: bioremediation, vol. 8. Enfield, NH, USA: Science Publishers. p. 233–68.

[70] van der Zee FP, Villaverde S, Garcia PA, Polanco FFdz. (2007). Sulfide removal by moderate oxygenation of anaerobic sludge environments. Bioresour Technol; 98:518–24.

[71] World Health Organization (1989). Health Guidelines for the Use of Waste water in Agriculture and Aquaculture. Technical Report Series- 778, Geneva: WHO.

[72] Walia R. (2007). Polishing of effluent from UASB reactor: ORP as a monitoring parameter, PhD thesis, Indian Institute of Technology, Roorkee, India.

[73] YingYu A, Yang FL, Bucciali B, Wong FS. (2009). Municipal wastewater treatment using a UASB coupled with cross-flow membrane filtration. J Environ Eng; 135(2):86–91.

[74] YingYu A, Bing W, Wong FS, Yang F. (2010). Post-treatment of up-flow anaerobic sludge blanket effluent by combining the membrane filtration process: fouling control by intermittent permeation and air sparging. Water Environ J; 24: 32–8.

[75] N. Matsuura, M. Hatamoto, H. Sumino, K. Syutsubo, T. Yamaguchi and A. Ohashi. (2010). Closed DHS system to prevent dissolved methane emissions as greenhouse gas in anaerobic wastewater treatment by its recovery and biological oxidation. Water Science & Technology—Wat Sci Technol , Vol 61, No 9, pp 2407–2415 .

Chapter 9

Determination of Anaerobic and Anoxic Biodegradation Capacity of Sulfamethoxasole and the Effects on Mixed Microbial Culture

Zeynep Cetecioglu, Bahar Ince, Samet Azman,
Nazli Gokcek, Nese Coskun and Orhan Ince

Additional information is available at the end of the chapter

http://dx.doi.org/10.5772/56049

1. Introduction

During last decades, concentration of human and veterinarian antibiotics in the environment, natural and engineered systems have been increased because of high amount production and consumption. This situation has aroused great concern due to the possibility of harmful effects on human, animals and plants [1,2]. Occurrence and fate of these compounds are one of the main issues because of their unknown potential risks and their effects on the environment. Approximately 500 tonnes of them are produced and consumed every year in the worldwide. Antibiotics are resistant to conventional biological treatment process and the wastewaters including these compounds are directly discharged to the receiving water bodies without efficient treatment. Hospitals and pharmaceutical industries are the main sources of high antibiotic concentration release to the environment [3]. Also sewage systems can transport these molecules and/or their metabolites since metabolization of them by humans and animals cannot be achieved completely [4]. During the transportation of antibiotics throughout treatment plants, elimination of these compounds can occur via biodegradation, photolysis and sorption to sludge but ultimate degradation of these compounds cannot be achieved in conventional treatment plants [4, 5, 6]. As a result of the introduction of metabolized and/or active antibiotics to the receiving water bodies caused an increase in the ratio of multi-antibacterial resistant pathogens [7].

Sulfamethoxasole (SMX) is a sulfonamide bacteriostatic antibiotic that is used to treat urinary tract infections. SMX inhibits the multiplication of bacteria, since they are competitive inhibitors of *p-amino benzoic acid* in the folic acid metabolism cycle [8]. Sulfonamide antibiotics, including SMX, have been found in the activated sludge processes and digested sludges in varying concentration from ng/L to μg/L levels [5, 8]. Behaviour of sulfonamide antibiotics has been reported as recalcitrant molecules thus sorption and desorption are the main pathways on antibiotic elimination from aquatic phases [9, 10]. Biodegradability of SMX has not been widely studied for anaerobic systems. There are few studies about anaerobic biodegradability characteristics of SMX in the literature [11, 12].

In this chapter, the aim is to reveal the anoxic and anaerobic biodegradability characteristics of SMX and the effects of this compound on microbial community. In this scope, biodegradation capacity and the effects on the microorganisms were investigated by destructive batch tests based on a modified version of Anaerobic Biodegradability of Organic Compounds-OECD 311 protocol [13] under three different electron acceptor conditions; nitrate reducing, sulfate reducing, and methanogenic conditions. Quantification of defined microbial groups was also carried out to determine the effects of SMX on abundance of microbial community.

1.1. Antibiotics

Antibiotics are among the most important groups of pharmaceuticals and chemotherapeutic agents that inhibit or terminate the growth of microorganisms, such as bacteria, fungi, or protozoa without affecting host [11, 12]. The term antibiotic used for drugs that block any of these microorganisms. Other terms as chemotherapeutics or antimicrobials are not synonymous because of their scopes; the term of antimicrobial is used for the medicine which is also effective against viruses and the expression "chemotherapeutical" referring to compounds used for the treatment of disease which kill cells, specifically microorganisms or cancer cells. The term "chemotherapeutical" may also refer to antibiotics (antibacterial chemotherapy).

The expression of antibiotic is originally used to describe any agent with biological activity against living organisms; however, "antibiotic" now refers to substances with antibacterial, anti-fungal, or anti-parasitical activity. During the years, this definition has been changed and now it includes also synthetic and semi-synthetic products. There are approximately 250 different compounds registered for use in medicine and veterinary application [16].

In this chapter, the term "antibiotic" refers only to drugs that kill or inhibit bacteria. Antibiotics that are sufficiently nontoxic to the host are used as chemotherapeutic agents in the treatment of infectious diseases of humans, animals and plants. They are extensively used for prevention and treatment of diseases caused by microorganisms in human and veterinary medicine as well as in aquaculture nowadays. Also, they are being still used as growth factor in livestock farming. Some compounds may be used for different purposes such as in growing fruit and in bee keeping other than human or veterinary medicine. The application purposes may vary from country to country.

Antibiotics are classified as their chemical structures and the mechanism of inhibition of microorganisms and they can be divided into subgroups such as β-lactams, quinolones, tetracyclines, macrolides, sulfonamides and others. The active compounds of antibiotics are often complex molecules, which may have different functionalities. In the environment, these molecules could be found as neutral, cationic, anionic, or zwitterionic forms. Because of the different functionalities within one molecule, their physicochemical and biological properties may change with pH levels [17].

1.2. Sulfamethoxazole

In this chapter, sulfamethoxazole (SMX) is selected as model compound. The systematic name of this compound is 4-amino-*N*-(5-methylisoxazol-3-yl)-benzenesulfonamide. Sulfamethoxazole and other sulfonamides have a similar structure to *p*-aminobenzoic acid and inhibit to the synthesis of nucleic acids in sensitive microorganisms by blocking the conversion of *p*-aminobenzoic acid to the coenzyme dihydrofolic acid, a reduced form of folic acid; dihydrofolic acid is obtained from dietary folic acid so sulfanomides do not have any influence on human cells. Their action is primarily bacteriostatic, although they may be bactericidal where concentrations of thymine are low in surrounding medium. The sulfonamides have a broad spectrum of action, but the development of widespread resistance has greatly reduced their usefulness, and susceptibility often varies widely even among nominally sensitive pathogens like Gram-positive and Gram-negative cocci.

There are several mechanisms of resistance including alteration of dyhydropteroate synthetase, the enzyme inhibited by sulfonamides, to a less sensitive form, or an alteration in folate biosynthesis to an alternative pathway; increased production of *p*-aminobenzoic acid; or decreased uptake or enhanced metabolism of sulfonamides.

Resistance may result from chromosomal alteration, or may be plasmid-mediated and transferable, as in many resistant strains of enterobacteria. High-level resistance is usually permanent and irreversible. There is complete cross-resistance between the different sulfonamides [18].

1.3. Consumption and occurrence

The yearly consumption of antibiotics worldwide is estimated between 500 tons [19]. Approximately 90% of the consumed antibiotics are excreted via urinary or fecal pathways from the human body after partial or no metabolism and they are transferred to the domestic sewage plants or directly to the environment. Conventional biological treatment of domestic sewage provides very low or no reduction for these compounds, which usually by-pass treatment and accumulate in the receiving waters.

Antibiotic consumption changes depending on the country and/or region however the situation is scarce and heterogenous. Country specific consumption for groups of antibiotics in DDDs can be found for Europe on the ESAC homepage [20]. Using patterns of different regions and countries are given Table 1. The relative importance of the different use patterns in different countries is still not known.

An increasing number of studies have been done to determine the source, occurrence, fate, and effects on the ecosystem of antibiotics. However, there is still a lack of understanding and knowledge of these compounds. So studies maybe focus on the strategies about stream segregation and at-source treatment of the concentrated streams appears.

Region/ Country	Total volume used in human medicine (ton/year)	Volume used in human medicine (gram per capita)	Thereof in hospitals (%)	Unuse medicaments	Measured in sewage up to (μg/L)	Measured in surface water up to (μg/L)	Reference
World wide	100000-200 000	N.D.	N.D.	N.D.	N.D.	N.D.	[19]
EU + Switzerland	8367	22.4	N.D.	N.D.	N.D.	N.D.	[21]
USA	4860	17	70	N.D.	1.9	0.73	[22, 23]
Canada	N.D.	N.D.	N.D.	N.D.	N.D.	0.87	[24]
Switzerland	34.2	4.75	20-40	N.D.	0.57	0.2	[25]
Germany	411	4.95	25	20-40	6	1.7	[16, 26]
Denmark	40	7.4	N.D.b	N.D.	5N.D.	N.D.	[27]
Austria	38	4.7	N.D.	20-30	N.D.	N.D.	[28]
Netherlands	40.9	3.9	20	N.D.	4.4	0.11-0.85	[29]
Italy	283	4.88	N.D.	N.D.	0.85-	0.25	[30]
Turkey	N.D.	31.4	N.D.	N.D.	N.D.	N.D.	[31]

Table 1. Country specific antibiotics consumption and occurrence data (N.D.: not defined)

1.4. Production and manufacturing

Pharmaceutical industries have minor importance on the sewage treatment plants. Only in some Asian countries, wastewaters from this industry contributes to the sewage and cause an increase in the concentration of single compound up to mg/L level [32, 33, 34]. Also in developed countries, manufacturing plants increases the total antibiotic concentrations in the domestical wastewater [35].

The main problem for this industry is that they still use the physicochemical treatment technologies in the plant to remove the compounds from their wastewater. However, this approach is expensive.

1.5. Elimination and treatment

In the literature, there are lots of studies focused on the fate of these compounds in conventional domestical wastewater treatment plants and also lab-scale applications in the innovative treatment methods. Elimination and/or treatment of these organic compounds are the results of biotic and abiotic processes. While biotic process is the biodegradation by microorganisms, abiotic processes are sorption, hydrolysis, oxidation-reduction, and photolysis.

1.5.1. Sorption

Before to assess the sorption characteristics of antibiotics, it is necessary to consider their physical and chemical parameters. Tolls [36] investigated the sorption behavior of these compounds in soil and the results showed that sorption mechanism of antibiotics could be very complex and difficult.

Additionally, binding to particles or the formation of complexes may prevent their detection. For example, tetracyclines are able to form complexes with double cations such as magnesiu-mor calcium [37]. Also humic substances cause the change in the surface properties and sites available for sorption and reactions. Gu and Karthikeyan [38] reported that there is a strong interaction between humic acids, hydrous Al oxide and tetracycline. Some studies showed that antibiotics used in medicine such as fluoroquinolones and macrolides can reach the terresrial environment by sewage sludge [38, 39].

Also sorption mechanism is a significant process for sulphonamides [36]. However, knowledge about the interaction of antibiotics with sludge and of sediments with sludge in activated sludge plants as well as the subsequent potential for their release back into the environment is still too sparse.

1.5.2. Photolysis

Photochemical process can be important in the surface waters and treatment plant effluents as another elimination process [40-43]. In the environment, photolysis process is not effective in turbid water or river and lakes, which are shadowed. So, the in the lab-scale experiments cannot reflect the photochemical process in the nature. Also, effectiveness of depletion process can differ under different environmental conditions such as pH, temperature, water hardness [44] and depends on type of matrix, location, season, latitude [45].

One of the problems about this type of process is that incomplete photo-transformation and photo-degradation can cause to more or less stable or toxic compounds although this does not necessarily have to happen [46-48].

The significance and extent of direct and indirect photolysis of antibiotics in the aquatic environment are different for each compound because some of them are light sensitive (e.g. quinolones, tetracyclines, sulphonamides, tylosin, nitrofuran antibiotics). However, not all compounds are photo-degradable [49]. Tetracyclines are senstive to photo-degradation. Samuelsen [50] investigated the sensitivity of oxytetracycline towards light in seawater as well as in sediments. The antibiotics proved to be stable in sediments rather than in seawater. As no mechanism of decomposition other than photolysis is known for them [51], the substance remains in the sediment for a long period, as shown by [52]. Boree *et al.* [53] showed that sulphanilic acid was found as a degradation product common to most of the sulpha drugs.

1.5.3. Hydrolysis and thermolysis

Another important pathway for the non-biotic decomposition of organic substances in the environment is hydrolysis. Some instability in water could be demonstrated for some tetra-cycylines [54]. In general, the hydrolysis rates for oxytetracycline increase with reascept to temperature at pH 7. The half-lives of oxytetracycline under investigation changed by differences in temperature, light intensity and flow rate from one test tank to another. However sulphonamides and quinolones are known as resistant antibiotic to hydrolysis.

1.5.4. Oxidation

Pharmaceutical industry wastewaters including antibiotic are well known for the difficulty of their elimination by conventional biological treatment methods and their important contribution to environmental pollution is due to their fluctuating and recalcitrant nature. For this reason, oxidation processes are usually applied.

The presence of carbon–carbon double bonds, aromatic bonds or nitrogen is a necessary essential for this application. However, the presence of these structural elements does not provided the fast and full degradation or even the complete degradation.

The effect of ozonation on the degradation of oxytetracycline in aqueous solution at different pH values (3, 7 and 11) was reported by Li *et al.* [55]. The study was designed that ozonation as a partial step in a combined treatment concept is a potential technique for biodegradability enhancement. It has been shown that COD removal rates increase with increasing pH as a consequence of enhanced ozone decomposition rates at elevated pH values. The results of bioluminescence data indicate that the initial by-products after partial ozonation (5–30 min) of oxytetracycline were more toxic than the parent compound [55].

Sulfamethoxazole was also efficiently degraded by ozonation [56]. An improvement in biodegradability by the increasing of BOD5/COD ratio from 0 to 0.28 was observed by the authors after 60 min of ozonation. The acute toxicity of the intermediates was checked and a slight acute toxicity increment in the first stage of ozonation was found. pH variation was found as important parameter on TOC and COD removal efficiencies. The complete sulfamethoxazole removal was achieved for an in photo-Fenton process [57]. Toxicity and inhibition tests pointed in the same direction: no toxic effect of oxidized intermediates was determined and also no inhibition was detected on activated sludge activity.

1.5.5. Biodegradation

Biodegradability of most antibiotics has been checked and it was found that they are not biodegradable under aerobic conditions until today [3, 11, 55, 58, 59]. Biodegradability characteristics have been weak for most of the compounds investigated in laboratory tests such as the OECD test series (301–303, 308) – even for some of the ß-lactams (Alexy *et al.*, 2004). Out of 16 antibiotics tested, only benzyl penicillin (penicillin G) was completely mineralized in a combination test (combination of the OECD 302 B and OECD 301 B tests; [11]).

Biodegradation for tetracycline was not observed during a biodegradability test (sequence batch reactor), and sorption was found to be the principal removal mechanism for tetracycline in activated sludge [61].

Some antibiotics occurring in soil and sediment proved to be quite persistent in laboratory testing as well as in field studies. Some of them were not biodegradable also under anaerobic conditions [12] others did [62]. Substances extensively applied in fish farming had long half-lives in soil and sediment, as reported in several investigations [63]; [64]; [65]; [66]; [67]; [68]; [69]). However, some substances were at least partly degradable ([70]; [71]; [72], [66]; [68]; [73]). Maki *et al.* [62] found that ampicillin, doxycycline, oxytetracycline, and thiamphenicol were significantly degraded, while josamycin remained at initial levels. Tylosin was biodegraded [42].

1.6. Problem definition and aim

The yearly consumption of antibiotics is 500 tons throughout the world according to the data of 2001. Approximately 90% of the consumed antibiotics after being partially metabolized or not being metabolized are excreted by the help of urea or feces from the body and transferred to the domestic sewage plants. These antibiotics are discharged into the receiving environment with no or low elimination after being treated in conventionally operated domestic sewage plants. While the concentration of these materials in domestic wastewaters and surface waters are in μg/l level, in pharmaceutical wastewater they are in 100-1000 mg/L level [74, 75, 76, 77]. As this low concentration in the surface wastewaters cause important problems in the ecosystem, it necessitates the removal of high antibiotic amount that are found in the pharmaceutical wastewaters. However, because the chemical removals of these materials are costly, biological treatment is essential. Antibiotics are the one of these compounds and the most often discussed pharmaceuticals because of their potential role in the spread and maintenance of (multi)resistance of bacterial pathogens. There are lots of studies that have been done in Europe and North America on the detection and removal of antibiotics in the receiving environment and the treatment plant [4, 5, 23, 24, 78-84]. However, the studies on the treatability of these antibiotics biologically are quite few [61, 85]. Also the scope of the studies done on the biodegradability potential of these materials is limited [11, 12, 86, 87]. Additionally, the studies on the microbial groups and species that are responsible for degradation have not been done, yet.

In this scope, determination of biodegradation characteristics of the refractory compounds and their toxic/inhibition effects on microbial community is substantial for environmental engineering. For this aim, the biodegradability of these sulfamethoxazole under anoxic and anaerobic conditions and also changes in microbial groups under the different conditions are explained in this chapter.

2. Materials and methods

2.1. Experimental approach

This study involves setting-up batch biodegradation test to investigate biodegradation characteristics of sulfamethoxasole (SMX) under anoxic and anaerobic conditions. The biodegradation test bottles were set up under nitrate reducing conditions (NRC), sulfate reducing conditions (SRC) and methanogenic conditions (MC). Experiment was carried out for 120 days. During the experiment, gas production was monitored daily. Destructive sampling was done in four different times (at 0th, 20th, 60th and 120th day). Wet chemical analysis (dissolved organic carbon [DOC], SMX measurements and electron acceptor measurements) and microbiological analysis (quantitative real-time PCR [Q-PCR]) were carried for four sampling times.

2.2. Set–up of batch biodegradation test bottles

In this study, two different seed sludges were used for setting-up of the batch tests. For NRC, the seed was taken from anoxic part of a domestic wastewater treatment plant in Istanbul whereas; test tubes for the SRC and MC were inoculated by anaerobic sludge from a full-scale UASB reactor treating alcohol distillery effluents.

The batch tests were constructed in 120 mL serum bottles, 100 mL of active volume, according to modified OECD 311 protocol [13]. The constituents of each experimental set for NRC, SRC and MC conditions are given in Table 2, 3 and 4, respectively. Also chemicals of the trace element solution and their amounts are given in Table 5. SMX was chosen as the model carbon source. The test tubes were set up as duplicates including positive and negative controls. Phenol was chosen as slowly biodegradable carbon source for positive control set. Negative control sets were constructed without any carbon source to determine endogenous decay. All sets were set-up in an anaerobic cabinet (Coy Laboratory Products, U.S.).

Experimental sets were destructed in 4 different sampling times. The first set was destructed immediately after all the test tubes were set-up, the other three sets were spoiled in day 20, day 60 and day 120. In each test tube, after inoculation 2000 mg/L TVS was maintained. Phenol and SMX concentrations were adjusted to 80±4.5 mg DOC/L and 280±1.0 mg DOC/L within the all experimental groups. The dissolved organic carbon (DOC) value of negative control bottles was 18.6±1.5 mg/L. All solutions were deoxygenated and adjusted to pH 7. Biodegradation test bottles were incubated at 20 °C and 35 °C for NRC and MC/SRC, respectively. All test bottles were stored at dark chambers to ensure occurring only biodegradation and sorption mechanisms during the experiment. The test tubes were shaken daily by hand.

CONSTITUENT	AMOUNT (g)
Anhydrous potassium dihydrogen phosphate (KH2PO4)	0,27
Disodium hydrogen phosphate dodecahydrate (Na2HPO4.12H2O)	1,12
Ammonium chloride (NH4Cl)	0,53
Potassium Nitrate (KNO3)	1
Calcium chloride dihydrate (CaCl2.2H2O)	0,075
Magnesium chloride hexahydrate (MgCl2.6H2O)	0,1
Iron (II) chloride tetrahydrate (FeCl2.4H2O)	0,02
Resazurin (oxygen indicator)	0,001
Sodium sulphide nonahydrate (Na2S.9H2O)	0,1
Stock solution of trace elements	10 ml
Add de-oxygenated water	to 1 liter

Table 2. Medium for nitrate reducing conditions (10 mM potassium nitrate)

CONSTITUENT	AMOUNT (g)
Anhydrous potassium dihydrogen phosphate (KH2PO4)	0,27
Disodium hydrogen phosphate dodecahydrate (Na2HPO4.12H2O)	1,12
Ammonium chloride (NH4Cl)	0,53
Potassium Sulfate (K2SO4)	1,8
Calcium chloride dihydrate (CaCl2.2H2O)	0,075
Magnesium chloride hexahydrate (MgCl2.6H2O)	0,1
Iron (II) chloride tetrahydrate (FeCl2.4H2O)	0,02
Resazurin (oxygen indicator)	0,001
Sodium sulphide nonahydrate (Na2S.9H2O)	0,1
Stock solution of trace elements	10 ml
Add de-oxygenated water	to 1 liter

Table 3. Medium for sulfate reducing conditions (10 mM potassium sulfate)

CONSTITUENT	AMOUNT (g)
Anhydrous potassium dihydrogen phosphate (KH2PO4)	0,27
Disodium hydrogen phosphate dodecahydrate (Na2HPO4.12H2O)	1,12
Ammonium chloride (NH4Cl)	0,53
Calcium chloride dihydrate (CaCl2.2H2O)	0,075
Magnesium chloride hexahydrate (MgCl2.6H2O)	0,1
Iron (II) chloride tetrahydrate (FeCl2.4H2O)	0,02
Resazurin (oxygen indicator)	0,001
Sodium sulphide nonahydrate (Na2S.9H2O)	0,1
Stock solution of trace elements	10 ml
Add de-oxygenated water	to 1 liter

Table 4. Medium for methanogenic conditions

CONSTITUENT	AMOUNT
Manganese chloride tetrahydrate (MnCl2.4H2O)50 mg	50 mg
Boric acid (H3BO3)	5 mg
Zinc chloride (ZnCl2)	5 mg
Copper (II) chloride (CuCl2)	3 mg
Disodium molybdate dihydrate (Na2MoO4.2H2O)	1 mg
Cobalt chloride hexahydrate (CoCl2.6H2O)	100 mg
Nickel chloride hexahydrate (NiCl2.6H2O)	10 mg
Disodium selenite (Na2SeO3)	5 mg
Add de-oxygenated water	to 1 liter

Table 5. Stock solution of trace elements

Headspace pressure was measured by hand-held pressure transducer (Lutron PM-9107, U.S.A.) every day. At each sampling time, biogas composition of the samples was determined via gas chromatography (Perichrom, France). DOC concentration of each sample was measured by Shimadzu ASI-V TOC analyser (Japan). Nitrate and sulfate concentrations were measured by DIONEX ICS 1500 ion chromatograph (U.S.A.). SMX measurements within the solid and liquid phase were carried by the protocol that is proposed previously by Karcı and Balcıoglu [88].

2.3. Calculation of mass balances

Theoretical CO_2 (Th CO_2) and Theoretical biogas (Th biogas), which were used for evaluation of biodegradation, were calculated according DOC, gas and ion chromatography results. Mass balances were calculated by the assumptions, which were described by Ritmann and Mc Carty [89]. Simplified mass balances were given in Equation 1-3 for NRC, SRC and MC, respectively.

$$SMX + NO_3^- \rightarrow Biomass + CO_2 + N_2 + H_2O \tag{1}$$

$$SMX + SO_4^{2-} \rightarrow Biomass + CO_2 + H_2S + HS^- + H_2O \tag{2}$$

$$SMX + H_2O \rightarrow Biomass + CO_2 + CH_4 \tag{3}$$

Ultimate biodegradation ratios were estimated by comparison of $ThCO_2$ and Th biogas production (which were assumed to be produced as a result of 100% biodegradation of tetracycline) were compared to actual CO_2 and biogas production within the batch tests, DOC elimination and SMX measurements.

2.4. Microbiological analyses

Genomic DNA (GDNA) was extracted from 0.5 g sludge using the Fast DNA Spin Kit for Soil (Qbiogene Inc., U.K.) following the manufacturer's instructions.

Q-PCR procedure recommended by Roche was followed and a Light Cycler Master Kit (Roche, Applied Science, Switzerland) was used to set up the reaction (2.0 µl master mix, 1.6 µl $MgCl_2$ 1.0 µl Primer F and R, 13.4 µl H2O, 1 µl sample). Absolute quantification analysis of the GDNA was carried out with a Light Cycler 480 Instrument (Roche Applied Science, Switzerland). Primers used in the quantification are given in Table 6.

Significant differences were determined according to independent sample t-test. Pearson correlation was used for the interactions between variables. All the statistical analyses were conducted by using SPSS (IBM, U.S.A) and $p<0.05$ level was used for significance.

http://dx.doi.org/10.5772/56049

Primer	Target Gene	Target Microorganisms	Respiration Conditions	References
Bac519f Bac907r	16S rRNA	Bacteria	All	90
Arc349f Arc806r	16S rRNA	Archaea	All	91
Met348f Met786r	16S rRNA	Methanogens	All	92
DSRp2060F	Sulfites reductase beta sub-unit (dsrB)	Sulfate Reducing Bacteria	Only sulfate reducing condition	93

Table 6. Primers and target groups for Q-PCR analysis

3. Results and discussion

3.1. Methane generation

Biogas generation in the test bottles operated under sulphate reducing and methanogenic conditions was observed daily. However methane content of the biogas in the test bottles were determined in each sampling time before the destruction of the test bottles as 0th, 20th, 60th and 120th days. Produced methane volume in sulfamethoxazole (SMX) and reference item (REF) fed bottles with non-carbon source (NC) fed bottles under sulphate reducing and methanogenic conditions were given in Figure 1 and 2, respectively.

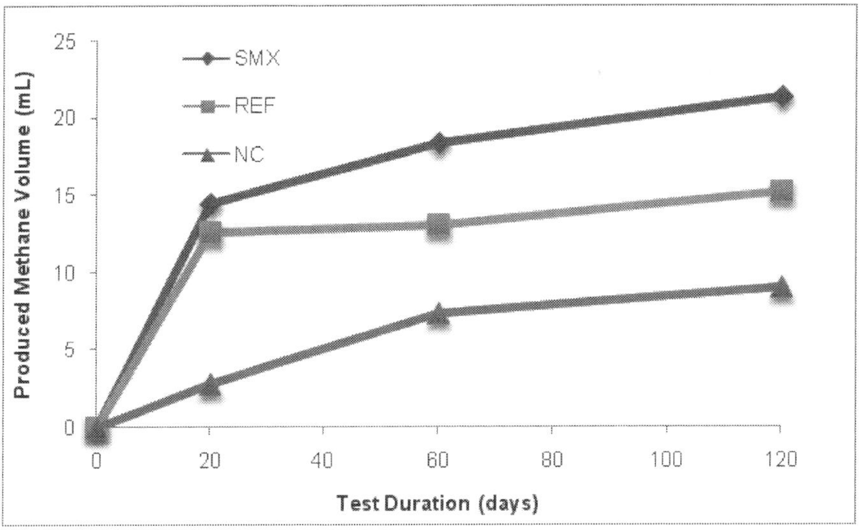

Figure 1. Methane production under sulfate reducing conditions

As seen in Figure 1, the maximum methane production associated with SMX was 21 mL while the maximum values were determined as 15 L and 9 mL in REF and NC test bottles, respectively, under sulphate reducing conditions. These values increased to 132 mL, 41 mL and 23 mL, respectively. This wide difference is expected as a result of sulphate inhibitory effect on methanogens. Another point to show the inhibition that most of the methane was produced during first 20 days under methanogenic conditions while methane production was slower under sulphate reducing conditions. Also it was known that sulphate reducers are much more versatile than methanogens and in environments where sulfate is present, sulfate-reducing bacteria compete with methanogenic consortia for common substrates. Compounds like propionate and butyrate, which require syntrophic consortia in methanogenic environments, are degraded directly by single species of sulfate reducing bacteria [94].

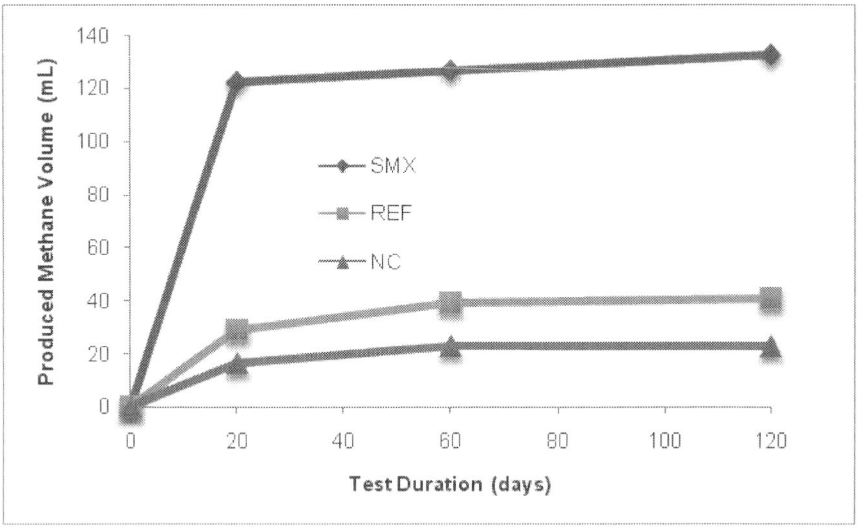

Figure 2. Methane production under methanogenic conditions

Positive and negative control groups were used to increase the reliability of the experiment. For positive control groups phenol was used as a carbon source. For all three electron-accepting conditions, phenol was biodegraded at the ratios between 74- 78% in 120 days, which indicated the ultimate biodegradation according to OECD protocol [13]. Measured CO_2 and biogas production within the negative control groups subtracted as blanks to reveal the actual biodegradation ratios. The CO_2 productions in the negative control test bottles reached a total of 4-12 mL in 120 days corresponding to 70 -100% of the theoretical CO_2 (Th CO_2) production while biogas production reached 40 mL corresponding to 100% of the Th biogas occurred via degradation of biomass completely.

3.2. Removal of dissolved organic carbon

Total organic carbon parameter was used to compare the biodegradation capacity of the antibiotic and reference item under nitrate reducing, sulphate reducing and methanogenic conditions. Also electron acceptors were measured in the test bottles. DOC removal was higher in the first 60 days in all electron-accepting condition. The removal between 60th-120th days, any significant changes were not observed.

In Figure 3, DOC and nitrate concentration changes in respect to time are given. ın the beginning of the experiment, nitrate concentration in each bottle was 250 mg/L. This concentration decreased to less than 10 mg/L in the first 20-day period of the experiment. Also decrease in DOC values was parallel to nitrate concentration except of SMX test bottles. The decrease in the DOC continued first 60 days while nitrate concentration was 1 mg/L.

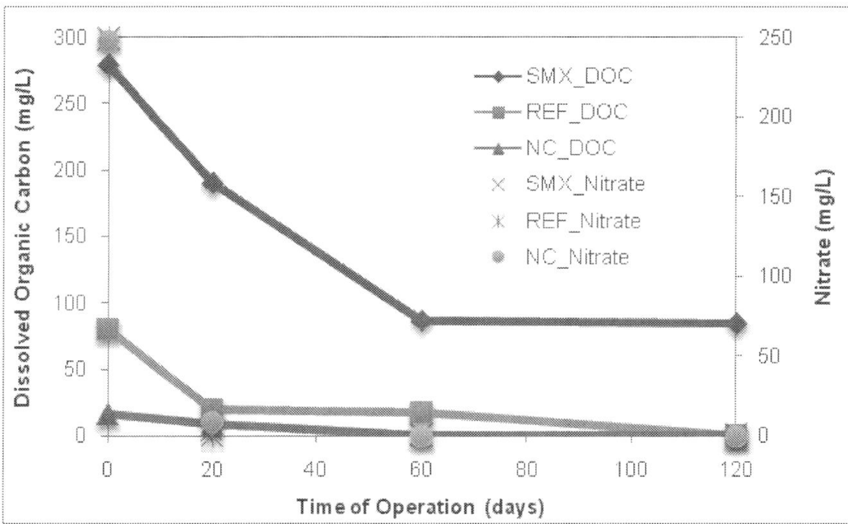

Figure 3. DOC and nitrate concentration in SMX, REF and NC bottles under nitrate reducing conditions

As seen in Figure 4, most of the DOC in the SMX bottle was consumed in the first 60 days. Also sulphate concentration decreased from 480 mg/L to 59 mg/L during same period.

In Figure 5, changes in DOC concentrations under methanogenic conditions are given. The results indicated that the removal of DOC mechanism was more quickly in the first 20 days. This pattern was also similar with the other respiration conditions. Also reference item was consumed in the same period. DOC concentration in SMX bottles decreased from 280 mg/L to 88 mg/L in the first 20 days. After this day, only 18 mg/L DOC was consumed and the final DOC concentration in SMX test bottles was determined as 70 mg/L.

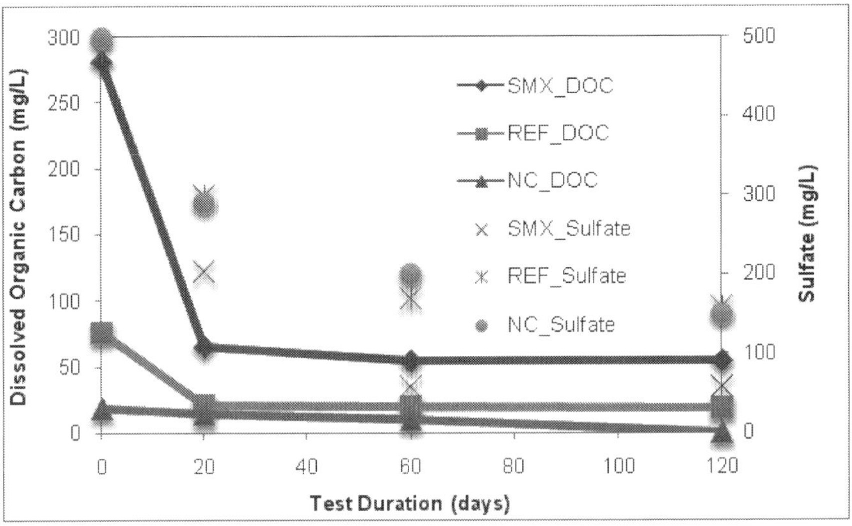

Figure 4. DOC and sulphate concentration in SMX, REF and NC bottles under sulphate reducing conditions

The most efficient DOC removal in SMX test bottles was observed under SRC and MC as 78/ and 74%, respectively. Under NRC, DOC removal was detected as 71%.

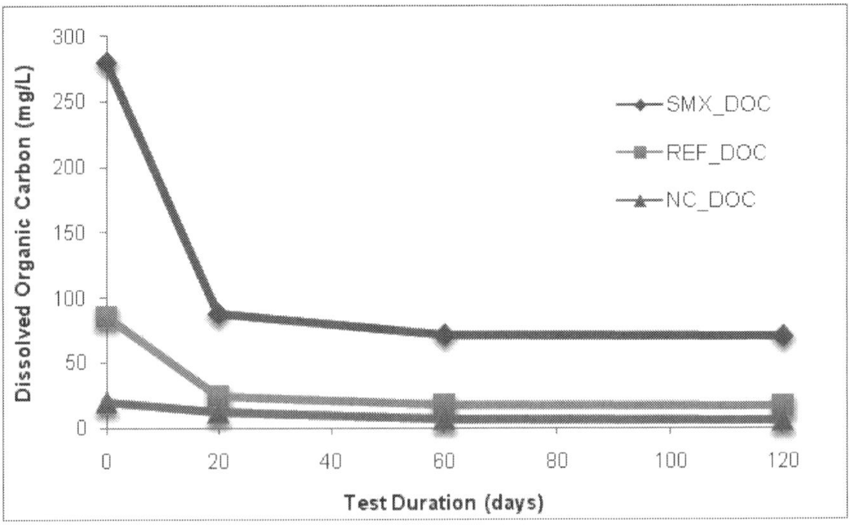

Figure 5. DOC concentration in SMX, REF and NC bottles under methanogenic conditions

In another study showed that SMX affected the propionic acid degradation and acetic acid utilization pathways in the higher concentrations [95]. Sponza and Demirden [96] also showed while sulfamerazine, which is another antibiotic from sulfonamid group, was being fed to the anaerobic system, an increase in VFA accumulation was observed with respect to rising of antibiotic concentration. Decreased utilization of butyrate and propionate is consistent with the fact that these substrates are used directly by bacteria, homoacetogens. SMX also has a bacteriostatic inhibition effect on folic acid production of especially gram positive and negative cocci [18]. VFAs are not directly used by methanogens, however different groups of syntrophic bacteria use specific VFAs.

3.3. Biodegradability of sulfamethoxazole and mass balance

SMX measurement was done for water and sludge matrix. The recovery was found as 92% after solid-phase extraction (SPE). Antibiotic measurement in the sludge showed that the SMX concentration in the sludge did not change in respect to time and it is found as 50,4±3 mg/L. This result indicated that velocity of biodegradation and sorption mechanisms are similar during the test. Antibiotic concentrations in water samples are given in Figure 6.

Antibiotic removal for three electro accepting conditions was same and it was detected as approx. 98% in water matrix. If the sorption mechanism takes into the consideration, the removal decreased to 70%. Most of the antibiotic was removed in the first 60 days. There is no significant change between 60th-120th days. Also the decrease in electron acceptor concentration under nitrate reducing conditions may be caused a negative impact on microbial activity [97]. However, it was clear that antibiotic removal was faster under methanogenic conditions. The results showed that 68% of SMX was removed under methanogenic conditions while ultimate SMX removal was 70%.

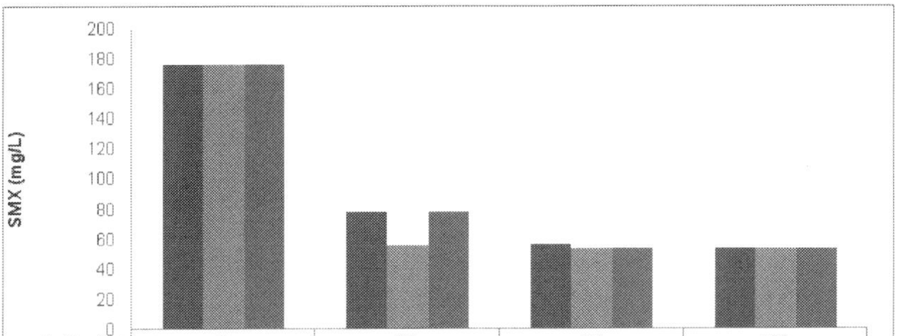

Test Duration (days)	0	20	60	120
SMX-N	176.00	78.27	56.59	53.23
SMX-M	176.00	55.75	53.30	53.30
SMX-S	176.00	78.28	53.43	53.34

Figure 6. Sulfamethoxazole concentration under nitrate reducing (SMX-N), sulphate reducing (SMX-S) and methanogenic (SMX-M) conditions

Figure 7 shows ultimate biodegradation ratios (evaluated according to gas production only derived from SMX biodegradation), sorption ratios according to SMX measurements within sludge and soluble microbial products (SP) and/or transformation products (TP) ratios that were calculated via DOC removal ratio compared with SMX biodegradation for each electron accepting condition throughout the operating period. SMX showed non-biodegradable behavior under SRC, NRC and MC according to OECD protocol [13].

SMX measurements within the sludge samples of the all experimental groups showed that 29% of the SMX sorbed to the solid media throughout the experiment time. Sorbed part of the SMX did not change for four sampling time. Stabile results indicated that sorption processes are more dominant rather than desorption processes since all serum bottles were shaken daily in order to increase the bioavailability of the carbon source. Yang et al. also confirmed the rapid sorption processes rather than biodegradation [10].

Under MC, biogas production showed that 23% of the SMX was mineralized. However, according to SMX and DOC measurements 40% of the SMX were removed from the liquid phase. This result indicated that parent compound transformed to SP and/or TP. 17% of the SMX was remained in liquid phase as its potential SP and/or TP. Gartiser et al. reported SMX as non-biodegradable compound (2.3%) as well [11]. Different results of two studies mainly emanated by the application of different methods and duration time of the experiments.

Under SRC, 32% of the SMX was ultimately biodegraded whereas; 8% of the parent compound transformed to SP and/or TP. Under NRC, 38% of the SMX was mineralized to CO_2 and 2% of the SMX converted to residual SP and/or TP. Biodegradation ratios within the conventional treatment plants which is reported by Hong et al. [98] complies with our results. In their study 40% of the SMX removed from liquid phase. Also in our study, anoxic biodegradation rate was the highest removal rate among the experimental groups. Overall elimination within three electron-accepting conditions was calculated as 69 %.

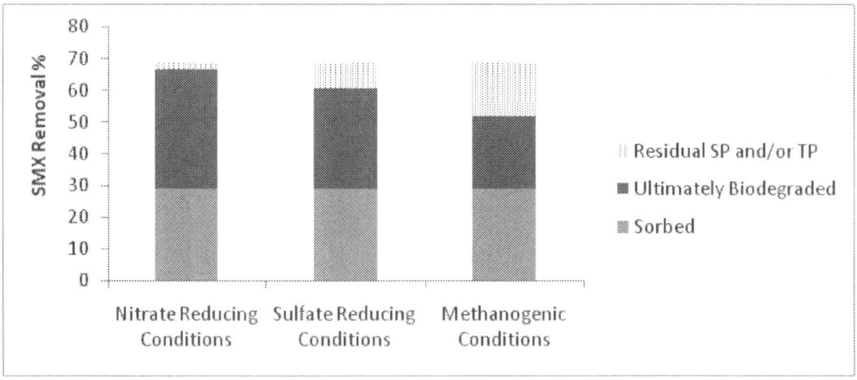

Figure 7. Biodegradation of SMX under different e-accepting conditions

3.4. Microbiological analyses

Q-PCR analyses were carried out for four sampling times. Four different taxonomic groups were quantified. These were; Bacteria, Archaea, methanogenic Archaea and Sulfate Reducing Bacteria (SRB). There was no significant change in the amount of these populations during 120 days (data not shown). However, under methanogenic conditions, biogas, antibiotic concentration and microbial quantification data indicate that there was a strong correlation between antibiotic concentration and amount of bacterial and methanogenic species. This correlation was a strong proof of the usability of SMX and showed that the bacterial and archaeal community continued to work together while this compound was only carbon source. Figure 8 and Table 7 show the changes within these groups under methanogenic conditions and their correlations with each other, respectively.

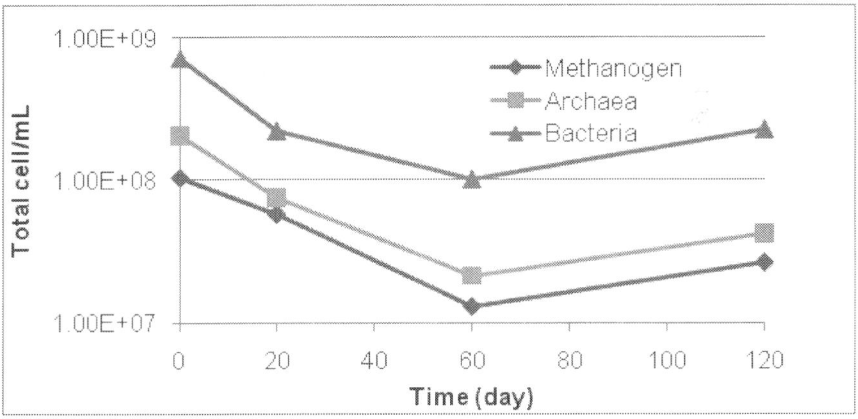

Figure 8. Microbial cell counts under methanogenic conditions

Table 7. Correlation analyses for methanogenic conditions (n=3, $p<0.05$)

As it can be seen in Table 7 for methanogenic conditions, there was a strong positive correlation between methanogenic count and bacterial count; also, they had a positive correlation with DOC and SMX concentration change over time. This observation shows that there may be syntrophic relationship between bacteria and methanogens in methanogenic conditions. In addition, SMX concentration is strongly correlated with bacterial and methanogenic count.

Insignificant change in microbial groups can be explained with two approaches. 1- SMX may have an inhibition on bacterial growth so SMX biodegradation delayed and microbial cells have faced with starvation as described by Gartiser et al. [12]. 2- Multi antibacterial resistant bacteria have been described by many authors [7, 99]. Based on that knowledge, bacterial populations in the batch tests might have gained resistance to SMX with time but would not able to use SMX efficiently. In this case, bacterial populations lowered their metabolic functions to survive rather than grow. Also changes in the population dynamics can be derived from the microbial interactions. Thus methanogens were not affected by SMX. Otherwise biogas production wouldn't have occurred or would have been inhibited because of their susceptibility to toxic compounds [100].

More detailed microbiological approach was applied in a parallel study [95]. The author tested the inhibition effect and biodegradability characteristic of same compound in long-term semi-continuous operation under anaerobic conditions. In that study, *Clostiridum spp.* was found in the system independently of operation time and SMX concentration in the system according to 16S rDNA clone library and denaturing gradient gel electrophoresis (DGGE) studies. It was also expected due to they are responsible to fermentation and some species especially produce the ethanol. Additionally, *Clostridium spp.* have the role on the starch degradation by exo-enzymes. Other OTUs, which were detected in the system, almost belong to the uncultured clones or unclassified bacterial cultured species. In addition to bacterial results, archaeal studies showed that acetoclastic methanogenic species disappeared in the last phases of operation in which SMX concentration increased. However the abundance of hydrogenotrophic methanogenswere higher than acetoclastic species and they were dominant during the operation.

Looking at this point, more detailed microbiological approach was needed to give the answers of two main questions: Which microbial groups are directly affected by SMX and which ones utilize this compound as a substrate? Next generation sequencing (NGS) based on DNA and also cDNA produced from total RNA represent more details about microbial community in each operation period. NGS is a novel sequencing technology for metagenomic studies. The main advantage of this technique is to sequence the mix GDNA directly without any preliminary study. By this means, the process does not cover the bias coming from polymerase chain reaction (PCR) and cloning. Additionally stable isotope probing (SIP) technique may be a good option to find the answer about which microbial groups utilize directly the SMX. In this technique, a labeled compound is given as substrate and then the produced GDNAs are monitored by labeled elements coming from utilized compound.

4. Conclusion

In the light of evaluations presented above, the significant findings of the study on the biodegradability characteristics of sulfamethoxazole under different electron accepting conditions may be outlined as follows:

The results suggested that the nature of the biodegradability characteristic of SMX are similar under nitrate reducing, sulphate reducing and methanogenic conditions and it was clear that biological treatment is suitable for this compound to remove from the wastewater during long retention times. However, methanogenic conditions should be selected because of obtaining biogas to use as energy source.

Microbial studies showed a syntrophic relationship between bacteria and methanogens in methanogenic conditions. Quantification of the main microbial groups has given general idea

http://dx.doi.org/10.5772/56049

about the effect of SMX and showed the next step to clarify this mechanism: To focus microbial kinetics in terms of metabolic expressions on mRNA level and also quantification of antibiotic resistance genes in the system, which is operated under methanogenic conditions, give more information about the removal mechanism of this compound. Also for detailed information about microbial community and changes in the community, NGS is a good option. SIP is the direct method to observe which microbial species utilize the SMX and its transformation products.

Acknowledgements

This study was funded by The Scientific and Technological Research Council of Turkey (TUBITAK), Project No: 109Y012.

Author details

Zeynep Cetecioglu[1], Bahar Ince[2], Samet Azman[1], Nazli Gokcek[1], Nese Coskun[1] and Orhan Ince

*Address all correspondence to: cetecioglu@itu.edu.tr

1 Istanbul Technical University, Environmental Engineering Department, Istanbul, Turkey

2 Bogazici University, Institute of Environmental Sciences, Istanbul, Turkey

References

[1] Li, B, & Zhang, T. Biodegradation and adsorption of antibiotics in the activated sludge process. *Environmental Science and Technology*. (2011). , 44, 3458-3473.

[2] Xiao, Y, Chang, H, Jia, A, & Hu, J Y. Trace analysis of quinolone and fluoroquinolone antibiotics from wastewaters by liquid chromatography electrospray tandem masss-pectrometry. Journal of Chromatography A, (2008).

[3] Kummerer, K. *Pharmaceuticals in the Environment: Sources, Fate, Effects and Risks*. 1st ed. Berlin, *Springer-Verlag*; (2001).

[4] Hirsch, R, Ternes, T, Haberer, K, & Kratz, K L. Occurrence of antibiotics in the aquatic environment. *Science of the Total Environment*. (1999).

[5] Golet, E M, Strehler, A, Alder, A. C, & Giger, W. Determination of fluoroquinolone antibacterial agents in sewage sludge and sludge-treated soil using accelerated sol-

vent extraction followed by solid-phase extraction. *Analytical Chemistry*. (2002). , 74(21), 5455-5462.

[6] Andreozzi, R, Caprio, V, Ciniglia, C, & De Champdore, M. Lo Giudice R, Marotta R, Zuccato E. Antibiotics in the environment occurrence in Italian STPs, fate, and preliminary assessment on algal toxicity of amoxicillin. *Environmental Science and Technology*. (2004). , 38(24), 6832-6838.

[7] Munir, M, Wong, K, & Xagoraraki, I. Release of antibiotic resistant bacteria and genes in the effluent and biosolids of five wastewater utilities in Michigan. *Water Science and Technology*. (2011). , 45, 681-693.

[8] Nieto, A, Borrull, F, Pocurull, E, & Marce, R M. Occurrence of pharmaceuticals and hormones in sewage sludge.*Environmental Toxicological Chemistry*.(2010). , 29(7), 1484-1489.

[9] Drillia, P, Dokianakis, S N, Fountoulakis, M S, Kornaros, M, Stamatelatou, K, & Lyberatos, G. (2005). *Journal of Hazardous Materials*., 122, 259-265.

[10] Yang, S, Lin, C F, Lin, A Y, & Hong, P A. Sorption and biodegradation of sulfonamide antibiotics by activated sludge: Experimental assessment using batch data obtained under aerobic conditions. *Water Research*.(2011). , 45, 3389-3397.

[11] Gartiser, S, Urich, E, Alexy, R, & Kummerer, K. Anaerobic inhibition and biodegradation of antibiotics in ISO test schemes. *Chemosphere*. (2007). , 66, 1839-1848.

[12] Gartiser, S, Urich, E, Alexy, R, & Kümmerer, K. Ultimate biodegradation and elimination of antibiotics in inherent tests. *Chemosphere*. (2007). , 67, 604-613.

[13] Anaerobic biodegradability of organic compounds in digested sludge- method by measurement of gas production (OECD 311)France; (2006). OECD- Organization for Economic Co-operation and Development.

[14] Korolkovas, A. (1976). Essentials of Medicine Chemistry. Wiley, New York.

[15] Foye, W. O, Lemke, T. L, & Williams, D. A. (1995). Principles of Medicinal Chemistry. MD. Williams&Wilkins, Baltimore.

[16] Kummerer, K, & Henninger, A. (2003). Promoting resistance by the emission of antibiotics from hospitals and households into effluents. Clin. Microbiol. Infec. , 9, 1203-1214.

[17] Cunningham, V. (2008). Special characteristics of pharmaceuticals related to environmental fate. In. Kummerer, K. (Ed.), Pharmaceuticals in the Environment. Sources, Fate, Effects and Risk, third ed. Springer, Berlin Heidelberg, , 23-34.

[18] Sweetman, S. C. Ed.), (2009). Martindale. The complete drug reference, 36th edition. Pharmaceutical Press, London UK.

[19] Wise, R. (2002). Antimicrobial resistance. priorities for action. J. Antimicrob. Chemoth., , 49, 585-586.

[20] http://www.esac.ua.ac.be/main.aspx?c=*ESAC2&n=1066l

[21] European Federation of Animal Health (FEDESA)(2001). Antibiotic use in farm animals does not threaten human health. FEDESA/FEFANA Press release, Brussels, 13 July.

[22] Union of Concerned Scientists(2001). Percent of all Antibiotics Given to Healthy Livestock. Press release, 8 January, Cambridge, MA

[23] Kolpin, D. W. Fur Kolpin long, E.T., Meyer, M.T., Thurman, E.M., Zaugg, S.D., Barber, L.B., Buxton, H.T., (2002). Pharmaceuticals, hormones, and others organic wastewater contaminants in U.S. streams, a national reconnaissance. Environ Sci Technol 36.1202-1211., 1999-2000.

[24] Miao, X. S, Bishay, F, Chen, M, & Metcalfe, C. D. (2004). Occurrence of Antimicrobials in the final effluents of waste water treatment plants in Canada. Environ Sci Technol 38., 3533-3541.

[25] Kummerer, K. (2004). Pharmaceuticals in the Environment, Ed. Kummerer, K., 2nd Ed. Springer, Verlag.

[26] Rönnefahrt, I. (2005). Verbrauchsmengen in der Bewertung des Umweltrisikos von Humanarzneimitteln, In. Umweltbundesamt (Hrsg) Arzneimittel in der Umwelt- Zu Risiken und Nebenwirkungen fragen Sie das Umweltbundesamt. Dessau, UBA texte 29/05.

[27] Kummerer, K. (2008). Antibiotics in the Environment. Pharmaceuticals in the Environment, Ed. Kummerer, K., 3rd Ed. Springer, Verlag., 3-35.

[28] Sattelberger, S. (1999). Arzneimittelruckstande in der Umwelt, Bestandsaufnahme und Problemstellung. Report des Umweltbundesamtes Österreich, Wien.

[29] Verbrugh, H. A, & De Neeling, A. J. Eds, (2003). Consumption of antimicrobial agents and antimicrobial resistance among medically important bacteria in the Netherlands. SWAB NETHMAP.

[30] Calamari, D, Zuccato, E, Castiglioni, S, Bagnati, R, & Fanelli, R. (2003). Strategic survey of therapeutic drugs in the rivers Po and Lambro in northern Italy, Environ Sci Technol., , 37, 1241-1248.

[31] Karabay, O. (2009). Türkiye'de antibiyotik kullanımı ve direnç nereye gidiyor?, ANKEM Dergisi, 23(2), 116-120.

[32] Larsson, D. G, De Pedro, C, & Paxeus, N. (2007). Effluent from drug manufactures contains extremely high levels of pharmaceuticals. J. Hazard. Mater. , 148, 751-755.

[33] Li, D, Yang, M, Hu, J, Ren, L, Zhang, Y, Chang, H, & Li, K. (2008). Determination and fate of oxytetracycline and related compounds in oxyteracycline production wastewater and the receiving river. Environ. Toxicol. Chem., , 27, 80-86.

[34] Li, D, Yang, M, Hu, J, Zhang, Y, Chang, H, & Jin, F. (2008). Determination of penicillin G and its degradation products in a penicillin production wastewater treatment plant and the receiving river. Water Res., , 42, 307-317.

[35] Thomas, K. V. (2008). The relevance of different point sources. Lecture given at the "First International Conference on Sustainable Pharmacy", April 2008, Osnabruck, Germany <http.//www.dbu.de/550artikel27307_135.html>., 24-25.

[36] Tolls, J. (2001). Sorption of veterinary pharmaceuticals in soils. a review. Environ. Sci. Technol., , 35, 3397-3406.

[37] Christian, T, Schneider, R. J, Färber, H. A, Skutlarek, D, Meyer, M. T, & Goldbach, H. E. (2003). Determination of antibiotic residues in manure, soil, and surface waters, Acta Hydroch. Hydrob., , 31, 36-44.

[38] Gu, C, & Karthikeyan, K. G. (2008). Sorption of the antibiotic tetracycline to humic-mineral complexes. J. Environ. Qual. , 37, 704-711.

[39] Trivedi, P, & Vasudevan, D. (2007). Spectroscopic investigation of ciprofloxacin speciation at the goethite-water interface. Environ. Sci. Technol., , 4, 3153-3158.

[40] Viola, G, Facciolo, L, Canton, M, & Vedaldi, D. Dall'Acqua, F., Aloisi, G.G., Amelia, M., Barbafina, A., Elisei, F., Latterini, L., (2004). Photophysical and phototoxic properties of the antibacterial fluoroquinolones levofloxacin and moxifloxacin. Chem. Biodivers., , 1, 782-801.

[41] Werner, J. J, Chintapalli, M, Lundeen, R. A, Wammer, K. H, Arnold, W. A, & Mcneill, K. (2007). Environmental photochemistry of tylosin. efficient, reversible photoisomerization to a less-active isomer, followed by photolysis. J. Agric. Food Chem., , 55, 7062-7068.

[42] Hu, D, & Coats, J. R. (2007). Aerobic degradation and photolysis of tylosin in water and soil. Environ. Toxicol. Chem. , 26, 884-889.

[43] Hu, Z, Liu, Y, Chen, G, Gui, X, Chen, T, & Zhan, X. (2011). Characterization of organic matter degradation during composting of manure-straw mixtures spiked with tetracyclines. Bioresource Technology, doi.j.biortech.2011.05.003.

[44] Werner, J. J, Arnold, W. A, & Mcneill, K. (2006). Water hardness as a

[45] Kallenborn, R, Fick, J, Lindberg, R, Moe, M, Nielsen, K. M, Tysklind, M, & Vasskog, T. (2008). Pharmaceutical residues in Northern European environments. consequences and perspectives. In. Kummerer, K. (Ed.), Pharmaceuticals in the Environment. Sources, Fate, Effects and Risk, third ed. Springer, Berlin-Heidelberg, , 61-74.

[46] Arslan-alaton, I, & Caglayan, A. E. (2006). Toxicity and biodegradability assessment of raw and ozonated procaine penicillin G formulation effluent, Ecotoxicol. Environ. Safe, , 63, 131-140.

[47] Gonzalez, O, Sans, C, & Esplugas, S. (2007). Sulfamethoxazole abatement by photo-Fenton toxicity, inhibition and biodegradability assessment of intermediates. J. Hazard. Mater. , 146, 456-459.

[48] Paul, T, Miller, P. L, & Strathmann, T. J. (2007). Visible-light-mediated TiO2 photocatalysis of fluoroquinolone antibacterial agents. Environ. Sci. Technol., , 41, 4720-4727.

[49] Turiel, E, Bordin, G, & Rodríguez, A. R. (2005). Study of the evolution and degradation products of ciprofloxacin and oxolinic acid in river water samples by HPLC-UV/MS/MS-MS. J. Environ. Monitor., , 7, 189-195.

[50] Samuelsen, O. B. (1989). Degradation of oxytetracycline in seawater at two different temperatures and light intensities, and the persistence of oxytetracycline in the sediment from a fish farm. Aquaculture, , 83, 7-16.

[51] Oka, H, Ikai, Y, Kawamura, N, Yamada, M, Harada, K, Ito, S, & Suzuki, M. (1989). Photodecomposition products of tetracycline in aqueous solution.*J. Agric. Food Chem.*, , 37, 226-231.

[52] Lunestad, B. T, & Goksøyr, J. (1990). Reduction in the antibacterial effect of oxytetracycline in sea water by complex formation with magnesium and calcium. *Dis. Aquat. Organ.*, , 9, 67-72.

[53] Boree, A. L, Arnold, W. A, & Mcneill, K. (2004). Photochemical fate of sulfa drugs in the aquatic environment. sulfa drugs containing five-membered heterocycylic groups, *Environ. Sci. Technol.*, , 38, 3933-3940.

[54] Halling-sørensen, B, Lykkeberg, A, Ingerslev, F, Blackwell, P, & Tjørnelund, J. (2003). Characterisation of the abiotic degradation pathways of oxytetracyclines in soil interstitial water using LC-MS-MS. *Chemosphere* 50 (10), 1331-1342.

[55] Li, K, Yediler, A, Yang, M, Schulte-hostede, S, & Wong, M. H. (2008). Ozonation of oxytetracycline and toxicological assessment of its oxidation by-products. *Chemosphere* , 72, 473-478.

[56] Dantas, R. F, Contreras, S, Sans, C, & Esplugas, S. (2007). Sulfamethoxazole abatement by means of ozonation. *J. Hazard Mater.* , 150, 790-794.

[57] Gonzalez, O, Sans, C, & Esplugas, S. (2007). Sulfamethoxazole abatement by photo-Fenton toxicity, inhibition and biodegradability assessment of intermediates*J. Hazard. Mater.* , 146, 456-459.

[58] Ingerslev, F, & Halling-sørensen, B. (2001). Biodegradability of metronidazole, olaqiondox, and tylosin, and formation of tylosin degradation products in aerobic soil/manure slurries*Ecotoxicol. Environ.* Safe , 48, 311-320.

[59] Ingerslev, F, Torang, L, Loke, M. L, Halling-sørensen, B, & Nyholm, N. (2001). Primary biodegradation of veterinary antibiotics in aerobic and anaerobic surface water simulation systems.*Chemosphere* , 44, 865-872.

[60] Gartiser, S, Urich, E, Alexy, R, & Kuummerer, K. (2007). Ultimate biodegradation and elimination of antibiotics in inherent tests.*Chemosphere,* , 67, 604-613.

[61] Kim, S, Eichhorn, P, Jensen, J. N, Weber, A. S, & Aga, D. S. (2005). Removal of antibiotics in wastewater. effect of hydraulic and solid retention timeson the fate of tetracycline in the activated sludge process. *Environ. Sci. Technol., 39,* 5816-5823.

[62] Maki, T, Hasegawa, H, Kitami, H, Fumoto, K, Munekage, Y, & Ueda, K. (2006). Bacterial degradation of antibiotic residues in marine fish farm sediments of Uranouchi Bay and phylogenetic analysis of antibiotic-degrading bacteria using 16S rDNA sequences. *Fisheries Sci.* , 72, 811-820.

[63] Jacobsen, P, & Berglind, L. (1988). Persistence of oxytetracyline in sediment from fish farms. *Aquaculture 70,* 365-370.

[64] Hansen, P. K, Lunestad, B. T, & Samuelsen, O. B. (1992). Effects of oxytetracycline, oxolinic acid and flumequine on bacteria in an artificial marine fish farm sediment. Can. *J. Microbiol.* , 38, 307-1312.

[65] Samuelsen, O. B, Torsvik, V, & Ervik, A. (1992). Long-range changes in oxytetracycline concentration and bacterial resistance towards oxytetracycline in fish farm sediment after medication. *Sci. Total Environ.,* , 114, 25-36.

[66] Samuelsen, O. B, Lunestad, B. T, & Fjelde, S. (1994). Stability of antibacterial agents in an artificial marine aquaculture sediment studied under laboratory conditions. *Aquaculture,* , 126, 183-290.

[67] Hektoen, H, Berge, J. A, Hormazabal, V, & Yndestad, M. (1995). Persistence of antibacterial agents in marine sediments. Aquaculture , 133, 175-184.

[68] Capone, D. G, Weston, D. P, Miller, V, & Shoemaker, C. (1996). Antibacterial residues in marine sediments and invertebrates following chemotherapy in aquaculture, *Aquaculture,* , 145, 55-75.

[69] Marengo, J. R, Brian, O, Velagaleti, R. A, & Stamm, R. R. J.M., (1997). Aerobic biodegradation of (14C)-sarafloxacin hydrochloride in soil. *Environ. Toxicol. Chem.,* , 16, 462-471.

[70] Donoho, A. L. (1984). Biochemical studies on the fate of monensin in animals and in the environment. *J. Anim. Sci.* , 58, 1528-1539.

[71] Gilbertson, T. J, Hornish, R. E, Jaglan, P. S, Koshy, K. T, Nappier, J. L, Stahl, G. L, Cazers, A. R, Napplier, J. M, Kubicek, M. J, Hoffman, G. A, & Hamlow, P. J. (1990). Environmental fate of ceftiofur sodium, a cephalosporin antibiotic. Role of animal excreta in its decomposition. *J. Agric. Food Chem.* , 38, 890-894.

[72] Samuelsen, O. B, Solheim, E, & Lunestad, B. T. (1991). Fate and microbiological effects of furazolidone in a marine aquaculture sediment. *Sci. Total Environ.,* , 108, 275-283.

[73] Thiele-bruhn, S. (2003). Pharmaceutical antibiotic compounds in soils- a review. *J. Plant. Nutr. Soil Sci.,* , 166, 145-167.

[74] Daughton, C. G, & Ternes, T. A. (1999). Pharmaceuticals and personal care products in the environment. agents of subtle change? *Environmental Health Perspectives,* , 107, 907-938.

[75] Stuer-lauridsen, F, Birkved, M, Hansen, L. P, Lutzhoft, H. C. H, & Halling-sorenson, B. (2000). Environmental risk assessment of human pharmaceuticals in Denmark after normal therapeutic use. *Chemosphere,* , 40, 783-793.

[76] Chelliapan, S, Wilby, T, & Sallis, P. J. (2006). Performance of an upflow anaerobic stage reactor (UASR) in the treatment of pharmaceutical wastewater containing macrolide antibiotics, *Water Res.,* 40 (3). , 507-516.

[77] Amin, M. M, Zilles, J. L, Greiner, J, Charbonneau, S, Raskin, L, & Morgenroth, E. (2006). Influence of the antibiotic erythromycin on anaerobic treatment of a pharmaceutical wastewater, *Environ. Sci. Technol.,* , 40, 3971-3977.

[78] Hartig, C, Storm, T, & Jekel, M. (1999). Detection and identification of sulphonamide drugs in municipal wastewater by liquid chromatography coupled with electrospray ionization tandem mass spectrometry. *J. Chromatogr.* A, , 854, 163-173.

[79] Alder, A. C, Mcardell, C. S, Golet, E. M, Ibrics, S, Molnar, E, Nipales, N. S, & Giger, W. (2001). Occurrence and fate fluoroquinolone, macrolide and sulfonamide antibiotics during wastewater treatment and in ambient waters in Switzerland. *In Pharmaceuticals and Personal Care Products in the Environment. Scientific and Regulatory Issues.* Daughton, C.G. and Jones-Lepp, T. (eds). Washington D.C.. Am. Chem. Soc., , 56-69.

[80] Golet, E. M, Alder, A. C, Hartmann, A, Ternes, T. A, & Giger, W. (2001). Trace determination of fluoroquinolone antibacterial agents in solid-phase extraction urban wastewater by and liquid chromatography with fluorescence detection. *Anal. Chem. 73*, 3632-3638.

[81] Lindsey, M. E, Meyer, M, & Thurman, E. M. (2001). Analysis of trace levels of sulfonamide and tetracycline antimicrobials, in groundwater and surface water using solid-phase extraction and liquid chromatography/mass spectrometry. *Anal. Chem.,* , 73, 4640-4646.

[82] Golet, E. M, Xifra, I, Siegrist, H, Alder, A. C, & Giger, W. (2003). Environmental exposure assessment of fluoroquinolone antibacterial agentsfrom sewage to soil. *Environ. Sci. Technol.* , 37, 3243-3249.

[83] Mcardell, C. S, Molnar, E, Suter, M. J. F, & Giger, W. (2003). Occurrence and fate of macrolide antibiotics in wastewater treatment plants and in the Glatt Valley Watershed, Switzerland. *Environ. Sci. Technol.,* , 37, 5479-5486.

[84] Gobel, A, Thomsen, A, Mcardell, C. S, Alder, A. C, Giger, W, Thesis, N, Löffler, D, & Ternes, T. (2005). Extraction and determination of sulfonamide and macrolide antimicrobials and trimethoprim in sewage sludge. *J. Chromatogr.* A , 1085, 179-189.

[85] Drillia, P, Dokianakis, S. N, Fountoulakis, M. S, Kornaros, M, Stamatelatou, K, & Lyberatos, K. G. (2005). On the occasional biodegradation of pharmaceuticals in the activated sludge process. The example of the antibiotic sulfamethoxazole. *J. Haz. Mat.* , 122, 259-265.

[86] Alexy, R, Scholl, A, & Kummerer, K. (2004). Elimination and degradability of 18 antibiotics studied with simple tests, *Chemosphere,* , 57, 505-512.

[87] Matamoros, V, Caselles-osorio, A, Garcia, J, & Bayona, J. M. (2008). *Sci. Tot. Env.,* 394. , 171-176.

[88] Karci, A, & Balciogli, A I. Investigation of the tetracycline, sulfonamide, and fluoroquinolone antimicrobial compounds in animal manure and agricultural soils in Turkey, *Science of the Total Environment.* (2009). , 407, 4652-4664.

[89] Ritmann, B E, & Mc Carty, P L. Environmental Biotechnology: Principles and Applications, 1st ed.,U.S.A., Mc GrawHill; (2001).

[90] Lane, D J. *S/23S rRNA sequencing, Nucleic acid techniques in bacterial systematics.* England, Wiley;(1991). , 1991, 205-248.

[91] Takai, K, & Horikoshi, K. Rapid detection and quantification of members of the archaeal community by quantitative PCR using fluorogenic probes. *Applied Environmental Microbiology.* (2000). , 66, 5066-5072.

[92] Sawayama, S, Tsukahara, K, & Yagishita, T. Phylogenetic description of immobilized methanogenic community using real-time PCR in a fixed-bed anaerobic digester. *Bioresource Technology.*(2006). , 97, 69-76.

[93] Geets, J, Borremans, B, Diels, L, Springael, D, Vangronsveld, J, Van Der Lelie, D, & Vanbroekhoven, K. DsrB gene-based DGGE for community and diversity surveys of sulfate-reducing bacteria. *Journal of Microbiological Methods.* (2005). , 66, 194-205.

[94] Stams, A. J. M. Oude Elferink, S. J.W.H., Westermann, P.Metabolic Interactions Between Methanogenic Consortia and Anaerobic Respiring Bacteria. In: Scheper, T. (ed.) Advances in Biochemical Engineering Biotechnology, (2003). , 31-56.

[95] Cetecioglu, Z. Evaluation of Anaerobic Biodegradability Characteristics of Antibiotics and Toxic/Inhibitory Effect on Mixed Microbial Culture. PhD Thesis. Istanbul Technical University, (2011).

[96] Sponza, D. T, & Demirden, P. (2007). Treatability of sulfamerazine in sequential up-flow anaerobic sludge blanket reactor (UASB)/completely stirred tank reactor (CSTR) processes. Separation and Purification Technology, 56., 108-117.

[97] Thomas, K. L, Lloyda, D, & Boddya, L. Effects of oxygen, pH and nitrate concentration on denitrification by Pseudomonasspecies, FEMS Microbiology Letters, (1994).

[98] Hong, C, & Ying, H J. LeZheng W and Bing S. Occurrence of sulfonamide antibiotics in sewage treatment plants. *Chinese Science Bulletin.* (2008)., 53(4), 514-520.

[99] Iwane, T, Urase, T, & Yamamoto, K. Possible impact of treated wastewater discharge on incidence of antibiotic resistant bacteria in river water. *Water Science and Technology.* (2001)., 43, 91-99.

[100] Dworkin, M, Falkow, S, Rosenberg, E, Schleifer, K H, & Stackebrandt, E. *The Prokaryotes,* 3nd ed., U.S.A, Springer; (2006).

Chapter 10

Biodegradation in Animal Manure Management

Matthieu Girard, Joahnn H. Palacios, Martin Belzile,
Stéphane Godbout and Frédéric Pelletier

Additional information is available at the end of the chapter

http://dx.doi.org/10.5772/56151

1. Introduction

Typical manure management strategies from intensive livestock feeding operations in Canada include the pre-storage of manure inside the animal buildings, long-term storage at the farm and finally field application of manure as fertilizer. Different biodegradation phenomena can occur in each of these steps, but naturally occurring biodegradation can cause harmful emissions. However, when used properly, biodegradation can also be beneficial and reduce pollution from animal wastes. This chapter will describe in detail the different processes involved in the biodegradation of manure, the emissions that are produced as well as how biodegradation can be used to treat both the manure and the emissions from manure management. The phenomena and systems described here can be applied to most livestock feeding operations (dairy and beef cattle, poultry, egg production, hog, etc.), but the specific examples and results will be provided for the swine industry.

2. Phenomena

2.1. Manure composition

Manure from animal husbandry contains a wide range of compounds that can be used by microorganisms for energy requirements or anabolic processes. In general, manure contains organic matter, nitrogen, phosphorous and potassium as well as numerous micronutrients (sulphur, copper and zinc for example). The specific concentrations of these components may vary according to several factors: building and storage management as well as the genetics of the animals, their growth stage and their diet. For example, experience has shown that hog diets supplemented with phytase had the effect of, among others, reducing the release of phosphorus in manure. Manure composition may also vary with water dilution when using water-saving drinkers in the building or a roof structure to cover the manure storage pit for example.

According to results from the analysis of various types of manure carried out by the Research and Development Institute for the Agri-Environment (IRDA), the typical swine manure composition for maternity, nursery and growing-finishing stages can be represented by the average values found in Table 1 [1]. The main difference between growth stages is related to the dry matter content. Indeed, manure produced by grower-finisher pigs is generally dryer and therefore more concentrated in nutrients.

Parameter	Unit	Growth stage		
		Maternity	Nursery	Growing-finishing
Dry matter	%	1.8	2.7	4.7
Total nitrogen	%	0.2	0.3	0.6
Ammonium nitrogen	mg/kg	1488	1545	2846
Phosphorus	mg/kg	593	762	1690
Potassium	mg/kg	1049	1964	3405
Copper	mg/kg	8.9	29.6	49.9
Zinc	mg/kg	41.0	208.7	82.9
Manganese	mg/kg	10.3	14.9	29.85
Calcium	mg/kg	697	701	1700
Magnesium	mg/kg	213	311	674

Table 1. Typical composition of swine manure for each growth stage (adapted from [1])

To provide a full description of manure composition, values for organic matter must also be considered. For swine manure, values of 19 to 51 gO_2/L as COD (chemical oxygen demand) are normally encountered [2,3]. The organic matter content depends essentially on the type of feed, manure management and manure age. After excretion, manure decomposes naturally; suspended solids contained in the manure are hydrolyzed into dissolved elements and biodegradation by microorganisms occurs. This decomposition of manure may be favored by appropriate conditions which depend on the proportion of elements contained in the slurry, the amount of oxygen, the pH and the temperature.

2.2. Aerobic biodegradation

Different microorganisms can grow by using various compounds found in manure both in the presence and in the absence of oxygen. The microorganisms can also be classified according to the compounds consumed.

Organic matter represents an important fraction of swine manure and includes many compounds that can be separated in four fractions: readily biodegradable (S_S), slowly biodegradable (X_S), inert soluble (S_I) and inert particulate (X_I). When considering biodegradation, only the biodegradable fractions (S_S and X_S) are taken into account. The S_S fraction is usually in soluble form and is composed of relatively small molecules such as volatile fatty acids (acetic, butyric and valeric acids), monosaccharides (sugar) and alcohols [3]. On the other hand, the X_S fraction is usually found as particles and is composed of high molecular weight organic polymers or dead biomass. This fraction of the organic matter cannot be directly assimilated by microorganisms and must first be hydrolyzed to S_S. The distribution of the organic matter can be quite variable among the different fractions and depends on many factors such as the type of feed and the manure storage time. For the S_S fraction, values from the literature vary from 8 to 30% of the total COD, from 30 to 60% for the X_S and from 10 to 60% for the inert fractions (S_I and X_I) [2-4]. Various types of microorganisms can degrade organic matter: bacteria, protozoa and fungi. As shown in equation 1, these microorganisms degrade the organic matter and release carbon dioxide (CO_2), water and biomass:

$$\text{Organic matter} + O_2 + \text{Nutrients} \rightarrow \text{Biomass} + CO_2 + H_2O \tag{1}$$

Nitrogen in manure can be found as ammonium (NH_4^+), trapped in organic molecules or as urea. Both the organic nitrogen and the urea must be broken down into NH_4^+ by microorganisms before they can be accessible. The aerobic oxidation of NH_4^+, nitrification, follows two distinct steps: the transformation of NH_4^+ into nitrite (NO_2^-) by *Nitroso* bacteria (*Nitrosomonas* for example) followed by the oxidation to nitrate (NO_3^-) performed by *Nitro* bacteria (*Nitrobacter* for example) [5]. The relative kinetics between the two steps are controlled by the temperature and will determine which compound is accumulated (NO_2^- or NO_3^-). The two separate steps of the nitrification reaction are presented in equations 2 and 3 while the combined reaction is given in equation 4 [6]:

$$NH_4^+ + 3/2\, O_2 \rightarrow NO_2^- + 2H_2 + H_2O \tag{2}$$

$$NO_2^- + 1/2\, O_2 \rightarrow NO_3^- \tag{3}$$

$$NH_4^+ + 2\, O_2 \rightarrow NO_3^- + 2H_2 + H_2O \tag{4}$$

Regarding the other compounds in manure (phosphorous, potassium and heavy metals), they can used by microorganisms for different microbiological processes or to synthesise certain compounds such as DNA and amino acids.

2.3. Anaerobic biodegradation

Without oxygen present, (e.g. when all the dissolved oxygen has been exhausted by aerobic respiration) several compounds are released by the anaerobic metabolism of microorganisms still utilizing nutrients in manure. By a complex series of reactions, the anaerobic biodegradation of manure produces different gases, mainly methane (CH_4), hydrogen sulfide (H_2S), ammonia (NH_3) and CO_2, as well as many intermediate compounds; the most noteworthy are volatile fatty acids and other odorous molecules. A study from the North Carolina State University identified a total of 331 compounds that cause odours from manure [7].

Biological decomposition during storage or during anaerobic digestion contributes to the transfer of nutrients, especially nitrogen and phosphorus, between different fractions and chemical forms in manure [8]. For nitrogen, anaerobic digestion can break down organic nitrogen and produce NH_4^+ and NH_3. If oxidised nitrogen compounds (NO_2^- or NO_3^-) are present, heterotrophic microorganisms can use these compounds as an electron acceptor and produce nitrogen gas (N_2). This process is called denitrification and requires a source of easily biodegradable organic carbon. It can also produce nitrous oxide (N_2O), a potent greenhouse gas, as a by-product. For phosphorus, anaerobic digestion contributes to moving some of the dissolved portion into the bodies of bacteria that carry out the anaerobic digestion process. All of the phosphorous present in the manure will still be present in the digester sludge [9]. Anaerobic digestion may also change the pH and the chemical form of salts and metals, such as iron, calcium and magnesium, which may affect the amount of suspended phosphates as a result of precipitation processes [8].

There is a huge interest in controlling anaerobic biodegradation for bioenergy production purposes. In fact, the anaerobic digestion of manure in an airtight container, under certain conditions, will form biogas, an energy source composed of a mixture of CH_4, CO_2 and trace amounts of other gases. Anaerobic digestion is a multi-stage process (Figure 1). Communities of hydrolytic bacteria break down complex organic matter from manure to simpler compounds (sugars, amino acids and fatty acids). Then, acid forming bacteria convert the simple compounds to alcohols and carbon acids (volatile fatty acids), as well as hydrogen, CO_2, NH_3, NH_4^+ and H_2S [10]. An amount of acetic acid is also produced at this stage, which along with hydrogen, can be used directly by methanogens. Other molecules, such as volatile fatty acids must first be catabolised to produce acetic acid, as well as CO_2 and H_2 that can be directly used by methanogens.

3. Biodegradation as a source of emissions

3.1. Buildings

Biodegradation processes begin before manure is even expelled from the animals. Intestinal microorganisms anaerobically degrade nutrients as they pass through an animal's digestive system. Once expelled, manure comes in contact with oxygen and aerobic microorganisms can become dominant after several hours. However, most animal housing systems use some sort of pre-storage for the manure inside the barn where anaerobic conditions can once again take over. Emissions from biodegradation of manure inside animal buildings are therefore a mix of anaerobic and aerobic products and are removed by the ventilation air. The emission rates

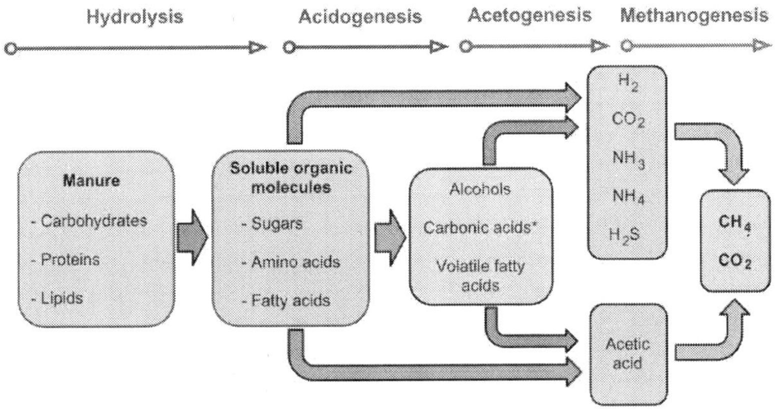

Figure 1. Process stages of anaerobic digestion (Modified from [10])

are affected by many factors such as ventilation flow rate, temperature, manure separation systems and animal activity among others.

Emissions from pig barns include a number of gases (CO_2, CH_4, and N_2O), dust particles (inhalable and breathable), bioaerosols (bacteria, viruses, endotoxins and fungi) and several other volatile compounds such as NH_3 and H_2S. In addition, an increasing importance is given to the odour nuisance associated with swine production. Thus, research in this area has become more important in recent years.

A baseline emission scenario of swine buildings was defined by [11] based on an inventory of gas, odour and dust emission data (Table 2). The resulting scenario provides a good overview of the magnitude of the emissions that are produced in swine production systems for the different growth stages (maternity, nursery and growing-finishing).

Odours, consisting of a complex mixture of several chemical compounds, are one of the major concerns in the emissions from the swine sector. Odours are expelled from barns by the ventilation system at 2.5 to 51.6 EOU/s/pig (EOU: European Odour Unit), depending on the growth stage. According to data, the nursery stage tends to emit fewer odours than the other stages. The use of odour reduction technologies in animal buildings, such as air cleaning technologies, could reduce the level of nuisance. In fact, downwind odours from confined feeding operations are considered to be a nuisance that may lead to a reduced quality of life by nearby residents.

NH_3 is produced by the degradation of urea in the urine on floors or still stored in the building. In a swine barn, average emissions range from 0.33 to 14 gNH_3/d/pig depending on the growth stage (Table 2). The rate of NH_3 emissions from buildings, storage structures and land spreading is favored when the liquid and solid fractions of the manure are not separated and

Parameter	Unit	Growth stage		
		Maternity	Nursery	Growing-finishing
			Europe	
Odours	EOU/s/pig	21.2 (16.3)	10 .69 (8.05)	13.75 (8.23)
NH_3	g/d/pig	14.2 (2.9)	0.94 (0.85)	7.75 (4.95)
CH_4	g/d/pig	57.7	10.7	12.42 (10.41)
CO_2	kg/d/pig	-	-	2.25 (087)
N_2O	g/d/pig	-	-	2.72 (3.26)
$PM_{2,5}$	mg/h/pig	-	6.4	6.9
PM_{10}	mg/h/pig	8.2 (2.55)	-	4.71 (2.50)
PM_{total}	mg/h/pig	50	-	-
			Canada	
Odours	EOU/s/pig	51.56 (53.45)	2.5 (0.69)	7.82 (8.19)
NH_3	g/d/pig	-	0.33 (0.14)	-
CH_4	g/d/pig	59.9 (45.13)	1.45 (2.37)	3.78 (3.76)
CO_2	kg/d/pig	5,29 (2,26)	0,55 (0,003)	1,71 (1,21)
N_2O	g/d/pig	0,0	0,007 (0,005)	0,04 (0,04)
$PM_{2.5}$	mg/h/pig	-	-	-
PM_{10}	mg/h/pig	-	-	-
PM_{total}	mg/h/pig	-	-	63 (4,12)

Confidence intervals in parentheses.

EOU : European odour unit

$PM_{2,5}$; PM_{10} and PM_{total}: particle matter smaller than 2.5, 10 and 100 micrometers respectively.

Table 2. Emissions scenario from swine buildings for different growth stages (adapted from [11])

when the manure: contains nitrogen from undigested food, has a high pH and has a high temperature. Moreover, a high contact area between the air and the manure as well as a high air movement at the surface increase NH_3 emissions.

In animal production, CH_4 comes from two sources, enteric fermentation in ruminants (cellulose digestion) and the decomposition of manure under anaerobic conditions. In the case of pig production, only the second source applies. At the building, where waste is handled in solid form in aerobic conditions, the production of CH_4 is minimal. Under anaerobic conditions, the production of CH_4 varies with the temperature and the composition of the manure. The emissions inventoried in Table 2 suggest that CH_4 is emitted from swine barns at 1.45 to 57.7 g/d/pig, with higher values at the maternity stage. The concern of CH_4 emissions from animal production systems is due to its high potential as a greenhouse gas and by the large quantities produced.

CO_2, produced by the metabolism of animals, is the most prominent gas in animal housing. Almost all CO_2 (96%) is produced by the respiration of animals; the rest comes from the decomposition of manure [12] and the combustion gases from heating systems. [13] showed that CO_2 emissions from pig manure can be estimated by multiplying the CH_4 emissions by a factor of 1.42 kg CO_2 per kilogram of CH_4.

N_2O emissions are not as high as the other gases expelled from animal buildings. For instance, according to Table 2, the maximal average N_2O emission was 2.7 g/d/pig. However, N_2O is also a major greenhouse gas and air pollutant. Considered over a 100-year period, it has 298 times more impact on climate change than CO_2 [14]. The formation of N_2O occurs during the processes of nitrification and denitrification over the course of manure management. In fact, it is during denitrification that N_2O is emitted, but to do so, nitrification must first take place. Under anoxic conditions, there is not enough oxygen for microorganisms who will take the oxygen they need from NO_3^-. Thus, the NO_3^- is then reduced to N_2. However, when the reaction is not complete (e.g. due to process kinetics or the sudden presence of dissolved oxygen), N_2O is emitted. It should be noted that this cannot occur under complete anaerobic conditions, since these microorganisms cannot survive.

The Environmental Protection Agency of the United States [15] defines particulate matter as a complex mixture of extremely small particles and liquid droplets. Moreover, according to [16], suspended particles in livestock buildings differ from other types of particles for three reasons: their concentration is usually 10 to 100 times greater than other indoor environments, they are vectors of odours and gases and they are biologically active, usually containing a wide variety of bacteria and microorganisms. The dust concentration in the air of buildings depends on several factors, such as relative humidity, temperature, level of animal activity, type and mode of feeding and presence of litter. Such particles have a significant impact on the health and well-being of both workers and animals. The consequences are mainly related to respiratory problems. However, in the inventory performed by [11] (Table 2), particular matter (PM) is the least documented parameter.

3.2. Storage

During manure storage, aerobic conditions are present at the surface of the manure, but after a shallow depth, anaerobic conditions prevail. Emissions from manure storage generally represent the intermediate and end products of anaerobic digestion: NH_3, CH_4, CO_2, H_2S and odours. The composition of the manure as well as the storage and weather conditions (temperature, pH, precipitation and wind) can affect the biodegradation of manure and will dictate the emission rates for each compound.

Typical gas emissions at the surface of manure tanks have been measured in the past using a special device. A sampling chamber, developed at IRDA, floats on the storage tank and takes gas samples in a confined space, swept by an airflow of 100 L/min. The gas concentrations are measured at the outlet of the chamber and the increase of the concentration compared to the ambient air is attributed to the emitting surface. A photo of the sample chamber floating on a manure storage tank is shown in Figure 2.

Figure 2. IRDA sampling device on a manure storage tank

Various research projects have been carried out using this instrument. Typical values for CH_4, CO_2 and N_2O are found in Table 3. Annual emissions represent the summation of the emissions over one year while daily averaged values represent this value distributed on a daily basis and the maximal daily values is the highest emission over one day during the year.

Parameter	Units	Gas		
		CH_4	CO_2	NH_3
Annual emissions	g/year/m²	7 940	19 096	530
Daily mean values	g/day/m²	22	52	1.5
Maximal daily values	g/day/m²	134	2662	6.0

Table 3. Gas concentrations from swine manure storage tanks

Generally, no N_2O is produced during swine manure storage [17] since anaerobic conditions prevail and the NH_4^+ cannot be oxidized to NO_3^-. N_2O is essentially generated once the slurry has been applied to agricultural land as a fertilizer where both aerobic and anoxic conditions can exist [18].

http://dx.doi.org/10.5772/56151

3.3. Fields

Measuring field emissions is a complex area of research and finding representative data for the phenomena which occur after manure application represents a challenge. That is why this section contains only an overview of data measured by different research groups.

Once manure has been applied to agricultural land as a fertilizer, microorganisms continue to degrade it and release additional compounds. Anaerobic conditions can remain for a short time after manure has been applied; therefore NH_3, CH_4, H_2S and odours are emitted directly following application. Other compounds, such as N_2O, require the combination of aerobic and anaerobic processes and can be produced for a long time after manure application. Certain field emissions (NH_3 and odours mainly) can be reduced by quickly incorporating manure into soil after application. Factors that influence these reactions are soil pH, exchange capacity and weather condition. It is therefore difficult to present typical values of gas emissions. An example of NH_3 emissions is however presented in Figure 3; emissions are presented on a daily basis following manure application.

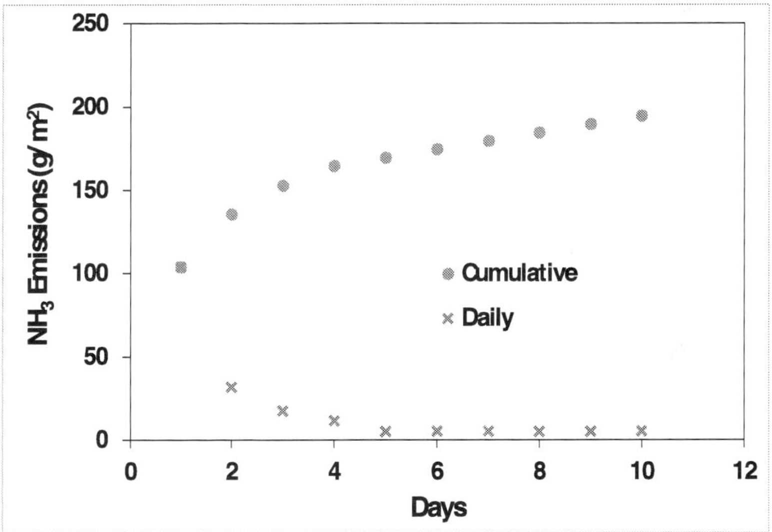

Figure 3. Gases emissions from the field fallowing application of manure (Adapted from [19])

Regarding the production of N_2O, emission values vary greatly. For example, cumulative values measured by [20] for a clay soil cultivated with silage corn varied between 0.255 and 0.873 g/m^2 for the entire growing season (May to October). Although it is only one example, it demonstrates the variation in the measurement of N_2O whereas several factors such as soil type and culture remained constant.

4. Biodegradation as treatment

4.1. Manure

Natural biodegradation phenomena in animal manure can cause harmful emissions, but certain biological processes, whether anaerobic or aerobic, can be used to treat manure. Aerobic biological processes for manure treatment can be relatively simple as in short-term manure aeration, which can remove up to 90% of the biodegradable organic compounds. This process can also significantly reduce odours (up to 96% as evaluated with volatile fatty acids) during manure storage for up to 190 days [21]. Biological processes using suspended biomass, such as aerated lagoons and activated sludge reactors developed for wastewater treatment, can be applied to treat manure [22]. Bioreactors using biomass fixed on a porous filter material can also be used to treat manure, but the solids must be removed prior to treatment in order to avoid clogging problems. The manure can be supplied to these systems from the bottom to obtain a submerged upflow system or from the top and trickle down the filter bed.

Laboratory-scale tests using upflow aerated biological filters showed good results for manure treatment: removal efficiencies of 88% for biodegradable organic matter and 94% for NH_4^+ with two 1.5 m^3 biofilters treating 8 m^3/d of flushed swine manure [23]. In another study, part of the effluent was recirculated to an anoxic reactor at the beginning of the process for complete nitrogen removal [24]. The bioreactor removed 72% of the organic matter as COD, 94% of the NH_4^+ as well as achieving a denitrification rate of 92%. This type of upflow biofilter is available commercially under the name Ekokan® Biofiltration Treatment System and removes between 90 and 98% of the NH_4^+ and between 40 and 70% of the biodegradable organic carbon from swine slurry pre-treated to remove solids [25]. The main disadvantage with this type of system is that the filter bed must be regularly backwashed to remove excess biomass.

In a trickling biofilter, the manure is supplied at the top and flows down through the filter bed. Trickling biofilters have been used for almost 100 years for wastewater treatment [5], but they have only recently been applied to manure treatment. Trickling biofilters can be quite simple consisting of a pile of filter material with passive aeration. However, the performance can be limited; results from preliminary tests showed removal efficiencies up to 56% for biodegradable organic matter and NH_4^+ [26]. Researchers in Québec (Canada) developed a highly engineered system using an enclosed biofilter with forced aeration, the Biosor® biofilter [27]. This type of system can provide a better performance with removal efficiencies up to 99% for the biodegradable organic matter and above 95% for the NH_4^+ [28,29]. Furthermore, a study on the nitrogen elimination mechanisms demonstrated that trickling biofilters can achieve simultaneous nitrification and denitrification which transforms the NH_4^+ directly to N_2 in one system [30]. By means of a mass balance, it was shown that 30% of the nitrogen was eliminated as N_2 and 10% as N_2O. For swine manure, loading rates between 0.017 and 0.035 m^3/m^2/d are generally recommended to avoid clogging problems [29,31]. Due to the high concentrations of nutrients in manure, these values are up to 2000 times lower than the loading rates recommended for wastewater treatment.

While biofiltration can be used to treat liquid manure, composting is a biological system that can treat solid materials to produce a biologically stable product rich in humic compounds [22]. In addition, composting can reach relatively high temperatures (40-60°C) which can reduce pathogenic microorganisms by up to 92% and improve the sanitary quality of the fertilizer produced [32]. Since swine manure is generally managed in liquid form, bulking agents must be added. These additives generally have a high carbon content, such as straw or sawdust, in order to improve the carbon to nitrogen mass ratio of the composting mix. A mass ratio between 25 and 30 is optimal, but values between 15 and 20 can be used to reduce bulking agent requirements. However, this increases reaction time by 30% [33,34]. A major disadvantage of composting is the loss of nitrogen, 10% as NH_3 and 3% as N_2O on average according to [35], which reduces the quality of the fertilizer. Since N_2O is a powerful greenhouse gas, emissions are particularly troubling. By adding nitrite-oxidizing bacteria, [36] were able to reduce N_2O emissions by 80%.

Aerobic bioreactors are usually operated at ambient temperatures with mesophilic microorganisms to maintain low operating costs, but reactors using a thermophilic biomass at temperatures between 50 and 75°C offer interesting advantages. This type of system reduces the quantity of pathogenic microorganisms to improve the sanitary quality of the manure. Furthermore, since nitrification ceases above 40°C, the nitrogen in manure is retained as NH_4^+ [37].

As previously described, anaerobic digestion can be used to treat manure and produce biogas for heat or energy requirements. This process also produces a good quality liquid fertilizer since nitrogen is mainly retained in the liquid. However, the nitrogen in the digestate (liquid effluent of the digester) is mainly NH_4^+ [38] and steps must be taken to reduce NH_3 losses (cover for storage reservoir, incorporation of digestate in soil after spreading, etc.). Manure can be fed to a digester either in batch, steady-state or semi-continuous modes. Batch systems are the simplest systems technically, but biogas production is irregular over time and the reaction rate is temperature dependant. Continuous systems are more complex, but provide consistent biogas production. Several parameters must be controlled for proper digester operation [39]:

- pH (between 6.8 and 7.4)
- dry matter content (maximum value of 14% for proper operation)
- NH_4^+ concentration (must be kept below 3 g/L to avoid inhibiting microorganisms)
- carbon to nitrogen ratio (helps optimise process and reduce reaction time).

A disadvantage specific to anaerobic digesters is the H_2S produced as a by-product which must generally be removed from the biogas to avoid deteriorating equipment.

4.2. Gaseous emissions

Biological treatment systems can also be used to treat gaseous emissions (NH_3, H_2S, greenhouse gases and odours) from manure management in order to improve air quality. Biological treatment of air is based on the capacity of microorganisms to transform organic and inorganic pollutants into non-toxic, odour free compounds.

Biological air treatment units (bioreactors) are an established technology, but research is ongoing for new media and reactor designs, microbial structure analysis and modeling of gas compounds removal [40]. Bioreactors can be used for reducing toxic VOC (volatile organic compound) emissions from industrial sources, but in agricultural applications, pollutant concentrations are lower and bioreactors must be simple, easy to operate and maintain and must meet investment and operating costs below those of the industrial sector.

[40-46] conducted detailed analyses from literature reviews providing interesting information on biological methods for air treatment: classification of biological reactors, analysis of the mechanisms involved in biological treatment, design of bioreactors and performance analysis.

The basic mechanisms for the biodegradation of pollutants are the same for all biological treatment systems: the contaminant is absorbed from the gas phase (contaminated air) to a liquid phase where biological degradation is initiated. For organic contaminants, oxidation reactions (and sometimes reductions) transform the contaminant in a mixture of CO_2, water vapour and biomass. The air pollutants (organic or inorganic) are used as a source of energy and/or carbon for the development of the microbial population. There are three types of bioreactors with different configurations, which can be used to achieve the transfer between the gas and the liquid phases and promote the microbial metabolic reactions: biofilters, biotrickling filters and bioscrubbers. Such equipment differ by the nature of the microbiological phase (microorganisms attached to the filter bed or suspended in the liquid) and by the circulation mode of the liquid (stationary or flowing) (Table 4).

Reactor	Microorganisms	Liquid phase
Biofilter	Fixed	Stationary
Biotrickling filter	Fixed	Flowing
Bioscrubber	Suspended	Flowing

Table 4. Classification of biological reactors for air treatment [47].

4.3. Biofilters

Biofiltration is the oldest and most widespread biotechnology for the treatment of gaseous emissions. The contaminated gases flow through a humid porous material, usually made of organic waste, where microorganisms capable of degrading the pollutants are present [48]. The microorganisms will grow attached to the material, thereby forming a wet biofilm wherein the air pollutants are absorbed and then degraded by the microorganisms. A liquid nutrient solution can be sprayed periodically over the filter bed to maintain proper moisture levels and to supplement certain nutrients if necessary. The moisture content of the filtration equipment and the maintenance of the biofilm are essential elements for maintaining the performance of this biological reactor. If a biofilter is not irrigated, moisture should be controlled by the humidity of the air fed to the device. This type of control on the moisture content of the filter media is not always effective and the variations in the humidity and temperature of the incoming air can affect the performance of the biofilter [41]. The filter material can also lose its porosity with time; it can even become clogged by excess biomass growth.

The most common type of biofilter is the open biofilter (Figure 4). This equipment can be exposed to atmospheric conditions and can be installed at ground level. Moreover, it usually uses packing materials readily available and affordable (soil or compost for example). The usual height of the filter bed of an open biofilter is between 1.0 and 1.5 m. Open systems are ideal for applications where space is not a constraint and they are known to be the least expensive solutions for odour control [41].

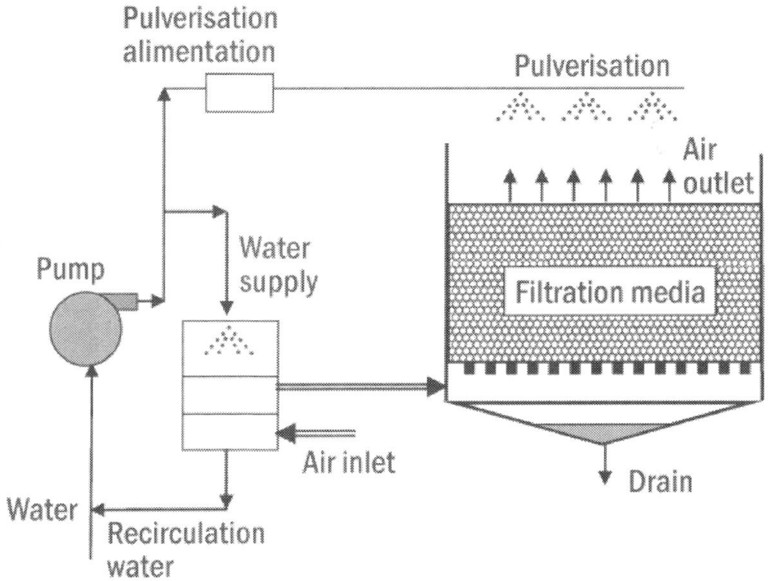

Figure 4. Diagram of an open biofilter system (Adapted from [42])

Closed biofilters (Figure 5) are generally more complex and may have either a circular or rectangular section. These air treatment systems allow a better control of some operating parameters (temperature, moisture, nutrients and pH) while being less sensitive to atmospheric conditions. The filter bed used in closed biofilters generally has a height that varies between 1.0 and 1.5 m and is composed of organic and/or inorganic materials. An air plenum at the inlet and the outlet of the biofilter is generally used for uniform air distribution. For most applications with a closed biofilter, downwards air circulation is more efficient than upward air flow due to a better control of filter bed moisture [47].

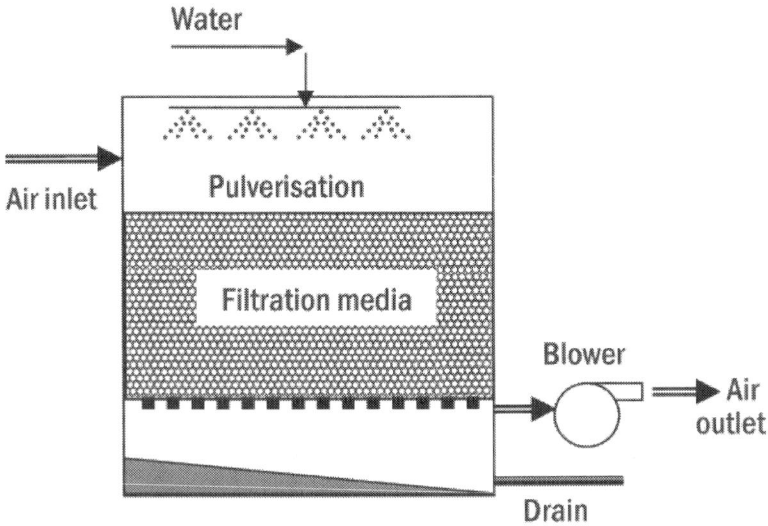

Figure 5. Diagram of a closed biofilter system (adapted from [41])

A study by [49] recommends maintaining the moisture content between 35 and 65% in the filter material. The average reductions of H_2S for low, medium and high relative humidity were 3, 72 and 87 %, respectively. Under the same conditions, the odour and NH_3 reductions were 42, 69 and 79% and 6, 49, and 81%, respectively. The optimal ratio of compost to wood chips recommended by the study for the treatment of air coming from swine buildings is 30% compost and 70% wood chips (on a weight basis).

[50] studied a pilot-scale biofilter treating swine ventilation air to determine the optimum operating conditions. The filter bed had a height of 0.5 m and was built using wood chips of at least 20 mm. The moisture content of the filter bed varied between 64 and 69%. Preliminary tests showed that the installation of a mechanical filter at the air inlet of the biofilter can reduce over 99% of airborne particles with an odour reduction of 19%. During the experiment, the system achieved a removal efficiency between 73 and 87% for NH_3. When the load of NH_3 was increased from 967 to 2 057 mg/h with a maximum volumetric load of 1 898 $m^3{}_{air}/m^3{}_{filter}/h$, the removal efficiency was reduced by 19%. The study recommended wood chips over 20 mm for biofilters that are used to treat air emitted from swine production facilities. The maximum recommended volumetric load is 1 350 $m^3{}_{air}/m^3{}_{filter}/h$ in order to ensure an odour removal efficiency greater than 90%. In summer operating conditions, the size of the biofilter was 0.148 m^2/pig. An efficient humidification system (humidifier at the air inlet and a spraying device above the bedding material) and an adapted air distribution system are determining factors for the design and the operation of treatment systems for high air flow rates.

In another study, [51] compared the effectiveness of two pilot-scale biofilters for the treatment of air from pig barns. The first biofilter used wood chips over 20 mm and the second one used wood chips with dimensions between 10 and 16 mm. The humidity in the filter bed was maintained at 69% and the volumetric load varied between 769 and 1847 $m^3{}_{air}/m^3{}_{filter}/h$ for a trial period of 63 days. Both biofilters reduced the odour in the range of 88 to 95%. The reduction of NH_3 was in the range of 64 to 92% for the first biofilter and 69 to 93% for the second. H_2S was reduced by 9 to 66% for the first biofilter while the results for the second ranged from an increase of 147% to a decrease of 51%. The pH was maintained between 6 and 8. Investigations show that there is a risk of forming anaerobic zones in the filter bed (second biofilter) which can release reduced sulphur compounds. The study concluded that biofiltration is an interesting technology for the removal of odour and NH_3 from the air emitted from swine production facilities.

[52] attempted to combine a strategy of minimum ventilation and biofiltration. A minimum air flow rate of 75 m^3/h/pig, corresponding to the conditions of summer nights, was established as a reference. The tests were carried out with a biofilter using wood chips with a filter bed height of 27 cm and an area of 80 m^2. The results showed an average removal efficiency of 44% for ammonia, 58% for H_2S and 54% for odours. The results are quite modest, but are partially due to reduced volumes of the treated air.

[53] studied the efficiency of biofilters in reducing NH_3 emitted from livestock buildings. The aim of the research was to test a filter bed composed of non-expensive organic and inorganic materials in combination with a diverse microbial population (multiculture). The tests were conducted on a bench-scale device with a closed-type reactor with a height of 0.5 m. The packing media was composed of peat (91% organic), vermiculite and perlite (ratio 3:1:1). In the second series of tests, the filter material was made from peat and polystyrene (3:1 ratio). The results of the study showed that the removal efficiency of NH_3 can be very high (99 to 100%) under conditions where the inlet concentration is 200 ppmv and flow rates are between 0.03 to 0.06 m^3/h.

A study on a pilot-scale plant by [54] demonstrated the performance of biofilters for odour reduction using different filter materials such as sand, soil, bark and wood mixtures. The reduction of odours analyzed by olfactometry was between 29 and 99.9% with odour concentrations at the inlet ranging between 143 100 and 890 000 OU/m^3. The study highlighted the presence of leachate resulting from wetting of the filter bed. This fluid plays a very important role in maintaining moisture, but it may also have other effects on the quality of the filter bed, such as washing, accumulation of large amounts of pollutants, interference with the airflow, formation of preferential paths, formation of anaerobic zones and release of NH_3 and H_2S. The study demonstrated the need for further research to clarify these aspects that have a direct influence on the performance and longevity of the biofiltration system.

In response to the questions raised regarding the accumulation of nitrogen compounds in the filter bed due to high inlet concentrations of NH_3, a study by Japanese researchers [55] tested the use of a new bacterium (Vibrio alginolyticus), which is able to effectively degrade high concentrations of NH_3. The study demonstrated the feasibility of using this marine bacterium for concentrations of ammonia between 120 and 2 000 ppmv with removal efficiencies greater than 85% for more than 60 days of operation.

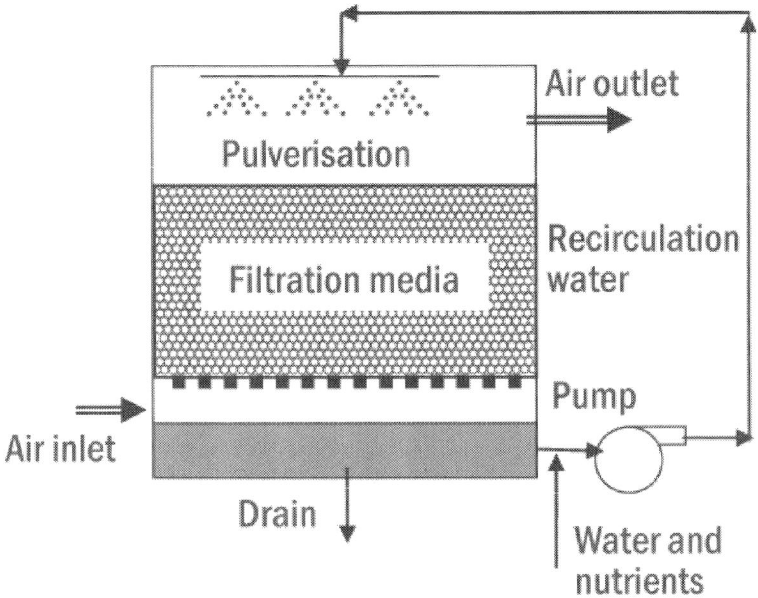

Figure 6. Diagram of a biotrickling filter (adapted from [42])

4.4. Biotrickling filters

Contrary to biofilters, biotrickling filters generally use an inorganic packing material with the liquid solution continuously recirculating over the filter bed (Figure 6). This technology offers many advantages: an easy control of key operating parameters (such as temperature, pH, nutrient supply and concentration of toxic compounds), low pressure drops and reduced space requirements by allowing high flow rates.

Air treatment by biotrickling filters is a relatively new technology and the majority of experimental results are from tests carried out with pilot-scale plants [44]. Various filter media, such as lava rock, random plastic packing, structured blocks of plastic and polyurethane foam have been used. The high porosity of these materials provides a minimal pressure drop on the airflow; higher airflow rates are therefore achievable. One of the main characteristics of biotrickling filters is the continuous flow of the liquid on the filter bed. It is therefore possible to improve process control by the addition of nutrients, adjustment of the pH and the temperature or by removing toxic by-products. For example, in the case of odour reduction and H_2S removal, production of sulfuric acid and reduction of pH and/or accumulation of sodium sulphate are the key controlling parameters. Biotriclickling filters also have other advantages over the other biological treatments in controlling air pollutants [56]: the height of the filter bed, the longer life of the filter media (above 10 years), the ease of control and the ability to treat air containing dust and grease.

The examples cited by [44] show that these reactors have good removal efficiencies for high concentrations of H_2S at low residence times (EBRT). Biotrickling filters seem a good option for the treatment of gases with high concentrations of H_2S and possibly for other sulfur compounds. Experiments on industrial applications have shown the potential of biofilters and biotrickling filters for the simultaneous removal of odour, H_2S and VOCs. From a total of eight cases of biotrickling filters used for the removal of H_2S and for inlet concentrations of 1 to 1000 mg/m^3, the reduction efficiencies varied from 95 to 99%. For the reduction of odours, the efficiencies obtained were 65 to 99%.

Biotrickling filters should be inoculated with a variety of microorganisms, since inorganic filter beds generally do not contain bacteria. The addition of nutrients can also become a tool for optimizing the performance of the reactor. The nutrient requirements depend on the type of pollutant to be treated, its concentration, the pollutant loading and the operating conditions of the reactor. However, excess nutrients can lead to an overproduction of biomass and eventually clog the reactor.

A cross-flow biotrickling filter was developed at the IRDA for the treatment of swine ventilation air. Pilot-scale tests were carried out with 6 units treating air from chambers housing 4 grower-finisher pigs. After a start-up phase between 9 and 20 days, the system was able to reduce emissions of NH_3 and odours by up to 68 and 82% respectively. Different operating conditions were tested (air residence time, liquid flow rate and type of packing material), but very little effect was observed since the system was probably over-sized. The biotrickling filters had no significant effect on CO_2 and CH_4 emissions.

4.5. Bioscrubbers

In a bioscrubber, each step of the treatment process is separate: the pollutants are first transferred to a liquid solution in an absorption unit and then the washing liquid is regenerated in a biological reactor which generally resembles an activated sludge reactor (Figure 7). Similar to the biotrickling filter, the operating conditions can easily be controlled in a bioscrubber and it is possible to optimize both the absorption and biodegradation of pollutants. [43] reported several types of absorbers such as the packed tower, the wet cyclone, the spray tower and the venturi scrubber.

The flow of air and water can either be counter-current, co-current or cross-current. The air velocity can vary between 1.5 m/s and 20 m/s in a spray tower, it can reach 25 m/s for a wet cyclone and between 40 and 50 m/s for a venturi. Bioscrubbers have the same space, flexibility and control than biotrickling filter, but they are only suitable for the treatment of highly water soluble compounds (Henry's coefficient below 0.01). Bioscrubbers have thereby increased the scope of application for the biological treatment of waste gases. The greatest advantage of bioscrubbers compared to biofilters and biotrickling filters is the ability to produce and maintain large amount of active microbial mass in smaller units. On the other hand, [57] considers that bioscrubbers and biotrickling filters present greater construction and operation complexity.

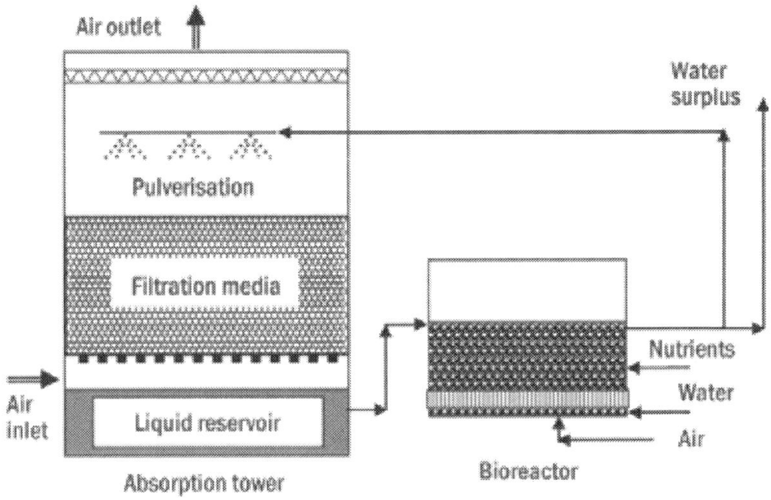

Figure 7. Diagram of a bioscrubber (adapted from [42])

[43] concluded that bioscrubbers offer the greatest efficiencies for the removal of H_2S, NH_3 and organic sulphur compounds. The performance analysis of bioscrubbers used in the industry for H_2S removal showed efficiencies over 98% for low, medium and high inlet concentrations (between 0 and 75 mg/m^3, 2 000 mg/m^3 and between 10 000 and 15 000 mg/m^3 respectively). The results cited for the reduction of odours show an efficiency of 80% for the reduction of organic sulfides with inlet concentrations between 4 000 and 22 000 OU.

There are many technologies available for the reduction of air contaminants and odours using biological reactions. Several systems have configurations that are similar to each other though different in terms of operating conditions. The difficulty of using one of the existing technologies for the treatment of air comes from the specific constraints of each application. Livestock buildings in general are characterized by a very large number of parameters that influence the application of air treatment.

Biofiltration has been the subject of many scientific publications and several units are currently installed throughout the world. Its efficiency has been demonstrated for reducing odours and to a lesser extent, for the reduction of NH_3 and H_2S emitted from barns. However, despite the advantage of being simple, the use of biofilters in livestock buildings is limited by several problems such as the accumulation of pollutants, the potential for clogging, the high pressure losses and the relatively rapid degradation of the filter bed. On the other hand, even if biotrickling filters and bioscrubbers have better features compared to biofiltration (e.g. rapid response to changes in operating conditions and longer life time), experimental research on these two technologies has just started and only specific solutions for specific applications and partial results are available. There are not many experimental studies on full-scale systems.

[45] highlighted that the technologies with the greatest potential for reducing the contaminants emitted from livestock buildings should, in all likelihood, come from the combination of different treatment systems. A combination of an air scrubber and a biotrickling filter or a combination of a biofilter and a biotrickling filter could provide greater capabilities than each technology used individually. However, there is no information in the literature on the efficiency of these different combinations of technologies.

5. Conclusion

This chapter outlined the importance of biodegradation associated with modern manure management practices. Depending on the type of management strategy and the production stage, both aerobic and anaerobic conditions can prevail, providing a wide range of emissions. The second part of the chapter explored the different technologies where biodegradation can be used to treat both the manure and the gaseous emissions. Biological treatment systems generally provide good removal efficiencies at relatively low costs and are well adapted to the agricultural sector.

Author details

Matthieu Girard*, Joahnn H. Palacios, Martin Belzile, Stéphane Godbout and Frédéric Pelletier

*Address all correspondence to: matthieu.girard@irda.qc.ca

Research and Development Institute for the Agri-Environment (IRDA), Québec (Qc), Canada

References

[1] Godbout, S. and Trudelle, M. 2002. Évaluation des performances techniques des séparateurs mécaniques à lisier et de leur rapport efficacité/coût. Final Report, IRDA, 84 pages.

[2] Andreottola, G., Bortone, G., and Tilche, A. 1997. Experimental validation of a simulation and design model for nitrogen removal in sequencing batch reactor. Water, Science and Technology, 35(1): 113-120.

[3] Boursier, H., Béline, F., and Paul, E. 2005. Piggery wastewater characterisation for biological nitrogen removal process design. Bioresource Technology, 96: 351-358.

[4] Aubry, G. 2008. Étude du comportement de l'azote dans un biofiltre à lit ruisselant traitant du lisier de porc. Ph.D. thesis, Department of Civil Engineering, Laval University, Quebec City, Qc. Canada.

[5] Metcalf and Eddy. 2003. Wastewater engineering: treatment and reuse - 4th edition. McGraw-Hill, New-York.

[6] Henze, M., Harremoës, P., la Cour Jansen, J., and Arvin, E. 2002. Wastewater treatment: biological and chemical processes – 3rd edition. Springer, New York, United States.

[7] Schiffman, S.S., J.L. Bennett, and J.H. Raymer. 2001. Quantification of odors and odorants from swine operations in North Carolina. Agric. Forest Meteor. 108(3): 213–240.

[8] Møller, H.B., S.G. Sommer and B.K. Ahring. 2002. Separation e ▪ ciency and particle size distribution in relation to manure type and storage conditions. Bioresource Technology 85, 189–196.

[9] Natural Resources Conservation Service. 2007. Manure chemistry - Nitrogen, Phosphorus, & Carbon. Manure management information sheet, number 7.

[10] Hamilton, D. W. 2009. Anaerobic Digestion of Animal Manures: Understanding the Basic Processes. Oklahoma Cooperative Extension Service. BAE-1747.

[11] Godbout, S., L. Hamelin, J. H. Palacios, F. Pelletier, S.P. Lemay and F. Pouliot. 2011. État de référence des émissions gazeuses et odorantes provenant des bâtiments porcins québécois. Final report. IRDA. 53 pages.

[12] Marquis, A. and P. Marchal. 1998. Qualité de l'atmosphère à proximité des bâtiments d'élevage. Cahiers d'études et de recherches francophones – Agricultures. 7(5): 377-385.

[13] Hamelin, L., M. Wesnæs, H. Wenzel and B.M. Petersen. 2010. Life cycle assessment of slurry management technologies II – emphasis on biogas production. Danish Ministry of the Environment, Environmental Protection Agency. http://www2.mst.dk/udgiv/publications /2010/978-87-92668-03-5/pdf/978-87-92668-04-2.pdf. Accessed January 30, 2013.

[14] IPCC. 2007. Fourth Assessment Report (AR4) by Working Group 1 (WG1), Chapter 2 "Changes in Atmospheric Constituents and in Radiative Forcing".

[15] EPA. Environmental Protection Agency. 2010. Particulate matter. http://www.epa.gov/pm/. Accessed January 30, 2013.

[16] Cambra-López, M., A.J.A. Aarnink, Y. Zhao, S. Calvet and A.G. Torres. 2010. Airborne particulate matter from livestock production systems: A review of an air pollution problem. Environmental Pollution, 158: 1–17.

[17] Chadwick, D.R., Sneath, R.W., Phillips, V.R., and Pain, B.F. 1999. A UK inventory of nitrous oxide emissions from farmed livestock. Atmospheric Environment, 33: 3345-3354.

[18] Velthof, G.L., Nelemans, J.A., Oenema, O., and Kuikman, P.J. 2005. Gaseous nitrogen and carbon losses from pig manure derived from different diets. Journal of Environmental Quality, 34: 698-706.

[19] Chantigny, M.H., MacDonald, J.D., Beaupré, C., Rochette, P., Angers, D.A., Massé, D.I., and Parent, L.-É. 2009. Ammonia volatilization following surface application of raw and treated liquid swine manure. Nutrient Cycling in Agroecosystems, 85(3), p. 275-286.

[20] Perron, M-H. 2010. Disponibilité de l'azote de biosolides de traitement de lisier de porc et de deux boues de papetière et émissions de N2O consécutives à leur épandage au champ. Master thesis. Laval University, Québec, Canada. 138 pages.

[21] Zhang, Z., and Zhu, J. 2006. Characteristics of solids, BOD5 and VFAs in liquid swine manure treated by short-term low-intensity aeration for long-term storage. Bioresource Technology, 97: 140-149.

[22] BAPE. 2003. L'état de la situation de la production porcine au Québec. Rapport d'enquête et d'audience publique du Bureau d'audiences publiques sur l'environnement, ISBN: 2-550-41393-8.

[23] Westerman, P.W., Bicudo, J.R., and Kantardjieff, A. 2000. Upflow biological aerated filters for the treatment of flushed swine manure. Bioresource Technology, 74: 181-190.

[24] Lanoue, M. 1998. Traitement du lisier de porc par procédé de biofiltration aérobie. M.Sc. thesis, Department of Chemical Engineering, Université de Sherbrooke, Sherbrooke, Qc. Canada.

[25] Westerman, P.W. and Arogo, J. 2004. Ekokan biofiltration technology performance verification. Available from the North Carolina State University's College of Agriculture and Life Sciences: http://www.cals.ncsu.edu/waste_mgt/smithfield_ projects/phase1report04/A.6EKOKAN%20final%20.pdf. [Accessed in January 2013].

[26] Sommer, S.G., Mathanpaal, G., and Dass, G.T. 2005. A simple biofilter for treatment of pig slurry in Malaysia. Environmental Technology, 26: 303-312.

[27] Buelna, G. 2000. Biofilter for purification of waste waters and method therefore. United States Patent Number 6,100,081.

[28] Dubé, R. 1997. Traitement du lisier de porc par biofiltration sur milieu organique : influence de l'aération. M.Sc. thesis, Department of Civil Engineering, Laval University, Quebec, Qc. Canada.

[29] Aubry, G., Lessard, P., Gilbert, Y., Le Bihan, Y., and Buelna, G. 2006. Nitrogen behaviour in a trickling biofilter treating pig manure, Water Science and Technology, IWA Conference BIOFILMS Systems VI, Amsterdam.

[30] Garzón-Zúñiga, M., Lessard, P., Aubry, G., and Buelna, G. 2005. Nitrogen elimination mechanisms in an organic media aerated biofilter treating pig manure. Environmental Technology, 26: 361-371.

[31] Garzón-Zúñiga, M., Lessard, P., Aubry, G., and Buelna, G. 2007. Aeration effect on the efficiency of swine manure treatment in a trickling filter packed with organic materials. Water, Science and Technology, 55(10): 135-143.

[32] Ros, M., García, C., and Hernández, T. 2005. A full-scale study of treatment of pig slurry by composting: kinetic changes in chemical and microbial properties. Waste Management, 26: 1108-1118.

[33] Huang, G.F., Wong, J.W.C., Wu, Q.T. and Nagar, B.B. 2004, Effect of C/N on composting of pig manure with sawdust. Waste Management, 24: 805–813.

[34] Zhu, N. 2007. Effect of low initial C/N ratio on aerobic composting of swine manure with rice straw. Bioresource Technology, 98: 9–13.

[35] Hassouna, M., Espagnol, S., Robin, P., Paillat J-M., Levasseur, P. and Li, Y. 2008. Monitoring NH3, N2O, CO2 and CH4 emissions during pig solid manure storage - effect of turning. Compost Science & Utilization. 16(4): 267-274.

[36] Fukumoto, Y., Suzuki, K., Osada, T., Kuroda, K., Hanajima, D., Yasuda, T., and Haga, K. 2006. Reduction of nitrous oxide emission from pig manure composting by addition of nitrite-oxidizing bacteria. Environmental Science and Technology, 40: 6787-6791.

[37] Juteau, P. 2006. Review of the use of aerobic thermophilic bioprocesses for the treatment of swine waste, Livestock Science, 102: 187– 196.

[38] Ortenblad, H. 2000. The use of digested slurry within agriculture. Available from: http://homepage2.nifty.com/biogas/cnt/refdoc/whrefdoc/d9manu.pdf [accessed in January 2013.]

[39] Ricard, M.-A., V. Drolet, A. Coulibaly, C.B. Laflamme et al. 2010. Développer un cadre d'analyse et identifier l'intérêt technico-économique de produire du biogaz à la ferme dans un contexte québécois. Report produced by the Centre de développement du porc du Québec (CDPQ), ISBN 978-2-922276-35-0, 110 pages.

[40] Deshusses, M., Z. Shareefdeen. 2005. Modeling of biofilter and biotrickling filters for odour and VOC control applications. Biotechnology for odour and air pollution control. ed. Z. Shareefdeen and A. Singh. Springer. Verlag, Berlin, Heidelberg.

[41] Devinny, J.S., M.A. Deshusses, T.S. Webster. 1999. Biofiltration for air pollution control. Lewis Publishers. Washington, DC, USA.

[42] Revah, S. and J.M. Morgan-Sagastume. 2005. Methods of odor and VOC control. Biotechnology for odour and air pollution control. ed. Z. Shareefdeen and A. Singh. Springer. Verlag, Berlin, Heidelberg.

[43] Singh, A., Z. Shareefdeen and O.P. Ward. 2005. Bioscrubber technology. Biotechnology for odour and air pollution control. ed. Z. Shareefdeen and A. Singh. Springer. Verlag, Berlin, Heidelberg.

[44] Iranpour, R., H.H.J. Cox, M.A. Deshusses and E.D. Schroeder. 2005. Literature review of air pollution control biofilters and biotrickling filters for odor and volatile organic compound removal. Environmental Progress. 24 (3).

[45] Lemay S.P., M. Martel, M. Belzile, D. Zegan, J. Feddes, S. Godbout, F. Pelletier. 2009. A systematic literature review to identify an air contaminant removal technology for swine barn exhaust air. Written for presentation at the CSBE/SCGAB 2009 Annual Conference Rodd's Brudenell River Resort, Prince Edward Island, Canada. 12-15 July 2009.

[46] Godbout S., S.P. Lemay, F. Pelletier, M. Belzile, J.P. Larouche, L.D. Tamini, J. H. Palacios and D. Zegan. 2010. Réduction des émissions gazeuses et odorantes aux bâtiments porcins : techniques simples et efficaces applicables à la ferme. Final Report. IRDA. 143p.

[47] Godbout, S., F. Pelletier, J.P. Larouche, M. Belzile, J.J.R. Feddes, S. Fournel, S.P. Lemay and J.H. Palacios. 2012. Greenhouse Gas Emissions from Non-Cattle Confinement Buildings: Monitoring, Emission Factors and Mitigation. Greenhouse Gases - Emission, Measurement and Management, Guoxiang Liu (Ed.), ISBN: 978-953-51-0323-3, InTech, DOI: 10.5772/31948.

[48] Delhoménie, M.-C, and Heitz, M. 2005. Biofiltration of air: a review. Critical Reviews in Biotechnology, 25(1-2): 53-72.

[49] Nicolai, R.E. and Janni, K.A. 2001. Biofilter media mixture ratio of wood chips and compost treating swine odors. Dept of Biosystems and Agricultural Engineering, University of Minnesota, USA.

[50] Sheridan, B. A., T. P. Curran and V. A. Dodd. 2002. Assessment of the influence of media particle size on the biofiltration of odorous exhaust ventilation air from a piggery facility. Bioresource Technology. 84: 129-143.

[51] Sheridan, B. A., T. Curran, V. Dodd and J. Colligan. 2002. Biofiltration of odour and ammonia from a pig unit - A pilot-scale study. Biosystems Engineering. 82 (4): 441-453.

[52] Hoff, S. J. and J. D. Harmon. 2006. Biofiltration of the critical minimum ventilation exhaust air. Workshop on Agricultural Air Quality. Washington, D.C., USA.

[53] Kalingan, A. E., C. M. Liao, J. W. Chen and S. C. Chen. 2004. Microbial Degradation of Livestock -Generated Ammonia Using Biofilters at Typical Ambient Temperatures. Journal of Environmental Science and Health. B 39 (1): 185-198.

[54] Luo, J. 2001. A pilot-scale study on biofilters for controlling animal rendering process odours. Water Science and Technology. 44 (9): 277-285.

[55] Kim, N. J., Y. Sugano, M. Hirai and M. Shoda. 2000. Removal of a high load of ammonia gas by a marine bacterium, Vibrio Alginolyticus. Journal of Bioscience and Bioengineering. 90 (4): 410-415.

[56] Deshusses, M. A. and D. Gabriel. 2005. Biotrickling filter technology. Biotechnology for odour and air pollution control. ed. Z. Shareefdeen and A. Singh. Springer. Verlag, Berlin, Heidelberg.

[57] Kraakman, A. 2005. Biotrickling and Bioscrubber Applications to control odour and air pollutants. Biotechnology for odour and air pollution control. ed. Z. Shareefdeen and A. Singh. Springer. Verlag, Berlin, Heidelberg.

Section 3

Biodegradation and Sustainability

Chapter 11

Methods for Separation, Recycling and Reuse of Biodegradation Products

Ganapati D. Yadav and Jyoti B. Sontakke

Additional information is available at the end of the chapter

http://dx.doi.org/10.5772/56241

1. Introduction

Thousands of chemicals and materials with varied properties and functionalities are manufactured and used for commercial and day-to-day applications, whose ultimate fate in the environment may not be known. During their manufacture and use, these substances are often discharged into the environment through different routes in air, water and land. Creation of tremendous quantities of solid waste of all kind and its effective disposal has posed innumerable problems that need technological breakthroughs. Many of these substances degrade slowly and exert toxic effects on plants and animals, thus causing large scale environmental degradation [1, 2]. Pollution by abandoned plastic articles is also a matter of great concern [3]. Industrial wastewaters associated with the manufacture of organic chemicals are voluminous and characteristically have concentrations ranging from a few ppm to a thousands of ppm. Biodegradation of such dissolved pollutants is an area of immense interest to various sectors. Emission of volatile organic compounds (VOCs) from various sources has detrimental effects on quality of air we breathe and on environmental phenomena. Biodegradation, either aerobic or anaerobic, can be an approach to cleave big molecules through a series of steps in to smaller molecules from a mosaic of chemicals and materials and some of them can be valorized as pollution abatement strategy and source of energy through biogas generation [2]. Biogas can be produced from nearly all kind of biomass, among which the primary agricultural sectors and various organic waste streams can be properly tapped as renewable source of energy. Untreated or poorly managed animal manure is a major source of air and water pollution. Nutrient leaching, mainly nitrogen and phosphorous, ammonia evaporation and pathogen contamination are some of the foremost threats [3]. A conservative estimate is provided by Steinfeld et al. [4] that the animal production sector is responsible for 18% of the overall green house gas emissions, measured in CO_2 equivalent and for 37% of the anthropogenic methane, which has 23 times the global warming potential of CO_2. Furthermore, 65% of anthropogenic nitrous oxide and 64% of anthropogenic ammonia emission originate from the worldwide animal production sector. Biogas production from anaerobic digestion of animal manure and slurries can be harnessed to alleviate greenhouse gas emissions in particularly ammonia and methane [5].

Plastics are bane and benefactor simultaneously. Over 230 million tons of plastic are produced annually. Plastics are used in all walks of life and provide improved insulation, lighter packaging, are found in cars, aeroplanes, railways, phones, computers, medical devices, etc. but appropriate disposal is often not properly addressed. On one hand, plastic waste and disposal is a hotly debated issue globally whereas on the other, it can contribute to reduce the carbon footprint. Many leading European countries recover more than 80% of their used plastics, by adopting an integrated waste and resource management strategy to address each waste stream with the best options [6]. Plastic sorting and separation, recycling, depolymerisation, cracking, and production of fuel are some of the strategies used to abate plastic pollution. Development of biopolymers is pursued vigorously. Biodegradation of plastics by microorganisms and enzymes appears to be the most effective process. When plastics are used as substrates for microorganisms, evaluation of their biodegradability should not only be based on their chemical structure, but also on their physical properties such as melting point, glass transition temperature, crystallinity, storage modulus, etc. [7-11].

This chapter has covered the mechanisms of biodegradation, biodegradation of a variety of industrial chemicals, plastics and other biomass, advances in anaerobic digestion technologies and biogas generation, plastic processing, biopolymer synthesis and degradation. Synthesis of biopolymers is covered. The scope for treating municipal organic solid waste, manure and polymers to generate biogas as a renewable energy option, and also as a pollution abatement strategy is discussed including technological aspects. The synthesis of biohydrogen, bioethanol, biobutanol and other biotransformation leading to valuable chemicals, which also involve breaking down of larger molecules, plastics and biomaterials are not addressed [7,10]. Biorefinery is a concept which is akin to petrorefinery, wherein biomass is converted into useful platform chemicals through extraction, controlled pyrolysis, fermentation, enzyme and chemical catalysis [12].

2. Mechanisms of biodegradation

Cellulose, lignocellulose and lignin are major sources of plant biomass and are polymeric substances; therefore, their recycling is indispensable for the carbon cycle [13]. Each of these polymer is degraded by a variety of microorganisms which produce scores of enzymes that work in tandem. The diversity of cellulosic and lignocellulosic substrates has contributed to the difficulties found in enzymatic treatment. Fungi are the best-known microorganisms capable of degrading these three polymers. Because the substrates are insoluble, both bacterial and fungal degradation occur exo-cellularly, either in association with the outer cell envelope layer or extra-cellularly. Microorganisms have two types of extracellular enzymatic systems, namely, the hydrolytic system, which produces hydrolases and is responsible for cellulose and hemicellulose degradation; and a unique oxidative and extracellular ligninolytic system, which depolymerizes lignin [13]. The man-made chemicals and materials are comprised of different entities and functional groups which need to be degraded effectively by microorganisms and no single microorganism is obviously capable of doing it [1,14].

Growth and co-metabolism are the two mechanisms of biodegradation. In the case of growth, organic substance is used as the sole source of carbon and energy, which leads to complete degradation (mineralization). Archaebacteria, prokaryotes and eukaryotes (like fungi, algae, yeasts, protozoa) play dominant role in mineralization [7]. On the contrary, co-metabolism encompasses the metabolism of an organic compound in the presence of a growth substrate which is used as the primary carbon and energy source. Thus, biodegradation processes and their rates differ greatly depending on the type of substrate and conditions such as temperature, pH, and aqueous phase solubility, but frequently the major final products of the degradation are carbon dioxide and methane [1,7,10].

2.1. Growth-associated degradation of aliphatic compounds

Growth-associated degradation produces CO_2, H_2O, and cell biomass. The cells act as the complex biocatalysts of degradation. Further, cell biomass may be mineralized after exhaustion of the degradable pollutants in a contaminated site. Bulk chemicals like aromatic hydrocarbons such as benzene, toluene, ethylbenzene, xylenes, and naphthalene are widely used as fuels, industrial solvents and feedstock for petrochemical industry. Phenols and chlorophenols are another class of chemicals, employed in a variety of industries. Since all micro-organisms make aromatic compounds such as aromatic amino acids, phenols, or quinines, in large amounts, many microorganisms have evolved catabolic pathways to degrade aromatic compounds. In general, man-made organic chemicals (xenobiotics) can be degraded by microorganisms, when the respective molecules are similar to natural compounds [7,10].

In general, benzene, condensed ring and related compounds are characterized by a higher thermodynamic stability than aliphatic compounds. Benzene oxidation begins with hydroxylation catalyzed by a dioxygenase leading to a diol (Scheme 1) which is then converted to catechol by a dehydrogenase.

Hydroxylation and dehydrogenation are also common in degradation routes of other aromatic hydrocarbons. The introduction of a substituent group onto the benzene ring renders alternative mechanisms possible to attack side chains or to oxidize the aromatic ring. Many aromatic substrates are degraded by a limited number of reactions such as hydroxylation, oxygenolytic ring cleavage, isomerization, and hydrolysis. The inducible nature of the enzymes and their substrate specificity enable bacteria such as *pseudomonads* and *rhodococci* with a high degradation activity, to acclimatize their metabolism to the effective utilization of substrate mixtures in polluted soils and also to grow at a high rate [10,15].

2.2. Co-metabolic degradation of organo-pollutants

Co-metabolism is a common phenomenon of microbial activities and the basis of biotransformation used in biotechnology to convert molecules in to useful modified forms. Microorganisms growing on a particular substrate also oxidize a second substrate. The co-substrate is not

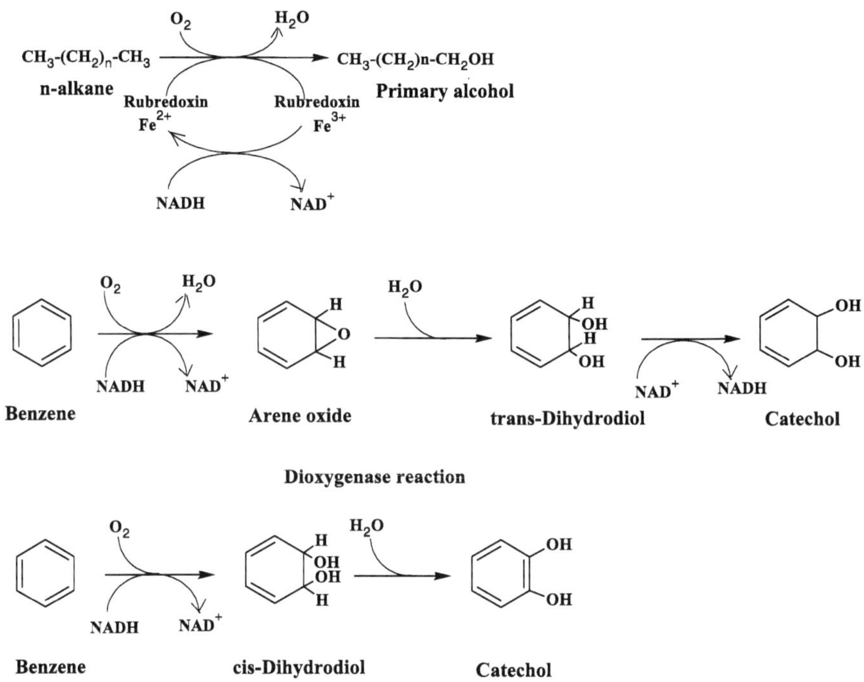

Scheme 1. Monooxygenase and dioxygenase reactions: In this mechanism, monooxygenase initially incorporates one O atom from O_2 into the xenobiotic substrate whereas the other is reduced to H_2O. On the contrary, dioxygenase incorporates both atoms into the substrate [15].

incorporated, but the product may be available as substrate for other organisms of a mixed culture. The rudiments of co-metabolic transformation are the enzymes of the growing cells and the synthesis of cofactors necessary for enzymatic reactions; for instance, of hydrogen donors (reducing equivalents, NADH) for oxygenases. Several aromatic substrates can be converted enzymatically to natural intermediates of degradation such as catechol and protecatechuate (Scheme 2) [15].

Co-metabolism of chloroaromatics is a general activity of bacteria in mixtures of industrial pollutants. The co-metabolic transformation of 2-chlorophenol leads to dead-end metabolites such as 3-chlorocatechol, which may be auto-oxidized or polymerized in soil to humic-like structures. Irreversible binding of dead end metabolites may fulfill the function of detoxification. The accumulation of dead-end products within microbes under selection pressure is the source for the evolution of new catabolic traits. Thus, recalcitrance of organic pollutants increases with increasing halogenation. Substitution of halogen as well as nitro and sulfo groups at the aromatic ring is accomplished by an increasing electrophilicity of the molecule. These compounds resist the electrophilic attack by oxygenases of aerobic bacteria. Compounds that persist under oxic condition are polychlorinated biphenyls (PCBs), chlorinated dioxins

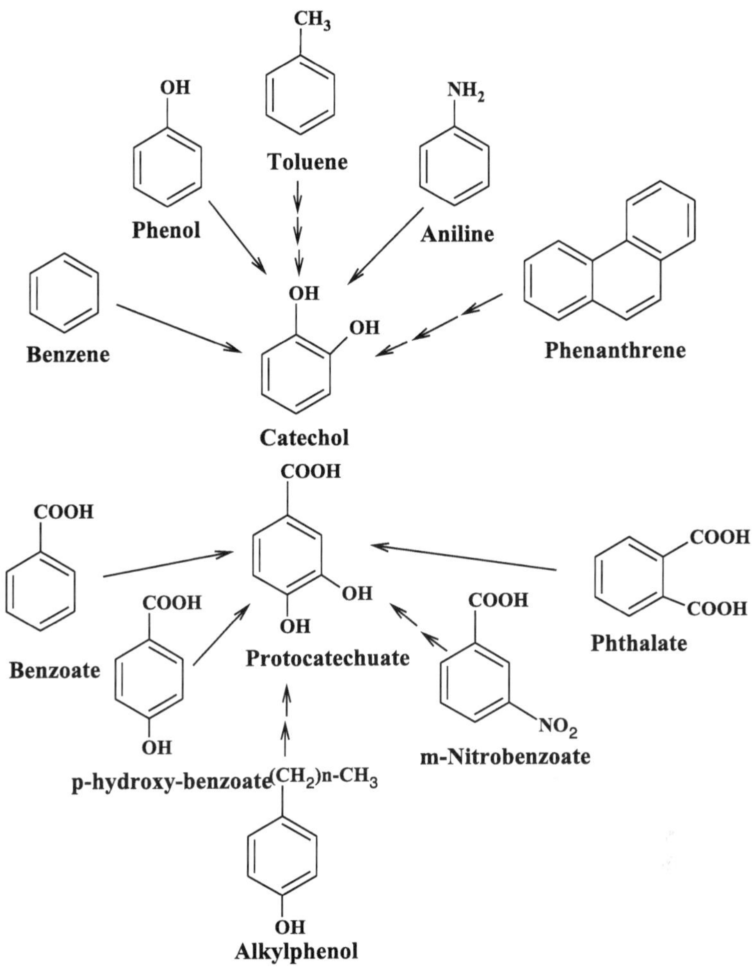

Scheme 2. Degradation of aromatic natural and xenobiotic compounds into two central intermediates, catechol and protocatechuate [15].

and some pesticides like DDT. To overcome the relatively high persistence of halogenated xenobiotics, reductive attack of anaerobic bacteria is of great value. Reductive dehalogenation achieved by anaerobic bacteria is either a gratuitous reaction or a new type of anaerobic respiration. The process reduces the degree of chlorination and, therefore, makes the product more accessible to mineralization by aerobic bacteria [7,15].

Reductive dehalogenation which is the first step of degradation of PCBs requires anaerobic conditions wherein organic substrates act as electron donors. PCBs accept electrons to allow the anaerobic bacteria to transfer electrons to these compounds. Anaerobic bacteria capable of catalyzing reductive dehalogenation seem to be relatively omnipresent in nature. Most dechlorinating cultures are a mixed consortia. Anaerobic dechlorination is always incomplete and the products are di- and monochlorinated biphenyls. These products can be metabolized further by aerobic microorganisms [2,7,15].

The rates of biodegradability of particular substrate is mainly related to accessibility of the substrate for enzymes and can be enhanced by several means as reviewed by van Lier et al. [16] such as (a) mechanical methods: the disintegration and grinding of solid particles present in sludge: releases cell compounds and creates new surface where biodegradation take place, (b) ultrasonic disintegration, (c) chemical methods: the destruction of complex organic compounds by means of strong, mineral acids or alkalis, (d) thermal pretreatment: thermal hydrolysis is able to split and decompose a significant part of the sludge solid fraction into soluble and less complex molecules, (e) enzymatic and microbial pre-treatment: a very promising method for the future for some specific substrates (e.g. cellulose, lignin etc.),(f) stimulation of anaerobic micro-organisms: some organic compounds (e.g. amino acids, cofactors, cell content) act as a stimulating agent in bacteria growth and methane production. Most of the above methods occur at the pre-methanation step and result in a better supply of methanogenic bacteria by suitable substrates.

3. Aerobic biodegradation

Many microorganisms grow under aerobic conditions. The so-called cellular respiration process (CSP) begins with aerobes which employ oxygen to oxidize substrates such as sugars and fats to derive energy. Before the onset of CSP, glucose molecules are degraded into smaller molecules in the cytoplasm of the aerobes. The smaller molecules then enter a mitochondrion, where aerobic respiration takes place. Oxygen is used to break down small entities into water and carbon dioxide, accompanied by release of energy. Aerobic degradation does not produce foul gases, unlike anaerobic process. The aerobic process leads to a more complete digestion of solid waste reducing build-up by more than 50% in most cases [1, 2, 7]. The major enzymatic reactions of aerobic biodegradation are oxidations catalyzed by oxygenases and peroxidases. Oxygenases are oxido-reductases that incorporate oxygen into the substrate as given in Scheme 1. Degradative organisms need oxygen at two metabolic sites, namely, at the initial attack of the substrate and at the end of the respiratory chain. Higher fungi possess a unique oxidative system for the degradation of lignin based on extracellular ligninolytic peroxidases and laccases [13]. This enzymatic system is important for the co-metabolic degradation of persistent organic pollutants. The predominant bacteria of polluted soils belong to a spectrum of genera and species (Table 1) [15].

The most important classes of organic pollutants in the environment are mineral oil constituents and halogenated petrochemicals, for the biodegradation of which the capacities of aerobic microorganisms are of great consequence. The most rapid and complete degradation of the majority of pollutants is brought about under aerobic conditions and these include petroleum hydrocarbons, chlorinated aliphatics, benzene, toluene, phenol, naphthalene, fluorine, pyrene, chloroanilines, pentachlorophenol and dichlorobenzenes. Many cultures of bacteria grow on these chemicals and are capable of producing enzymes which degrade them into non-toxic species. [7,15].

Gram negative bacteria	Gram positive bacteria
Pseudomonas species	*Nocardia* species
Xanthomonas species	*Mycobacteria* species
Alcaligenes species	*Corynebacterium* species
Flavobacterium species	*Arthobacter* species
Cytophaga group	*Bacillus* species

Table 1. Predominant bacteria in soil samples polluted with aliphatic and aromatic hydrocarbons, polycyclic aromatic hydrocarbons, and chlorinated compounds [15]

There are several essential attributes of aerobic microorganisms degrading organic pollutants amongst which metalobic processes top the list. The chemicals must be accessible to the degrading organisms. For example, hydrocarbons are immiscible in water and their degradation requires the production of biosurfactants in order to have effective biodegradation [14]. The initial intracellular attack of organic pollutants is an oxidative process and therefore, the activation and incorporation of oxygen is the main enzymatic reaction catalyzed by oxygenases and peroxidases. Peripheral degradation pathways convert organic pollutants step by step into intermediates of the central intermediary metabolism, such as the tricarboxylic acid cycle. Biosynthesis of cell biomass from the central precursor metabolites (acetyl-CoA, succinate, pyruvate) is required [14,15]. Sugars needed for various biosyntheses and growth must be synthesized by gluconeogenesis. The predominant degraders of organo-pollutants in the oxic zone of contaminated areas are chemo-organotropic species that are able to use a large number of natural and xenobiotic compounds as carbon sources and electron donors for the generation of energy. Although many bacteria are able to metabolize organic pollutants, a single bacterium does not possess the enzymatic capability to degrade all or even most of the organic pollutants from a heterogeneous mixture originating from particular industries. Thus, mixed microbial communities have the most powerful biodegradative potential. The genetic information of more than one organism is necessary to develop a system which could be used on industrial scale to degrade the complex mixtures of organic compounds present in contaminated areas. The genetic potential and certain environmental factors such as temperature, pH, and available nitrogen and phosphorus sources govern the rate and the extent of degradation [14].

4. Anaerobic biodegradation

Among biological treatments, anaerobic digestion is frequently the most economical process, due to the high energy recovery linked to the process and its limited environmental impact.

Anaerobic biodegradation results when the anaerobic microbes are predominant over the aerobic microbes. Here oxygen does not serve as the final electron acceptor or reactant. Manganese and iron ions, and substances like sulfur, sulfate, nitrate, carbon dioxide, some organic intermediates and pollutants are reduced by electrons originating from oxidation of organic compounds [7]. The common example of anaerobic process is the biodegradable waste in landfill. Paper and other materials degrade more slowly over longer periods of time. Biogas, coming from anaerobic digestion, mainly consists of methane and can be collected efficiently and used for eco-friendly power generation as has been demonstrated on larger scale [3, 16]. Anaerobic digestion is widely used, as part of an integrated waste management system, to treat wastewater sludge and biodegradable waste because it provides volume and mass reduction of the input material. It reduces the emission of landfill gas into the atmosphere [17-20]. Anaerobic digestion is a renewable energy source because the process produces methane and CO_2-rich biogas suitable for energy production helping to replace fossil fuel requirement. Also, the nutrient-rich solids left after digestion can be used as fertilizer [16,21].

There are four major biological and chemical steps of anaerobic digestion: hydrolysis, acidogenesis, acetogenesis, and methanogenesis [17,18]. The mechanism commences with bacterial hydrolysis of the organic matter to break down insoluble organic polymers such as carbohydrates and make them available for other bacteria. Acetogenic bacteria convert the sugars and amino acids into carbon dioxide, hydrogen, ammonia, and organic acid. Methanogens then ultimately transform these products in to methane and carbon dioxide [19].

4.1. Advances in anaerobic digestion technologies

Thermophilic anaerobic digestion of manure [20] and assessment of biodegradability of macropollutants [21] have demonstrated the prowess of anaerobic digestion which is now a general method used to stabilize municipal wastewater treatment residuals [22,23]. The so-called phased or staged anaerobic digestion is a recent technology for digestion facilities which include four different configurations of reactors: staged mesophilic digestion, temperature-phased digestion, acid/gas phased digestion, and staged thermophilic digestion [24]. Phased or staged configurations are multiple reactor digestion systems. Phased anaerobic digestion is defined as a digestion system having two or more tanks, each with exclusive operating conditions that support unique biomass populations, which may be acid-forming, methane-forming, thermophilic, or mesophilic organism populations. Effective digestion is achieved by manipulating operational parameters such as solids retention time (SRT) and temperature. Temperature phased digestion system is found better than the other systems during each study phase by having higher volatile solids reduction (VSR), higher methane production, and lower residual biological activity [24,25].

On industrial scale, anaerobic digestion of solid waste is considered as a mature technology [16,26]. Around 60% of the plants are reported in Europe to operate at the mesophilic range (40% thermophilic) with continued increase in capacity over the years in most European countries. Yields from the biomethanization process are very much dependent on operating conditions and the kind of substrate used. Digestion of grey wastes or residual refuse after source separation, has caught attention of industry and some of the solutions considered are landfilling or incineration [23]. However, anaerobic digestion is a better option since it gives number of advantages such as greater flexibility, the possibility of additional material recovery (up to 25%), and a more efficient and ecological energy recovery. In this case the low-calorific organic fraction is digested, the high-calorific fraction is treated thermally and the non-energy fractions can be recovered and reused. It is predicted that this residual refuse will be treated by anaerobic digestion [16, 23].

A very high growth potential is expected for the anaerobic digestion of organic fraction of municipal solid waste (OFMSW). Around 50% of MSW is landfilled, with a content of around 30% of organic fraction (without considering paper and cardboard). The growth potential for this technology is very important to reduce greenhouse gases emission as agreed at the Kyoto Summit [23]. Further, the consolidation of anaerobic digestion as a mainstream technology for the OFMSW should occur since the digested residue can be considered quite stable organic matter with a very slow turnover of several decades given adequate soil conditions. Thus, the natural imbalance in CO_2 can be adjusted by restoring or creating organic rich soil. The removal of CO_2 constitutes an extra benefit that would place anaerobic digestion as one of the most relevant technologies in this field. The degradation of chlorinated compounds need to be examined in greater depth, as anaerobic treatment offers high potential in this area [28].

Several novel reactors with high mass transfer rates, such as fluidized bed reactors, expanded granular sludge bed (EGSB) reactors [29-32], and membrane bioreactors [33] with different configurations have been used, in which hydraulic retention times (HRT) are uncoupled from the solids retention time (SRT) to make anaerobic technology economical alternative for conventional wastewater treatment systems. The upflow anaerobic sludge blanket (UASB) reactors [30] and/or related systems are mostly applied, wherein spontaneous formation of granular conglomerates of the anaerobic organisms occurs, leading to anaerobic sludge with an extremely low sludge volume index and optimal settling properties [21]. Besides, several large scale biogas plants have been built which combine waste from agriculture, industry and households and produce both biogas and a liquid fertiliser which is re-circulated back on agri-land. The combination of anaerobic digestion with other biological or physico-chemical processes has led to the development of optimised processes for the combined removal of organic matter, sulphur and nutrients. In conjunction with anaerobic digestion which removes mainly carbon, other processes are used to remove nitrogen and phosphorus (with oxic phase), which mainly use micro-organisms and also physico-chemical processes. For the treatment of municipal wastewater, the ANANOX process [34] takes advantage of sulphate reduction to sulphide to provide an electron donor for the denitrification process [35-37]. The integration of the nitrogen cycle in anaerobic digestion could be maximised with the application of the ANAMMOX process that makes use of particular micro-organisms that are able to oxidise ammonium to N_2 gas with nitrite as electron acceptor [38,39].

5. Biodegradation of industrial organic pollutants

Knowledge of fate of chemicals discharged in the environment, the life cycle analysis and the mechanisms by which they degrade are of great importance in designing biodegradation systems since many of the industrial chemicals are toxic, recalcitrant and bioaccumulating in organisms [40-42].

5.1. Volatile Organic Compounds (VOCs)

There are two classes of VOCs that are responsible for a large number of land and groundwater contamination: (i) petroleum hydrocarbons (PHCs) such as gasoline, diesel, and jet fuel, and (ii) chlorinated hydrocarbon (CHC) solvents such as the dry cleaning agents such as tetra-chloroethylene, perchloroethylene (PCE) and the degreasing solvents such as trichloroethylene (TCE), 1,1,1-trichloroethane (TCA), and PCE.

PHCs biodegrade readily under aerobic medium, whereas CHCs characteristically biodegrade much more slowly and under anaerobic conditions [43]. Because PHC biodegradation is relatively rapid when oxygen is present, aerobic biodegradation can usually limit the concentration and subsurface migration of petroleum vapours in unsaturated soils. Further, CHC biodegradation can produce toxic moieties, such as dichloroethylene and vinyl chloride, while petroleum degradation usually produces carbon dioxide, water, and sometimes methane or other simple hydrocarbons. A second primary difference is density of pollutant. PHC liquids are lighter than water and immiscible. PHCs can float on the groundwater surface (water table), whereas chlorinated solvents being heavier than water sink through the groundwater column to the bottom of the aquifer. These major differences in biodegradability and density lead to very different subsurface behaviour that often reduces the potential for human exposure.

5.1.1. Petroleum Hydrocarbons (PHCs)

It is known that microorganisms capable of aerobically degrading PHCs are present in nearly all subsurface soil environments [44-49]. Effective aerobic biodegradation of PHCs hinges on the soil having adequate oxygen and water content to provide a habitat for sufficient populations of active microorganisms. If oxygen is present, these organisms will generally consume available PHCs. Furthermore, aerobic biodegradation of petroleum compounds can occur relatively quickly, with degradation half lives as short as hours or days under some conditions [50]. Some petroleum compounds can also biodegrade under anaerobic conditions; however, above the water table, where oxygen is usually available in the soil zone, this process is insignificant and often much slower than aerobic biodegradation. Aerobic biodegradation consumes oxygen and generates carbon dioxide and water. This leads to a characteristic vertical concentration profile in the unsaturated zone in which oxygen concentrations decrease with depth and VOCs including PHCs and methane from anaerobic biodegradation and carbon dioxide concentrations increase with depth [51,52].

5.1.2. Chlorinated Hydrocarbon (CHC) Solvents

Chlorinated solvents such as tetrachloroethylene (TCE), 1,1,2,2-tetrachloroethane, carbon tetrachloride, and chloroform are released as waste products by spills, land-filling, and discharge to sewers during manufacture and their use as solvents in a variety of cleaning processes or as vehicles for solid slurries. TCE is a major pollutant of the industry. It is biodegraded under anaerobic conditions through hydrogenolysis that sequentially produces isomers of 1,2-dichloroethylene (1,2-DCE), vinyl chloride (VC), and ethylene. Some labs have also reported ethane [53,54], methane [55], and carbon dioxide [56] as degradation products.

In addition to anaerobic degradation through reductive dechlorination (hydrogenolysis), TCE and other chlorinated VOCs can be susceptible to co-metabolic oxidation by aerobic microorganisms that have oxygenases with broad substrate specificity. Methanotrophs are microorganisms that primarily oxidize methane for energy and growth using methane monooxygenase (MMO) enzymes and are a group of aerobic bacteria transform TCE through co-metabolic oxidation [57-59]. In contrast to reductive dechlorination, where the degradation rate generally decreases as the degree of chlorination of the aliphatic hydrocarbon decreases, the less-chlorinated VOCs such as 1,2-DCE and VC are more straightforwardly and quickly degraded through aerobic oxidation reactions than the higher chlorinated compounds such as TCE [60]. Methane-oxidizing bacteria are known to convert TCE to its epoxide, which then breaks down immediately in water to form dichloroacetic acid, glyoxylic acid, or one-carbon compounds such as formate or CO. The two carbon acids accumulate in the water phase, while formate and CO are further oxidized by methanotrophic bacteria to CO_2. Hence, coupling of anaerobic and aerobic degradation processes has been recommended as the best possible bioremediation method for chlorinated VOCs such as TCE [60-62].

5.2. Quinoline

Quinoline occurs commonly in coal tar, oil shale, and petroleum, and is used as an intermediate and solvent in many industries [63,64]. Due to its toxicity and repulsive odor, quinoline-containing waste is detrimental to human health and environmental quality. The study of quinoline- degrading bacteria not only helps to reveal the metabolic mechanism of quinoline, but also benefits the bio-treatment of quinoline-containing wastewater. Although different genera of bacteria may produce different intermediates, almost all of them transform quinoline into 2-hydroxyquinoline in the first step [63, 65]. A quinoline-degrading bacteria strain, *Pseudomonas* sp. BW003, was isolated from the activated sludge in a coking wastewater treatment plant. *Pseudomonas* strains degrade quinoline via the 2-hydroxyquinoline and 2,8-hydroquinoline pathway, and then transform 2,8-hydroquinoline into 8-hydrocumarin, which is then transformed into 2,3-dihydroxyphenylpropionic acid, and finally to CO_2 and H_2O (Scheme 3) [66-69]. Quinoline-N is transformed into ammonia-N, as reported in few genera of bacteria. Thus, quinoline pollution can be eliminated by applying such degrading bacteria in the treatment with bio-augmentation [70-72].

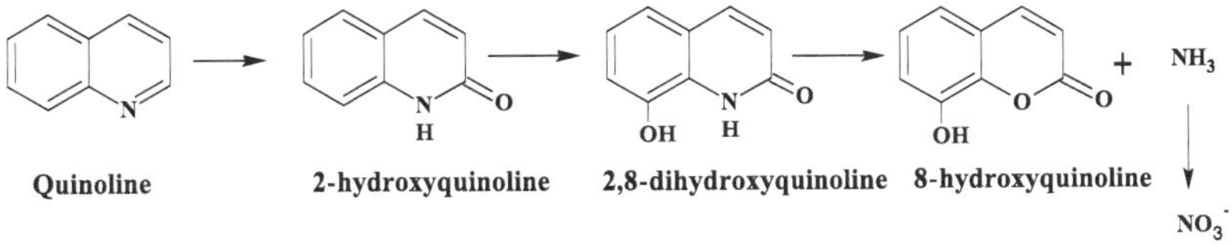

Scheme 3. Degradation products of quinoline [63]

5.3. Phenols

Phenols are harmful to organisms at low concentrations and classified as hazardous pollutants because of their potential to harm human health. They exist in different concentrations in wastewaters originated from coking, synthetic rubber, plastics, paper, oil, gasoline, etc. Biological treatment, activated carbon adsorption and solvent extraction are some of the most widely used methods for removing phenol and family compounds from wastewaters [73-76]. Biological treatment is economical, practical, promising and versatile approach for it leads to complete mineralization of phenol. Many aerobic bacteria are capable of using phenol as the sole source of carbon and energy [77]. In recent years, the strain of *Pseudomonas putida* has been the most widely used to degrade phenol. Under aerobic conditions, phenol may be converted by the bacterial biomass to CO_2; other intermediates such as benzoate, catechol, *cis*-cis-muconate, β-ketoadipate, succinate and acetate are formed during the biodegradation process [77, 78]. *p*-Nitrophenol (PNP) is one of the most widely used nitrophenolic compounds in industry and finds important applications in agriculture, polymers, pigment and pharmaceutical industries. However, PNP is highly toxic for both the environment and humans and its efficient removal from the environment is required. Hydroquinone (HQ), 4-nitrocatechol (4-NC) and 1,2,4-benzenetriol (1,2,4-BT) are the metabolic intermediates of the PNP biodegradation [80,81].

Chlorinated phenols are common and encountered even in relatively pristine environments [82,83]. These compounds are formed during the bleaching of pulp with chlorine [82-84]. As the pulp accounts only for about 40-45% of the original weight of the wood, these effluents are heavily loaded with organics [85]. Chlorophenols are also used as fungicides and may be formed from hydrolysis of chlorinated phenoxyacetic acid herbicides. Chlorophenols, part of the *adsorbable organic halides* (AOX), are present in bleaching effluents at concentrations ranging from 0.1 to 2.6 ppm [86]. Aqueous effluents from industrial operations such as polymeric resin production, oil refining and coking plants also contain chlorophenolic compounds. Pentachlorophenol (PCP) is the second most heavily used pesticide in the US. As compared to phenol, chlorophenolic compounds are more persistent in the environment. Toxicity and bioaccumulative potential of chlorophenols increases with the degree of chlorination and with chlorophenol lipophilicity. Haloaromatic compounds are degraded via the formation of halocatechols as intermediates which are subsequently cleaved by dioxygenases, by the mechanism delineated earlier. Dehalogenation then occurs by the elimination of the hydrogen halide, with subsequent double bond formation on the aliphatic intermediate [87]. In anaerobic environments, the biodegradation of chlorinated aromatics takes place through reductive dehalogenation leading to the formation of less toxic and more biodegradable compounds. Reductive dechlorination of 2,4-dichlorophenol is followed by carboxylation, ring fission and acetogenesis, and methanogenesis which finally led to the complete mineralization of 2,4-DCP, which is also biodegraded to 4-chlorophenol in anaerobic sediments. Similarly, biodegradation of PCP under anaerobic conditions occurs through reductive dechlorination [88].

5.4. Fluoro benzenes

Toluene degrading enzymes can transform many 3-fluoro-substituted benzenes to the corresponding 2,3-catechols with the concomitant release of inorganic fluoride. The substrates that induce 2,3-dioxygenase are 3-fluorotoluene, 3-fluorotrifluorotoluene, 3-flurohalobenzene,

3-fluoronisole, and 3-fluorobenzonitrile. While 3-fluorotoluene and 3-fluoronisole produce only deflorinated catechols, other substrates led to catechol products both with and without the toluene substituent [89].

5.5. Polycyclic Aromatic Hydrocarbons (PCAHs)

PCAHs are toxic, mutagenic and resist biodegradation [90]. Many strategies have been developed to treat them, including volatilization, photooxidation, chemical oxidation, bioaccumulation, and adsorption on soil particles [91]. Soil clean-up may be achieved using different remediation technologies, among which bioremediation is an effective and low-cost alternative that has garnered widespread use [92]. Two processes have been found to increase the activity of microorganisms during bioremediation: bio-stimulation and bio-augmentation. Bio-stimulation involves the addition of nutrients and/or a terminal electron acceptor to increase the meager activity of indigenous microbial populations. Bio-augmentation involves the addition of external microbial strains (indigenous or exogenous) which have the ability to degrade the desired toxic compounds [93]. The added specific PCAHs degrading strain, which has a competitive capacity to become dominant species with indigenous microbial strains or grow simultaneously with indigenous microbial strains, may greatly enhance the rate of PCAHs biodegradation [94,95]. The ability to degrade PCAHs depends on the complexity of their structure and the extent of enzymatic adaptation by bacteria. In general PCAHs with 2 or 3 aromatic rings are readily degraded, but those with 4 or more are usually recalcitrant and genotoxic. Such examples of PCAHs are acenaphthene, acenaphthylene, anthracene, naphthalene, fluorene, phenanthrene, chrysene, pyrene, etc. The major metabolites are 4-phenanthroic acid and 4-hydroxyperinapthenone. Cinnamic and phthalic acids are ring fission products [96].

Naphthalene is carcinogenic and persistent organic pollutant [97]. Bacteria such as *Pseudomonas putida, Rhodococcus opacus, Mycobacterium* sp., *Nocardia otitidiscaviarum,* and *Bacillus pumilus* are known to biodegrade naphthalene [98-102]. Some metabolites of naphthalene, such as salicylic acid, 1-naphthol and *o*-phthalic acid could be degraded and support cell growth (Scheme 4). Phenanthrene was used as a model compound for PCAH degradation which shows 1-hydroxy 2-naphthoic acid (1H2NA) as intermediate biodegradation product [103].

5.6. Plasticizers

Plasticizers are polymeric additives, used to impart flexibility to polymer materials. The biodegradation of some plasticizers can lead to the formation of metabolites with increased persistence and toxicity relative to the original compounds [104-106]. Use of plasticizers has grown considerably, both with respect to product variety and production volume [107]. Phthalates are the most widely used plasticizers. Presence of phthalates and their metabolites in rats, mice, human plasma and liver are related to adverse health effects such as endocrine disruption and peroxisome proliferation [108,109]. The high production volumes of phthalates and their incomplete biodegradation have led to the presence of these compounds and a number of toxic and stable metabolites in surface waters, groundwater, air, soil and tissue of living organisms [104, 110-113]. Such findings have led to stricter environmental regulations

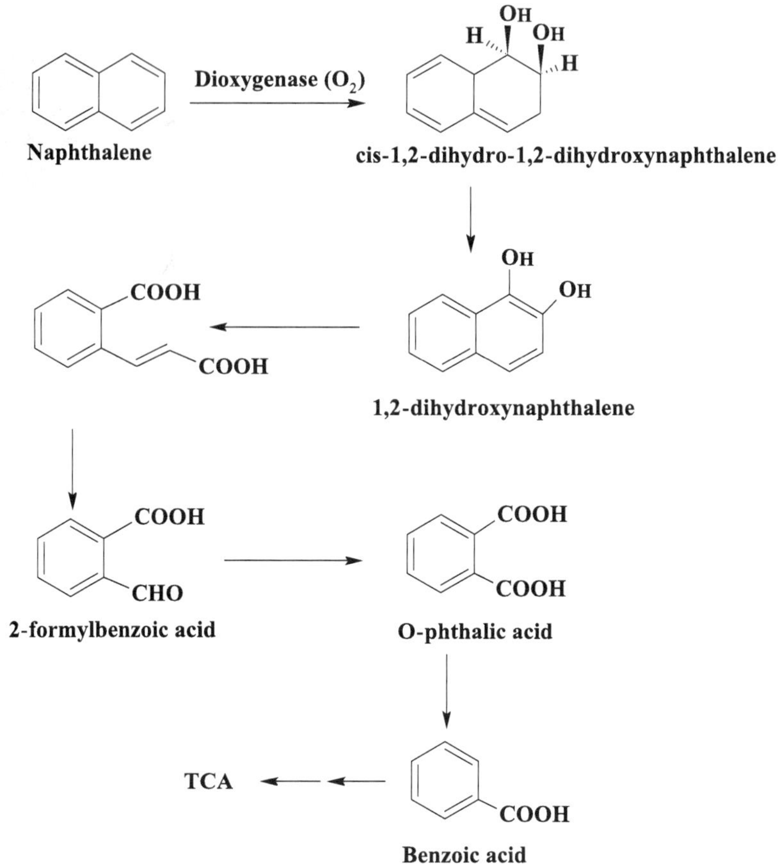

Scheme 4. Proposed pathway for the degradation of naphthalene [103]

and consequently have broadened the criteria used to evaluate plasticizers. Consequently, dibenzoates have been approved by the European Chemical Agency as alternatives to phthalates [114]. However, the degradation of dipropylene glycol dibenzoate (D(PG)DB) and diethylene glycol dibenzoate (D(EG)DB) by common soil microorganisms such as *Rhodotorula rubra* and *Rhodococcus rhodochrous* can lead to the formation and accumulation of monobenzoate metabolites [115,116] that exhibit high acute toxicity [115]. Other compounds including 1,5-pentandiol and 1,6- hexanediol dibenzoates were reported to produce less stable metabolites and have also been tested as potential alternatives to commercial dibenzoate plasticizers [116-118]. Scheme 5 shows the biodegradation products of dibenzoates by *R. Rhodochrous*, which include 2-[2-(benzoyloxy)propoxy] propanoic acid, 1,3-propanediol monobenzoate and 3-(benzoyloxy) propanoic acid [119].

5.7. Plastics

Over the years, plastics have brought economic, environmental and social advantages. Today's material world uses tremendous quantities of plastics of all hue and origins. However, their wide spread use has also increased plastic waste, which brings its own economic, environmental and social problems. The redesign of plastic products, whether individual polymer or product structure, could help alleviate some of the problems associated with plastic waste. Redesign could have an impact at all levels of the hierarchy established by the European Waste Framework Directive: prevention, re-use, recycle, recovery and disposal [120].

Polyethylene, polypropylene and polystyrene, and water-soluble polymers, such as polyacrylamide, polyvinyl alcohol and polyacrylic acid are bulk polymers used in a variety of industries and products. Some of the plastics are not biodegradable and deleterious to the environment due to their accumulation. Plastics can be disposed of by incineration or recycling, but incineration is very difficult, dangerous and expensive whereas recycling is a long process and not very efficient. Some plastics still cannot be recycled or incinerated due to pigments, coatings and other additives during manufacture of materials. Making biodegradable and ecofriendly plastics will avoid accumulation, recycling and incineration [121].

5.7.1. Polyvinyl alcohol

Polyvinyl alcohol (PVA) is water-soluble but also has thermoplasticity. In addition to its use as a water-soluble polymer, for instance, as a substituent for starch in industrial processes, it can also be molded in various shapes, such as containers and films. PVA can therefore be used to make water-soluble and biodegradable carriers, which may be useful in the manufacture of delivery systems for chemicals such as fertilizers, pesticides, and herbicides. Among the vinyl polymers produced industrially, PVA is the only one known to be mineralized by microorganisms [122]. Extensive use of PVA, in textile and paper industries generates considerable amount of contaminated wastewaters [121]. In aqueous solution, the biodegradation mechanism of PVA involves the random endocleavage of the polymer chains. The initial step is associated with the specific oxidation of methane-carbon bearing the hydroxyl group, as mediated by oxidase and dehydrogenase type enzymes, to give β-hydroxyketone as well as 1,3-diketone moieties. The latter groups are able to facilitate the carbon-carbon bond cleavage as promoted by specific β-diketone hydrolase, leading to the formation of carboxylic and methyl ketone end groups [123,124]. Most of the PVA-degrading microorganisms are aerobic bacteria belonging to *Pseudomonas, Alcaligenes,* and *Bacillus* genus. A very moderate PVA biodegradation was reported [125-128].

5.7.2. Polyhydroxyalkanoates

Polyhydroxyalkanoates (PHAs) are degraded to CO_2 and water in aerobic conditions and methane in anaerobic conditions by microbes found in soil, water and other various natural habitats. PHAs are the only proposed replacement polymers that are completely biodegradable [129]. The structures of these polymers have a very similar structure of the petroleum-derived thermoplastics [130].

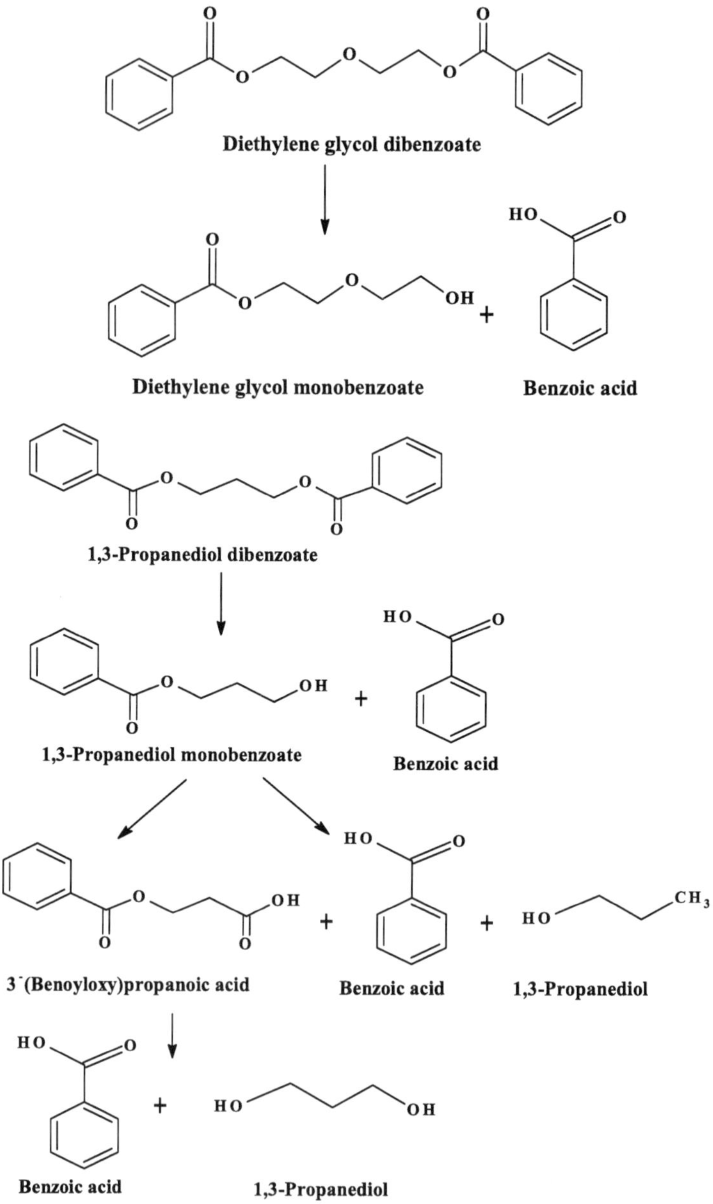

Scheme 5. Proposed biodegradation pathways of diethylene glycol dibenzoate and 1,3-propanediol dibenzoate [116]

PHAs also possess similar physical properties as plastics including the ability to be molded, made into films, and also into fibers. Efforts are underway to identify bacteria, which produce various kinds of PHAs [129] as well as the production of these polyesters or create certain kinds of PHAs by changing the kind of bacteria [130] and/or the substrates given to the bacteria and genetically enhancing bacteria [131].

6. Prospective of anaerobic digestion and biogas energy

The foregoing analysis shows that anaerobic digestion technologies have matured so far to treat several organic micro-pollutants, halogenated compounds, substituted aromatics, azo-linkages, nitro-aromatics and the like in industrial effluents and also for municipal effluents containing industrial loads. A very high growth potential is envisaged for the anaerobic digestion of organic fraction of municipal solid waste [27]. Novel reactor and control systems ought to be developed for different purposes depending on the source of pollutants or biomass. Anaerobic digestion of sewage sludge followed by recycling on agricultural land is currently the largest world-wide application of anaerobic processes. Treatment of sludge and slurries targeted at the production of safe end products can be tackled with niche anaerobic technologies [16]. It is predicted that major future process developments will come from the deployment of pre- and post treatment processes, including physical, chemical and biological processes, for the reclamation of the products from the wastewater treatment system. Wastewater treatment for reuse will be effective if anaerobic digestion is adopted for mineralizing organic matter. Hence, anaerobic digestion has the potential to play a major role in closing water, raw materials, and nutrient cycles in industrial processes [37]. Further development is required on the community on-site treatment of domestic sewage under a wide range of conditions, opting for the reuse of the treated water in agriculture and making use of the mineralized nutrients for fertilization purposes. An upstream integration of the anaerobic process with industrial primary production processes under extreme conditions of temperature, pH, salinity, toxic and recalcitrant compounds, and variable load is envisaged in future [39].

There is an emphasis worldwide on renewable energy system among which biogas produced from any biological feedstocks including primary agricultural sectors and from various organic waste streams will come in to prominence in near future [22]. It is estimated [3] that at least 25% of all bioenergy in the future can originate from biogas, produced from wet organic materials like animal manure, slurries from cattle and pig production units as well as from poultry, fish and fur, whole crop silages, wet food and feed wastes, etc. Anaerobic digestion of animal manure offers several environmental, agricultural and socio-economic benefits throughout such as improved fertilizer quality of manure, considerable reduction of odors and inactivation of pathogens and more importantly production of biogas production, as clean, renewable fuel, for multiple utilizations [16]. This biogas can be upgraded to natural gas to mix with the existing natural gas grid which will be cost effective. The potential development of biogas from manure co-digestion includes the use of new feedstock types such as by-products from food processing industries, bio-slurries from biofuels processing industries as well as the biological degradation of toxic organic wastes from pharmaceutical industries, etc. [3,16,22]. This will also call for better reactor systems and careful process control to increase the biogas yield, which will be more attractive if coupled with less capital and operating costs. Integration of biogas production into the national energy grids will eventually be commercially viable since the biogas from anaerobic co-digestion of animal manure and suitable organic wastes would overcome the major environmental and veterinary problems of the animal production and organic waste disposal.

7. Plastic waste separation, reprocessing and recycle

In 2009, around 230 million tonnes of plastic was produced; ~25 % which was used in the European Union [131]. About 50 % plastic is used for single-use disposable applications, such as packaging, agricultural films and disposable consumer items [132]. Although plastics consume approximately 8 % world oil production: 4 % as raw material for plastics and 3-4 % as energy for manufacture [132], supplies are being depleted. Bioplastics make up only 0.1 to 0.2 % total plastics [115]. It is estimated that plastics reduce 600 to 1300 million tonnes of CO_2 through the replacement of less efficient materials, lighter and fuel efficient vehicles, housing sector, contribution to insulation, preservation of food, packaging, use in wind power rotors and solar panels [133]. However, plastic littering and pollution of land and sea have been matters of great concern which should be strategically and technologically solved. Plastics recovery, in addition to increased diversion from disposal, results in significant energy savings (~50-75 MBtu/ton of material recycled) compared with the production of virgin materials, which also leads to reductions in greenhouse gas emissions due to avoided fuel use. Limiting the plastics that enter landfills can lower the costs associated with waste disposal. It is believed that the recycled plastic will fetch as much as 70 % of the typical price for virgin plastics [136].

7.1. Waste reduction hierarchy

The motto of waste reduction by plastics is by following the principles of (i) prevention, (ii) reuse (iii) recycle, (iv) recovery, and (v) disposal [119].

i. *Prevention* – Using minimum and as less types of plastic in the product by clever product redesign.

ii. *Reuse* – Products could be designed for reuse by facilitating the dismantling of products and replacement of parts. This could involve standardizing parts across manufacturers [137].

iii. *Recycle* – Some types of plastics are easier to recycle than others, for example polyethylene terephthalate (PET). By using fewer types and colors (or colorless) of plastics the recycling process becomes easier. The use of "intelligent" or smart polymers that undergo changes under certain conditions could also reduce disassembly time [138]. For example, a polymer that changes shape when subject to magnetic or electric fields could aid the disassembly of electronic goods.

http://dx.doi.org/10.5772/56241

iv. Recovery – Energy can be recovered from plastics in waste-to-energy plants. By designing products to consider the possibility of energy recovery, plastic may have a greater end-of-life use.

v. *Disposal* – Biodegradable plastics are less persistent in the environment than traditional plastics, but need specific and suitable end-of-life treatment.

7.2. Bioplastics

Since disposal is one of the important aspect, bioplastics are being favored. There are three main categories of bio-based plastics: (i) Natural polymers from renewable sources, such as cellulose, starch and plant-based proteins, (ii) Polymers synthesised from monomers derived from renewable resources. For example, polylactic acid (PLA) is produced by the fermentation of starch, corn or sugar, (iii) Polymers produced by microorganisms. For example, PHA (polyhydroxyalkanoate) is produced by bacteria through fermentation of sugar or lipids [139].

Biodegradable plastics are not by definition bio-based and bio-based plastics are not always biodegradable, although some fall into both categories, such as PHAs. The term *bioplastics* is often used to refer to both bio-based and biodegradable plastics. The main applications of bioplastics are disposable plastic bags, packaging and loose fill packaging (beads and chips), dishes and cutlery, electronic casings and car components. However, bioplastics cannot substitute all types of plastic; particularly certain types of food packaging that require gas permeability [135]. Development of novel biodegradable plastic is a solution for the plastic disposal problem since plastics are immiscible in water and are thermo-elastic polymeric materials. Biodegradability of plastics is governed by both their chemical and physical properties. Other factors affecting degradability are the forces associated with covalent bonds of polymer molecules, hydrogen bonds, van der Waals forces, coulombic forces, etc. Enzymatic degradation is an effective way. Lipase and esterase can hydrolyze fatty acid esters, triglycerides and aliphatic polyesters. These lipolytic enzymes have an important role in the degradation of natural aliphatic polyesters such as cutin, suberin and esteroid in the natural environment and animal digestive tract.

As stated earlier, biodegradable plastics decompose in the natural environment from the action of bacteria. Biodegradation of plastics can be achieved through the action of micro-bacteria and fungi in the environment to metabolize the molecular structure of plastic films to produce an inert humus-like material that is less harmful to the environment, along with water, carbon dioxide and/or methane. They may be composed of either bioplastics or petro-plastics. The use of bio-active compounds compounded with swelling agents ensures that, when combined with heat and moisture, they expand the plastic's molecular structure and allow the bio-active compounds to metabolize and neutralize the plastic [140]. Compostable plastics are biodegradable and meet certain criteria, such as rate of biodegradation and impact on the environment. Degradable plastics include biodegradable and compostable plastics, but also plastics that degrade by chemical and physical processes such as the action of sunlight. Purely biodegradable plastics are different from oxy-biodegradable plastics, which contain small amounts of metal salts to speed up degradation. It has been suggested that this process be called "oxo-fragmentation" to avoid confusion [139,140].

It is possible to produce polymers biologically, e.g., PHA grown in genetically modified corn plant leaves, PLA (polylactic acid) produced by the fermentation of sugars extracted from plants, PHA produced by bacteria. Bioplastics could also help alleviate climate change by reducing the use of petroleum for the manufacture of traditional plastics. It is claimed that CO_2 emissions released at the end-of-life of bio-based plastics are offset by absorption of CO_2 during the growth of plants for their production [141].

7.3. Sorting plastic materials

The technical difficulties and high cost associated with separating plastics have limited recycling in the past. Consumer goods often contain as many as 20 different types of plastic as well as non-plastic materials such as wood, rubber, glass, and fibers. There is upsurge of new products and pigment types, which can pose a challenge to the recycling infrastructure. Consequently, the cost of producing virgin materials is often less than the cost of collecting and processing post-consumer plastics. Used plastic material will contain more than one base polymer, and resins with a variety of additives, including coloring agents and thus technologies for cleaning and separating the materials are an important part of most plastics recycling systems. A particular concern for recycled plastics is their use as food containers requiring stringent regulations to avoid contamination [140].

Separation of different types of polymers from each other is many times a desired part of plastics recycling processes which are classified as macrosorting, microsorting, or molecular sorting.

7.3.1. Macrosorting

Macrosorting involves the sorting of whole or nearly whole objects such as separation of PVC bottles or caps from PET bottles, separation of polyester carpet from nylon carpet, and sorting of automobile components by resin type. Various devices are now commercially available to separate plastics by resin type, which typically rely on differences in the absorption or transmission of certain wavelengths of electromagnetic radiation, or color or resin type. Particularly for recycling of appliances, carpet, and automobile plastics, several IR spectra based equipment are used [135].

7.3.2. Microsorting

Microsorting is a size-reduction process to reduce the plastic material in to small pieces which is then separated by resin type or color; for instance, separation of high-density polyethylene (HDPE) base cups from PET soft drink bottles using a sink-float tank. More modern separation processes, such as the use of hydrocyclones, also rely primarily on differences in the density of the materials for the separation. A number of other characteristics have also been used as the basis for microsorting systems, including differences in melting point and in triboelectric behavior. In many of these systems, proper control over the size of the plastic flakes is

http://dx.doi.org/10.5772/56241

important in being able to reliably separate the resins. Some systems rely on differences in the grinding behavior of the plastics combined with sieving or other size-based separation mechanisms for sorting. Sometimes cryogenic grinding is used to facilitate grinding and to generate size differences [135].

Three new separation technologies, developed by MBA Polymers, Argonne National Laboratory, and Recovery Plastics International (RPI), could break down these barriers and increase plastics recycling [138].

7.3.2.1. Automated separation

According to the process developed by MBA Polymers, plastic scraps from computers and other electronics are first ground into small pieces. Magnets and eddy-current separators then extract ferrous and non-ferrous metals. Paper and other lighter materials are removed with jets of air. Finally, a proprietary sorting, cleaning, and testing process involving various technologies, allows the separation of different types of plastics and compound them into pelletized products comparable to virgin plastics [138].

7.3.2.2. Froth flotation

Argonne National Laboratory (ANL) has developed a process to separate acrylonitrile-butadiene styrene (ABS) and high-impact polystyrene (HIPS) from recovered automobiles and appliances. The froth flotation process separates two or more equivalent-density plastics by modifying the effective density of the plastics. There is a careful control of the chemistry of the aqueous "froth" so that small gas bubbles adhere to the solid surface and facilitate the plastic to float to the top [135].

7.3.2.3. Skin flotation

Recovery Plastics International (RPI) has developed an automated process capable of recovering up to 80 % plastics found in automobile shredder residue (ASR). The process starts with the separation of light lint materials, followed by the removal of rocks and metals, granulation, washing, and, finally, automated skin flotation separation. This final step adds a skin of plasticizer to make the surface of the targeted plastic hydrophobic. Air bubbles then can attach to the plastic, allowing it to float above denser, uncoated pieces. It is estimated this new skin flotation technology could be used to treat about one-third of the estimated 7 million tons of ASR disposed off each year [141].

7.3.3. Molecular sorting

Molecular sorting deals with sorting of materials whose physical form has been completely disrupted, such as by dissolving the plastics in solvents using either a single solvent at several temperatures or mixed solvents, followed by reprecipitation. There is a need to control emissions and to recover the solvents, without any residual solvent in the recovered polymer to avoid leaching in stored material. There are at present no commercial systems using this approach. Some research effort has focused on facilitating plastics separation by incorporating chemical tracers into plastics, particularly packaging materials, so that they can be more easily identified and separated.

It has become obvious that many of the difficulties of recycling plastics are related to difficulties in separating plastics from other wastes and in sorting plastics by resin type. Design of products can do a lot to either aggravate or minimize these difficulties [134,135]. The concept of green product embeds recycling at the design stage itself.

7.4. Plastic reprocessing and recycling

For plastics recycling to be effective, it is necessary to have (i) a system for collecting the targeted materials, (ii) a facility capable of processing the collected recyclables into a form which can be utilized to make a new product and, (iii) new products made in whole or part from the recycled material must be manufactured and sold.

Processing of recyclable plastics is necessary to transform the collected materials into raw materials for the manufacture of new products. Three general categories of processing can be identified: (1) physical recycling, (2) chemical recycling, and (3) thermal recycling, wherein the particulars of the processing are often specific to a given plastic or product.

7.4.1. Physical processes

Physical recycling, the most popular option, covers size and shape alteration, removing contaminants, blending in additives if desired, and similar approaches that change the appearance of the recycled material, but do not alter its basic chemical structure. Plastic containers, for example, are processed including grinding, air classification to remove light contaminants, washing, gravity-separation, screening, rinsing, drying, and often melting and pelletization, accompanied by addition of colorants, heat stabilizers, or other ingredients, depending on type of plastic [132].

7.4.2. Chemical reactions

Chemical recycling of plastics deals with chemical reactions using catalysis or solvents such as methanol, glycols or water leading to depolymerization or breaking polymers into monomers or useful chemicals, or fuels [134]. The products of the reaction then can be separated and reused to produce either the same or a related polymer. An example is the glycolysis process sometimes used to recycle polyethylene terephthalate (PET), in which the PET is broken down into monomers, crystallized, and repolymerized. Condensation polymers, such as PET, nylon, and polyurethane, typically much more amenable to chemical recycling than are addition polymers such as polyolefins, polystyrene, and polyvinyl chloride (PVC). Most commercial processes for depolymerization and repolymerization are restricted to a single polymer, which is usually PET, nylon 6, or polyurethane. Methanolysis is another common reaction using methanol [134].

7.4.3. *Thermal cracking*

Thermal cracking or recycling also involves cracking of the chemical structure of the polymer using heat in the absence of sufficient oxygen for combustion. At these elevated temperatures, the polymeric structure breaks down. Thermal recycling can be applied to all types of polymers. However, the typical yield is a complex mixture of products, even when the feedstock is a single polymer resin. If reasonably pure compounds can be recovered, products of thermal recycling can be used as feedstock for new materials. When the products are a complex mixture which is difficult to separate, they are most often used as fuel. There are relatively few commercial operations today which involve thermal recycling of plastics. Some European nations have such feedstock recycling facilities. Many plastic resin companies use fluidized bed cracking to produce a waxlike material from mixed plastic waste [134-136, 139]. The product, when mixed with naptha, can be used as a raw material in a cracker or refinery to produce feedstocks such as ethylene and propylene. In certain case, syn gas can be produced and used in Fisher-Tropsch synthesis to produce valuable chemicals.

In landfill, both synthetic and naturally occurring polymers do not get the necessary exposure to UV and microbes to degrade. The discarded plastics occupy space and none of the energy put into making them is being reclaimed. Reclaiming the energy stored in the polymers can be done through incineration, but this can cause environmental damage by release of toxic gases into the atmosphere. Therefore, recycling is a viable alternative in getting back some of this energy in the case of some polymers. With ever increasing petroleum prices, it would be financially viable to recycle polymers rather than produce them from raw materials [141].

8. Conclusions

The modern society needs thousands of chemicals and materials of all sorts which are produced annually and used in all sectors of economy. However, their fate in the environment is of great concern since some are toxic, recalcitrant and bioacumulating and hence their discharge into the environment must be regulated. Better understanding of the mechanism of biodegradation has a high ecological significance that depends on indigenous microorganisms to transform or mineralize the organic contaminants. Thus, biodegradation processes differ greatly depending on conditions, but frequently the main final products of the degradation are carbon dioxide and/or methane. Microorganisms have enzyme systems to degrade and utilize different hydrocarbons as a source of carbon and energy. Slow and partial biodegradation of chlorophenolic compounds under aerobic as well as anaerobic natural environment has been observed. Aerobic degradation takes place via formation of catechols while anaerobic degradation occurs via reductive dechlorination. Acclimatization of biomass to chlorophenols markedly enhances their ability to degrade such compounds, both by reducing the initial lag phase as well as by countering biomass inhibition. Aerobic processes as well as anaerobic processes partially remove chlorophenols. However, enhanced removal efficiency can be obtained by operating anaerobic and aerobic treatment processes in combination. Thus microbial degradation can be a key component for clean-up strategy of organopollutants and plastics.

Renewable energy system among which biogas produced from biological feedstocks will play a major role in energy sector. Anaerobic digestion of animal manure, slurries from cattle and pig production units as well as from poultry, fish and fur, whole crop silages, wet food and feed wastes, etc offers several environmental, agricultural and socio-economic benefits by improved fertilizer quality of manure, considerable reduction of odors, inactivation of pathogens and production of biogas production, as clean and renewable fuel. This biogas can be upgraded to natural gas to inject in to the existing natural gas grid which will be cost effective. Biogas from anaerobic co-digestion of animal manure and suitable organic wastes would overcome the major environmental and veterinary problems of the animal production and organic waste disposal.

The recycling of plastics is environmentally beneficial because plastics reduce millions of tonnes of CO_2 emissions through the replacement of less efficient materials, development of lighter and fuel efficient transport systems, housing material, energy saving insulation, food preservation and storage, energy efficient packaging, use in wind power rotors and solar panels. Processing of recyclable plastics is necessary to transform the collected materials into raw materials for the manufacture of new products. Bioplastics offer a very good solution to environmentally deleterious materials. Biodegradation of plastics can be achieved through the action of micro-bacteria and fungi.

Author details

Ganapati D. Yadav* and Jyoti B. Sontakke

*Address all correspondence to: gdyadav@yahoo.com; gd.yadav@ictmumbai.edu.in

Department of Chemical Engineering, Institute of Chemical Technology, Matunga, Mumbai, India

References

[1] Alexander M. Biodegradation and Bioremediation. Academic Press: New York; 1999.

[2] Lily Y, Young LY, Cerniglia CE. Microbial Transformation and Degradation of Toxic Organic Chemicals. Wiley-Liss Inc. New York, NY; 1995.

[3] Holm-Nielsen JB, Al Seadi T, Oleskowicz-Popiel P. The future of anaerobic digestion and biogas utilization. Bioresource Technology 2009; 100: 5478–5484.

[4] Steinfeld H, Gerber P, Wasenaar T, Castel V, Rosales M, de Haan C. Livestock's long shadow. Environmental Issues and Options. Food and Agriculture Organization (FAO) of United Nations; 2006.

[5] Nielsen LH, Hjort-Gregersen K, Thygesen P, Christensen J. Samfundsøkonomiske analyser af biogasfllesanlg. Rapport 136; 2002. Fødevareøkonomisk Institut, København (Summary in English).

[6] http://www.unesco.org/new/en/natural-sciences/ (accessed December 2012)

[7] Van Agteren MH, Keuning S, Janssen D, Handbook on Biodegradation and Biological Treatment of Hazardous Organic Compounds. Kluwer, Dordrecht, The Netherlands; 1998.

[8] Tokiwa Y, Calabia BP, Ugwu CU, Aiba S. Biodegradability of Plastics. International Journal of Molecular Science 2009; 10: 3722–3742.

[9] Griffin GJL. Chemistry and Technology of Biodegradable Polymers, Springer, London; 1994.

[10] Schmidt M., editor. Synthetic Biology: Industrial and Environmental Applications, Wiley-Blackwell; 2012.

[11] NIIR Board of Consultants and Engineers (Ed.), Medical, Municipal and Plastic Waste Management Handbook. National Institute of Industrial Research, New Delhi; 2009.

[12] Stuart PR, El-Halwagi MM., editor. Integrated Biorefineries: Design, Analysis and Optimization, CRC Press; 2012.

[13] Perez JJ, Munoz-Dorado J, de la Rubia TJ, Martınez J. Biodegradation and biological treatments of cellulose, hemicelluloses and lignin: an overview. International Microbiology 2002; 5: 53-63.

[14] Berna JL, Cassani G, Hager CD, Rehman N, Lopez I, Schowanek D, Steber J, Taeger K, Wind T. Anaerobic Biodegradation of Surfactants-Scientific Review. Tenside Surfactants Detergents 2007; 44: 312-347.

[15] Fritsche W, Hofrichter M. Aerobic Degradation by Microorganisms: Principles of Bacterial Degradation. In: Rehm HJ, Reed G, Puhler A, Stadler A. (eds.) Biotechnology, environmental processes II, vol IIb. Wiley-VCH, Weinhein. p145-167.

[16] Lier JB van, Tilche A, Ahring BK, Macarie H, Moletta R, Dohanyos M, Hulshoff Pol LW, Lens P, Verstraete W. New perspectives in anaerobic digestion. Water Science and Technology 2001; 43(1): 1-18.

[17] Rozzi A, Remigi E. Anaerobic biodegradability: Conference Proceeding. In: 9th World Congress, Anaerobic digestion 2001, Workshop 3 Harmonisation of anaerobic activity and biodegradation assays, 9-2-2001, Belgium.

[18] Dolfing J, Bloemen GBM. Activity measurement as a tool to characterize the microbial composition of methanogenic environments. Journal of Microbiological Methods 1985; 4: 1-12.

[19] Soto M, Mendez R, Lema JM. Methanogenic activity tests. Theoretical basis and experimental setup. Water Research 1993; 27: 850–857.

[20] Angelidaki I, Ahring BK. Thermophilic anaerobic digestion of livestock waste: the effect of ammonia. Applied Microbiology Biotechnology 1993; 38: 560–564.

[21] Angelidaki I. Sanders W. Assessment of the anaerobic biodegradability of macropollutants. Reviews in Environmental Science and Biotechnology 2004; 3: 117–129.

[22] Holm-Nielsen JB, Oleskowicz-Popiel P, 2007. The future of biogas in Europe: Visions and targets until 2020. In: Proceedings: European Biogas Workshop-Intelligent Energy Europe, 14–16 June 2007, Esbjerg, Denmark. Mata-Alvarez J, Macé S, Llabrés P, Anaerobic digestion of organic solid wastes. An overview of research achievements and perspectives. Bioresource Technology 2000; 74: 3-16.

[23] Reusser S, Zelinka G. Laboratory-scale comparison of anaerobic-digestion alternatives. Water Environment Research 2004; 76(4): 360-379.

[24] David C, Inman DC. Comparative studies of alternative anaerobic digestion technologies. M.S. (Environ. Eng.) Thesis, Virginia Polytechnic Institute and State University; 2004.

[25] Riggle D. Acceptance improves for large-scale anaerobic digestion. Biocycle 1998; 39 (6): 51-55.

[26] Mata-Alvarez, J., Tilche, A., Cecchi, F., editor. The treatment of grey and mixed solid waste by means of anaerobic digestion: future developments. Proceedings of the Second International Symposium on Anaerobic Digestion of Solid Wastes, Barcelona, vol. 2. Graphiques 92, 15-18 June, 1999. p302-305.

[27] Christiansen N. Hendriksen HV, Jarviene KT, Ahring B. Degradation of chlorinated aromatic compounds in UASB reactors. Water Science and Technology 1999; 31: 249-259.

[28] Man AWA de, Last ARM van der, Lettinga G. The use of EGSB and UASB anaerobic systems or low strength soluble and complex wastewaters at temperatures ranging from 8 to 30°C. Proceedings of the 5th International Symposium on Anaerobic Digestion. Bologna, Italy, 1988. p197-211.

[29] Driessen WJBM, Habets LHA, Groeneveld N. New developments in the design of Upflow Anaerobic Sludge Bed reactors. 2nd Specialised IAWQ conference on Pretreatment of Industrial Wastewaters, October 16-18,1996.

[30] Zoutberg GR, Been P de. The biobed EGSB (Expanded Granular Sludge Bed) systems covers short comings of the upflow anaerobic sludge blanket reactors in the chemical industry. Water Science and Technology 1997; 35(10): 183-188.

[31] Collins AG, Theis TL, Kilambi S, He L, Pavlostathis SG. Anaerobic treatment of low strength domestic wastewater using an anaerobic expanded bed reactor. Journal of Environmental Engineering 1998: 652-659.

[32] Nagano A, Arikawa E, Kobayashi H. Treatment of liquor wastewater containing high strength suspended solids by membrane bioreactor system. Water Science and Technology 1992; 26(3-4): 887-895.

[33] Garuti G, Dohanyos M, Tilche A. Anaerobic-aerobic wastewater treatment system suitable for variable population in coastal are: the ANANOX process. Water Science and Technology 1992; 25(12):185-195.

[34] Oude Elferink SJWH, Visser A, Hulshoff Pol LW, Stams AJM. Sulfate reduction in methanogenic bioreactors. FEMSMicrobiology Reviews 1994; 15: 119-136.

[35] Boopathy R, Kulpa CF, Manning J. Anaerobic biodegradation of explosives and related compounds by sulfate-reducing and methanogenic bacteria: a review. Bioresource Technoology 1998; 63(1): 81-89.

[36] Houten RT van, Lettinga G. Biological sulphate reduction with synthesis gas: microbiology and technology. In: Wijffels RH., Buitelaar RM., Bucke C., Tramper J. (eds.) Progress in Biotechnology. Elsevier, Amsterdam, The Netherlands; 1996. pp. 793-799..

[37] Jetten MSM, Strous M, Pas-Schoonen KT Van de, Schalk J, Van Dongen UGJM, Van De Graaf AA, Logemann S, Muyzer G, Van Loosdrecht MCM, Kuenen JG. The anaerobic oxidation of ammonium. FEMS Microbiology Reviews 1999; 22: 421-437.

[38] Lier JB van, Lettinga, G. Appropriate technologies for effective management of industrial and domestic wastewaters: the decentralised approach. Water Science and Technology 1999; 40 (7): 171-183.

[39] Abrahamsson K, Klick S. Degradation of Halogenated Phenols in Anoxic Marine Sediments. Marine Pollution Bulletin 1991; 22: 227-233.

[40] Suflita JM, Horowitz A, Shelton DR, Tiedje JM. Dehalogenation: A Novel Pathway for the Anaerobic Biodegradation of Haloaromatic Compounds. Science 1982; 218: 1115-1117.

[41] Annachhatre AP, Gheewala SH. Biodegradation of Chlorinated Phenolic Compounds. Biotechnology Advances 1996; 14(1): 35-56.

[42] Howard PH. Handbook of Environmental Degradation Rates. Lewis Publishers: Chelsea MI; 1991.

[43] Zobell CE. Action of microorganisms on hydrocarbons. Bacteriological Reviews 1946; 10(1-2): 1-49.

[44] Atlas RM. Microbial degradation of petroleum hydrocarbons: an environmental perspective. Microbiological Reviews 1981; 45(1): 180–209.

[45] Wilson JT, Leach LE, Henson M, Jones JN. In situ biorestoration as a ground water remediation technique. Ground Water Monitoring Review 1986; 6: 56–64.

[46] Leahy JG, Colwell RR. Microbial degradation of hydrocarbons in the environment. Microbiological Reviews 1990; 54(3): 305–315.

[47] Bedient PB, Rifai HS, Newell CJ. Ground Water Contamination: Transport and Remediation. PTR Prentice-Hall Inc. Englewood Cliffs, NJ; 1994.

[48] EPA (Environmental Protection Agency). Monitored Natural Attenuation of Petroleum Hydrocarbons. Remedial Technology Fact Sheet. EPA/600/F-98/021. Office of Research and Development, Washington, DC; May 1999.

[49] DeVaull G. Indoor vapor intrusion with oxygen-limited biodegradation for a subsurface gasoline source. Environmental Science and Technology 2007; 41(9): 3241–3248.

[50] Roggemans S, Bruce CL, Johnson PC. Vadose Zone Natural Attenuation of Hydrocarbon Vapors: An Empirical Assessment of Soil Gas Vertical Profile Data. API Technical Bulletin No. 15. American Petroleum Institute, Washington, DC, 2002.

[51] http://apiep.api.org/environment (accessed December 2012)

[52] EPA (Environmental Protection Agency). Petroleum Hydrocarbons And Chlorinated Hydrocarbons Differ In Their Potential For Vapor Intrusion. Office of Underground Storage Tanks, Washington, D.C. 20460. 2011. p1-11.

[53] www.epa.gov/oust (accessed December 2012)

[54] Belay N, Daniels L. Production of Ethane, Ethylene, and Acetylene from Halogenated Hydrocarbons by Methanogenic Bacteria. Applied Environmental Microbiology 1987; 53(7): 1604-1610.

[55] de Bruin WP, Kotterman MJJ, Posthumus MA, Schraa G, Zehnder AJB. Complete Biological Reductive Transformation of Tetrachloroethene to Ethane. Applied Environmental Microbiology 1992; 58(6): 1996-2000.

[56] Bradley PM, Chapelle FH. Methane as a Product of Chloroethene Biodegradation under Methanogenic Conditions. Environmental Science Technology 1999; 33(4): 653-656.

[57] Vogel TM, McCarty PL. Biotransformation of Tetrachloroethylene to Trichloroethylene, Dichloroethylene, Vinyl Chloride, and Carbon Dioxide under Methanogenic Conditions. Applied Environmental Microbiology 1985; 49: 1080-1083.

[58] Little, CD, Palumbo AV, Herbes SE, Lidstrom ME, Tyndall RL, Gilmer PJ. Trichloroethylene Biodegradation by a Methane-Oxidizing Bacterium. Applied Environmental Microbiology 1988; 54(4): 951-956.

[59] Tsien H, Brusseau GA, Hanson RS, Wackett LP. Biodegradation of Trichloroethylene by Methylosinus trichosporium OB3b. Applied Environmental Microbiology 1989; 55(12): 3155-3161.

[60] Wilson JT, Wilson BH. Biotransformation of Trichloroethylene in Soil. Applied Environmental Microbiology 1985; 49(1): 242-243.

[61] Pfaender FK. Biological Transformations of Volatile Organic Compounds in Groundwater. In: Ram NM, Christman RF, Cantor KP (eds.) Significance and Treatment of Volatile Organic Compounds in Water Supplies. Lewis Publishers: Chelsea, MI 1990. p205–226.

[62] Bouwer EJ. Bioremediation of Organic Contaminants in the Subsurface. In: Mitchell R. (Eds.) Environmental Microbiology. John Wiley & Sons: New York 1992. p287–318.

[63] Lorah MM, Olsen LD, Capone DG, Baker JE. Biodegradation of Trichloroethylene and Its Anaerobic Daughter Products in Freshwater Wetland Sediments. Bioremediation Journal 2001; 5(2): 101–118.

[64] Fetzner S. Bacterial degradation of pyridine, indole, quinoline, and their derivatives under different redox conditions. Applied Environmental Microbiology 1998; 49: 237–250.

[65] Shukla OP. Microbial transformation of quinoline by a Pseudomonas species. Applied Environmental Microbiology 1986; 51: 1332–1442.

[66] Kaiser JP, Feng YC, Bollag JM. Microbial metabolism of pyridine, quinoline, acridine, and their derivatives under aerobic and anaerobic conditions. Microbiological Reviews 1996; 60: 483–498.

[67] Carl B, Arnold A, Hauer B, Fetzner S. Sequence and transcriptional analysis of a gene cluster of Pseudomonas putida 86 involved in quinoline degradation. Gene 2004; 331: 177–188.

[68] Kilbane JJ, Ranganathan R, Cleveland L, Kayser KJ, Ribiero C, Linhares MM, Selective removal of nitrogen from quinoline and petroleum by Pseudomonas ayucida IGTN9m. Applied Environmental Microbiology 2000; 66: 688–693.

[69] Shukla OP. Microbiological degradation of quinoline by Pseudomonas stutzeri: the coumarin pathway of quinoline catabolism. Microbios 1989; 59: 47–63.

[70] Shukla OP. Microbiological transformation of quinoline by Pseudomonas sp. Applied Environmental Microbiology 1986; 51: 1332-1442.

[71] O'Loughlin EJ, Kehrmeyer SR, Sims GK, Isolation, characterization, and substrate utilization of a quinoline-degrading bacterium. International Biodeterioration and Biodegradation 1996; 38: 107–118.

[72] Sugaya K, Nakayama O, Hinata N, Kamekura K, Ito A, Yamagiwa K, Ohkawa A. Biodegradation of quinoline in crude oil. Journal of Chemical Technology Biotechnology 2001; 76: 603–611.

[73] Sun Q, Bai Y, Zhao C, Xiao Y, Wen D, Tang X. Aerobic biodegradation characteristics and metabolic products of quinoline by a Pseudomonas strain. Bioresource Technology 2009; 100: 5030-5036.

[74] Annadurai G, Juang R, Lee DJ. Microbial degradation of phenol using mixed liquors of Pseudomonas putida and activated sludge. Waste Manage 2002; 22: 703–710.

[75] Mohan D, Chander S. Single component and multi-component adsorption of phenols by activated carbons. Colloids and Surfaces A: Physicochemical &. Engineering Aspects 2001; 177: 183–196.

[76] Dursun G, Cicek HC, Dursun AY. Adsorption of phenol from aqueous solution by using carbonised beet pulp. Journal of Hazardous Materials B 2005; 125: 175–182.

[77] Patterson JF. Industrial Wastewater Treatment Technology, Second ed., Butterworths, London, 1985.

[78] Tepe O, Dursun AY. Combined effects of external mass transfer and biodegradation rates on removal of phenol by immobilized Ralstonia eutropha in a packed bed reactor. Journal of Hazardous Materials 2008; 151: 9-16.

[79] Knoll G, Winter J. Anaerobic degradation of phenol in sewage sludge: benzoate formation from phenol and carbon dioxide in the presence of hydrogen. Applied Environmental Microbiology 1987; 25(4): 384–391.

[80] El-Naas MH, Al-Muhtaseb SA, Makhlouf S. Biodegradation of phenol by Pseudomonas putida immobilized in polyvinyl alcohol (PVA) gel. Journal of Hazardous Materials 2009; 164: 720–725.

[81] Carrera J, Martín-Hernández M, Suárez-Ojeda ME, Pérez J. Modelling the pH dependence of the kinetics of aerobic p-nitrophenol biodegradation. Journal of Hazardous Materials 2011; 186: 1947–1953.

[82] Ye J, Singh A, Ward O. Biodegradation of nitroaromatics and other nitrogen containing xenobiotics. World Journal Microbiology Biotechnology 2004; 20: 117–135.

[83] Abrahamsson K, Klick S. Degradation of Halogenated Phenols in Anoxic Marine Sediments. Marine Pollution Bulletin 1991; 22: 227-233.

[84] Hakulinen R, Woods S, Ferguson J, Benjamin M. The Role of Facultative Anaerobic Microorganisms in Anaerobic Biodegradation of Chlorophenols. Water Science & Technology 1985; 17: 289-301.

[85] Jain V, Bhattacharya SK, Uberoi V. Degradation of 2,4-Dichlorophenol in Methanogenic Systems. Environmental Technology 1994; 15: 577-584.

[86] Leuenberger C, Giger W, Coney R, Graydon JW, Molnar-Kubica E. Persistent Chemicals in Pulp Mill Effluents. Water Research 1985; 19: 885-894.

[87] Sierra-Alvarez R, Field JA, Kortekaas S, Lettinga G. Overview of the Anaerobic Toxicity caused by Organic Forest Industry Wastewater Pollutants. Water Science Technology 1994; 29: 353-363.

[88] Wood JM. Chlorinated Hydrocarbons: Oxidation in the Biosphere. Environmental Science & Technology 1982: 16: 291A-297A.

[89] Annachhatre AP, Gheewala SH. Biodegradation of Chlorinated Phenolic Compounds. Biotechnology Advances 1996; 14 (1): 35-56.

[90] Cheremisinoff NP. Biological Degradation of hazardous Waste. In: Biotechnology for Waste and Wastewater Treatment. Noyes Publications: Westwood, New Jersey, USA. 1996. p37-110.

[91] Smith MJ, Lethbrideg G, Burns RG. Bioavailability and biodegradation of polycyclic aromatic hydrocarbons in soils. FEMS Microbiology Letters 1997; 152: 141–147.

[92] Yuan SY, Wei SH, Chang BV. Biodegradation of polycyclic aromatic hydrocarbons by a mixed culture. Chemosphere 2002; 41: 1463–1468.

[93] Ulrici W. Contaminated soil areas, different countries and contaminants, monitoring of contaminants, In: Rehm HJ., Reed G., Puhler A., Stadler P. (Eds.) Environmental Processes II Soil Decontamination Biotechnology: A Multi Volume Comprehensive Treatise, In: J. Klein (Ed.), Second Ed., vol. 11b, Wiley–VCH,Weihheim, FRG, 2000. p5-42.

[94] Odokuma LO, Dickson AA, Bioremediation of a crude oil polluted tropical rain forest soil. Global Journal of Environmental Science 2003; 2: 29–40.

[95] Cheung KC, Zhang JY, Deng HH, Ou YK, Leung HM, Wu SC, Wong MH. Interaction of higher plant (jute), electrofused bacteria and mycorrhiza on anthracene biodegradation. Bioresource Technology 2008; 99: 2148–2155.

[96] Somtrakoon K, Suanjit S, Pokethitiyook P, Kruatrachue M, Lee H, Upatham S. Enhanced biodegradation of anthracene in acidic soil by inoculated Burkholderia sp. VUN10013. Current Microbiology 2008; 57: 102–107.

[97] Li X, Lin X, Li P, Liu W, Wang L, Ma F, Chukwuka KS. Biodegradation of the low concentration of polycyclic aromatic hydrocarbons in soil by microbial consortium during incubation. Journal of Hazardous Materials 2009; 172: 601–605.

[98] Santos EC, Rodrigo JS, Jacques Bento FM, Peralba MDCR, Selbach PA, Enilso LSS, Camargo FAO. Anthracene biodegradation and surface activity by an iron-stimulated Pseudomonas sp. Bioresource Technology 2008; 99: 2644–2649.

[99] Zeinali M, Vossoughi M, Ardestani SK. Naphthalene metabolism in Nocardia otitidiscaviarum strain TSH1, a moderately thermophilic microorganism. Chemosphere 2008; 72: 905–909.

[100] Hwang G, Park SR, Lee CH, Ahn IS, Yoon YJ, Mhin BJ. Influence of naphthalene biodegradation on the adhesion of Pseudomonas putida NCIB 9816-4 to a naphthalene-contaminated soil. Journal of Hazardous Materials 2009; 171: 491–493.

[101] Gennaro PD, Rescalli E, Galli E, Sello G, Bestetti G, Characterization of Rhodococcus opacus R7, a strain able to degrade naphthalene and o-xylene isolated from a polycyclic aromatic hydrocarbon-contaminated soil. Research in Microbiology 2001; 152: 641–651.

[102] Calvo C, Toledo FL, González-López J. Surfactant activity of a naphthalene degrading Bacillus pumilus strain isolated from oil sludge. Journal of Biotechnology 2004; 109: 255–262.

[103] Kelley I, Freeman JP, Evans FE, Cerniglia CE. Identification of metabolites from degradation of naphthalene by a Mycobacterium sp. Biodegradation 1990; 1: 283–290.

[104] Lin C, Gan L, Chen ZL. Biodegradation of naphthalene by strain Bacillus fusiformis (BFN). Journal of Hazardous Materials 2010; 182: 771–777.

[105] Staples CA, Peterson DR, Parkerton TF, Adams WJ. The environmental fate of phthalate esters: a literature review. Chemosphere 1997; 35: 667–749.

[106] Nalli S, Cooper DG, Nicell JA. Biodegradation of plasticizers by Rhodococcus rhodochrous. Biodegradation 2002; 13: 343–352.

[107] Nalli S, Cooper DG, Nicell JA. Metabolites from the biodegradation of di-ester plasticizers by Rhodococcus rhodochrous. Science of the Total Environment Journal 2006; 366: 286–294.

[108] Rahman M, Brazel CS. The plasticizer market: an assessment of traditional plasticizers and research trends to meet new challenges. Progress in Polymer Science 2004; 29: 1223–1248.

[109] Tickner JA, Schettler T, Guidotti T, McCally M, Rossi M. Health risks posed by use of di-2-ethylhexyl phthalate (DEHP) in PVC medical devices: a critical review. American Journal of Industrial Medicine 2001; 39: 100–111.

[110] Onorato TM, Brown PW, Morris P. Mono-(2-ethylhexyl)phthalate increase spermatocyte mitochondrial peroxiredoxin 3 and cyclooxygenase 2. Journal of Andrology 2008; 29: 293–303.

[111] Horn O, Nalli S, Cooper DG, Nicell JA. Plasticizer metabolites in the environment. Water Research 2004; 38: 3693–3698.

[112] Nalli SS, Horn OJ, Grochowalski AR, Cooper DG, Nicell JA. Origin of 2- ethylhexanol as a VOC. Environmental Pollution Journal 2006; 140: 181–185.

[113] Barnabé S, Beauchesne I, Cooper DG, Nicell JA. Plasticizers and their degradation products in the process streams of a large urban physicochemical sewage treatment plant. Water Research 2008; 42: 153–162.

[114] Beauchesne I, Barnabé S, Cooper DG, Nicell JA. Plasticizers and related toxic degradation products in wastewater sludges. Water Science & Technology 2008; 57: 367–374.

[115] Deligio T. Phthalate Alternative Recognized by ECHA. 2009.

[116] http://www. plasticstoday.com/articles/phthalate-alternative-recognized-echa/ (accessed December 2012)

[117] Gartshore J, Cooper DG, Nicell JA. Biodegradation of plasticizers by Rhodotorula Rubra. Environmental Toxicology & Chemistry 2003; 22: 1244–1251.

[118] Pour AK, Cooper DG, Mamer OA, Maric M, Nicell JA. Mechanism of biodegradation of dibenzoate plasticizers. Chemosphere 2009; 77: 258–263.

[119] Firlotte N, Cooper DG, Maric M, Nicell JA. Characterization of 1,5-pentanediol dibenzoate as a potential green plasticizer for poly(vinyl chloride). Journal of Vinyl Additive Technology 2009; 15: 99–107.

[120] Pour AK, Mamer OA, Cooper DG, Maric M, Nicell JA. Metabolites from the biodegradation of 1,6-hexanediol dibenzoate, a potential green plasticizer, by Rhodococcus rhodochrous. Journal of Mass Spectrometry 2009; 44: 662–671.

[121] Pour AK, Roy R, Coopera DG, Maric M, Nicell JA. Biodegradation kinetics of dibenzoate plasticizers and their metabolites. Biochemical Engineering Journal 2013; 70: 35-45.

[122] http://ec.europa.eu/environment/waste/framework/index.htm (accessed December 2012)

[123] Shimao M. Biodegradation of plastics. Current Opinion in Biotechnology 2001; 12: 242–247.

[124] Chiellini E, Corti A, D'Antone S, Solaro R. Biodegradation of poly(vinylalcohol) based materials. Progress in Polymer Science 2003; 28: 963-1014.

[125] Sakai K, Hamada N, Watanabe Y. Studies on the poly(vinyl alcohol)-degrading enzyme. Part VI. Degradation mechanism of poly(vinyl alcohol) by successive reactions of secondary alcohol oxidase and β-diketone hydrolase from Pseudomonas sp. Agricultural & Biological Chemistry 1986; 50: 989-996.

[126] Suzuki T. Degradation of poly(vinyl alcohol) by microorganisms. Journal of Applied Polymer Science Applied Polymer Symposium 1979; 35: 431-437.

[127] Hatanaka T, Kawahara T, Asahi N, Tsuji M. Effects of the structure of poly(vinyl alcohol) on the dehydrogenation reaction by poly(vinyl alcohol) dehydrogenase from Pseudomonas sp. 113P3. Bioscience Biotechnology Biochemistry 1995; 59: 1229-1231.

[128] Bloembergen S, David J, Geyer D, Gustafson A, Snook J, Narayan R. Biodegradation and composting studies of polymeric materials. In: Doi Y, Fukuda K. (Eds.) Biodegradable plastics and polymers. Amsterdam: Elsevier; 1994. p601-609.

[129] David C, De Kesel C, Lefebvre F, Weiland M. The biodegradation of polymers: recent results. Angewandte Makromolekulare Chemie 1994; 216: 21-35.

[130] Chiellini E, Corti A, Sarto GD, D'Antone S. Oxo-biodegradable polymers e Effect of hydrolysis degree on biodegradation behaviour of poly(vinyl alcohol). Polymer Degradation and Stability 2006; 91: 3397-3406.

[131] Khanna S, Srivastava AK, Recent Advances in microbial polyhydroxyalkanoates. Process Biochemistry 2005; 40: 607-619.

[132] Ghatnekar MS, Pai JS, Ganesh M. Production and recovery of poly-3-hydroxybutyrate from Methylobacterium sp.V49. Journal of Chemical Technology and Biotechnology 2002; 77: 444-448.

[133] DeMarco S. Advances in polyhydroxyalkanoate production in bacteria for biodegradable plastics. MMG 445. Basic Biotechnology eJournal 2005; 1: 1-4.

[134] Mudgal S, Lyons L, Bain, J. Plastic Waste in the Environment – Final Report for European Commission DG Environment. BioIntelligence Service; 2010. http://www.ec.europa.eu/environment/ (accessed December 2012)

[135] Hopewell J, Dvorak R, Kosior E. Plastics recycling: challenges and opportunities. Philosophical Transactions of the Royal Society B 2009; 364: 2115-2126.

[136] Plastics Europe. An analysis of European Plastics production, demand and recovery for 2009. Plastics - the Facts 2010.

[137] http://www.plasticseurope.org/ (accessed December 2012)

[138] The Encyclopedia of Polymer Science and Technology, 4th Edition, John Wiley and Sons, New York; 2012.

[139] Selke SE. Plastics recycling In: Harper CA. (Ed.), Handbook of plastics, elastomers and composites, 4th edition, McGraw-Hill, New York; 2002. p693–757.

[140] Fact sheet, Recycling the hard stuff. U.S. Environmental Protection Agency, Solid Waste and Emergency Response, 2002 EPA 530-F-02-023 Washington, D.C. http://www.docstoc.com/docs (accessed December 2012)

[141] Hendrickson CT, Matthews DH, Ashe M, Jaramillo P, McMichael FC. Reducing environmental burdens of solid-state lighting through end-of-life design. Environmental Research Letters 5. 2010. Doi: 10.1088/1748-9326/5/1/014016.

[142] Cerdan C, Gazulla C, Raugei M, Martinez E, Fullana-i-Palmer P. Proposal for new quantitative eco-design indicators: a first case study. Journal of Cleaner Production 2009; 17: 1638-1643.

[143] Plastic Waste: Redesign and Biodegradability. Science for Environmental Policy, Future Brief, 2001; 1: 1-8.

[144] Tokiwa Y, Calabia BP, Ugwu CU, Aiba S. Biodegradability of Plastics. International Journal of Molecular Science 2009; 10: 3722–3742.

[145] Jackson S, Bertényi T. Recycling of Plastics. ImpEE Project. 2006. p1-27

[146] http://www-g.eng.cam.ac.uk/ (accessed December 2012)

Chapter 12

Biodegradation and Mechanical Integrity of Magnesium and Magnesium Alloys Suitable for Implants

S. González, E. Pellicer, S. Suriñach, M.D. Baró and J. Sort

Additional information is available at the end of the chapter

http://dx.doi.org/10.5772/55584

1. Introduction

Most conventional orthopedic implants used for repairing joint and bone fractures consist of metallic biomaterials with polycrystalline microstructure that exhibit high hardness, good corrosion resistance and excellent fatigue and wear resistance. Usually, once the patient has recovered from a traumatic injury, a revision surgery is necessary in order to remove the implant from the body and avoid problems associated with osteopenia, inflammation of adjacent tissues or sarcoma. Alternatively, to avoid post-extraction of the implant, intensive efforts are being made in recent years to develop new classes of so-called "biodegradable implants", composed of non-toxic materials that become reabsorbed by the human body after a reasonable period of time. These implants are usually based on polymeric materials. However, polymeric implants are often rather costly and exhibit relatively low mechanical strength. Sometimes organic polymers can also react with human tissues, leading to osteolysis. For these reasons, it is highly desirable to develop cost-effective biodegradable metallic alloys, with better mechanical performance than polymers.

Although biodegradation is usually associated with the breakdown of organic matter into simple chemicals through the action of microorganisms, metals can also undergo biodegradation. Although corrosion should be generally avoided in the engineering field, it is advantageous for certain applications such as biodegradable implants. Since the 18th century, when Au, Ag and Pt elements were used for the fabrication of biomaterials [1], a large number of alloys have been developed so far. Some of the most employed metallic biomaterials for permanent implants are austenitic steels [2], Co-Cr-Mo [3], titanium and Ti-6Al-4V alloys [4] due to their biocompatibility and adequate mechanical behavior. To avoid post-extraction of these materials, intensive efforts have been made in recent years to develop the so-called "biodegradable implants". The materials of choice for biodegradable metallic implants are iron-based [5] and Mg-based alloys [6] owing to their relatively fast biodegradability. From the point of view of the mechanical performance, Mg alloys are preferred because their stiffness (i.e., Young's modulus) is closer to that of human bone [7].

Since "biodegradable implants" become reabsorbed by the human body after a certain period of time, they should be composed of biocompatible alloying elements. For this reason the potential cytotoxicity of the constituent elements of an implant material has to be seriously considered at an early stage of material development. For example, elements such as Ni, Al, Cr and V are not suitable to be in contact with human tissues [8]. Their substitution by non-toxic elements such as Zn and Ca has permitted the fabrication of biocompatible Mg-based alloys with potential use as biomaterials. However, the problem with some Mg alloys is their exceedingly high corrosion rates in physiological conditions, which makes their biodegradability to be faster than the time required to heal the bone [9]. For this reason it is important to decrease their degradation rate, and to keep their mechanical integrity until the bone heals. Another drawback of magnesium and its alloys is that corrosion is accompanied by intense hydrogen evolution. This gas can be accumulated in pockets next to the implants or can form subcutaneous gas bubbles.

This book chapter deals with the fundamental aspects of corrosion of magnesium based alloys in bodily fluids and reviews the various techniques that can be used to tune their degradation rate. The time-dependent evolution of their mechanical properties during the biodegradation process is also outlined.

2. Basic aspects of corrosion

Corrosion is a surface phenomenon greatly influenced by different media-related factors (chemical, electrochemical and physical) in which the material is placed. The corrosion behavior of Mg in aqueous environments proceeds by an electrochemical reaction with water to yield magnesium hydroxide $Mg(OH)_2$ and hydrogen gas [10]:

$$\text{Anodic reaction: } Mg \rightarrow Mg^{2+} + 2e \qquad (1)$$

$$\text{Cathodic reaction: } 2H_2O + 2e^- \rightarrow H_2(g) + 2OH^- \qquad (2)$$

$$\text{Overall reaction: } + 2H_2O \rightarrow Mg(OH)_2 + H_2 \qquad (3)$$

The hydroxide anions generated through the cathodic reaction cause an increase of the pH of the solution [11] (eq. (2)). The formation of a magnesium hydroxide $Mg(OH)_2$ layer onto the

Mg surface can further protect the metal from ongoing corrosion provided that the electrolyte pH and/or the presence of chloride anions or other species induce breakage of the passive film. According to the potential-pH Pourbaix diagram for magnesium in pure water at 25°C (Fig. 1), a passivation region exists for pH values above 10.4 [12] (alkaline environment) where the $Mg(OH)_2$ layer is stable. In neutral or acid environments (pH lower than 10.4) this layer is unstable. The diagram also shows that the immunity region of the diagram is below the region of water stability. However, bodily fluids are more aggressive than pure water. Body fluids are complex saline solutions containing ingredients such as proteins, blood serum, etc [13]. The most common fluids to carry out in-vitro tests and thereby to predict the degradation rate of magnesium and its alloys are Hank's balanced salt solution (HBSS), phosphate buffered solution (PBS) and simulated body fluid (SBF). All of them are acellular isotonic solutions (i.e., solutions with the same salt concentration as blood and cells) to make the sample, cell or tissue stable during an experiment.

HBSS [14] is mainly composed of chloride, sodium, potassium, magnesium and calcium ions. However, there are varieties of ingredients which can consist of glucose, potassium chloride (KCl), potassium dihydrogen phosphate (KH_2PO_4), sodium dihydrogen phosphate (NaH_2PO_4), and sodium chloride (NaCl). Additional ingredients can include hydrated magnesium sulfate ($MgSO_4$ $7H_20$) and sodium bicarbonate ($NaHCO_3$).

PBS as its name implies [15] is a buffer solution consisting of a mixture of a weak acid and its conjugate base or a weak base and its conjugate acid. It aims to maintain a neutral pH in order not to destroy the cell or tissue sample and to maintain the osmolarity of the cells. The main ingredients are sodium phosphate and sodium chloride (NaCl) but in some recipes potassium phosphate and potassium chloride (KCl) are added.

SBF is a solution that has an inorganic ions concentration and pH almost equal to that of human extracellular fluid (i.e., the human blood plasma). The ions concentration in SBF is: Na^+ (142.0), K^+ (5.0), Mg^{2+} (1.5), Ca^{2+} (2.5), Cl^- (148.8), $HCO3^-$ (4.2) and $PO4^{2-}$ (1.0) mmol/dm^3 and it is buffered at pH 7.25 [16].

Chloride ions are able to dissolve the $Mg(OH)_2$ layer [17] yielding the soluble $MgCl_2$ salt [18], according to the following reaction:

$$Mg\left(OH\right)_2 + 2Cl^- \leftrightarrow MgCl_2 + 2OH^- \tag{4}$$

Chloride ions are thus detrimental for the corrosion resistance of passive systems. Yet, other studies point out to opposite effects. For example, chloride ions were found to improve surface stability of Mg-Y-RE alloy in artificial plasma solution [19]. Other species can also degrade the protective passive characteristics of $Mg(OH)_2$ layer. Baril and Pébère found that the addition of increasing concentrations of $NaHCO_3$ to a deaerated Na_2SO_4 media leads to an accelerated corrosion of magnesium due to dissolution of MgO and $Mg(OH)_2$ films [20]. On the contrary, certain anionic species like HCO_3^- have beneficial effects and can be added to the electrolyte to increase the stability of the corrosion

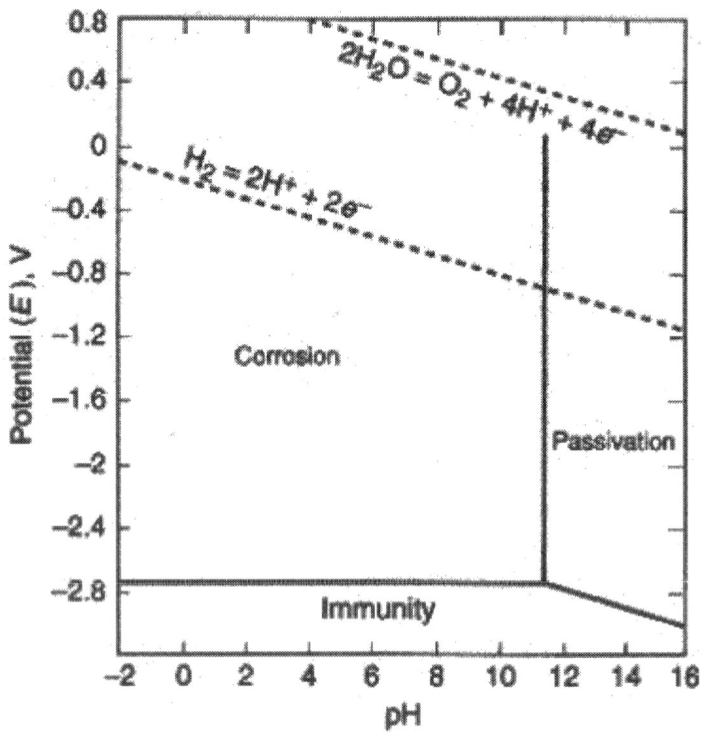

Figure 1. Potential-pH Pourbaix diagram for Mg in water at 25 °C.

products.The presence of dissolved O_2 appears not to play a major role in the corrosion of magnesium when immersed in saline solutions or fresh water [21].

3. Hydrogen evolution

One of the major drawbacks of Mg as biomaterial is the formation of H_2 gas when it is in contact with body tissues. The evolved H_2 bubbles from magnesium implants can be accumulated and form gas pockets that may lead to necrosis of the neighboring tissues and delay healing of the surgery region [22]. However, if the H_2 gas is generated slowly enough it can be transported away from the implant and can thus be tolerated by the body. According to Song [22] a hydrogen release rate in the human body of 0.01 ml/cm^2/day can be tolerated. Dissolution of Mg and concomitant hydrogen evolution can be retarded by either purification of Mg or through appropriate alloying. Fig. 2 shows the average rate of hydrogen evolution (ml/cm^2/day) for commercially pure Mg (CP-Mg) and different Mg alloys [22]. The highest release of hydrogen stands for CP-Mg, about 26 ml/cm^2/day, and decreases with the addition of certain elements. For example, it decreases to 1.502 ml/cm²/day for ZE41 alloy (4 wt. % Zn, 1 wt. % RE), to 0.280 ml/cm²/day for Mg1.0Zn (1.0 wt. % Zn), to 0.068 ml/cm²/day for AZ91 (9 wt. % Al, 1 wt. % Zn) and to 0.012 ml/cm²/day for Mg2Zn0.2Mn (2 wt. % Zn, 0.2 wt. % Mn).

By measuring the hydrogen evolution rate the corrosion rate associated with magnesium is directly obtained since the release of one mol of H_2 implies the consumption of one mole of Mg according to eq. (3) [23]. The rate of H_2 gas evolution for Mg in Hank's solution at 37°C and different pH values was studied by Ng et al. [23] over a period of 7 days. They reported that the hydrogen evolution rate decreases with the increase of the solution pH. However, the volume of H_2 gas evolved over the time at a given pH (between 5.5 and 6.8) practically does not change. The same authors reported that the average H_2 evolution rate initially drops very fast from 153.3 to 1.079 ml/cm²/day when the pH rises from 5.5 to 7.4 but it slows down at pH 8.0 (0.534 ml/cm²/day) [23]. This was attributed to the accumulation of corrosion products that covered the sample surface, forming a progressively thicker layer with pH. Similarly, Zainal Abidin et al. [24] suggested that the formation of a partially protective film on Mg2Zn0.2Mn and ZE41 samples after long immersion times in Hank's solution was responsible for the decrease of the corrosion rate and concomitant H_2 evolution.

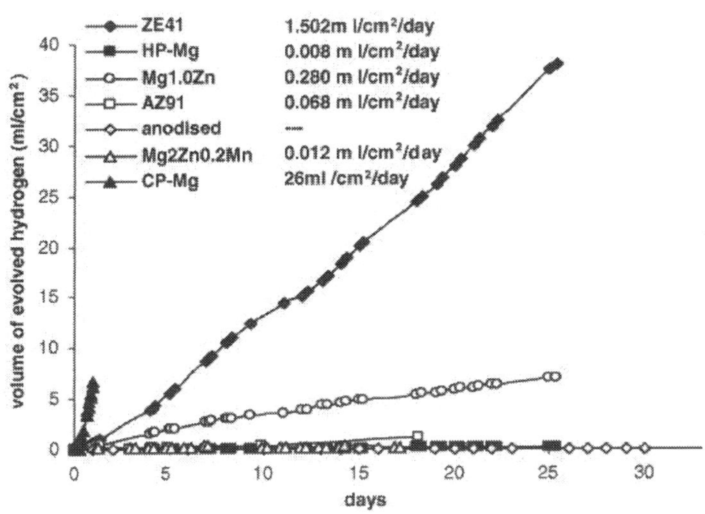

Figure 2. Hydrogen evolution in SBF and their average rates for various Mg-based alloys. Reprinted from Song G [22], page 3, with permission from Elsevier.

It is important to stress that magnesium shows an unusual electrochemical phenomenon known as "negative difference effect" (NDE) [25], which basically consists of an increase of the H_2 evolution rate at more positive potentials. For most metals, hydrogen evolution decreases with an increase of the applied potential or current density [26].

4. Corrosion of Mg and Mg alloys

When Mg and its alloys are used as biomaterials for implant applications they can be subjected to a combination of corrosion and stress (erosion, fatigue, etc). Since galvanic and pitting corrosion are the most common corrosion types of Mg and Mg alloys, this chapter primarily focuses on them:

4.1. Galvanic corrosion

Galvanic corrosion is an electrochemical process that occurs when two metals having different electrochemical potentials are in close contact with a common electrolyte. Of these two metals, the one that is more active in the galvanic series corrodes preferentially. Fig. 3 shows the galvanic series of different alloys listed in the order of the potential they exhibit in flowing seawater [27]. The black boxes of Fig. 3 correspond to the potentials in low-velocity or poorly aerated water. The reference potential is the Standard Calomel Electrode (SCE).

Although the composition of seawater differs slightly from that of saline body fluid and thus the corrosion potential is not expected to be exactly the same, Fig. 3 already gives a rough idea of the activity of different metals and alloys. The most positive (noble) material will be protected against corrosion at the expense of the material with more negative potential. Since the electrochemical potential of Mg and its alloys is located at the most negative side of this series (i.e., below -1.6 V), almost all the other metals in contact with it will be cathodically protected. Therefore, Mg will undergo galvanic corrosion; i.e., galvanic couples between the Mg metal or its alloy and the other metal will result in the dissolution of the former. The driving force for the galvanic corrosion depends on the difference between the potential (i.e., nobility) of both materials.

Regrettably, the corrosion of Mg alloys not only occurs when they are in close contact with other metals but also within the material itself. Mg alloys do not normally have a uniform microstructure, composition and crystalline orientation. This lack of uniformity is sufficient to promote the occurrence of galvanic couples [28]. The galvanic effect depends on a variety of factors; the crystal orientation of the matrix phase (i.e., the continuous phase of pure Mg into which the second phase/s is/are embedded), the alloying element concentrations in the matrix phase, the type and concentration of secondary phases along grain boundaries and the type and concentration of impurity particles in the matrix phase [28]. In the following, the main features having an influence of the corrosion rate of Mg are summarized:

- **Crystal orientation of the matrix phase:** polycrystalline pure Mg matrix immersed in neutral 0.01 M NaCl solution is more stable and corrosion resistant when grains possess a basal orientation [29]. This behavior can be explained considering that densely packed crystallographic planes (i.e., basal planes) normally have a higher atomic coordination and thus a lower dissolution tendency than non-compact planes [30]. For this reason, by controlling surface texture, one can improve the corrosion resistance of the material. For example, by rolling an AZ31 alloy it is possible to orient most of the crystallographic basal planes of the grains parallel to the rolling surface and thus decrease the corrosion rate of the rolled surface [31].

- **Alloying element concentrations:** the corrosion behavior of Mg phase can be tuned as a function of the concentration of elements in solid solution. Depending on the nature and distribution of these elements within the matrix phase, the occurrence of micro-galvanic cells can be either mitigated or favored. For example, an Al-containing Mg matrix phase becomes more passive as the Al content increases and consequently the corrosion rate decreases [32]. In as-cast Mg-Al alloys, the aluminum content in solid solution can vary from 1.5 wt. % at the grain center to about 12 wt. % at the grain boundary due to segregation during solidification [33]. Since Al has higher potential (Fig. 3) than Mg, corrosion mainly occurs at the interior of Mg grains. On the contrary, in Zr-containing Al-free Mg alloys, the central areas of the grains (which are enriched in Zr) do not corrode while the grain boundaries severely corrode.

- **Type and concentration of secondary phases along grain boundaries:** Mg intermetallic phases are typically nobler than the Mg matrix. As a consequence, they act as micro-galvanic cathodes and the dissolution of the Mg matrix is accelerated. Yet, in some cases the intermetallic phases can stop the corrosion process. Hence, they actually play a dual role in the corrosion of Mg alloys [34]. Namely, the presence of a finely and continuously distributed secondary phase can stop the corrosion process while the presence of small amount of discontinuous secondary phase particles will accelerate it [25,35].

- **Type and concentration of impurity particles within the matrix phase:** the corrosion resistance of Mg alloys can be improved by limiting the concentration of critical impurities. However, not all the impurity elements have the same effect on the corrosion behavior. Some of them have little influence while others are very detrimental to the corrosion resistance. For example, Zn and Ca, which are frequently employed in the biomaterials field [6,36], have moderate accelerating effects on corrosion rates. Contrarily, Ni, Fe, Cu and Co are deleterious due to their low solid-solubility limits in Mg and their ability to act as cathodic sites [37]. The corrosion rate also depends on the impurity concentration. Each impurity has a tolerance limit. For impurity concentrations lower than the tolerance limit, there is no significant influence on the corrosion rate, whereas above this limit the corrosion rate sharply increases (Fig. 4) [37]. There is a rough relationship between the solubility of some elements in Mg alloys and their critical concentrations [38]. For example, the tolerance limit of Fe in Mg corresponds to the solubility of Fe in Mg [39].

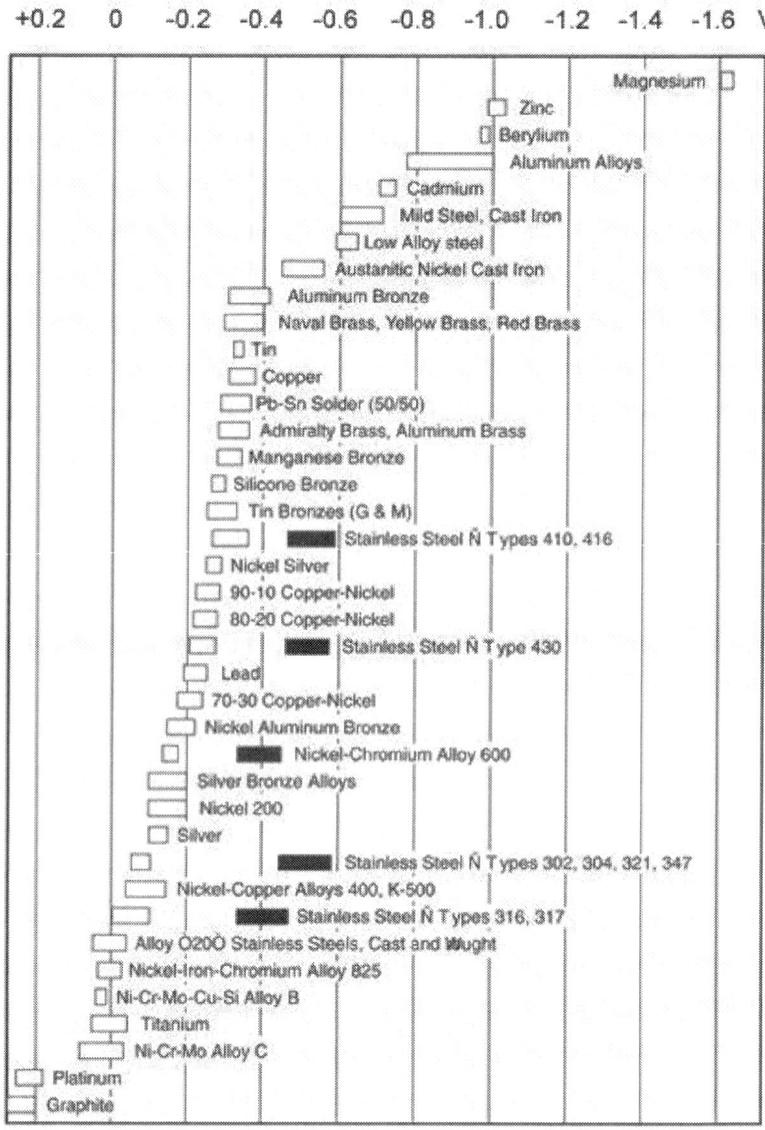

Figure 3. The galvanic series of metals, semi-metals and alloys of industrial interest showing their potentials (in volts) in flowing sea water, arranged from the most noble (bottom) to the most active (top) material. The values are referred to saturated calomel half-cell reference electrode. Adapted and reprinted from Amtec Consultants [27]

http://dx.doi.org/10.5772/55584

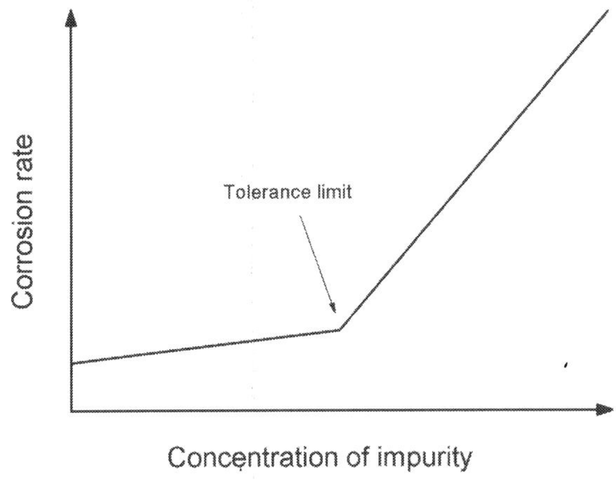

Figure 4. Schematic picture showing the dependence of the impurity concentration on the corrosion rate of Mg. The tolerance limit sets the threshold between the region for which an increase of the impurity concentration hardly affects the corrosion rate (left) and the region for which a further increase of the impurity concentration abruptly increases the corrosion rate.

- **Amorphous versus crystalline microstructure:** When a liquid is cooled below its liquidous temperature, it either crystallizes or, if crystallization is suppressed, it forms an amorphous solid. The microstructure and constituency of a material can be altered on purpose by means of rapid solidification processing at quenching rates of 10^5-10^6 K/s [40]. The development of novel Mg alloys with higher glass-forming ability has permitted to obtain amorphous materials with lower critical cooling rates. Since amorphous materials usually exhibit better corrosion and wear resistance than their crystalline counterparts, they can be potentially used for biomedical applications. Moreover, by controlling the crystallization events from the early stages of solidification, it is possible to tune the microstructure (i.e., nature and size of crystalline phases) and, in turn, optimize the corrosion performance of the material. This issue will be deeply tackled in section 5.1.2.

The micro-galvanic corrosion is also dependent on the solution in which the alloy is immersed. In a 3 % NaCl solution, the secondary phases present in the AZ91, ZE41 and Mg2Zn0.2Mn alloys can accelerate the corrosion rate. On the contrary, they do not play an important role when these alloys are immersed in Hank's solution [24]. The driving force for micro-galvanic corrosion between α-Mg and the secondary phases can be alleviated with the formation of a protective surface film on ZE41 and Mg2Zn0.2Mn during long immersion times in Hank's solution [24].

4.2. Pitting corrosion

It is a type of corrosion in which there is an intense localized attack on sample surface that leads to the formation of small holes in the metal. Mg alloys are prone to pitting corrosion when the passivation layer (which consists of $Mg(OH)_2$ [41] or a mixture of MgO and $Mg(OH)_2$ [42,43]) breaks down locally. When this occurs, the corrosion can be initiated at these local sites that act as small anodic surface areas. As aforementioned (see section: Basic Aspects of Corrosion), this protective coating may be damaged in physiological solutions because it is sensitive to both the chloride ion concentration and the solution pH. Physiological environments also contain phosphates, carbonates and sulfates that have different effects on the degradation behavior of Mg. Sulfate ions appear to stimulate the corrosion of Mg [44] while phosphate ions can delay pitting corrosion. Finally, carbonate ions can favor surface passivation and inhibit chloride-induced pitting corrosion due to the precipitation of stable magnesium carbonates into the pits. The presence of ions in different concentrations may explain why corrosion in Hank's solution is more severe than in simulated blood plasma [45]. Nevertheless, immersion of AZ91, ZE41 and Mg2Zn0.2Mn alloys in Hank's solution favors the formation of a more protective surface film than in 3% NaCl solution.

The large influence of the electrolyte composition on the corrosion behavior could explain why the same alloy exhibits different modes of corrosion depending on the environment. For example, a commercial AZ91 alloy immersed in 1M NaOH solution for 1 h and then re-immersed in 0.01M NaCl for 3h exhibited pitting [46] whereas the same alloy immersed in Minimum Essential Medium (MEM) at 37°C and 5 % CO_2 exhibited general corrosion mode [46]. Pitting can be also initiated by small surface defects such as scratches [47]. While galvanic corrosion is caused by local change of composition, pitting appears to be mainly influenced by the formation of a partially protective film. In fact, for the AZ91, ZE41 and Mg2Zn0.2Mn alloys, which are two-phase Mg alloys, their corrosion rates in Hank's solution are similar to that of HP-Mg despite the tendency of the second phase to accelerate the corrosion rate [24]. The formation of a more protective and compact film on the surface of the Mg-Nd-Zn-Zr alloy than on AZ31 alloy is responsible for the slower corrosion rate on the former alloy in Hank's solution at 37°C for 240h [48]. After immersion, deep pits were detected on the surface of the AZ31 alloy while that of the Mg-Nd-Zn-Zr alloy remained smooth.

The corrosion rate of Mg alloys is also influenced by the flowing conditions of the physiological environment (i.e., under dynamic physiological conditions the corrosion rate would slow down compared with steady conditions). This behavior is attributed to the fact that if the Hank 's solution is flowing, the absorption of Cl^- ions on the surface of the protective layer would be hindered. This phenomenon could explain the difference between in vitro and in vivo corrosion of degradable Mg alloys. For example, corrosion tests of AZ91D and LAE442 alloys in physiological solution indicate that both alloys corrode about four orders of magnitude slower in vivo than in vitro [49].

5. Tuning the biodegradation rate

The main limitation of Mg alloys for their use as implant materials is their exceedingly high corrosion rate in physiological conditions (i.e., pH = 7.4-7.6 and large chloride concentrations), which causes their biodegradability to be faster than the time required to heal the bone [50].

For this reason it is important to decrease the degradation rate of Mg alloys to remain implanted in the human body for at least 12 weeks [51]. Moreover, although the human body strives to keep a constant value of the pH, the presence of a fast corroding Mg implant can lead to local alkalinization that would unfavorably affect the pH close to the implant. Song suggested that a pH higher than 7.8 can have a poisoning effect [23]. These drawbacks associated with an exceedingly fast degradation rate suggest the need to control the biodegradation rate of Mg alloys.

So far, two kinds of methods have been used to slow down the corrosion rates of Mg alloys:

a. Surface coatings or surface modification:

In a broad sense, coatings can be divided in two classes: conversion coatings and deposited coatings. Conversion coatings consist of protective layers prepared using chemical (immersion in chemical baths to form calcium phosphate-containing layers, fluoride-containing layers, etc) or electrochemical processes (passivation, anodization, etc) [52]. Likewise, deposited coatings can be divided into metallic [53-56], organic [57] and inorganic [58,59]. The corrosion resistance of Mg and Mg alloys can be also improved through surface modification using various techniques such ion implantation [60], surface cladding and melting with laser or electron beam, [61], plasma surface modification [62], surface amorphization [6], etc.

b. By controlling the composition and the microstructure:

Although there are numerous review articles dealing with the growth of coatings [52] and surface modification procedures for Mg alloys [63], none of them deeply focuses, to the best of our knowledge, on the different means to tune the corrosion rate of these materials based on the control of their microstructure and composition. For this reason, the following section focuses on this subject.

5.1. Compositional and microstructural control:

5.1.1. Microstructural modification and thermal treatment

Microstructural control is an effective means to tune the strength and corrosion resistance of Mg alloys. More grain boundaries that act as corrosion barriers [64,65] are formed when the grain size is reduced. The microstructure can be refined using different severe plastic deformation (SPD) methods such as extrusion and equal channel angular extrusion (ECAP). Subsequent heat treatments further allow controlling the microstructure in order to tune the mechanical and corrosion performance. There are, in fact, multiple combinations of SPD and/or heat treatments to optimize the microstructure. For example, an alloy can be first heat treated and then plastically deformed or an alloy can be simply heat treated from the as-cast condition (i.e., without undergoing plastic deformation).

An example of microstructural optimization through heat treatments is the effective control of the corrosion behavior of the as-cast (F) Mg3Nd0.2Zn (wt. %) (NZ) and Mg3Nd0.2Zn0.4Zr (wt. %) (NZK) alloys through solution heat treatment (T4) and solution heat treated and artificially aged (T6) in 5 % NaCl solution. The T4 treatment is carried out at 540°C for 6 h followed by water quench at 25°C. After this solution treatment the alloys are artificially aged in an oil bath at 200°C for 16h (T6) [66]. Immersion tests indicate that the highest corrosion rates stand for the as-cast samples: 1.353 mg/cm^2/day and 0.203 mg/cm^2/day for NZ and NKZ alloys, respectively. The heat treatments increase the corrosion resistance in the following order: F < T6 < T4. The lowest corrosion rates values are obtained at T4 conditions (0.266 mg/cm^2/day for NZ alloy and 0.11 mg/cm^2/day for NZK alloy) and only increase slightly at T6 conditions. The change in the corrosion rate after the heat treatments is ascribed to micro-structural modifications. In the as-cast condition the microstructure of both alloys consist of α-Mg matrix and an eutectic $Mg_{12}Nd$ compound inhomogeneously distributed within the matrix. Because of the discontinuous distribution, the $Mg_{12}Nd$ acts as a microgalvanic cathode and, so, it accelerates the corrosion of the matrix. The authors reached this conclusion by comparing the role of $Mg_{12}Nd$ phase on the corrosion behavior of NZ and NZK alloys with the role of β phase (i.e., $Mg_{17}Al_{12}$) on the corrosion of AZ alloys [27]. Song and Atrens proposed that when the β phase is discontinuously distributed within the material, the corrosion rate of AZ alloys increases (see section 5.1.3) and thus the same behavior is expected for the $Mg_{12}Nd$ phase. However, when the NZ and NZK alloys are subjected to T4 or T6 treatments, the $Mg_{12}Nd$ compound dissolves into the matrix and microgalvanic couples are no longer present. The slightly higher corrosion rates detected at T6 to T4 is attributed to the precipitation of very small Nd-rich precipitates. Consistently, the corrosion morphologies reveal that the localized attack zones are more severe in the as-cast than in the T4 condition. Also, in the T6 condition the attack is slightly more severe than in T4 condition.

For a ZE41 alloy (4 wt. % Zn, 1 wt. % RE) the corrosion behavior improves after heat treating for 5 days at 500°C [67]. This improvement is again related to microstructural changes that occur during heat treatment. The Mg_7Zn_3RE phase present in the material before heating partly redissolves, which explains the increasing concentration of Zn and RE in the matrix. Similarly, the corrosion resistance of the as-cast Mg10Gd3Y0.4Zr (wt. %) alloy increases with solution treatments due to the dissolution into the α-Mg matrix of the (Gd+Y)-rich eutectic present in the as-cast condition [68]. The improvement of the corrosion resistance greatly depends on the thermal treatment. Namely, it is highest for a T4 solution treatment (500°C for 6 h) than for any of the T6 solution treatments (oil bath at 250°C for 0.5, 16, 193 and 500 h). The reason lies in that an increasing ageing time increases the volume fraction of secondary phase that act as cathodes and thereby ultimately increases the corrosion rate.

The microstructure of alloys can be optimized if the temperature is properly controlled during the dynamic recrystallization in an extrusion process. The corrosion behavior in SBF at 37°C of Mg3Nd0.2Zn0.4Zr (wt. %) NZK alloy initially solution-treated at T4 conditions (at 540°C for 10 h and then water quenched to room temperature) is effectively modified by controlling the extrusion temperature (250°C, 350°C and 450°C) [69]. Both immersion and electrochemical tests indicate that the corrosion rate in the extruded condition at 250°C, 350°C and 450°C is much slower than in the T4 state. Moreover, the corrosion resistance increases with the decrease of the extrusion temperature and so does the grain size.

Deformation processing can also have an effect on the redistribution of solutes within the microstructure and ultimately affect the corrosion behavior. When a ZK60 (6 wt.% Zn, 0.5 wt.

% Zr) alloy is processed by an integrated extrusion combined with ECAP, it is observed that Zn-Zr and Mg-Zn intermetallics become fractured and redistributed within the microstructure. Electrochemical and immersion tests in NaCl electrolytes indicate that grain refinement and redistribution of Zr and Zn solutes improve the corrosion resistance [70].

The corrosion behavior also depends on microstructural effects such as twins, dislocations, etc., caused by deformation processing. For example, the corrosion resistance in 3.5 % NaCl of AZ31B magnesium alloy has been studied in the initially hard rolled condition and after heat treating at 200, 300, 400 and 500°C for 3 h in an inert atmosphere of argon and subsequent quenching in water to room temperature [71].

The initial average grain size of 35 μm in the as-received condition increases with the heat treating temperature to 50 μm at 200°C (HT 200), 65 μm at 300°C (HT 300), 90 μm at 400°C (HT 400) and to 250 μm at 500°C (HT 500). In the HT300 conditions, the microstructure is untwined because a high density of twins are eliminated and so the intra-granular corrosion is the least. However, in the as-received and HT200 conditions the deformation twins and thus the dislocation density is higher. This can explain the more serious corrosion of the HT200 microstructure compared with the HT300 microstructure despite the fact that the HT200 microstructure is finer and thus the physical corrosion barrier is larger. In other words, twins accelerate the corrosion. From the physical metallurgy viewpoint, in the as rolled conditions (i.e., after plastic deformation), the amount of twins is the largest but they are progressively annihilated as temperature increases. For this reason potentiodynamic polarization curves show that the corrosion rate increases as the microstructure becomes more twinned.

Not only the presence of twins but also the distribution and density of dislocations are correlated with the corrosion behavior. The AZ31 alloy plastically deformed by ECAP at 350°C with a pressing speed of 350 mm/min exhibit twins and a higher density of dislocations than after being extruded at 350°C with an extrusion ratio of 10.24 (in this case twins were not observed) [65]. From corrosion studies in 3.5 % NaCl saturated with $Mg(OH)_2$ at pH 10.5, the authors concluded that the corrosion rate of AZ31 alloy decreases after extrusion but it increases after ECAP, suggesting that the twins and/or presence of higher density of dislocations decisively affect the corrosion rate.

The corrosion behavior of a AZ31 Mg alloy with different grain sizes immersed in two different solutions, NaCl and phosphate-buffer solution (PBS) has been studied by other researchers. The microstructure is refined by ECAP with a first pass of 250°C and successively heat treated to 300°C and rolled [72]. The best corrosion behavior is attained by the alloy having finest grains after long-term immersion in PBS [72]. This behavior is related to the formation of a mixed compact protective layer of P-containing compounds together with magnesium hydroxide that promotes protection against the chloride ions. The superior corrosion behavior of the fine-grained AZ31 alloy over the coarser one is attributed to the formation of a layer of corrosion products that provides better protection against the diffusion of aggressive ions to the surface of the material [72].

Although these results suggest that the corrosion performance can be tuned by controlling the microstructure, other factors such as the chemical composition plays a more important role.

For example, Liao et al. [73] observed that the fine grained AZ31B alloy exhibits a lower corrosion resistance than the AM60 alloy with coarser grains.

5.1.2. Amorphous and partly amorphous alloys

As aforementioned, the grain size can be tuned by controlling the cooling rate. For certain compositions such as AZ91 alloy [74] rapid cooling is an effective technique to obtain fine grain sizes. For other Mg-based compositions a sufficiently fast cooling rate can lead to the formation of glassy materials. Moreover, rapid cooling allows to extend the solubility of alloying elements in Mg alloys and to form a homogeneous single-phase structure (i.e., metallic glass) with a very different corrosion behavior than that of crystalline Mg alloys [75]. Typically, amorphous materials possess stronger corrosion and chemical resistance than their crystalline counterparts due to the absence of grain boundaries, segregated phases, secondary particles and also due to chemical homogeneity [76]. For this reason different Mg-based glassy materials have been studied over the years. For example, glassy $Mg_{60+x}Zn_{35-x}Ca_5$ ($0<x<7$ at. %) ribbons of 50 μm in thickness can be successfully obtained by melt spinning [36]. Immersion tests of these ribbons in SBF lead to the formation of corrosion layers that are different from those found in Zn-poor and Zn-rich alloys. For the Zn-rich alloys(above 28 at. % Zn), the Zn-rich oxygen-containing passivating layer that is formed on the surface of the ribbon is responsible for the more noble behavior of these alloys as compared to the Zn-poor alloys [36]. Moreover, a high Zn content appears to reduce hydrogen evolution. In fact, due to the extended solubility of Zn in the amorphous structure of the Mg-Zn-Ca system, the $Mg_{60}Zn_{35}Ca_5$ glass only exhibits marginal hydrogen evolution during in vitro and in vivo degradation [36].

Through the addition of different alloying elements to the Mg-Zn-Ca alloys family, the corrosion behavior can be tuned as well. Small Pd additions are enough to decrease the glass forming ability of glassy $Mg_{72}Zn_{23}Ca_5$ alloys and to shift the corrosion potential towards more positive values [6]. Cytotoxic tests do not show the presence of death cells, which confirm that these alloys might have potential use as implants [77]. Cytocompatibility tests also show that metallic glass $Mg_{66}Zn_{30}Ca_4$ and $Mg_{70}Zn_{25}Ca_5$ samples have higher cell viability and exhibit more positive corrosion potential than that of as-rolled crystalline pure Mg [78].

It is well known that glassy materials can be used as precursors of crystalline phases by controlling the crystallization temperature and/or time. Since the corrosion behavior depends on the structure (i.e., amorphous vs. crystalline) of the material, the extent of crystallization can be controlled to tune the corrosion rate. For example, glassy $Mg_{67}Zn_{28}Ca_5$ ribbons exhibit an increase of the corrosion resistance in simulated body fluid with the increase of annealing temperature up to a maximum and then the resistance decreases rapidly for higher temperatures. The best corrosion resistance of these ribbons is attained at 160°C, when the microstructure is constituted by a metastable crystalline $Mg_{102.08}Zn_{39.6}$ phase embedded in an amorphous matrix [76]. This behavior was explained considering that the electrochemical activity of this phase is similar to that of its amorphous matrix. However, the newly formed phases at 225°C are more active and worsen the corrosion resistance of the alloy [76].

To determine the effect that alloying elements have on the corrosion resistance of rapidly solidified magnesium alloys, different binary Mg-based glassy alloys were studied by using electrochemical techniques in pH 9.2 sodium borate [79]. These studies concluded that the corrosion rate of magnesium is decreased for larger contents of aluminium. Similarly, low concentrations of zinc and lithium decrease the corrosion rate below that of pure magnesium [79]. These results indicate that composition has an important influence on the corrosion rate of glassy Mg alloys, as it has also been observed in crystalline alloys.

5.1.3. Influence of alloying elements on corrosion performance

As was explained on section 4.1, the corrosion rate of magnesium alloys depends on the nature and concentration of impurities. The corrosion resistance can be improved either by purifying Mg or through appropriate additions of alloying elements. Mg alloys are basically classified [34] in two groups: 1) those that contain Al as primary alloying element and 2) those that do not contain Al and have small amounts of Zr to refine the microstructure. Al is generally considered as a beneficial element to improve the corrosion resistance [80]. For small contents, Al remains in solid solution, but above certain concentration β-$Mg_{17}Al_{12}$ secondary particles precipitate.

The β-phase is very stable in NaCl solutions and it is inert to corrosion due to the formation of a passive thin film on its surface. However, β-phase is also an effective cathode, which can explain the dual role of these precipitates in the corrosion of AZ alloys according to Song and Atrens [25]. A fine and continuous distribution of β-phase is recommended to increase the corrosion resistance. For example, Lunder et al. reported that Al additions higher than 8 wt. % increase the corrosion resistance of Mg-Al alloys [81]. An improvement of the corrosion resistance with the Al content is also found in AZ91, AZ61 and AZ31 alloys in 5 % NaCl [82]. However, Song et al. reported an increase of the corrosion rate in NaCl in the following order AZ501 < AZ21 < AZ91 [83].

For the second family of alloys, small additions of Zr refine the microstructure of Mg and improve the corrosion resistance [84]. Since Zr can easily combine with impurities, especially Fe and Ni, it can form insoluble precipitates that settle out during melting. This purification effect of Zr enhances the corrosion resistance of Mg [85]. Depending on the composition, a minimum concentration of Zr is required to observe such effect. For example, for Zr contents from 0 to 0.42 wt. % the corrosion resistance of Mg10Gd3Y (10 wt. % Gd, 3 wt. %) deteriorates, whereas for higher Zr contents, ranging from 0.42 to 0.93 %, the corrosion resistance improves. The distinct behavior is attributed to the differences in size and distribution of the Zr-rich particles [86]. An exceedingly larger amount of Zr can lead to precipitation of Zr within the matrix, which becomes detrimental for the corrosion performance of the alloy [34].

The addition of 1 wt. % of Al, Ag, In, Mn, Sn, Zn and Zr elements decrease the volume of evolved hydrogen gas, and thus decrease the corrosion rate, of Mg when immersed either in SBF or in Hank's solution [87]. On the contrary, the addition of 1 wt. % Y or Si have a deleterious effect on the corrosion performance.

Ca is an essential element to the body since it is a major component of human bones. For this reason, it has been used over the years to fabricate biocompatible Mg-based alloys. The concentration of Ca has, though, to be carefully controlled to avoid the precipitation of Mg_2Ca particles (that takes place for Ca contents ranging from 0.8 to 5 wt. % [88] or from 1 to 3 wt. % [89] depending on the system under study). These Mg_2Ca particles form micro-galvanic cells within the Mg matrix and accelerate preferentially the dissolution of the latter, worsening the corrosion resistance of the binary Mg-xCa alloy. For 1.5 wt. % Ca, a protective oxide layer of MgO and CaO is formed after heating to 500°C for 1h [90]. The influence of Ca on the corrosion behavior not only depends on its amount but also on the composition of the Mg-based alloy to which it is added. For example, the addition of 13 wt. % Ca increases the corrosion rate of AZ91D alloy (37 wt. % Al, 0.5 wt. % Zn) [91]. An improvement of the biocorrosion resistance is also detected when 0.2-0.4 wt. % Ca is added to a Mg-Si alloy since it refines the grain size and modifies the morphology of Mg_2Si phase [92]. The same holds when 1.6 wt. % Zn is added to Mg-Si alloy due to the modifications on the Mg_2Si phase morphology derived from the addition; namely from a coarse eutectic structure to a small dot or short bar shape [92]. Zn is an essential element to the human body and capable of decreasing the corrosion rate of pure Mg in small amounts. For example, the corrosion rate (measured in terms of volume of evolved hydrogen) of CP-Mg decreases from 26 ml/cm^2/day to 0.280 ml/cm^2/day with the addition of 1 wt.% Zn [22]. The addition of 6 wt. % zinc shifts the corrosion potential toward more cathodic values and decreases the in-vitro degradation rate of high purity Mg in SBF [93]. However, concentrations above the equilibrium solid solubility of Zn in Mg (i.e., 6.2 wt. %) [94] can lead to an increase of the corrosion rate in 3 % NaCl due to the formation of β-Mg_7Zn_3 phase in the magnesium matrix [95]. The introduction of Mn can help to decrease the corrosion rate of Mg-Zn alloys. Ahmed et al. [96] reported that adding Mn to a Mg-based alloy containing 4 to 8 wt. %. Zn decreases the dissolution rate of Mg. The corrosion rate of Mg2Zn0.2Mn (2 wt. % Zn, 0.2 wt. % Mn) in Hank's solution is also smaller than that of Mg1Zn (1 wt. % Zn) [22].

Other atypical alloying elements such as Y, Ce, Ti and Sc were reported to improve the corrosion performance when alloyed with Mg at a level below the solubility limit [97].

6. Biodegradation and mechanical integrity

The use of Mg alloys as weight-bearing implants requires that the material should have sufficient strength not only at the moment of being implanted but also when the alloy degrades over the time while remaining in contact with body fluids. It is important that implants keep their strength at least until the bone heals. For this reason different studies have been carried out to evaluate the mass loss and evolution of the strength over the implantation or immersion time [77].

According to Pietak et al. [98] the best technique to assess the degradation of Mg alloys is measuring the mechanical integrity as a function of the incubation time. Nevertheless, the measurement of the mass change [46] has been more frequently used. However, this procedure has several shortcomings due to the association of non-soluble degradation products that precipitate on the sample and obscure the mass loss [98].

The mechanical integrity can be evaluated using various mechanical tests (three-point bending, tensile tests, nanoindentation, etc). These tests can be performed under physiological conditions or in air.

From nanoindentation tests, Pellicer et al. [77] studied the evolution of the Young's modulus (E_r), hardness (H), and H/E_r ratio (which is an indirect measure of the wear resistance) of amorphous $Mg_{72}Zn_{23}Ca_5$ after different immersion times in HBSS as shown in Fig. 5. The same study was carried out on crystalline $Mg_{70}Zn_{23}Ca_5Pd_2$ alloy and the results were systematically compared. While the stiffness of both compositions decreases with the immersion time, hardness exhibits a more complex dependence, especially for the $Mg_{72}Zn_{23}Ca_5$ alloy. Namely, an increase is observed after short-term immersion, which was mainly attributed to the fast dissolution of Mg and the concomitant enrichment in Zn (Zn is mechanically harder, so solution hardening takes place). For longer immersion times, the dissolution progresses and the alloy not only undergoes surface chemical change but the surface is also physically modified with to the formation of flaws such as pores and corrugations, which cause a decrease of hardness [77]. The values of H/E_r increase from 0.053 for the as-cast $Mg_{72}Zn_{23}Ca_5$ alloy to 0.1 for $Mg_{72}Zn_{23}Ca_5$ after 2h immersion in HBSS at 37°C, thus indicating that the effect of porosity on the Young's modulus for short-immersion times is more noticeable than on hardness. These results are consistent with those observed in many other metallic alloys [99].

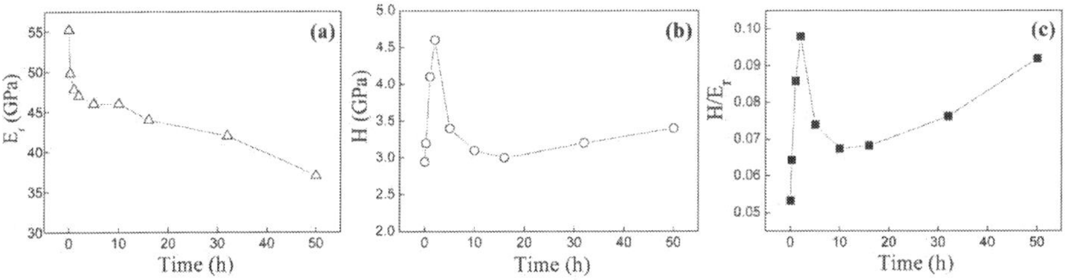

Figure 5. Dependence of (a) reduced Young's modulus, E_r, (b) hardness, H and (c) H/E_r ratio on the immersion time in HBBS at 37°C for $Mg_{72}Zn_{23}Ca_5$ alloy. Adapted and reprinted from Pellicer et al. [77], page 8, with permission from John Wiley&Sons.

To evaluate the mechanical integrity of AZ91Ca alloy (i.e., a calcium-containing magnesium alloy) in SBF at 36.5°C, slow strain rate tensile tests at $1.2x10^{-7}$ s^{-1} were carried out [91]. The AZ91Ca alloy shows lower elongation (3.5±0.2 %) and lower ultimate tensile strength (106±10 MPa) in the SBF than in air (4.6±0.3 % and 126±5 MPa, respectively). The decrease of the mechanical performance is, however, small and thus the alloy is not highly susceptible to corrosion in SBF. From electrochemical experiments it is observed that the AZ91Ca alloy exhibits improved corrosion resistance compared to the AZ91 alloy, which can be attributed to the formation of calcium phosphate on the surface of the AZ91Ca alloy. This surface film has higher stability than the film formed on AZ91 alloy for this reason not only the general corrosion resistance but also the pitting corrosion resistance improve.

Krause et al. [100] compared the evolution of the mechanical behavior of Mg0.8Ca (8 wt.% Ca), LAE442 (4 wt. % Li, 4 wt. % Al, 2 wt. % RE) and WE43 (4 wt. % Y, 3 wt. % RE) alloys implanted in rabbits for 3 and 6 months by three-point bending tests. All the samples exhibit biodegradation as can be deduced from the loss in volume with implantation period. The MgCa0.8 alloy degrades slowly during the first 3 months but its corrosion rate accelerates during the following 3 months. The LA442 alloy exhibits slower degradation rate than the Mg0.8Ca and WE43 alloys. The difference of degradation rate is responsible for the distinct mechanical performance of the alloys over the time. Thee-point bending test results indicate the following trend in the initial strength: LAE442 (255.67±5.69 N) > WE43 (238.05 ±21.68 N) > Mg0.8Ca (178.76±25.15 N). However, after 3 months the strength trend changes so that it decreases in the following order: WE43 (185.59±15.64 N) > LAE442 (153.21±18.45 N) > Mg0.8Ca (115.42±9.66 N). After 6 months the strength follows this sequence: LAE442 (134.68±14.68 N) > WE43 (122.23±23.65 N) > Mg0.8Ca (52.90±5.96 N). From the results of the maximal applied force it can be deduced that the LAE442 alloy degrades faster during the first 3 months and slower between 3 and 6 months. The degradation rate of Mg0.8Ca and WE43 alloys is different; it decreases in a linear manner over the time [100]. The ductility of the alloys was also assessed from three-point bending tests by measuring the bending displacement but concluding results could not be obtained due to high scattering. The Mg0.8Ca alloy exhibits the highest loss and the LAE442 the lowest loss in volume after 6 months.

The evolution of the bending strength of Mg6Zn (6 wt. % Zn) alloy with the immersion time in physiological saline solution (0.9 % NaCl) [88] is similar to that of implanted LAE442 alloy [100]. For short immersion times (i.e., 3 days) the degradation rate of Mg6Zn is very fast (0.20±0.05 mm/year) and exhibits a large weight loss but it becomes slower (0.07±0.02 mm/year) for longer immersion times (i.e., 30 days). The bending strength of the alloy decreases rapidly with an initially small weight loss but then decelerates as the percentage weight loss increases. This behavior was attributed to the formation of surface defects such as corrosion holes during degradation.

Plates of ZEK100 (1 wt. % Zn, 0.1 wt. % Zr, 0.1 wt. % RE (rare earth)) were mechanically tested in vitro after 14 (2 weeks), 28 (4 weeks) and 42 days (6 weeks) of immersion with a constant laminar flow rate in HBSS via four-point bending tests [101]. The bending strength decreases from immersion week 2 to week 4 but increases again after 6 weeks. The lowering of the bending strength is attributed to dissolution of the plate whereas the increase at longer times can be explained by precipitation of calcium phosphates from the solution on the surface of the plate. This behavior was supposed to be caused by a decrease of the implant volume during the first 4 weeks and an increase for longer times up to 8 weeks.

The mechanical behavior of ZEK100 alloy was also tested via three-point bending after being implanted in rabbit tibiae for 3 and 6 months [102]. The corrosion rate increases from 0.067 mm/year after 3 months to 0.154 mm/year after 6 months. The volume of the implant tends to reduce with the increase of the implantation time. This can explain why the initial maximum force of 241 N (the maximum force at breakage) decreases to 153 and 100 N after 3 and 6 months, respectively.

Figure 6 shows a comparative summary of the mechanical properties (compressive yield stress, $\sigma_{y,C}$, and Young's modulus, E) of various families of materials that can be used as bioabsorbable implants, such as metallic alloys, biodegradable polymers and ceramics. From the correlation $H \approx C\ \sigma_{y,C}$ (where C is the so-called constraint factor and is normally close to 3 for crystalline metallic alloys and slightly higher for metallic glasses [103]) the mechanical hardness is observed to be directly proportional to $\sigma_{y,C}$.

The yield stress of Mg-Zn-Ca bulk metallic glasses is relatively large (Figure 6), thus indicating that they are one of the hardest biodegradable materials reported in the literature. Moreover, the values of Young's modulus of glassy Mg-Zn-Ca are closer to that of cortical bone (E_{bone} = 3– 20 GPa) than most crystalline Mg-based alloys (they are also mechanically softer) and synthetic hydroxyapatites. Considering that the $Mg_{70}Zn_{23}Ca_5Pd_2$ alloy is fully crystalline, its hardness is also generally larger than that of most Mg-Zn-Ca crystalline alloys. Compared with most Fe-based biodegradable alloys, Mg-based bulk metallic glasses are generally harder. Moreover, Fe-based alloys are typically ferromagnetic at room temperature (except the antiferromagnetic FeMn), which precludes their use in nuclear resonance imaging techniques for diagnostics purposes. The Young's modulus of $Mg_{72}Zn_{23}Ca_5$ becomes closer to that of Ca-based or Sr-based biodegradable metallic glasses after long-term immersion in HBSS as well as to that of polymeric materials reinforced with glassy fibers. At the same time, the hardness of $Mg_{72}Zn_{23}Ca_5$ alloy is higher than that of all these materials.

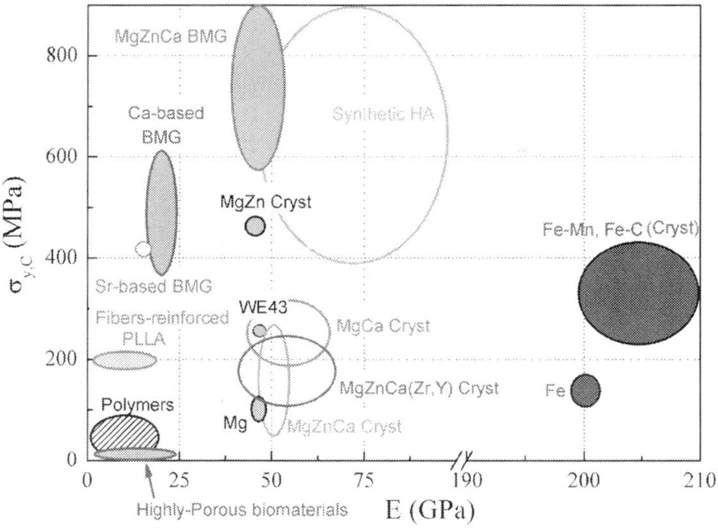

Figure 6. Comparison of the mechanical properties (compressive yield stress, $\sigma_{y,C}$, versus Young's modulus, E) in linear scale for different families of biodegradable implant materials, including metallic alloys (Mg-based, Ca-based, Sr-based, or Fe-based), ceramics (e.g., synthetic hydroxyapatites) and polymers. Adapted and reprinted from Pellicer et al. [77], page 13, with permission from John Wiley&Sons.

7. Summary and conclusion

Magnesium and its alloys are suitable materials for biomedical applications due to their low weight, high specific strength, stiffness close to bone and good biocompatibility. Specifically, because magnesium exhibits a fast biodegradability, it has attracted an increasing interest over the last years for its potential use as "biodegradable implants". However, the main limitation is that Mg degrades too fast and that the corrosion process is accompanied by hydrogen evolution. In these conditions, magnesium implants lose their mechanical integrity before the bone heals and hydrogen gas accumulates inside the body. To overcome these limitations different methods have been pursued to decrease the corrosion rate of magnesium to acceptable levels, including the growth of coatings (conversion and deposited coatings), surface modification treatments (ion implantation, plasma surface modification, etc) or via the control of the composition and microstructure of Mg alloys themselves.

For tuning efficiently the composition and microstructure it is first necessary to understand two of the most common types of corrosion that Mg and Mg alloys exhibit: galvanic and pitting corrosion. Galvanic corrosion develops because magnesium almost always behaves anodically in contact with other metals. Galvanic couples are usually encountered when the concentration of the alloying element surpasses their corresponding maximum solid solubility in magnesium. The alloying element then segregates during solidification or annealing and an inhomogeneous microstructure is formed. The extent of the galvanic effect depends on a number of factors such as the crystal orientation of the magnesium matrix, the type of secondary phases and impurity particles, the solution in which the alloy is immersed and the grain size. The concentration and distribution of secondary phases is also important for the corrosion behavior. A fine and continuous distribution of secondary phases typically improves the corrosion performance.

Mg alloys are susceptible to form a passivation layer of $Mg(OH)_2$ or a mixture of $Mg(OH)_2$ and MgO in aqueous solutions. Due to the presence of chloride ions in physiological fluids, the protective coating may be destroyed and localized attack (i.e., pitting corrosion) initiates. Physiological environments also contain carbonates, phosphates, sulfates and other ingredients that have different effects on the corrosion behavior of magnesium. Before magnesium alloys can be used as real implants it is necessary to evaluate the biodegradability and mechanical performance over the immersion (in-vitro) or implantation (in vivo) time.

Acknowledgements

This work has been partially financed by the 2009-SGR-1292 and MAT2011-27380-C02-01 research projects. S. G. acknowledges the Juan de la Cierva Fellowship from the Spanish Ministry of Science and Innovation. M.D.B. was partially supported by an ICREA Academia award.

http://dx.doi.org/10.5772/55584

Author details

S. González[1], E. Pellicer[1], S. Suriñach[1], M.D. Baró[1] and J. Sort[2]

*Address all correspondence to: Sergio.Gonzalez@uab.es

1 Departament de Física, Facultat de Ciències, Universitat Autònoma de Barcelona, Barcelona, Spain

2 Institució Catalana de Recerca i Estudis Avançats (ICREA) and Departament de Física, Facultat de Ciències, Universitat Autònoma de Barcelona, Barcelona, Spain

References

[1] Bhat SV. Biomaterials. Kluwer Adademic Publishers. Boston, MIT, USA; 2002. p265.

[2] Sumita M, Hanawa T, Teoh SH. Development of nitrogen-containing nickel-free austenitic stainless steels for metallic biomaterials-review. Matererials Science and Engineering C 2004; 24: 753-760.

[3] Puleo DA. Biochemical surface modification of Co-Cr-Mo. Biomaterials 1996; 17: 217-222.

[4] Geetha M, Singh AK, Asokamani R, Gogia AK. Ti-based biomaterials, the ultimate choice for orthopaedic implants. A review. Progress in Materials Science 2009; 54: 397-425.

[5] Moravej M, Mantovani D. Biodegradable metals for cardiovascular stent application: interests and new opportunities. International Journal of Molecular Sciences 2011; 12: 4250-4270.

[6] González S, Pellicer E, Fornell J, Blanquer A, Barrios L, Ibañez E, Solsona P, Suriñach S, Baró MD, Nogués C, Sort J. Improved mechanical perfomance and delayed corrosion phenomena in biodegradable Mg-Zn-Ca alloys through Pd-alloying. Journal of the Mechanical Behavior of Biomedical Materials 2012; 6: 53-62.

[7] Hanbook of Biomaterials Properties. In: Black J, Hasting GW (ed.). Chapman and Hall. London; 1998.

[8] Zhou Z, Liu X, Liu Q, Liu L. Evaluation of the potential cytotoxicity of metals associated with implanted biomaterials (I). Preparative Biochemistry and Biotechnology 2009; 39: 81-91.

[9] Li XN, Gu ZJ, Lou SQ, Zheng YF. The development of binary Mg–Ca alloys for use as biodegradable materials within bone. Biomaterials 2008; 29: 1329–1344.

[10] Godard HP, Jepson WB, Bothwell MR, Kane RL. The Corrosion of Light Metals. In: John Wiley&Sons (ed.). New York; 1967.

[11] Wang L, Shinohara T, Zhang B-P. Influence of deaerated condition on the corrosion behavior of AZ31 magnesium alloy in dilute NaCl solutions. Matererials Transactions 2009; 50: 2563-2569.

[12] Pourbaix M. Atlas of Electrochemical Equilibrium in Aqueous solutions. In: 2nd Ed. NACE. Houston; 1974.

[13] ASM Handbook. Volume 13A Corrosion: Fundamentals, Testing and Protection. In: Cramer SD, Covino BS, Jr. ASM International; 2003.

[14] Lonza Walkersville Inc. http://vgn.uvm.edu/outreach/documents/Hanksbufferedsalinesolution.pdf (accessed 27 November 2012).

[15] Medicago AB. Phosphate buffered saline specification sheet; 2010.

[16] Kokubo T, Kushtani T, Sakka S, Kitsugi T, Yamamuro T. Solutions able to reproduce *in vivo* surface-structure changes in bioactive glass-ceramic A-W3. Journal of Biomedical Materials Research 1990; 24: 721-734.

[17] Tunold R, Holtan H, Berge MBH, Lasson A, Steen-Hansen R. The corrosion of magnesium in aqueous solution containing chloride ions. Corrosion Science 1977; 17: 353-365.

[18] Hara N, Kobayashi Y, Kagaya D, Akao N. Formation and breakdown of surface films on magnesium and its alloys in aqueous solution. Corrosion Science 2007; 49: 166-175.

[19] Quach NC, Uggowitzer PJ, Schmutz P. Corrosion behavior of an Mg-Y-RE alloy used in biomedical applications studied by electrochemical techniques. Chemie 2008; 11: 1043-1054.

[20] Baril G, Pébère N. The corrosion of pure magnesium in aerated and deaerated sodium sulphate solutions. Corrosion Science 2001; 43: 471-484.

[21] Handbook of Corrosion data. In: Craig BD and Anderson D (ed.). Materials data series. Materials Park, OH: ASM International; 1995.

[22] Song GL. Control of biodegradation of biocompatible magnesium alloys. Corrosion Science 2007; 49: 1696-1701.

[23] Ng WF, Chiu KY, Cheng FT. Effect of pH on the in vitro corrosion rate of magnesium degradable implant material. Materials Science and Engineering C 2010; 30: 898-903.

[24] Zainal Abidin NI, Martin D, Atrens A. Corrosion of high purity Mg, AZ91, ZE41 and Mg2Zn0.2Mn in Hank's solution at room temperature. Corrosion Science 2011; 53: 862-872.

[25] Song GL, Atrens A. Corrosion mechanisms of magnesium alloys. Adv. Eng. Mater. 1999; 1: 11-33.

[26] Song GL, Atrens A, StJohn DH, Wu X, Nairn J. The anodic dissolution of magnesium in chloride and sulphate solutions. Corrosion Science 1997; 39: 1981-2004.

[27] Amtec Consultants. Experts in Coatings & Corrosion; 2011.

[28] Song GL. Corrosion characteristics of Mg alloys: NACE DOD Conference, July 2011, Palm Springs, CA.

[29] Song GL, Xu Z. Crystal orientation and electrochemical corrosion of polycrystalline Mg. Corrosion Science 2012; 63: 100-112.

[30] Suárez MF, Compton RG. Dissolution of magnesium oxide in aqueous acid: an atomic force microscopy study, Journal of Physical Chemistry B 1998; 102: 7156–7162.

[31] Song GL, Mishra R, Xu Z. Crystallographic orientation and electrochemical activity of AZ31 Mg alloy. Electrochemistry Communications 2010; 12: 1009-1012.

[32] Song GL, Bowles A, StJohn DH. Corrosion resistance of aged die cast magnesium alloy AZ91D. Materials Science and Engineering A 2004; 366: 74-86.

[33] Dargusch MS, Dunlop GL, Pettersen K. Mg alloys and their applications. Wolfsburg, Germany: Werkstoff-Information GmbH; 1998 p277-282.

[34] Song GL. Recent progress in corrosion and protection of magnesium alloys. Advanced Engineering Materials 2005; 7: 563-586.

[35] Liu C, Xin Y, Tang G, Chu PK. Influence of heat treatment on degradation behavior of bio-degradable die-cast AZ63 magnesium alloy in simulated body fluid. Materials Science and Engineering A 2007; 456: 350-357.

[36] Zberg B, Uggowitzer PJ, Löffler JF. MgZnCa glasses without clinically observable hydrogen evolution for biodegradable implants. Nature Materials 2009; 8: 887-891.

[37] ASM Specialty Handbook - Magnesium and Magnesium Alloys. In: Avedesian MM and Baker H (ed.). ASM international. Materials Park, OH; 1999.

[38] Roberts CS. Chapter Mg alloy systems. In: Mg and its alloys, John Wiley & Sons; 1960. p42-80.

[39] Liu M, Uggowitzer PJ, Nagasekhar AV, Schmutz P, Easton M, Song G-L, Atrens A. Calculated phase diagrams and the corrosion of die cast Mg-Al alloys. Corrosion Science 2009; 51: 602-619.

[40] Wang WH, Dong C, Shek CH. Bulk metallic glasses. Materials Science and Engineering R 2004; 44: 45-89.

[41] Badawy WA, Hilal NH, El-Rabiee M, Nady H. Electrochemical behavior of Mg and some Mg alloys in aqueous solutions of different pH. Electrochimica Acta 2010; 55: 1880-1887.

[42] Phillips RC, Kish JR. On the self-passivation tendency of Mg-Al-Zn (AZ) alloys in aqueous solutions. ECS Transactions 2012; 41: 167-176.

[43] Yao HB, Li Y, Wee ATS. An XPS investigation of the oxidation/corrosion of melt-spun Mg. Applied Surface Science 2000; 158: 112-119.

[44] Xu Y, Huo K, Tao H, Tang G, Chu PK. Influence of aggressive ions on the degradation behavior of biomedical magnesium alloy in physiological environment. Acta Biomaterialia 2008; 4: 2008-2015.

[45] Yang L, Zhang E. Biocorrosion behavior of magnesium alloy in different simulated fluids for biomedical application. Materials Science and Engineering C 2009; 29: 1691-1696.

[46] Kirkland NT, Lespagnol J, Birbilis N, Staiger MP. A survey of bio-corrosion rates of magnesium alloys. Corrosion Science 2010; 52: 287-291.

[47] Chen J, Wang J, Han E-H, Ke W. In situ observation of pit initiation of passivated AZ91 magnesium. Corrosion Science 2009; 51: 477-484.

[48] Zong Y, Yuan G, Zhang X, Mao L, Niu J, Ding W. Comparison of biodegradable behaviors of AZ31 and Mg-Nd-Zn-Zr alloys in Hank's physiological solution. Materials Science and Engineering B 2012; 177: 395-401.

[49] Witte F, Fischer J, Nellesen J, Crostack H-A, Kaese V, Pisch A, Beckmann F, Windhagen H. In vitro and in vivo corrosion measurements of magnesium alloys. Biomaterials 2006; 27: 1013-1018.

[50] Li ZJ, Gu XN, Lou SQ, Zheng YF. The development of binary Mg-Ca alloys for use as biodegradable materials within bone. Biomaterials 2008; 29: 1329-1344.

[51] Staigner MP, Pietak AM, Huadmai J, Dias G. Magnesium and its alloys as orthopedic biomaterials: A review. Biomaterials 2006; 27: 1728-1734.

[52] Hornberger H, Virtanen S, Boccaccini AR. Biomedical coatings on magnesium alloys - A review. Acta Biomaterialia. 2012; 8: 2442-2455.

[53] Zhong C, Liu F, Wu Y, Le J, Liu L, He M, Zhu J, Hu W. Protective diffusion coatings on magnesium alloys: A review of recent developments. Journal of Alloys and Compounds 2012; 520: 11-21.

[54] Fukumoto S, Sugahara K, Yamamoto A, Tsubakino H. Improvement of corrosion resistance and adhesion of coating layer for magnesium alloy coated with high purity magnesium. Mater. Trans. 2003; 44: 518-523.

[55] Zhang E, Xu L, Yang K. Formation by ion plating of Ti-coating on pure Mg for biomedical applications. Scripta Materialia 2005; 53: 523-527.

[56] Cui X, Jin G, Li Q, Yang Y, Li Y, Wang F. Electroless Ni-P plating with a phytic acid pretreatment on AZ91D magnesium alloy. Materials Chemistry and Physics 2010; 121: 308-313.

[57] Hu R-G, Zhang S, Bu J-F, Lin C-J, Song G-L. Recent progress in corrosion protection of magnesium alloys by organic coatings. Progress in Organic Coatings 2012; 73: 129-141.

[58] Feil F, Fürbeth W, Schütze M. Purely inorganic coatings based on nanoparticles for magnesium alloys. Electrochimica Acta 2009; 54: 2478-2486.

[59] Boccaccini AR, Keim S, Ma R, Li Y, Zhitomirsky I. Electrophoretic deposition of biomaterials. In: J.R. Soc. Interface (ed.). 2010; 7: S581-S613.

[60] Wu G, Xu R, Feng K, Wu S, Wu Z, Sun G, Zheng G, Li G, Chu PK. Retardation of surface corrosion of biodegradable magnesium-based materials by aluminum ion implantation. Applied Surface Science 2012; 258: 7651-7657.

[61] Wang C, Li T, Yao B, Wang R, Dong C. Laser cladding of eutectic-based Ti-Ni-Al alloy coating on magnesium surface. Surface and Coatings Technology 2010; 205: 189-194.

[62] Yang J, Cui F-Z, Lee IS, Wang X. Plasma surface modification of magnesium alloy for biomedical application. Surface and Coating Technology 2010; 205: S182-S187.

[63] Zeng R, Dietzel W, Witte F, Hort N, Blawert C. Progress and challenge for magnesium alloys as biomaterials. Avanced Biomaterials 2008; 10: B3-B14.

[64] Hamu GB, Eliezer D, Wagner L. The relation between severe plastic deformation microstructure and corrosion behavior of AZ31 magnesium alloy. Journal of Alloys and Compounds 2009; 468: 222-229.

[65] Liu CL, Xin YC, Tang GY, Chu PK. Influence of heat treatment on degradation behavior of biodegradable die-cast AZ63 magnesium alloy in simulated body fluid. Materials Science and Engineering A 2007; 456: 350-357.

[66] Chang J-W, Fu P-H, Guo X-W, Peng L-M, Ding W-J. The effects of heat treatment and zirconium on the corrosion behavior of Mg-3Nd-0.2Zn-0.4Zr (wt. %) alloy. Corrosion Science 2007; 49: 2612-2627.

[67] Neil WC, Forsyth M, Howlett PC, Hutchinson CR, Hilton BRW. Corrosion of heat treated magnesium alloy ZE41. Corrosion Science 2011; 53: 3299-3308.

[68] Peng L-M, Chang J-W, Guo X-W, Atrens A, Ding W-J, Peng Y-H. Influence of heat treatment and microstructure on the corrosion of magnesium alloy Mg-10Gd-3Y-0.4Zr. Journal of Applied Electrochemistry 2009; 39: 913-920.

[69] Zhang X, Yuan G, Mao L, Niu J, Fu P, Ding W. Effects of extrusion and heat treatment on the mechanical properties and biocorrosion behaviors of a Mg-Nd-Zn-Zr alloy. Journal of the Mechanical Behavior of Biomedical Materials 2012; 7: 77-86.

[70] Orlov D, Ralston KD, Birbilis N, Estrin Y. Enhanced corrosion resistance of Mg alloy ZK60 after processing by integrated extrusion and equal channel angular pressing Acta Materialia 2011; 59: 6176-6186.

[71] Aung N, Zhou W. Effect of grain size and twins on corrosion behavior of AZ31B magnesium alloy. Corrosion Science 2010; 52: 589-594.

[72] Alvarez-Lopez M, Pereda MD, del Valle JA, Fernandez-Lorenzo M, Garcia-Alonso MC. Corrosion behavior of AZ31 magnesium alloy with different grain sizes in simulated biological fluids. Acta Biomaterialia 2010; 6: 1763-1771.

[73] Liao J, Hotta M, Yamamoto N. Corrosion behavior of fine-grained AZ31B magnesium alloy. Corrosion Science 2012; 61: 208-214.

[74] Ning Z, Cao P, Wang H, Sun J, Liu D. Effect of cooling conditions on grain size of AZ91 alloy. Journal of Materials Science and Technology 2007; 23: 645-649.

[75] Scully JR, Gebert A, Payer JH. Corrosion and related mechanical properties of bulk metallic glasses. Journal of Materials Research 2007; 22: 302-313.

[76] Wang Y, Tan MJ, Pang J, Wang Z, Jarfors AWF. In vitro corrosion behaviors of Mg67Zn28Ca5 alloy: from amorphous to crystalline. Materials Chemistry and Physics 2012; 134: 1079-1087.

[77] Pellicer E, González S, Blanquer A, Barrios L, Ibañez E, Solsona P, Suriñach S, Baró MD, Nogués C, Sort J. On the biodegradability, mechanical behavior, and cytocompatibility of amorphous Mg72Zn23Ca5 and crystalline Mg70Zn23Ca5Pd2 alloys as temporary implant materials. J. Biomed. Mater. Research A 2013; 101A: 502–517.

[78] Gu X, Zheng Y, Zhong S, Xi T, Wang J, Wang W. Corrosion of, and cellular responses to Mg-Zn-Ca bulk metallic glasses. Biomaterials 2010; 31: 1093-1103.

[79] Makar GL, Kruger J. Corrosion studies of rapidly solidified magnesium alloys. Journal of the Electrochemical Society 1990; 137: 414-421.

[80] Baliga CB, Tsakiropoulos P. Development of Corrosion Resistant Magnesium Alloys: Part II. Structure of the Corrosion Products formed on the Surfaces of Rapidly Solidified Mg-16Al Alloys. Materials Science and Technology 1993; 9: 513-519.

[81] Lunder O, Lein JE, Aune TK, Nisancioglu K. The role of magnesium aluminum (Mg17Al12) phase in the corrosion of magnesium alloy AZ91. Corrosion 1989; 45: 741–748.

[82] Corrosion resistance of aluminum and magnesium alloys: understanding, performance and testing. In: Ghali E, Revie W (ed.). Wiley series in corrosion. John Wiley & Sons; 2010.

[83] Song GL, Atrens A, Wu X, Zhang B. Corrosion behavior of AZ21, AZ501 and AZ91 in sodium chloride. Corrosion Science 1998; 40: 1769-1791.

[84] Song GL, StJohn D. Corrosion performance of magnesium alloys MEZ and AZ91. International Journal of Cast Metals Research 2000; 12: 327-334.

[85] Song GL, StJohn D. The effect of zirconium grain refinement on the corrosion behavior of magnesium-rare earth alloy MEZ. Journal of Light Metals 2002; 2: 1-16.

[86] Sun M, Wu G, Wang W, Ding W. Effect of Zr on the microstructure, mechanical properties and corrosion resistance of Mg-10Gd-3Y magnesium alloy. Materials Science and Engineering A 2009; 523: 145-151.

[87] Gu X, Zheng Y, Cheng Y, Zhong S, Xi T. In vitro corrosion and biocompatibility of binary magnesium alloys. Biomaterials 2009; 30: 484-498.

[88] Kim W-C, Kim J-G, Lee J-Y, Seok H-K. Influence of Ca on the corrosion properties of magnesium for biomaterials. Materials Letters 2008; 62: 4146-4148.

[89] Li Z, Gu X, Lou S, Zheng Y. The development of binary Mg-Ca alloys for use as biodegradable materials within bone. Biomaterials 2008; 29: 1329-1344.

[90] You BS, Park WE, Chung IS. The effect of calcium additions on the oxidation behavior in magnesium alloys. Scripta Materialia 2000; 42: 1089-1094.

[91] Kannan MB, Raman RKS. In vitro degradation and mechanical integrity of calcium-containing magnesium alloys in modified simulated body fluid. Biomaterials 2008; 29: 2306-2314.

[92] Zhang E, Yang L, Xu J, Chen H. Microstructure, mechanical properties and biocorrosion properties of Mg-Si(-Ca, Zn) alloy for biomedical application. Acta Biomaterialia 2010; 6: 1756-1762.

[93] Zhang S, Zhang X, Zhao C, Li J, Song Y, Xie C, Tao H, Zhang Y, He Y, Jiang Y, Bian Y. Research on an Mg-Zn alloy as a degradable biomaterial. Acta Materialia 2010; 6: 626-640.

[94] Quan Y, Chen Z, Gong X, Yu Z. CO_2 laser beam welding of dissimilar magnesium-based alloys. Materials Science and Engineering A 2008; 496: 45-51.

[95] Kattamis TZ. Lasers in Metallurgy. In: Mukherjee K. and Mazumder J. (ed.). The Metals Society of AIME. Warrendale, PA; 1981. p1–10.

[96] Ahmed S, Edyvean RGJ, Sellars CM, Jones H. Effect of addition of Mn, Ce, Nd and Si additions on rate of dissolution of splat quenched Mg-Al and Mg-Zn alloys in 39 % NaCl solution. Materials Science and Technology 1990; 6: 469-474.

[97] Südholz AD, Birbilis N, Bettles CJ, Gibson MA. Corrosion behavior of Mg-alloy AZ91E with atypical alloying additions. Journal of Alloys and Compounds 2009; 471: 109-115.

[98] Pietak A, Mahoney P, Dias GJ, Staigner MP. Bone-like matrix formation on magnesium and magnesium alloys. Journal of Materials Science: Materials in Medicine 2008; 19: 407-415.

[99] Tancret F, Osterstock F. Modelling the toughness of porous sintered glass beads with various fracture mechanisms. Philosophical Magazine 2003; 83: 125–136.

[100] Krause A, von der Höh N, Bormann D, Krause C, Bach F-W, Windhagen H, Meyer-Lindenberg A. Degradation behavior and mechanical properties of magnesium implants in rabbits tibiae. Journal of Materials Science 2010; 45: 624-632.

[101] Waizy H, Weizbauer A, Modrejewski C, Witte F, Windhagen H, Lucas A, Kieke M, Denkena B, Behrens P, Meyer-Lindenberg A, Bach F-W, Thorey F. In vitro corrosion of ZEK100 plates in Hank's Balanced Salt Solution. BioMedical Engineering OnLine 2012; 11: 12.

[102] Huehnerschulte TA, Angrisani N, Rittershaus D, Bormann D, Windhagen H, Meyer-Lindenberg A. In vivo corrosion of two novel magnesium alloys ZEK100 and AX30 and their mechanical suitability as biodegradable implants. Materials 2011; 4: 1144-1167.

[103] Fornell J, Concustell A, Suriñach S, Li WH, Cuadrado N, Gebert A, Baró MD, Sort J. Yielding and intrinsic plasticity of Ti-Zr-Ni-Cu-Be bulk metallic glass. Int. J. Plast. 2009; 25: 1540–1559.

Chapter 13

Biodegradation of Nitrogen in a Commercial Recirculating Aquaculture Facility

S. Sandu and E. Hallerman

Additional information is available at the end of the chapter

http://dx.doi.org/10.5772/55841

1. Introduction

1.1. Need for biodegradation of nitrogen species in aquaculture systems

Commercial production of fish involves high levels of feeding. While digestive breakdown of lipids and carbohydrates yields water and carbon dioxide as waste products, digestion of proteins also yields nitrogenous compounds. In teleost (i.e., bony) fishes, these nitrogenous wastes are excreted predominately as ammonia. Total ammonia-nitrogen (TAN) consists of ionized ammonia (NH_4^+-N) and un-ionized ammonia (NH_3-N), the latter of which can prove toxic to fish. The fraction of TAN in the unionized form is dependent upon the pH and temperature of the water (Losordo 1997, Lekang 2007) and to a lesser degree its salinity (Diaz et al. 2012). At pH values less than 7.5, most ammonia is in the ionized form, and high levels of TAN can be tolerated. At higher pH, however, levels of un-ionized ammonia become problematic. Hence, biodegradation of ammonia is critical for the success of fish culture. Nitrifying bacteria, including *Nitrosomonas* sp., utilize NH_3-N as the energy source for growth, producing nitrite, NO_2-N. While nitrite-nitrogen is not as toxic as un-ionized ammonia-nitrogen, it can prove harmful to fish. The most common mode of toxicity is anoxia, as nitrite-nitrogen crosses the gills into the circulatory system and converts hemoglobin to methemoglobin, rendering it unable to bind and transport oxygen to the tissues (Palachek and Tomasso 1984, Svobodova et al. 2005). Other nitrifying bacteria, including *Nitrobacter* sp., utilize nitrite as their energy source, producing nitrate, NO_3-N. Nitrate-nitrogen concentrations are not generally of concern to aquaculturists, as most species can tolerate levels as high as 200 mg/L (Russo and Thurston 1991). Nitrate rarely reaches such high levels, as it is removed from the system by water exchanges and by passive denitrification in anaerobic pockets within the production or filtration systems (van Rijn 1996, Tal et al. 2006) or in denitrification reactors (Hamlin et al. 2008, Sandu et al. 2011).

Controlled degradation of nitrogenous wastes in filtration units is a major consideration in design and operation of commercial recirculating aquaculture systems. Among the technologies available (Crab et al. 2007), biological filtration is most commonly used. Biological filters are designed to provide abundant surface area for the attachment of complex microbial communities (Schreier et al. 2010) rich in *Nitrosomonas* and *Nitrobacter* species (Chen et al. 2006, Itoi et al. 2007, van Kessel et al. 2010). The nitrification capacity of the water treatment system is often the factor that limits production in a recirculating aquaculture system (Lemarie et al. 2004, Eschar et al. 2006, Diaz et al. 2012).

1.2. Utility of a nitrogen budget

The production efficiency of an aquaculture system can be evaluated through analysis of the conversion of nitrogen to fish biomass and to biodegradation pathways (Thoman et al. 2001). Nitrogen dynamics can be quantified by a mass balance equation, most simply as the difference between nitrogen in the feed supply and nutrients subsequently fixed as fish biomass. A nitrogen budget can quantify nitrogen fixation in fish biomass at various fish stocking densities (Suresh and Kwei 1992; Siddiqui and Al-Harbi 1999), nutrient release into the water column as dissolved and particulate excretion of fish (Krom and Neori 1989), and deposition of nitrogen into pond sediment (Acosta-Nassar et al. 1994). By estimating total nitrogen budgets for a particular species and culture system, we can evaluate the efficacy of water treatment processes (Porter et al. 1987). Hence, a nitrogen budget provides information crucial for the design and optimization of a production system, feeding strategies, and water and effluent treatment processes.

1.3. Biodegradation of nitrogenous wastes in a tilapia production system

Blue Ridge Aquaculture (BRA) in Martinsville, Virginia, USA is a large commercial facility that produces 1300 metric tons of hybrid tilapia *Oreochromis sp.* per year. To our knowledge, it is the largest recirculating aquaculture facility in existence. Before our study, little information existed about nitrogen budgets in commercial-scale fish production facilities, especially those using freshwater recirculating systems. By deriving a nitrogen budget, we can quantify the forms and proportions of nitrogen ingested as food as it becomes bound in tilapia biomass, excreted as metabolites, biodegraded by microorganisms, lost as gas by denitrification, or released in effluent. Knowledge of the nitrogen budget can help optimize operations, improving facility efficiency and maximizing production. Using Blue Ridge Aquaculture as our study system, our objectives were to: (1) examine nitrogen dynamics for the grow-out systems, (2) relate the nitrogen budget to water quality, (3) evaluate biofilter loading and nitrogen removal efficiency, and (4) predict maximum system carrying capacity. All abbreviations used in this chapter are shown in Table 1.

a = mole fraction of unionized ammonia nitrogen (decimal fraction)
ACR = areal conversion rate (mg/m^2-d)
$A_{NH3\text{-}N}$ = concentration of unionized ammonia nitrogen (mg/L)

http://dx.doi.org/10.5772/55841

A_{TAN} = maximum allowable concentration of total ammonia nitrogen (mg/L)

BRA = Blue Ridge Aquaculture

C_{TAN} = total ammonia nitrogen concentration in fish tank (mg/L)

C_{TANe}= total ammonia nitrogen concentration in the effluent from filters (mg/L)

C_{TANi} = total ammonia nitrogen concentration in the supply water (mg/L)

E_a = efficiency of rotating biological contactor for removal of ammonia nitrogen (percent)

FA = amount of feed (kg)

FB = fish biomass (kg)

FC = feed conversion factor (decimal fraction)

FP = protein content of feed (decimal fraction)

FR_{MTAN} = maximum feeding rate (kg/d)

LC50 = lethal concentration of a compound to 50% of the individuals in a population

LN = nitrogen load (g N/kg fish produced)

L_{TAN} = ammonia loading (g/hr)

N_{DENIT} = nitrogen gas removed by denitrification (mg/L)

N_{DIN} = dissolved inorganic nitrogen (mg/L)

N_{feed} = nitrogen fixed in feed (g/kg feed)

N_{fish} = nitrogen fixed in fish (g/kg fish produced)

N_{mort} = nitrogen fixed in dead fish (g/kg fish removed)

N_{NO2^-} = nitrite nitrogen (mg/L)

N_{NO3^-} = nitrate nitrogen (mg/L)

N_{NH3vol} = nitrogen removed by ammonia volatilization (mg/L)

$NO_3\text{-}N_{pass}$ = nitrate removed passively by denitrification (mg/L)

$NO_3\text{-}N_{exch}$ = nitrate removed by exchange of water (mg/L)

N_{TAN} = total ammonia nitrogen (mg/L)

N_{TON} = total organic nitrogen (mg/L)

PC = protein content of feed (decimal fraction)

$P_{NO3\text{-}N}$ = partitioning of nitrate nitrogen (g/kg)

P_{TAN} = production rate of ammonia nitrogen (g/kg)

Q = flow rate through system (m^3/min or L/h)

Q_f = recirculation flow rate (m^3/min or L/h)

RAS = recirculating aquaculture system

R_{TAN} = ammonia removal rate (g/h)

S = surface area (m^2)

SBM_{MTAN} = maximum biomass that could be sustained by system (kg fish)
$TAN_{exchange}$ = ammonia removed by water exchange (mg/L)
$TAN_{pass+vol}$ = ammonia removed by passive nitrification and ammonia volatilization (mg/L)
$TAN_{RBC\ nitrification}$ = ammonia removed by nitrification in rotating biological contactor (mass/volume)
t = time
TKN = total Kjeldall nitrogen (g)
TNI = total nitrogen input (kg/day)
TNR = total nitrogen recovered (kg/day)
TNUA = total nitrogen unaccounted for (kg/day)

Table 1. Abbreviations and associated units.

Tilapias are a group of fishes of great importance to world aquaculture (Costa-Pierce and Rakocy 1997, Fitzsimmons 1997, Lim and Webster 2006). Tilapias adapt readily to a range of production systems ranging from traditional extensive pond systems to high-input intensive pond systems to super-intensive recirculating aquaculture systems. Like all fishes, tilapias are sensitive to concentrations of nitrogenous wastes. The 48-hour LC_{50} value for NH_3 for Jordan tilapia *Oreochromis aureus* was 2.40 mg/L (Redner and Stickney 1979). The 48-hour LC_{50} value for hybrid red tilapia *O. mossambicus x O. niloticus* fry was 6.6 mg/L (Daud et al. 1988), although the threshold lethal concentration was 0.24 mg/L. The 24-hour LC_{50} value for un-ionized ammonia for *O. niloticus* was 1.46 mg/L (Evans et al. 2006) Sublethal effects of NH_3-N include tissue damage, decreased growth, increased feed conversion ratio, acute stress response, increased disease susceptibility, and reduced reproductive capacity (Russo and Thurston 1991, Yildiz et al. 2006, El-Sherif and El-Feky 2008, Benli et al. 2008). Tilapias also exhibit sensitivity to elevated nitrite concentrations. The 96-hour LC_{50} for nitrite-nitrogen for *O. aureus* was 16.2 mg/L at pH 7.2 and 22 mg/L chloride (Palachek and Tomasso 1984). Acute nitrite toxicity for *O. niloticus* varied with chloride levels and with fish size, with smaller fish proving more tolerant (Atwood et al. 2001, Wang et al. 2006). Nitrite-nitrogen levels should be kept below 5 mg/L within tilapia culture vessels (Losordo 1997). Knowledge of these toxicity values is useful for setting criteria for the design or evaluating the performance of filters for biodegradation of nitrogenous wastes in aquaculture systems.

2. Methods

2.1. Culture systems

The BRA facility includes systems for broodstock holding, fish breeding, egg incubation/ hatching, fingerling rearing, and food-fish production. The main building houses 42 recirculating aquaculture systems for grow-out to market size (Figure 1) that were the focus of our study. Each grow-out system (Figure 2) includes a fish production tank, a sedimentation basin

http://dx.doi.org/10.5772/55841

for solids removal, a rotating biological contactor (RBC) for microbial biodegradation including nitrification, and an oxygenation unit. Each fish production system is rectangular in shape, built from concrete, holds 215 m^3 of water, and consists of a fish-rearing tank (119 m^3), a multi-tube clarifier sedimentation basin (37 m^3), an air-driven rotating biological contactor (59 m^3 basin volume, 13,366 m^2 surface area per shaft), and an underground U-tube oxygenation system. The total volume of the grow-out unit is 9030 m^3. The water surface is at the same level in the fish tank, sedimentation and RBC compartments, and water passes freely from one section into another through large pipes or apertures. A pump receives water from the rotating biological contactor compartment and pushes it through U-tubes and then to the far end of the fish production tank, driving the recirculation. The filtration rate is 3.8 m^3/min, and the system turnover time is about once per hour.

Figure 1. Commercial-scale tilapia grow-out systems at Blue Ridge Aquaculture. The grow-out units are to the right of the catwalk and sedimentation basins to the left. Photograph courtesy of Blue Ridge Aquaculture.

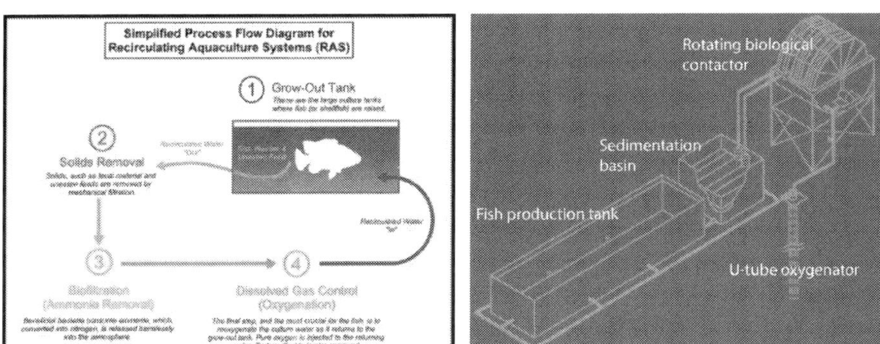

(A) (B)

Figure 2. A) Conceptual diagram and (B) and engineering drawing of a single recirculating tilapia grow-out system at Blue Ridge Aquaculture. Diagram courtesy of Blue Ridge Aquaculture.

BRA practices partial water exchange daily for controlling solids, dissolved organics and nutrient accumulation in fish grow-out tanks. Water is exchanged daily from the system in the interval between 2:00 p.m. and 8:00 a.m. Management practice is to completely flush the sedimentation basin after each instance that 227 kilograms of feed has been administered to a particular production unit. The exchange rate averages 22.3% per day, but the daily percentage varies among production units as a function of the size of fish, water quality requirements, and the amount of feed delivered to the system. The exchange water originates from wells, and is supplemented with municipal tap water when necessary. Exchange water replaces that used to remove settled particulate material, and thereby dilutes dissolved organic materials, dissolved nutrients, and salts.

Fish are fed commercially-prepared pelleted diets containing 36 or 40% minimum crude protein and 8-16% lipid levels, varying with the age of the fish. The feed is distributed hourly to the tanks over the 24-hour period. Fish production is managed so that 21-27 metric tons of 600-g fish reaches marketable size each week for shipment to a live market.

2.2. System boundaries

For the purpose of this study, the 42 recirculating aquaculture systems for grow-out were delimited as a unique system for purposes of quantifying the nitrogen budget. In certain contexts as set out below, N dynamics were quantified in greater detail in four individual systems. Broodstock holding and spawning facilities, a hatchery, and two greenhouses for fingerling production contain only a small part of the facility fish biomass, volume and exchange flow (i.e., they handle 3.0% of the fish biomass and 4.4% of the total nitrogen input). Because of their small contribution, the nitrogen budgets for these systems are not presented here, but can be found in Sandu (2004).

2.3. Inputs, outputs and nitrogen pools

The nitrogen budget is expressed as a mass-balance equation of all nitrogen forms, with total inputs plus generation equal to total outputs plus consumption. We found no measurable amounts of dissolved inorganic nitrogen in the replacement water. Hence, feed provided to fish was the sole nitrogen source in the form of organic nitrogen (N_{feed}). Multiplication of N_{feed} by the total amount of feed provided the mass of total nitrogen input (*TNI*). The nitrogen budget was accounted for in five known pools:

1. Nitrogen fixed in fish biomass as organic nitrogen, N_{fish},
2. Nitrogen fixed in dead fish biomass as organic nitrogen, N_{mort},
3. Dissolved inorganic nitrogen, N_{DIN}, which included N_{TAN}, N_{NO2}, and N_{NO3},
4. Total organic nitrogen in effluent, N_{TON}, and
5. Nitrogen gas removed from the system by passive denitrification, N_{denit}, and by ammonia volatilization, $N_{NH3\ vol}$.

All transformations among pools were assumed to be in a dynamic equilibrium over a defined period of time. We accounted for the mass fractions of nitrogen from Pools 1 to 4 (i.e., the measurable pools) as total nitrogen recovered (*TNR*), while the difference between total nitrogen input and total nitrogen recovered constituted pool 5, the mass fraction of total nitrogen unaccounted for (*TNUA*).

2.4. Analytical techniques

Analyses of fish and of feed for protein content followed Thiex et al. (2002), who indicated that by dry weight, 16% of protein is nitrogen. Samples were processed at the Forage Testing Laboratory, Virginia Polytechnic Institute and State University, Blacksburg, Virginia. Analyses for inorganic dissolved nitrogen forms (TAN, NO_2^--N, and NO_3^--N) were conducted on site using a Hach DR2400 spectrophotometer (Hach Company, Loveland, Colorado). Total Kjeldall nitrogen (TKN) was determined using macro-Kjeldall Standard Method 4500 – N_{org} B (APHA et al., 1998). Samples were acidified below pH 2 using H_2SO_4, refrigerated with ice, and transported to the Department of Civil and Environmental Engineering at Virginia Polytechnic Institute and State University, Blacksburg, Virginia, for analysis. Temperature and pH were measured directly on site using an Acorn Meter (Kit Model pH 6, Oakton, Vernon Hills, Illinois). Alkalinity was determined on-site using the Hach Permachem® Method. Dissolved oxygen (DO) was measured using a YSI (Model 550, Yellow Springs, Ohio) instrument. We calculated total organic nitrogen as the difference between TKN and total ammonia nitrogen (TAN).

2.5. Nitrogen budget determination

Under steady-state conditions, fish biomass does not fluctuate significantly over time (i.e., harvest equals growth), and the daily rations of feed are constant. Under these assumptions, we derived the nitrogen budget by determining the nitrogen input with feed and the output of nitrogenous compounds in known pools. We quantified daily amounts of nitrogen in feed, fish, and mortalities using information on feed consumption, fish production, and mortalities provided by BRA management. We measured the components of dissolved inorganic nitrogen and total organic nitrogen pools directly. We extrapolated mean values to the entire exchange volume from a day to determine the mass of nitrogen recovered in these forms. We assumed that the amount of nitrogen missing from the balance was lost by passive denitrification and by ammonia volatilization.

We considered both types of feed used in the system (with 36% or 40% standard protein content) to determine nitrogen fixed in feed, N_{feed}. We collected samples from three different points in storage silos for nitrogen content determination. We calculated N_{feed} as a composite using the equation:

$$N_{feed} = \Sigma\left(FA\text{x}PC\text{x}0.16\right) \tag{1}$$

where *FA* = amount of feed, *PC* = protein content of the feed, and 0.16 = concentration of nitrogen in protein (Thiex et al. 2002). We determined *PC* by laboratory analyses because protein content may differ from that claimed by the feed producer. We obtained the total mass of nitrogen originating from the feed input, *TNI*, by multiplying N_{feed} by the amount fed, *FA*.

To determine fixation of nitrogen in fish, N_{fish}, we analyzed protein content in triplicate samples of muscle tissue from fish from three size-classes. We estimated the proportions of fish in each size-class as 5% juveniles (i.e., newly introduced to the system from the hatchery), 60% intermediate, and 35% marketable size. With data on protein content of each fish size-class, we determined N_{fish} as a composite using the equation:

$$N_{\text{fish}} = \Sigma\left(FBxFPx0.16\right) \tag{2}$$

where *FB* = biomass of fish, and *FP* = protein content of the fish.

About 3.5% of the fish production (by number) was lost as mortalities. We assumed that nitrogen fixed in dead fish, N_{mort}, had the same nitrogen content as N_{fish}. In order to determine the biomass of N_{mort}, we collected mortalities daily from the production system for a two-week period, sorted them by size, and weighed them. We used these data to determine N_{mort} using equation 2.

Nitrogen load, L_N, entered the water column as ammonia and as organic nitrogen bound in feces. We quantified L_N as all nitrogen from feed that was not accounted for as living or dead fish as using the equation:

$$L_N = \left[N_{\text{feed}} - \left(N_{fish} + N_{\text{mort}}\right)\right] \ / \ FB \tag{3}$$

Hence, L_N quantified the amount of nitrogen that sustained the nitrogen cycle throughout the system, supplying all effluent nitrogen pools.

We quantified total organic nitrogen as the difference between TKN and TAN from the effluent. We obtained values for TKN, TAN, NO_2^--N, and NO_3^--N by analyzing seven samples collected from the effluent discharge pipe at 3-hour intervals between 2:00 p.m. and 8:00 a.m. because effluent originated from the production system only during that interval. We repeated the tests twice (on different days) and averaged the results. We estimated daily production of these nitrogen forms by multiplying the average concentration (mg/L) by the volume of wastewater released from the system during a one-day period.

All nitrogen in feed that was not recovered as living or dead fish or as total organic nitrogen represented the dissolved inorganic fraction that entered the water as TAN. Hence, we determined ammonia production as:

$$P_{TAN} = N_{\text{feed}} - \left(N_{\text{fish}} + N_{\text{mort}} + N_{TON}\right) \ / \ FA \tag{4}$$

The sum of TAN, NO_2^--N, and NO_3^--N found in the effluent represented the fraction of nitrogen recovered as dissolved inorganic nitrogen, N_{DIN}. The summation of N_{DIN}, N_{fish}, N_{mort}, and N_{TON} provided the value for total nitrogen recovered, *TNR*. We determined total nitrogen unaccounted for, *TNUA*, by subtracting total nitrogen recovered, *TNR*, from total nitrogen input, *TNI*.

2.6. RAS carrying capacity, RBC design, TAN and NO_3^--N removal

We used a simplified version of a model proposed by Losordo and Westers (1994) to determine the carrying capacity of the production system; that is, we considered only the parts of the model concerning maximum system carrying capacity with respect to TAN. Modeling of the flow rate through biofilters was unnecessary because the flow rate was fixed among all recirculating aquaculture systems at 3.78 m^3/min.

Four recirculating aquaculture systems chosen for intensive study held different age-groups of fish from juvenile to marketable size in order to represent the overall population in the facility. We knew total fish biomass, fish size, feeding rate, crude protein content of feed, daily percent body weight fed, flow rate through the system, and daily rate of exchange for each selected system. We measured other parameters, such as TAN, NO_3^--N, NO_2^--N, pH, temperature, and dissolved oxygen, using standard methods (APHA et al., 1998). We performed these analyses on composite samples collected from the fish-rearing tanks or from the rotating biological contactor's influent and effluent at four-hour intervals. By sampling from appropriate locations, we determined the effects of fish tanks, biofilters or sedimentation basins on each parameter. The experiments extended between consecutive water exchanges. We scaled the data to 24-hour intervals and determined mean and variance for each water quality parameter.

We determined the maximum system carrying capacity with respect to TAN as follows. We calculated the maximum allowable TAN concentration, $A_{TAN,}$ as:

$$A_{TAN} = A_{NH3-N} / a \tag{5}$$

where a = the mole fraction of unionized ammonia nitrogen as determined by pH and temperature (Huguenin and Colt, 1989). We calculated maximum feed rate, FR_{mTAN}, by assuming that the TAN concentration of a fish tank equals A_{TAN}, as:

$$FR_{mTAN} = \left[A_{TAN} x Q_f x E_a + Q\left(C_{TAN} - C_{TANi}\right)\right] \ / \ \left(0.092 x PC\right) \tag{6}$$

where Q_f = the recirculating flow rate, or flow rate to the RBC, known to be 227,100 L/hr, and 0.092 = model constant coefficient. We determined the efficiency of the rotating biological contactor for removal of ammonia nitrogen, E_a as:

$$E_a = \left[\left(C_{TAN} - C_{TANe}\right) / C_{TAN}\right] \text{ x } 100 \tag{7}$$

We estimated the maximum biomass that could be sustained within the system, SBM_{mTAN} as:

$$SBM_{mTAN} = FR_{mTAN} / \%BW \tag{8}$$

where $\%BW$ = the feeding rate, expressed as a percent of body weight per day.

P_{TAN} is the rate of production of TAN in the system by metabolism of fish and microbial degradation of uneaten feed. We estimated P_{TAN} as a function of the feed rate and the percentage of protein in feed:

$$P_{TAN} = \left(FA * PC * 0.102\right) / t \tag{9}$$

where t = the period of time from the onset of feeding to the next feeding.

This equation is based on the following assumptions and empirical estimates:

a. 16% of feed protein is nitrogen,

b. 80% of the nitrogen is assimilated,

c. unassimilated nitrogen in fecal matter is removed rapidly from the tank,

d. 80% of assimilated nitrogen is excreted, and

e. all of the TAN is excreted during t hours.

The coefficient 0.102 represents the product of values suggested by assumptions *a* through *d* (i.e., 0.16 x 0.8 x 0.8 = 0.102).

We determined the mass flow rate of TAN to a rotating biological contactor, or ammonia loading, L_{TAN}, from known (Q_f) and experimentally determined (C_{TANf}) parameters as:

$$L_{TAN} = Q_f \text{x} C_{TANf} \tag{10}$$

We determined the ammonia removal rate, R_{TAN}, as:

$$R_{TAN} = \left(C_{TANf} - C_{TANe}\right) \text{x} Q_f \tag{11}$$

The fraction (R_{TAN} x 100) / P_{TAN} represents the percentage of TAN that was removed by means other than the rotating biological contactor.

We estimated the nitrification performance of a rotating biological contactor as areal conversion rate, *ACR*, representing the amount of TAN oxidized by a unit of surface area in 24 hours:

$$ACR = R_{TAN} / S \tag{12}$$

http://dx.doi.org/10.5772/55841

where S, the surface area of an RBC, was 13,336 m^2.

The mass balance quantifying the partitioning of P_{TAN} removal was:

$$P_{TAN} = TAN_{\text{pass + vol}} + TAN_{\text{RBC nitrification.}} + TAN_{exchange} \quad (13)$$

We used a similar approach to determine NO_3^--N partitioning using the equation:

$$P_{NO3^- -N} = NO_3^- - N_{\text{pass.}} + NO_3^- - N_{\text{exch}} \quad (14)$$

2.7. Statistical analysis

We used linear regressions to determine the relationship between daily TAN production (P_{TAN}) and TAN removal efficiency per pass (E_a), and between fish biomass and percent P_{TAN} transformed by passive denitrification in the four systems tested.

3. Results

3.1. Nitrogen budget

We derived the nitrogen budget for the entire production system for mean conditions of 28.4ºC, pH 7.14, and alkalinity 119.0 mg/L as $CaCO_3$. For annual production of 1300 metric tons of fish biomass, BRA administers 2210 metric tons of feeds. These amounts correspond to 6054.8 kg feed consumed per day and 3561.6 kg fish weight gain per day. Of the feed utilized, 95% (5752.0 kg) was nominally 36% protein and 5% (302.8 kg) 40% protein content. However, laboratory analyses showed that the actual protein contents of the two feeds were somewhat lower, 35.0±0.2% and 39.8±0.2%, respectively. The estimated percentages of feed types and the laboratory-determined protein concentrations were used to determine the nitrogen fixed in feed, N_{feed} = 56.38 g/kg feed. By extrapolating N_{feed}, we determined a total nitrogen input of TNI = 341.381 kg/day.

Laboratory analyses showed that the three size-classes of fish from small to large had 18.04±0.16, 20.75±0.02 and 22.26±0.74% protein content, respectively. From these data, we determined that the nitrogen fixed in fish was N_{fish} = 33.83 g/kg produced. Extrapolating to the daily biomass of fish produced, the total nitrogen assimilated in fish was 120.49 kg/day.

Loss of fish represented 3.5% of the total production by number, with weighing of dead fish indicating losses of 2, 1, and 0.5% from the respective size-classes. This was the equivalent of 30.6 kg fish/day or 1.03 kg total N_{mort}/day, representing 0.86% of the total nitrogen assimilated. Hence, 35.3% of nitrogen from feed was assimilated in fish flesh (34.4% harvested and 0.86% removed with mortalities), and 64.7% was unassimilated or excreted in different forms. This latter term included nitrogen in uneaten feed that we accounted for in the overall budget as

N_{TON}. The nitrogen excreted, L_N, was 62.0 g/kg fish produced. Subsequently, the cumulative daily nitrogen loading for the entire system, L_N, was 221.3 kg.

Analyses of the effluent wastewater (estimated at 2017 m^3/day) indicated, on average, 2.88 mg/L TAN, 1.09 mg/L NO_2^--N, 49.3 mg/L NO_3^--N, and 32.05 mg/L TON. Extrapolated to the entire effluent volume, the overall flows were 5.8 kg N_{TAN}/day (representing 1.70% of total nitrogen input, *TNI*), 2.2 kg N_{NO2}/day (0.64% *TNI*), 99.4 kg N_{NO3}/day (29.1% *TNI*), and 64.6 kg N_{TON}/day (18.9% *TNI*). Determination of total organic nitrogen, N_{TON}, allowed estimation of P_{TAN} = 25.81 g/kg feed. The recovered fraction of dissolved inorganic nitrogen, N_{DIN}, resulted from the summation:

$$1.70\% N_{TAN} + 0.64\% N_{NO2} + 29.13\% N_{NO3} = 31.47\%$$

Total nitrogen recovered, *TNR*, was determined as a percentage of *TNI* as:

$$85.69\% TNR = 34.43\% N_{\text{fish}} + 0.86\% N_{\text{mort}} + 1.70\% N_{TAN} +$$
$$+ 0.64\% N_{NO2} + 29.13\% N_{NO3} + 18.93\% N_{TON}$$

We then estimated total nitrogen unaccounted for, *TNUA*, as 14.3% of *TNI*. Hence, the subsequent nitrogen mass balance for the production system was:

$$341.381 \text{ kg} TNI / \text{day} = 292.529 \text{ kg} TNR / \text{day} + 48.852 \text{ kg} TNUA / \text{day}$$

Table 2 summarizes the daily nitrogen budget for the production system. The relatively low value of total nitrogen unaccounted for, *TNUA*, was presumably due to nitrogen lost as nitrogen gas produced by denitrification and as ammonia lost to volatilization. Passive denitrification was likely the primary pathway because recirculated fish culture water passed through the sedimentation basin numerous times. As discussed below, the sediment blanket and associated thick biofilm in the multi-tube clarifier created anoxic conditions favorable for microbially mediated denitrification.

Units	Nitrogen pool							
	TNI	H_{fish}	N_{mort}	N_{TAN}	N_{NO2}	N_{NO3}	N_{TON}	*TNUA*
Kg	341.38	119.45	1.03	5.81	2.20	99.44	64.65	48.85
%	100.00	34.99	0.30	1.70	0.64	29.12	18.94	14.31

Table 2. Daily nitrogen budget for the grow-out system at Blue Ridge Aquaculture.

3.2. Carrying capacity, RBC design, TAN and NO_3^--N removal

The carrying capacity model indicated that recirculating aquaculture systems at Blue Ridge Aquaculture could support biodegradation of up to 3.15 mg TAN/L. This value corresponds to 0.025 mg/L maximum allowable unionized ammonia (A_{TAN}) at conditions of pH 7.0 and temperature of 30ºC (Huguenin and Colt 1989); our average values of these parameters for the four recirculating aquaculture systems monitored in greater detail were pH 7.09 and 27.8ºC. At 0.025 mg/L TAN, a recirculating system should be able to receive a maximum feeding rate of $FR_{max\ TAN}$ = 269.8 kg feed/day, which would support a fish biomass of $SBM_{max\ TAN}$ = 10,287.4 kg fish/system. Estimates of these parameters for each selected RAS are presented in Table 3. Comparison with actual feeding rates at the time of experiment (Table 4) showed that system loadings were 56.7 - 91.5% of the maxima estimated (Table 3, Figure 3). Over the four tanks examined in detail, TAN removal efficiency per pass, E_a, averaged 54.4%. We determined the rate of TAN production (P_{TAN}, Table 3). We determined P_{TAN} per kg of feed consumed by dividing these values by the daily amount of feed introduced into a system: i.e., 40.6 g/kg feed for feed with 40% crude protein content, and 36.7 g/kg feed for feed with 36% crude protein content. We found a positive, linear relationship between P_{TAN} (which also was proportional to the feeding rate) and E_a (slope = 0.0013, r^2 = 0.72), thereby showing that the rotating biological contactors efficiently removed various loadings of ammonia. None of the RBCs tested were working at maximum capacity.

Parameter	Units	RAS Tested				
		A12	A11	B16	A18	Average
Maximum feed rate (FR_{maxTAN})	kg/day	240.4	286.1	261.6	290.9	269.8
Maximum system biomass (SBM_{maxTAN})	kg	4202.5	11443.0	9871.0	15633.0	10287.4
Actual BW as % of SBM_{maxTAN}	%	56.66	91.52	66.54	76.77	72.87
TAN tank concentration	mg/L	1.77	2.32	2.04	2.10	2.06
TAN conc. in RBC influent ($CTAN_i$)	mg/L	1.77	2.32	2.04	2.10	2.06
TAN conc. in RBC effluent ($CTAN_e$)	mg/L	0.84	1.01	0.99	0.90	0.94
TAN removal efficiency per pass (E_a)	%	52.39	56.47	51.47	57.28	54.40
P_{TAN}/kg feed	g	40.6	36.7	36.7	36.7	37.7
Daily TAN production (P_{TAN})	g/day	5522.4	9626.4	6397.9	8161.9	7427.1
Ammonia loading (L_{TAN})	g/hr	402.19	526.87	463.28	478.50	467.71
Ammonia removal rate (R_{TAN})	g/hr	210.75	297.50	238.46	274.11	255.20
Areal conversion rate (ACR)	mg TAN/m^2-d	378.4	534.2	428.2	429.2	442.5
ACR at SBW_{maxTAN}	mg TAN/m^2-d	667.8	583.7	643.4	641.1	634
Mass TAN introduced by exchange	g/day	39.47	73.10	50.79	90.93	63.57
P_{TAN} introduced with water exchange	%/day	0.71	0.76	0.79	1.11	0.84
*Total TAN removed by water exchange	%/day	0.08	0.34	0.22	0.35	0.25

*Daily TAN percentage removal by water exchange, assuming that exchange water is treated using the treatment train tested by Sandu (2004) with 1.60 mg/L TAN.

Table 3. Experimentally determined and predicted parameters for estimation of maximum system carrying capacity with regard to TAN for tested units.

Parameter	Units	RAS Tested				
		A12	A11	B16	A18	Average
Water exchange rate	% volume/day	11.5	21.3	14.8	18.4	16.5
Flow rate through system (*Q*)	L/hr	1028.0	1903.7	1322.7	1645.8	1475.1
Fish size	g/fish	43	192	245	424	226
Fish biomass	kg	2381.0	10473.0	6568.5	12002.0	7856.1
Feeding rate (*FR*)	kg/day	136.0	262.0	174.0	222.3	198.6
Feed protein content (*FP*)	%	40	36	36	36	37
Percent body weight fed	kg feed/kg fish-d	5.72	2.50	2.65	1.85	3.18

Table 4. Characteristics of the recirculating aquaculture systems selected for evaluation.

Figure 3. Towards the end of a tilapia production cycle, stocking densities approach system carrying capacity. Photograph courtesy of Blue Ridge Aquaculture.

The mass flow-rate of TAN to a rotating biological contactor, L_{TAN}, averaged 467.7 g/hr, which was removed at an average rate of R_{TAN}= 255.2 g/hr. Per-system values are presented in Table 3. The ratio of R_{TAN} to P_{TAN} showed that rotating biological contactors removed an average of 84.0% of total ammonia nitrogen from the selected systems. From the difference, 1.1% of TAN was recovered from exchanged water and 15.0% remained unaccounted for, probably transformed to NO_2^--N and NO_3^--N by passive nitrification or lost by volatilization of ammonia. Data in Table 3 show that fish biomass in the system was positively correlated with the percentage of total ammonia nitrogen transformed by passive nitrification (slope = 0.0015, r^2 = 0.69); although the correlation was not strong, it shows that systems with higher biomass had lower water quality and larger microbial populations, including nitrifiers that promoted in-situ biodegradation of ammonia.

The rotating biological contactors removed between 378.4 and 534.2 mg TAN/m^2/day (442.5 mg TAN/m^2/day on average, Table 3). The areal conversion rate, *ACR*, increased with the loading of total ammonia nitrogen. Average *ACR* under conditions of maximum system biomass was estimated at 634.0 mg TAN/m^2/day. We note that the difference between existing *ACR* and predicted maximum *ACR* is consistent with that between the existing fish biomass and predicted maximum fish biomass.

We derived a daily nitrogen budget partitioning the total ammonia nitrogen removal from each RAS (Table 5). On average among systems, 84.0% of TAN was removed by rotating biological contactors, 14.9% by passive nitrification and ammonia volatilization, and only 1.1% was removed by periodic water exchange.

System	[1]P_{TAN}		[2]$TAN_{pass+vol}$		[3]$TAN_{RBC\ nitrification}$		[4]$TAN_{exchange}$	
	g	%	g	%	g	%	g	%
A12	5522.4	100	421.30	7.63	5057.41	91.58	43.69	0.79
A11	9626.4	100	2380.50	24.73	7139.90	74.17	106.00	1.10
B16	6397.9	100	610.22	9.54	5722.92	89.45	64.76	1.01
A18	8161.9	100	1463.66	17.93	6578.49	80.60	119.75	1.47
Average	7427.2	100	1108.51	14.93	6235.09	83.95	83.55	1.12

[1] TAN production over a 24-hour period.

[2] TAN removed by passive nitrification and by ammonia volatilization.

[3] TAN removed by nitrification in RBC.

[4] TAN removed with exchanged water.

Table 5. Partitioning of total ammonia nitrogen removal for each recirculating aquaculture system studied.

We conducted tests on the same recirculating aquaculture systems to determine the fate of NO_3^--N following its production by nitrification. We regarded $P_{NO3^- -N}$ as approximately equal to P_{TAN} by assuming that TAN lost from the systems by water exchange and volatilization was negligible. Data on $P_{NO3^- -N}$, water exchange rates, and NO_3^--N concentrations before and after water exchange allowed us to determine the total mass of NO_3^--N in the systems at these times and the amounts of NO_3^--N lost by water exchange and passive denitrification. That is, we derived a daily mass balance quantifying $P_{NO3^- -N}$ removal pathways from each RAS (Table 6). Results indicated that NO_3^--N accumulation was in the range of 9.1 – 17.2 mg/L in each RAS over a 24-hour period. On average, 44.1% of NO_3^--N was removed by water exchange, and the difference of 55.9% was removed by passive denitrification. NO_3^--N in effluent could be subject to microbial denitrification if water reuse is implemented (Sandu et al. 2008).

Parameter	Units	RAS Tested				
		A12	A11	B16	A18	Average
Daily NO_3^--N production ($P_{NO3^- -N}$)	g	5522.4	9626.4	6397.9	8161.9	7427.9
NO_3^--N conc. before exchange	mg/L	57.3	57.3	50.9	49.1	53.6
System mass NO_3^--N before exchange	g	12290.85	12290.85	10918.05	10531.95	11507.92
NO_3^--N conc. after exchange	mg/L	40.5	40.1	38.9	40.0	39.9
System mass NO_3^--N after exchange	g	8687.25	8601.45	8344.05	7872.15	8376.22
NO_3^--N and removed by exchange	g/day	3603.6	3689.4	2574.0	2659.8	3132.45

Parameter	Units	RAS Tested				
		A12	A11	B16	A18	Average
$P_{NO3^{-}-N}$ removed by exchange	%/day	65.25	38.33	40.23	32.58	44.10
NO_3^{-}-N lost by passive denitrification	g/day	1918.8	5937.0	3823.9	5502.1	4295.45
$P_{NO3^{-}-N}$ lost by passive denitrification	%/day	34.75	61.67	59.77	67.42	55.90

Table 6. Dynamics and partitioning of $P_{NO3^{-}-N}$ removal for each recirculating aquaculture system studied.

NO_2^{-}-N always remained at concentrations lower than 0.3 mg/L in the fish tanks. Its concentration increased slightly as water passed through the sedimentation basin, but decreased again to concentrations lower that those in fish tanks after contact with the RBC, creating an equilibrium concentration. Because NO_2^{-}-N concentrations were generally stable and below levels considered a threat to fish, we pursued no further determination of NO_2^{-}-N dynamics.

4. Discussion

We quantified nitrogen fixation and biodegradation through the recirculating tilapia production system at Blue Ridge Aquaculture, a large commercial production facility. The 34.4% of total nitrogen input assimilated by the fish indicated excellent nitrogen utilization relative to other production systems. For example, Suresh and Kwei (1992) found that less than 20% of nitrogen was assimilated by tilapia using feed with 22% crude protein content and much lower fish stocking densities than those at BRA. Using feed with 34% crude protein content, Siddiqui and Al-Harbi (1999) reported 21.4% nitrogen assimilation by red tilapia. Although Suresh and Kwei (1992) found decreasing nitrogen assimilation with increasing fish density, Refstie (1977), Rakcocy and Allison (1981), and Vijayan and Leatherland (1988) reported the opposite finding. We attribute the high nitrogen assimilation in our study to higher protein content in feeds used at BRA, well-managed water quality, and to production of selectively bred fish (Hallerman 2000). Also, most earlier studies reported higher mortality rates, diminishing total nitrogen accumulated in fish.

The small amounts of nitrogen recovered as TAN and NO_2^{-}-N likely were due to biodegradation in rotating biological contactors, which oxidized them effectively to NO_3^{-}-N. Most of the nitrogen recovered as total organic nitrogen (18.93%) was probably due to feces, noting that feed was consumed by fish almost instantly at distribution, and that only fine particulates could escape as wasted feed. Assuming that some organic nitrogen in feces dissolved upon contact with water, our results with tilapia, which accounted for nitrogen from the entire organic pool, broadly agree with those of Porter et al. (1987, who found 10% fecal nitrogen) and Thoman et al. (2001, who recovered 14% nitrogen from suspended solids) for other species.

For total nitrogen unaccounted for (14.31%), removal of N_2 gas through passive denitrification is the most reasonable explanation. Although denitrification may seem surprising given the relatively high dissolved oxygen in the recirculating systems, development of anoxic microsites in sediment provides likely sites for denitrification (Brandes and Devol 1997). Anoxic microsites could arise in fish tanks where particles accumulated, or more likely, in the sedimentation basin, where a blanket of sediments developed for 19 – 36 hours before removal. We observed that large amounts of gases rapidly collected beneath the water surface in the sedimentation basin; however, samples we collected were contaminated with oxygen, precluding evaluation of biologically-generated nitrogen production. A thick biofilm on the tanks' walls also could have provided anoxic microsites, contributing to NO_3^--N removal. This explanation was supported by our results for Tank A12, where fish were harvested and the biofilm removed from the walls less than two weeks before our monitoring began. The time for regrowth of the biofilm to a thickness that could allow denitrification was limited. Subsequently, less than 35% of NO_3^--N production was removed by passive denitrification from this particular system, considerably less than in the other three systems monitored. In-situ denitrification has been reported by other authors. For example, Bovendeur et al. (1987) found that 40 – 80% of TAN oxidized by nitrification then was reduced by denitrification. Thoman et al. (2001) attributed 9 – 21% losses of systems' nitrogen to denitrification. The 56% removal of NO_3^--N by passive denitrification in our study represented an important, positive outcome, because it could reduce by more than half the investment necessary for nitrogen removal should the effluent be treated and reused as suggested by Sandu et al. (2008, 2011).

Our results indicated that despite high fish densities maintained at BRA, the systems are not being operated at their maximum carrying capacity. Our results showed that an average of 73% of the recirculating systems' productive potential was utilized, although utilization approached 92% in systems holding fish close to harvest size. In particular, much productive potential can be realized in systems holding smaller fish for long periods. By better distributing fish biomass among systems via more frequent grading, net production could be increased within existing space. Our suggestion for increased production is supported by the excellent average removal efficiency for rotating biological contactors (54.4%) at a recirculation rate of almost one pass per hour, and by an average areal conversion rate of 442.5 mg TAN/m^2/day, which maintained an average TAN of 2.06 mg/L in fish tanks. Up to 2830 mg TAN/m^2/day can be removed by a rotating biological contactor (Rogers and Klemeston 1985), suggesting that the biofilters could function successfully under the maximum conditions of 3.15 mg/L TAN and 634 mg TAN/m^2/day areal conversion rate that we predicted. Additionally, reusing water using a treatment train such as that described by Sandu (2004) and Sandu et al. (2008, 2011) with 1.6 mg/L TAN, only 0.84% of total ammonia nitrogen produced would be reintroduced to the recirculating systems. This additional loading would be removed easily by the rotating biological contactors, without significant increase of TAN throughout the systems.

5. Conclusion

Routine aquaculture production generates waste products for which controlled biodegradation in treatment units is a major consideration in design and operation of recirculating aquaculture systems. Biodegradation of nitrogenous wastes is critical, especially for un-ionized ammonia and nitrite, which are toxic to fish. We quantified the dynamics of nitrogen through a large commercial recirculating aquaculture facility producing hybrid tilapia *Oreochromis* sp. Our nitrogen budget evaluated total ammonia nitrogen (TAN) production and removal in biofilters, quantifying the fate of nitrate-nitrogen (NO_3^--N) and determining the systems' maximum carrying capacity under steady-state conditions. Most of the recovered nitrogen was in fish, nitrate-nitrogen, and total organic nitrogen pools, with relatively small proportions as total ammonia nitrogen, mortalities, and nitrite-nitrogen, totaling 86%. The remaining 14% of the nitrogen budget unaccounted for likely was lost by passive denitrification to nitrogen gas and by volatilization of ammonia. Our nitrogen biodegradation model predicts that the systems could operate safely at up to 3.15 mg/L total ammonia nitrogen. Under current production conditions, system loading was 57-92% of the maximum fish biomass that could be supported. The biofilters' areal conversion rate could be increased by half under conditions of maximum biomass loading. NO_2^--N was not a parameter of concern, always remaining below 0.3 mg /L. Our results showed that microbial biodegradation of fish wastes was more than adequate and that fish production could be increased within the existing farm infrastructure, especially by more frequent grading of fish in order to stock production systems at densities approaching carrying capacity. With denitrification, discharged culture water could be reused to realize savings in operating costs. Beyond the narrow interest in our study system, our approach can be applied more broadly to other fish culture systems.

Acknowledgements

We are grateful for the support of the Commercial Fish and Shellfish Technologies program and the Department of Fish and Wildlife Conservation at Virginia Tech University. Blue Ridge Aquaculture graciously allowed access to facilities and production records. Julie Petruska trained S.S. water quality testing procedures. The expertise of Nancy Love was indispensable in experimental design and analysis.

Author details

S. Sandu and E. Hallerman*

*Address all correspondence to: ehallerm@vt.edu

Department of Fish and Wildlife Conservation, Virginia Polytechnic Institute and State University, Blacksburg, VA, USA

References

[1] Acosta-Nassar, M.V.; Morrell, J.M., & Corredor, J.E. (1994). The nitrogen budget of a tropical semi-intensive freshwater fish culture pond. *Journal of the World Aquaculture Society*, Vol. 25, No. 2, pp. 261-270, ISSN 0893-8849

[2] Atwood, H.L.; Fontenot, Q.C., Tomasso, J.R., & Isely, J.J. (2001). Toxicity of nitrite to Nile tilapia: effect of fish size and environmental chloride. *North American Journal of Aquaculture*, Vol. 63, No. 1, pp. 49-51, ISSN 1522-2055

[3] APHA (American Public Health Association), American Water Works Association & Water Environment Federation (1998). *Standard Methods for the Examination of Water and Wastewater, 20th edition*. American Public Health Association, ISBN 0-875530235-7, Washington, DC

[4] Benli, A.C.K.; Koksal, G., & Ozkul, A. (2008). Sublethal ammonia exposure of Nile tilapia (*Oreochromis niloticus* L.): effects on gill, liver, and kidney histology. *Chemosphere*, Vol. 72, No. 9, pp. 1355-1358, ISSN 0045-6535

[5] Bovendeur, J.; Eding, E.H., & Henken, A.M. (1987). Design and performance of a water recirculation system for high-density culture of the African catfish, *Clarias gariepinus* (Burchell 1822). *Aquaculture*, Vol. 63, No. 1-4, pp. 329-353, ISSN 0044-8486

[6] Brandes, J.A. & Devol, A.H. (1997). Isotopic fractionation of oxygen and nitrogen in coastal marine sediments. *Geochimical et Cosmochimical Acta*, Vol. 61, pp. 1798-1801, ISSN 0016-7037

[7] Chen, S.; Ling, J., & Blancheton, J. (2007). Nitrification kinetics of biofilm as affected by water quality factors. *Aquacultural Engineering*, Vol. 34, No 3, pp.179-197, ISSN 0144-8609

[8] Colt, J. (2006). Water quality requirements for reuse systems. *Aquacultural Engineering*, Vol. 34, No. 3, pp. 143-156, ISSN 0144-8609

[9] Costa-Pierce, B.A. & Rakocy, J.E. eds. (1997). *Tilapia Aquaculture in the Americas*. World Aquaculture Society, Baton Rouge, LA. 522 pp. in two volumes, ISBN 1-88807-01-6 and 1-88897-04-0

[10] Crab, R.; Avnimelech, Y., Defoird, T., Bossier, P., & Verstraete, W. (2007). Nitrogen removal techniques in aquaculture for sustainable production. *Aquaculture*, Vol. 270, No. 1-4, pp. 1-14, ISSN 0044-8486

[11] Daud, S.K.; Hasbollah, D., & Law, A.T. (1988). Effects of unionized ammonia on red tilapia (*Oreochromis mossambicus/O. niloticus* hybrid) fry. Pages 411-413 in Pullin, R.S.V., Bhukaswan, T., Tonguthai, K., and Maclean, J.L. *The Second International Symposium on Tilapia in Aquaculture*. ICLARM Conference Proceedings 15. Thailand Department of Fisheries and International Center for Living Aquatic Resources Management, ISBN 971-1022-58-3, Manila

[12] Diaz, V.; Ibanez, R., Gomez, P., Urtiaga, A.M., & Ortiz, I. (2012). Kinetics of nitrogen compounds in a commercial marine recirculating aquaculture system. *Aquacultural Engineering*, Vol. 50, No. 1, pp. 20-37, ISSN 0144-8609

[13] El-Sherif, M.S.; & El-Feky-A.M. (2008) Effect of ammonia on Nile tilapia (*O. niloticus*) performance and some hematological and histological measures. *8th International*

Symposium on Tilapia in Aquaculture, Cairo, Egypt, October 12-14, 2008, http://ag.arizona.edu/azaqua/ista/ISTA8/FinalPapers/PDF%20Files/39%20mohamed %20shreif12.pdf

[14] Eshchar, M.; Lahav, O., Mozes, N., Peduel, A, & Ron, B., (2006). Intensive fish culture at high ammonium and low pH. *Aquaculture,* Vol. 255, No. 1-4, pp. 301-313, ISSN 0044-8486

[15] Evans, J.J.; Pasnik, D.J., Brill, G.C., & Klesius, P.H. (2006) Un-ionized ammonia exposure in Nile tilapia: toxicity, stress response, and susceptibility to *Streptococcus agalactiae. North American Journal of Aquaculture,* Vol. 68, No. 1, pp. 23-33, ISSN 1522-2055

[16] Fitzsimmons, K., ed. (1997). *Tilapia aquaculture: Proceedings from the Fourth International Symposium on Tilapia in Aquaculture.* Natural Resource, Agriculture, and Engineering Service, Cooperative Extension, 152 Riley-Robb Hall, Ithaca, NY. 808 pp. in two volumes, ISBN 0-935817-58-1

[17] Hallerman, E.M. (2000). Genetic improvement of fishes for commercial recirculating aquaculture systems: a case study involving tilapia. In Libey, G.S. & Timmons, M.B., eds. Proceedings of the Third International Conference on Recirculating Aquaculture, Roanoke, VA, July 20-23, 2000.

[18] Hamlin, H.J.; Michaels, J.T., Beaulaton, C.M., Graham, W.F., Dutt, W., Steinbach, P., Losordo, T.M., Schrader, K.K., and Main, K.L. (2008). Comparing denitrification rates and carbon sources in commercial upflow denitrification biological filters in aquaculture. *Aquacultural Engineering,* Vol. 38, No. 2, pp. 79-92, ISSN 0144-8609

[19] Huguenin, J.E. & Colt, J. (1989). *Design and operating guide for aquaculture seawater systems.* Elsevier Interscience, ISBN 0-444-50577-6, Amsterdam. 264 pp

[20] Itoi, S.; Ebihara, N., Washio, S, and Sugita, H. (2007). Nitrate-oxidizing bacteria, *Nitrospira,* distribution in the outer layer of the biofilm from filter materials of a recirculating water system for the goldfish *Carassius auratus. Aquaculture,* Vol. 264, No. 1-4, pp. 297-308, ISSN 0044-8486

[21] Krom, M.D. & Neori, A. (1989). A total nitrogen budget for an experimental intensive fishpond with circularly moving seawater. *Aquaculture,* Vol. 83, No. 1-2, pp. 345-358, ISSN 0044-8486

[22] Krom, M.D.; Porter, C., & Gordin, H. (1985). Nutrient budget of a marine fish pond in Eilat, Israel. *Aquaculture,* Vol. 51, No. 1, pp 65-80, ISSN 0044-8486

[23] Lekang, O.I. (2007). Ammonia removal. Chapter 9 in *Aquaculture Engineering.* Blackwell Publishing, ISBN 978-1-4051-2610-6, Oxford

[24] Lemarie, G.; Dosdat, A., Coves, D., Dutto, G., Gasset, E., & Person-Le Ruyet, J. (2004). Effect of chronic ammonia exposure on growth of European seabass (*Dicentrarchus labrax*) juveniles. *Aquaculture,* Vol. 229, No. 1-4, pp. 479-491, ISSN 0044-8486

[25] Lim, C. & Webster, C.D., eds. (2006). *Tilapia Biology, Culture, and Nutrition*. The Haworth Press, Binghamton, New York, NY, USA, ISBN 13: 978-1-56022-318-4

[26] Losordo, T.M. (1997). Tilapia culture in intensive recirculating systems. Pages 185-211 in Costa-Pierce, B.A. & Rakocy, J.E. *Tilapia Aquaculture in the Americas, volume 1*. World Aquaculture Society, ISBN 1-88807-01-6, Baton Rouge, Louisiana, USA

[27] Losordo, T.M., and Westerman, P.W. (1994). An analysis of the biological, economic, and engineering factors affecting the cost of fish production in recirculating aquaculture systems. *Journal of the World Aquaculture Society* Vol. 25, No. 2, pp. 193-203. ISSN 1749-7345

[28] Losordo, T.M. & Westers, H. (1994). System carrying capacity and flow estimation. In: M.B. Timmons & T.M. Losordo, editors. *Aquaculture water reuse system: Engineering design and management*. Developments in Aquaculture and Fisheries Sciences, vol. 27. Elsevier Science, ISBN 9780444895851, Amsterdam. Pp. 9-60

[29] Palacheck, R.M. & Tomasso, J.R. (1984). Toxicity of nitrite to channel catfish (*Ictalurus punctatus*), tilapia (*Tilapia aurea*), and largemouth bass (*Micropterus salmoides*): evidence for a nitrite exclusion mechanism. *Canadian Journal of Fisheries and Aquatic Sciences* Vol. 41, No. 12, pp. 1739-1744, ISSN 1205-7533

[30] Porter, C.B.; Krom, M.D., Robbins, M.G., Brickell, L., & Davidson, A. (1987). Ammonia excretion and total N budget for gilthead seabream (*Sparus aurata*) and its effect on water quality conditions. *Aquaculture,* Vol. 66, No. 3-4, pp. 287-297, ISSN 0044-8486

[31] Refstie, T. (1977). Effects of density on growth and survival of rainbow trout. *Aquaculture,* Vol. 10, No. 3, pp. 231-242, ISSN 0044-8486

[32] Racocy, J. & Allison, R. (1981). Evaluation of a closed recirculating system for the culture of tilapia and aquatic macrophytes. In: Fish Culture Section, *Bioengineering Symposium for Fish Culture,* American Fisheries Society, Bethesda, Maryland, USA, Pages 296-307

[33] Redner, B.D. & Stickney, R.R. (1979). Acclimation to ammonia by *Tilapia aurea*. *Transactions of the American Fisheries Society* Vol. 108, No. 4, pp. 383-388, ISSN 1548-8659

[34] Rogers, G.L. & Klemetson, S.L. (1985). Ammonia removal in selected aquaculture water reuse biofilters. *Aquacultural Engineering,* Vol. 4, No. 2, pp. 135-154, ISSN 0144-8609

[35] Russo, R.C. & Thurston, R.V. (1991). Toxicity of ammonia, nitrite and nitrate to fishes. Pages 58-89 in D.E. Brune and J.R. Tomasso, eds. *Aquaculture and water quality*. Advances in World Aquaculture 3. World Aquaculture Society, Baton Rouge, Louisiana, USA.

[36] Sandu, S.I. (2004). Evaluation of ozone treatment, pilot-scale wastewater treatment plant, and nitrogen budget for Blue Ridge Aquaculture. Ph.D. dissertation, Virginia Polytechnic Institute and State University, Blacksburg, Virginia, USA.

[37] Sandu, S.; Brazil, B., & Hallerman, E. (2008). Efficacy of a pilot-scale wastewater treatment plant upon a commercial aquaculture effluent: I. Solids and carbonaceous compounds. *Aquacultural Engineering*, Vol. 39, No. 1, pp. 78-90, ISSN 0144-8609

[38] Sandu, S.; Brazil, B., & Hallerman, E. (2011). Efficacy of pilot-scale wastewater treatment upon a commercial recirculating aquaculture facility effluent. Pages 141-158 in B. Sladonja, ed. *Aquaculture and the Environment: A Shared Destiny*. Intech. ISBN 978-953-307-749-9, Rijeka, Croatia.

[39] Schreier, H.J.; Mirzoyan, N., & Saito, K. (2010). Microbial diversity of biological filters in recirculating aquaculture. *Current Opinion in Biotechnology*, Vol. 21, No. 3, pp. 318-325, ISSN 0958-1669

[40] Siddiqui, A.Q. & Al-Harbi, A.H. (1999). Nutrient budget in tanks with different stocking densities of hybrid tilapia. *Aquaculture*, Vol. 170, No. 3-4, pp. 245-252, ISSN 0044-8486

[41] Suresh, A.V. & Kwei, L.C. (1992). Effect of stocking density on water quality and production of red tilapia in a recirculating water system. *Aquacultural Engineering*, Vol. 11, No. 1, pp. 1-22, ISSN 0144-8609

[42] Svobodova, Z.; Machova, J., Poleszczuk, G., Huda, J., Hamackova, J., & Kroupova, H. (2005). Nitrite poisoning of fish in aquaculture facilities with water-recirculating systems. *Acta Veterinaria Brno*, Vol. 74, pp. 129-137, ISSN 0001-7213

[43] Tal, Y.; Watts, J.E.M., and Schreier, H.J. (2006). Anaerobic ammonium-oxidizing (annamox) bacteria in associated activity in fixed-film biofilters of a marine recirculating aquaculture system. *Applied and Environmental Microbiology*, Vol. 72, No. 4, pp. 2896-2904, ISSN 0099-2240

[44] Tchobanoglous, G. & Schoeder, E.D. (1985). *Water Quality: Characterization, Modeling, Modification*. Addison-Wesley Publishing Company. ISBN: 10-0201054337, Reading, MA

[45] Thiex, N.J.; Manson, H., Anderson, S., & Persson, J. A. (2002). Determination of crude protein in animal feed, forage, grain, and oilseeds by using block digestion with a copper catalyst and steam distillation into boric acid: collaborative study. *Journal of the Association of Official Agricultural Chemists*, Vol. 85, pp. 309-317, ISSN 0095-9111

[46] Thoman, E.S.; Ingall, E.D., Davis, D.A., & Arnold, C.R. (2001). A nitrogen budget for a closed, recirculating mariculture system. *Aquacultural Engineering*, Vol. 24, No. 3, pp. 195-211, ISSN 0144-8609

[47] van Kessel, M.A.J.H.; Harhangi, H.R., van de Pas-Schoonen, K., van de Vossenberg, J., Flik, G., Jetten, M.S.M., Klaren, P.H.M., & Op den Camp, H.J.M. (2010). Biodiversi-

ty of N-cycle bacteria in nitrogen removing bed biofilters for freshwater recirculating aquaculture systems. *Aquaculture*, Vol. 306, No. 1-4, pp. 177-184, ISSN 0044-8486

[48] van Rijn, J. (1996). The potential for integrated biological treatment systems in recirculating fish culture: A review. *Aquaculture* Vol. 139, No. 3-4, pp. 181-201, ISSN 0044-8486

[49] Vijayan, M.M. & Leatherland, J.F. (1988). Effects of stocking density on the growth and stress response in brook charr *Salvelinus fontinalis*. *Aquaculture*, Vol. 75, No. 1-2, pp. 159-170, ISSN 0044-8486

[50] Wang, Y.; Zhang, W., Li, W., & Xu, Z. (2006). Acute toxicity of nitrite on tilapia (*Oreochromis niloticus*) at different external chloride concentrations. *Fish Physiology and Biochemistry*, Vol. 32, No. 1, pp. 49-54, ISSN 0920-1742

[51] Yildiz, H.Y.; Koksal, G., Borazan, G., & Benli, C.K. (2006). Nitrite-induced methemoglobinemia in Nile tilapia, *Oreochromis niloticus*. *Journal of Applied Ichthyology*, Vol. 22, No. 5, pp. 427-426, ISSN 1439-0426

Chapter 14

Aerobic Biodegradation Coupled to Preliminary Ozonation for the Treatment of Model and Real Residual Water

P. Guerra, J. Amacosta, T. Poznyak, S. Siles,
A. García and I. Chairez

Additional information is available at the end of the chapter

http://dx.doi.org/10.5772/56011

1. Introduction

1.1. Sequence of water treatment methods

Residual and waste water have become a problem of paramount importance in modern societies [1]. Recently, the number of proposals to solve this issue has incremented importantly [2]. Several methods were proposed since thirty years ago using a wide variety of physical, biological and chemical principles. Biological treatments are cheap and environmentally friendly [3]. Nevertheless, they require a long time to eliminate pollutants and they are limited by the toxicity and initial concentration of the water sample that must be treated [4]. On the other hand, chemical treatments are capable to promote the faster decomposition for a wide range of toxic compounds [5]. Despite this adequate performance to decompose organics dissolved in water, they are hundreds or thousands of times more expensive than pure biological methods [6-11].

1.2. Ozonation followed by biodegradation

Just some years ago, the attractive features of both methods (biological and chemical) have attracted attention to develop more advanced schemes to manage more toxic and complex pollutant mixtures [12-13]. Indeed, remarkable results have come from a sequence of treatments (usually called trains of treatments) using the combination of several individual options. Regularly, the treatment trains are using a sequence defined by a physical method followed by the biological scheme and finally, one last chemical process completes the treatment.

However, this arrangement does not always work efficiently when the initial pollutant mixture has complex composition or they are very toxic. The stage that is usually compromised by this aspect is the biological one [14-15].

Several mixed processes have been recently proposed including a process based on chemical oxidative compounds plus biological based decomposition course [12-18]. Among others, oxygen injected with high pressure, ozonation [13], catalytic and photocatalytic processes and others have been tested to perform the chemical decomposition [16-18]. Most of these treatment methods have important advantages but also have important drawbacks. These problems can be classified into two main areas: the first one contains all economic aspects associated with high cost required to implement these treatments, the second one includes all troubles associated to the resources needed to complete the transformation from very toxic compounds to simpler ones that can be considered as no toxicity and no hazardous [19]. Nevertheless, these drawbacks may be solved by biologically based treatments.

Nowadays, a different way of thinking has emerged to improve the efficiency of waste and residual water: changing the order of treatments to include a chemical pre-treatment before the biological process. The idea is to reduce the complexity as well as the toxicity of the organic mixture of chemical methods. Theoretically, this condition must have a positive effect on the microorganism's efficiency to decompose the simpler and less toxic organics.

Beltran et al. [13] reported that combined ozonation and aerobic treatment increased the removal efficiency from 82% or 76% for the COD or the total phenolic content, respectively. Benitez et al. [20] demonstrated the COD removal for wine vinasses containing organic matter and aromatic compounds was enhanced (from 27.7% to 39.3%), when the combined ozonation and biological process was used. Aparicio et al. [21] reported the use of combined wastewater treatment set up in a resin-producing factory. After biological treatment of the ozonated effluent, the organic carbon and nitrogen removal was increased from 27 to 97% and from 27 to 80%, respectively.

The possible benefits coming from the combination of pre-treatment with ozone and a sequential biodegradation are almost evident; however, there are still several questions about this procedure. For example, what time is adequate to move the organics mixture from the ozonation reactor to the biological one or what conditions should be set-up for both reactions still remain as open problems. Another important issue that must be explained is what conditions must fulfill the microorganism strains to handle the pollutant mixture produced by the preliminary ozonation. This is an important aspect conditioned to the composition of the mixture supplied to microorganisms that can modify the organics elimination by biodegradation. Moreover, there is just a few of works describing what type of microorganisms is responsible for the elimination of residual compounds after ozonation [22]. In recent reports, [23-29] catechol, hydroquinone and several low weight organic acids have been recognized like the main byproducts obtained after ozonation of phenol and its chlorinated derivatives. Nevertheless, what relative concentration of each byproduct is the most adequate to construct the combined process including ozonation and biological reaction has not been determined yet.

1.3. Phenol and its chlorinated derivatives as artificial wastewater

Phenol and its chlorinated derivatives are simple examples of how the biodegradation can be efficient or not for closely related pollutants [26, 27]. It has been broadly reported the efficient decomposition of phenol by biological strands. Many methods of eliminating phenol and its derivatives using chemical and biological systems have been studied [23-24]. Biological systems are environmentally friendly, low-cost technologies that can be successfully used to remove phenols by using different microbial strains but with pure cultures.

However, when any chlorophenol is exposed to the same microorganisms, the toxicity of this compound reduces the decomposition efficiency to 20 or 30 % compared to the same conditions observed in phenol treatment.

On the other hand, most of advanced oxidation processes can remove the chlorine atom from the chlorophenols in few minutes or even seconds [5]. Therefore, the possible sequential treatment based on ozone followed by biodegradation can use the advantages offered by this couple of water treatment schemes. Indeed, phenol's ozonation generates simpler organic acids that can be assimilated by microorganisms [14-15].

1.4. Lignin and its derivatives as real wastewater

A more complex situation arises when the pollutants in residual water are toxic and also with complex structure. As an example, pulp and paper industry wastewater mainly contains lignin and its derivatives (chlorinated phenolic compounds, resin acids, dioxins and dioxin-like compounds and many others) [29-32].

Lignin is a three-dimensional biopolymer which confers the resistance to stress, protects the plant from the microbial enzymatic hydrolysis and also acts as a binder of the fibers of cellulose and hemicellulose in the wood [33,34]. Is formed by the coupling of three monolignols (paracoumaryl alcohol, coniferyl alcohol and sinapyl alcohol) and with some functional groups (ROH, ϕOH, ROMe, RR'C=O, RCOOH, RSO_3R', etc.). The extractable products like fatty acids, phenols, terpenes, steroids, waxes, tannins and resinic acids, also found in the wastewaters, confer the physicochemical properties of each plant such as color, smell, strength, hardness [37-39].

Lignin and its derivatives have shown to be a very complex, toxic mixture with mutagenic and teratogenicity activities. Biological treatment of wastewater with these residuals has partial efficiency [39-41].

1.5. Motivation and contribution of this study

In this chapter, a combined method to treat residual water is proposed. The treatment is based on the preliminary action of ozone followed by the biological treatment using a microorganism consortium. Two water samples were used to evaluate the combined treatment: a model mixture of chlorophenols and residual water obtained from the paper industry after the bleaching step from the Kraft process. This selection was done to illustrate the efficiency of the combined process using ozone and biodegradation in a row.

2. Materials and methods

2.1. Model and real residual water

Model solutions were artificially prepared with 4-Chlorophenol (4-CPh) or 2,4-Dichlorophenol (2,4-DCPh) (120 mg/L) as model solutions. All these chemical products have 99% purity.

For the real residual water, the sample was obtained from a Kraft process in the bleaching step; collected at 4 °C and sterilized in autoclave under the temperature of 121 °C and a pressure of 15 pounds. The mixture was characterized by simple analytical methods based on UV/VIS spectroscopy.

These two polluted water samples were treated by ozonation, aerobic biodegradation and the combination of both processes (ozonation followed by biodegradation). The biodegradation was developed using a microbial consortium acclimated to the particular composition of carbon source remaining in the reactor after/before ozonation [42-44].

2.2. Ozonation procedure

The ozonation treatment was carried out in a semi-batch glass reactor (250 mL). Ozone concentration at the reactor's input was 30 mg/L. The maximum ozonation time was 60 min for both the real and artificial wastewater. The ozone/oxygen mixture was injected through a ceramic porous at the inferior part of the reactor with a flow of 0.5 L/min. The ozone was produced by the ozone generator HTU500G 'G' (corona discharge type, "AZCO" INDUSTRIES LIMITED, Canada). The Ozone Analyzer BMT 963 "S" (BMT Messtechnik, Berlin) provides the ozone detection in the gas phase at the reactor output. This information was used to perform the ozone monitoring, to control the ozonation degree and to study the ozone decomposition. The ozone concentration was sampled by a data acquisition system implemented in a regular computer (Figure 1).

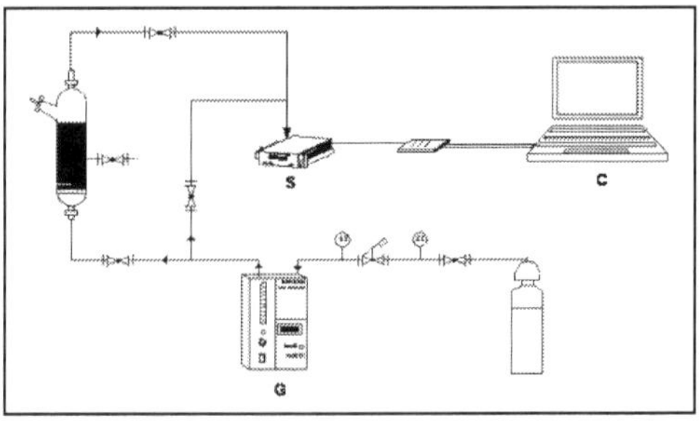

Figure 1. Simple scheme of the ozonation setup including the reactor where the ozonation is carried out. The ozone concentration produced in (G) is monitored in the UV sensor (S). A data acquisition board is connected to a personal computer to register the ozone concentration.

The ozonation of was carried out using two different pHs: 7 (for model and real water) and 12 (only model solution). Ozonation of residual water samples was carried out for diluted solutions (1:10). This change promotes the reduction of reaction time and helped to decrease foaming [16].

2.3. Microbial culture, mineral media

For the biological treatment, different microbial consortia were cultivated and acclimated during 6 months (by a fill-and-draw procedure) to the specific carbon source (model chlorophenols and real water solutions with and without previous ozonation).

The mineral media used for all the experiments contains (g/L): 3.0 $(NH_4)_2SO4$, 0.6 KH_2PO_4, 2.4 K_2HPO_4, 1.5 $MgSO_4$•7H2O, 0.15 $CaSO_4$, and 0.03 $FeSO_4$. A mixed microbial culture from a biofilter used to remove aromatic compounds and gasoline vapors (Dr Revah's Laboratory, Universidad Autonoma Metropolitana Iztapalapa, Mexico) was independently adapted for three months to phenol (100 mg/L) and to a mixture of oxalic and formic acids (100 mg/L each) in mineral media.

The mixed culture was cultivated in an Erlenmeyer flask of 1 L with 500 mL of mineral media. These compositions were inoculated with 50 mL of the microbial mixture. Reactors were kept at ambient temperature and shaken in an orbital shaker at 200 RPM. The mineral media was also kept invariable for these experiments. In all the studied samples, the biomass amount and the organic degradation as well were periodically measured in triplicate.

The cultures were harvested between 24 and 30 h, corresponding to the exponential growth phase and then used for the model and real water treatment (phenol and chlorophenols mainly) and their corresponding ozonation products (catechol, hydroquinone and organic acids mainly) [46].

For the biological treatment, different microbial consortia (set of several microorganisms with different species) were cultivated and acclimated during 6 months (by a fill-and-draw procedure) to the specific carbon source (model chlorophenols and real water solutions with or without previous ozonation). In this study, the biological media is composed of a complex consortium of the microbial population previously identified [22] by the extraction of DNA samples using an Easy-DNATM Kit (Invitrogen, USA) [44].

Inoculums of the corresponding microbial consortium were added into the batch reactor containing the model solution or the real water (with or without previous ozonation). Reactors were kept at ambient temperature and shaken in an orbital shaker at 200 RPM. Chlorophenolsenols (from model solution) and real water's components concentration, as well as ozonation products concentration in reactors and the biomass amount were periodically measured by triplicate.

2.4. Analytical methods

Several analytical methods were used in order to characterize, identify and quantify the samples. UV Spectroscopy (Lambda 2S, Perkin Elmer) was used for monitoring the global behavior of ozonation (λ=260 and 210 nm) and biodegradation (λ=210 nm) of real water as well as the microbial growth (OD600).

The control of chlorophenols or the components of real water decomposition, as well as the intermediates and final products formed in the ozonation step was made by high performance liquid chromatography (HPLC), (Series 200, Perkin Elmer) equipped with UV-VIS detector series 200. Two wavelengths were periodically monitored (210 nm and 270 nm). Analytical details are shown in Table 1.

Analysis conditions	Compounds	
	Phenols	**Organic acids and real water**
Column	Platinum C-18 (Alltech), 250 x 4.6mm	Prevail Organic Acid (Grace), 150 x 4.6mm
Mobile Phase	60:40 (water : methanol)	KH_2PO_4 25 mMol in water (pH = 2.5)
λ(nm)	210	
Flow rate	1mL / min	
Sample volume	30μL	

Table 1. HPLC analysis conditions

The study was made on raw material and samples of study, both in the stage of ozonation of biodegradation. Identification and qualitative determinations were made taking into account the retention times of components and the quantitative analysis by integration of signals, in relation to the corresponding calibration curve.

3. Results

For both kind of waters, model solution or real water, three processes were evaluated: ozonation, biodegradation (without ozonation) and the combined treatment (ozonation followed by biodegradation). Those are described below.

3.1. Ozonation

3.1.1. Model solution preliminary ozonation

Chlorophenols (CPhs) decomposition was faster at pH 12 (8 and 5 minutes) than pH 7 (15 and 8 minutes) for 4-CPh and 2,4-DCPh, respectively. Some by-products like catechol, hydroquinone, oxalic and formic acids were formed. All these are some of the products identified in CPhs ozonation [5]. Besides, some other ones were observed but they could not be identified, however, they were monitored by HPLC and referred as non- identified phenolic compounds and organic acids. During ozonation, both identified and non-identified phenolic compounds were rapidly decomposed, while oxalic and formic acids were mostly accumulated during the whole reaction period. The maximum concentration detected for the different ozonation conditions were previously published [5]. All these compounds constituted the carbon source for adapted bioprocess applied at the next step. Then, the percentage of CPhs decomposition and the by-products accumulation/decomposition was considered to stop the ozonation and carry out the biodegradation step.

3.1.2. *Real residual water ozonation*

In the case of real residual wastewater ozonation, a significant decrease of organic compounds concentration was followed in the UV spectra (Figure 2). Lignin derivatives was followed in the UV region of λ=260 nm and organic acids in the region of λ= 210 nm.

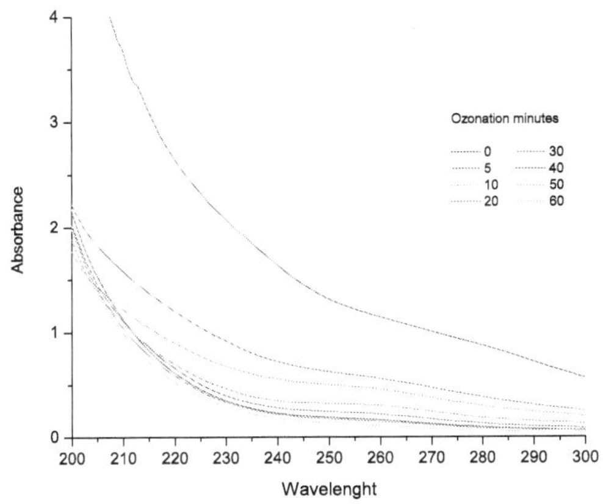

Figure 2. Variation of UV-VIS spectrum in the ozonation process.

A significant decrease in the region of lignin derivatives (94%) and in the corresponding organic acids (83%), which also tends to decrease in absorbance is depicted. At the end of ozonation no longer variation in the UV spectra is observed. Therefore, the susceptible organic matter to be ozonated has been completely reacted with ozone.

The identification of organic compounds in the original sample and also during the ozonation was done by the HPLC technique. The decomposition dynamics during the reaction with ozone was determined. Moreover quantification of accumulated products is gotten. Main byproducts were hydroquinone, catechol and simple organic acids such as maleic acid and several unidentified compounds.

The main recalcitrant accumulated product during ozonation was oxalic acid. Indeed, its presence was observed from the beginning to the end of ozonation. Its relative importance for the sequent biodegradation motivated its quantification by HPLC (Figure 3).

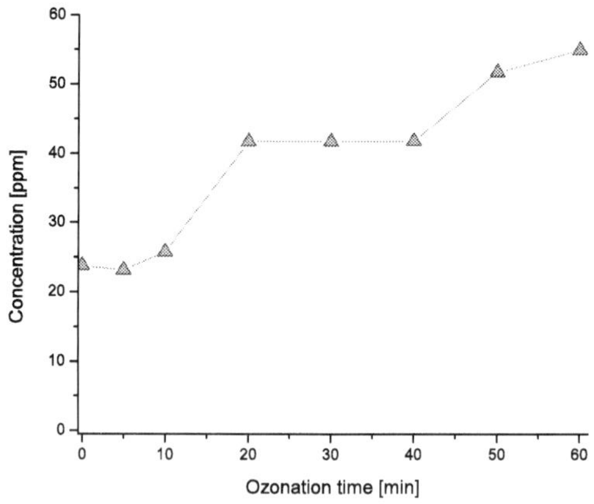

Figure 3. Quantification of oxalic acid during the ozonation of real wastewater.

As it can be seen, oxalic acid was contained in the real residual wastewater before any treatment (23 mg/L), but after the first 30 minutes 43 mg/L of this acid was detected. After 60 minutes of ozonation the acid concentration was increased up to 55 mg/L.

3.2. Biodegradation

3.2.1. Biological treatment of model solution (without ozonation)

Poor degradation of CPhs was observed in the case of non ozonated samples (30% and 40% for 4-CPh and 2,4-DCPh, respectively) after 10 days. No matter the microorganisms were previously acclimated to CPhs, the initial concentration was toxic enough to inhibit biodegradation, which is one of the biological treatments principal disadvantages. The biodegradation of ozonation products identified during ozonation was also tested. The minority compounds were eliminated obeying the following order of elimination: phenol>catechol>hydroquinone. For final products, a mixture of oxalic and formic acids with a concentration of 100 mg/L of each one was also tested. Microorganisms were able to eliminate both compounds during two days. These results are very important because they demonstrated that: highly toxic substrates, which cannot be eliminated by bioprocess, but they can be easily degraded by ozone and transform them into several compounds.

These results demonstrated that highly toxic substrates which cannot be eliminated by bioprocess are in the model solution, but they can be easily degraded by ozone and transform them into several compounds. Additionally, an acclimated consortium whit capability to eliminate the ozonation products was produced. So, it was expected that ozonation products were easier to be eliminated by biodegradation than original CPhs. Therefore, one can expect that combined treatment is more efficient than individual ones.

3.2.2. Biological treatment of real residual water (without ozonation)

The real residual water biodegradation showed a similar behavior to model solution one. It showed a partial decomposition of the organic compounds up to 20 % after five days due to the complexity and heterogeneity of this sample without ozonation. Therefore, organic matter degradation is poor (Figure 4).

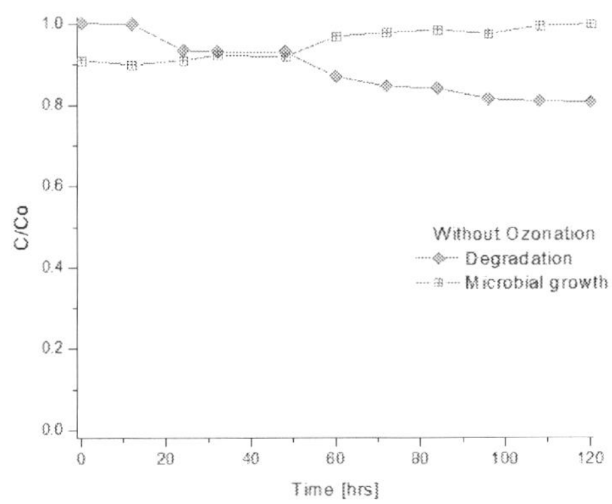

Figure 4. Growth dynamics of biomass and decomposition of organic matter without pre-treatment with ozone.

Nevertheless, microbial growth was obtained but probably because of the consumption of oxalic acid present in non ozonated sample. Taking into account that oxalic acid (which could be eliminated by microorganisms) is in real water (before ozonation) and this compound is formed further in ozonation, it is expected that combined treatment was successful. In general, by comparing in Figure 4 the growth / degradation dynamics appears in normalized form to compare them. Both, the increase of the Optical Density measured at 600 nm (corresponding to microorganisms grow), and the decrease for the consumption of the source of carbon are shown together.

First 20 minutes are characterized by a partial decomposition of the organic compounds up to 10 %. However, after 50 minutes, an additional decomposition around 10% is obtained but no more pollutants elimination was achieved until the end of the experiment (120 hours).

As it was expected, the organic matter decomposition is not significant. Only 20% of the total organic matter was eliminated due to the complexity and heterogeneity of the wastewater sample. As expected, microbial growth was not considerable because the pollutants nature as well as their concentration in the sample.

3.3. Combined treatment

3.3.1. Biological treatment of model solution after ozonation step

Taking into account the results obtained in previous sections, several ozonation conditions were chosen to test combined treatment. Three principal aspects were considered to choose pre ozonation conditions: the reduction (or complete elimination) of CPhs concentration, the accumulation of phenolic compounds and the final production of organic acids. As phenolic compounds are formed since the beginning but rapidly decomposed during ozonation, its presence was evaluated during biodegradation, in order to consider the needy to continue ozonation till remove them. Table 2 shows these pre ozonation conditions with corresponding ozonation products concentration.

Identified Compounds	Concentration (mg/L)						
	Ozonation time (min)						
	4-CPh				2,4-DCPh		
	pH 7			pH 12	pH 7		pH 12
	10	15	30	5	5	8	5
4-CPh	9	-	-	18	-	-	-
2,4-DCPh	-	-	-	-	18	-	-
Oxalic Acid	10	15	27	30	6	9	48
Formic Acid	154	137	43	39	36	50	54

Table 2. Ozonation conditions for the sequential treatment

In order to measure the biodegradation of pre-ozonated samples, the total area registered in HPLC chromatograms (under each condition reported in Table 2) was integrated. A normalized version of this area was used to evaluate the evolution of pollutants decomposition (before biodegradation). For the model sample, total degradation of two kinds of substrates (phenolic compounds and organic acids) was monitored. We could corroborate that the amount of each substrate (phenolics or organic acids) accumulated in ozonation has a very important influence during biodegradation step.

Figures 5, 6, 7 and 8 show the biodegradation on combined treatment of organic acids and phenolic compounds. For both kinds of substrates, during the first two days of biodegradation, it is remarkable the influence of ozonation conditions (Table 2). Indeed, it is a direct relationship between the formation-decomposition dynamics of by-products ozonation and the biodegradation facility of different kind of substrates.

Remembering the separated processes describe above, during ozonation, the phenolic compounds tend to accumulate during the first stage of treatment and after that, they tend to decomposed (Figures 5 and 6 for 4-CPh and 2,4-DCPh respectively). On the other hand, organic

http://dx.doi.org/10.5772/56011

acids are formed since the beginning and they tend to accumulate (being the oxalic acid the most representative acid at the end of ozonation). On the other hand during biodegradation the oxalic and formic acids demonstrate to be easier to eliminate than phenolic compounds (Figures 7 and 8 for 4-CPh and 2,4-DCPh respectively). Under this perspective, it is very clear why during the first days of biodegradation in combined treatment; the phenolic compounds are more degraded if the ozonation time was higher (under pH 7). Without a doubt, ozonation partially degraded these compounds, during biodegradation and then, they are less complicated to be eliminated by microorganisms. This last idea is confirmed by the decomposition dynamics observed for biodegradation of 4-CPh and 2,4-DCPh ozonated under pH 12. Similar behavior was observed for organic acids biodegradation.

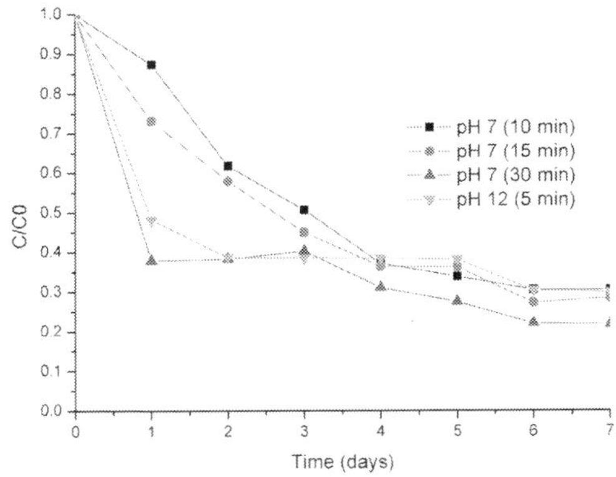

Figure 5. Biodegradation of phenolic compounds accumulated during 4-CPh ozonation

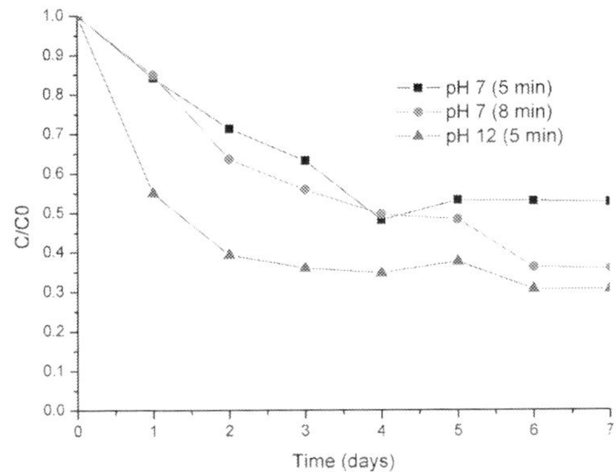

Figure 6. Biodegradation of phenolic compounds accumulated during 2,4-DCPh ozonation

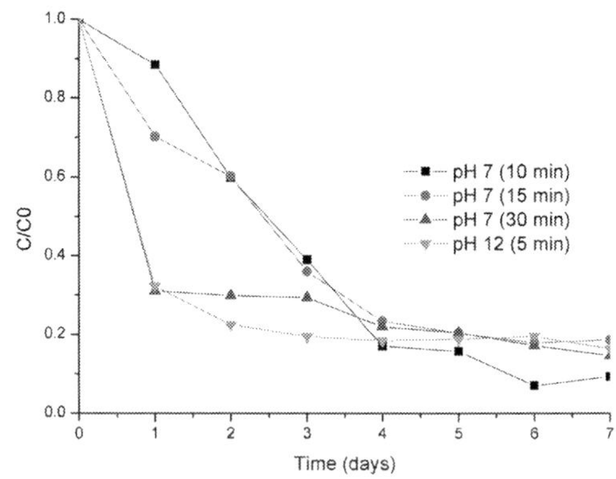

Figure 7. Biodegradation of organic acids accumulated during 4-CPh ozonation

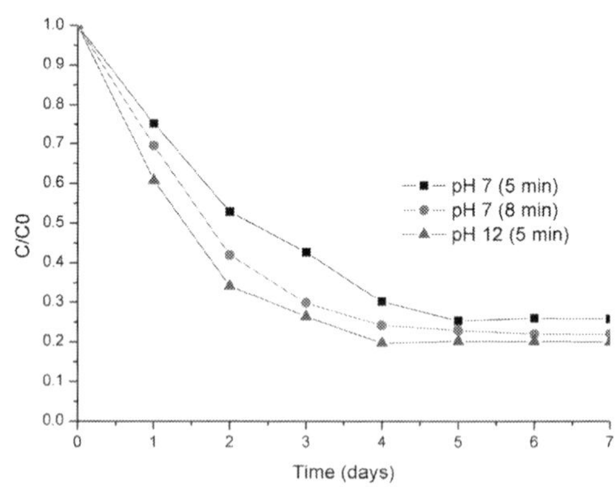

Figure 8. Biodegradation of organic acids accumulated during 2,4-DCPh ozonation

Biodegradation profiles are similar for phenolic compounds and organic acids. These organic acids are easier to eliminate than phenolic compounds. The 4-CPh decomposition degree for each substrate was between 81-90% and 70-78% for organic acids and phenolic compounds, respectively. On the other hand, 2,4-DCPh have decomposition degree between 74-80% and 47-69% for organic acids and phenolic compounds correspondingly. Table 3 shows a summary of elimination efficiency obtained for each pre ozonation condition.

It is important to notice that no matter the ozonation conditions, resulting by-products (phenolic compounds as well as organic acids) were metabolized by the microbial population since the beginning of the biotreatment, some of them faster than the others, but thank to

http://dx.doi.org/10.5772/56011

Compound	pH	Ozonation time (min)	% Biodegradation	
			Organic Acids	Phenolic Compounds
4-CPh	7	10	90	70
		15	81	71
		30	85	78
	12	5	86	70
2,4-DCPh	7	5	74	47
		8	78	64
	12	5	80	69

Table 3. Percentage biodegradation of 4-CPh and 2,4-DCPh pre-ozonated.

previous acclimation the microbial consortia is able to consume ozonated substrates. It is remarkable that residual 4-CPh (not fully eliminated during ozonation) remaining in ozonated samples during 10 and 5 minutes under pH 7 and 12, respectively was eliminated in biodegradation during the first day of treatment.

Analyzing the individual biodegradation profile of each compound formed during previous ozonation, a serial degradation is inferred. This means than some compounds were preferably consumed (in the earlier days of biodegradation) because of their energetic wealth or ease of degradation. When those organics were depleted, microbial consortium was able to metabolize all others (data not shown).

As it was previously mentioned, a microbial consortium was acclimated to specific carbon source, in this case, ozonation by-products. So, microbial population had developed specific abilities to degrade ozonation products.

Figure 9 shows the global biodegradation behavior in combined treatment for 4-CPh ozonated 10 minutes at pH 7. Substrates elimination and microbial growth are parallel. They were presented in a normalized way as diminution of the HPLC area (phenolic compounds and organic acids) and optical density increase.

The correlation between the optical density measured at 600 nm and the pollutants decomposition suggests organic matter integration in the biomass concentration. Between the day 0 and 1, poor degradation is obtained and in agreement, poor microbial growth. Between day 1 and 4, the major organic acids depletion and an important one for phenolic compounds is observed, so a second growth step appears as a result of metabolism of these substrates. Between day 4 and 7 the third growth step is observed as a result of metabolism of residual substrates (phenolic compounds and organic acids consumed until the end). It is remarkable that biodegradation trends of both substrates are simultaneous. However, from the second day, consortium shows an evident preference for organic acids (90% removal) over the phenolic compounds (70% elimination). Those dynamics were similar for all the combined treatments considered in this study (data not shown).

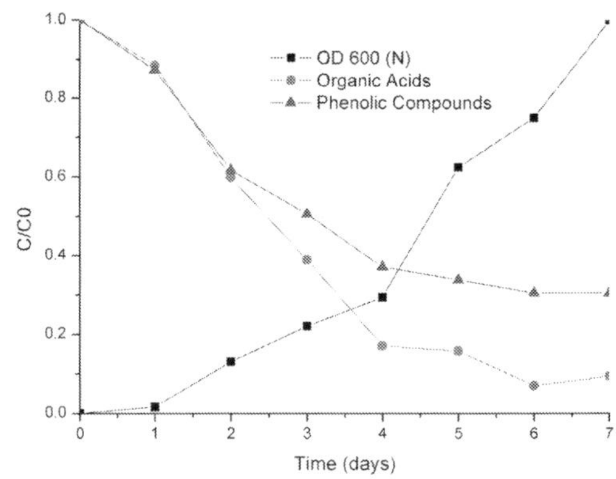

Figure 9. Substrates degradation and microbial growth during biodegradation of pre-ozonated 4-CPh (pH 7, 10 min).

Microbial growth in ozonated substrates was followed by optical density at 600 nm. This analysis was done when different pre-ozonation times (10, 15 and 30 minutes for pH is 7.0 and 5 minutes for pH is fixed to 12.0) were considered.

When 4-CPh is ozonated, microbial growth was faster when pH is 7.0 and ozonation time is fixed to 30 minutes. This accelerated biomass accumulation is associated to the major pollutants decomposition. Moreover, when pH was fixed to 12 and the reaction time was 5 minutes, the lower biomass velocity growing was achieved. If pH was fixed to 7.0, when reaction time was 15 minutes, the lag phase was delayed more than any others (Figure 10). This is explained by the accumulation of phenolic compounds under this reaction conditions. This is confirmed by the faster biomass accumulation when ozonation time was 10 minutes and phenolic compounds were not so higher than the previous case.

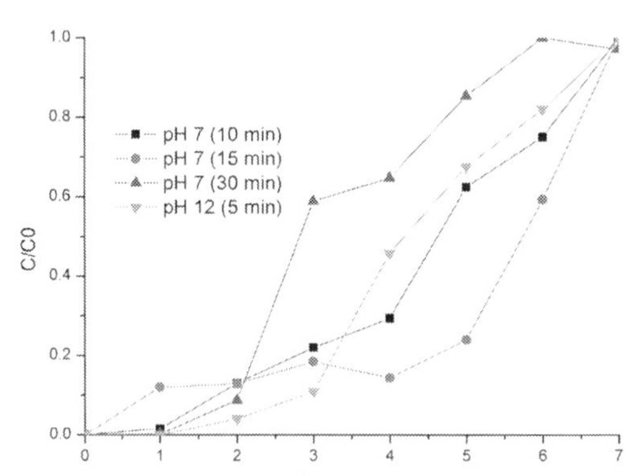

Figure 10. Biomass growth in biodegradation of organic acids accumulated during 4-CPh ozonation

http://dx.doi.org/10.5772/56011

When 2,4-DCPh was ozonated, the faster biomass accumulation was obtained when pH was 7.0 and the ozonation time was 8.0 minutes. Once again this condition coincides to the case when lower phenolic derivatives were observed in the reactor. Indeed, when pH was 7.0 with reaction time was 5 minutes and when pH was 12 and reaction time was 5, a lower biomass accumulation was determined. This is in agreement to the previous case, because under these two cases higher phenolic concentrations were observed. As one can understand, when phenolic compounds were at these levels, no important organic acids were in the reactor. This is a contrary situation to the case when phenolics compounds were at their minimum (among other studied cases) concentrations.

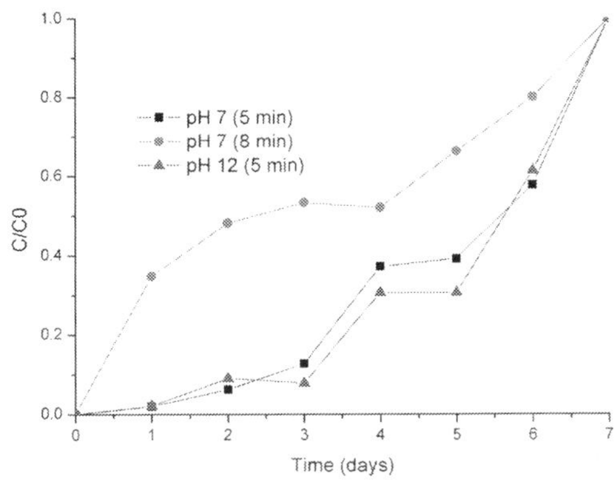

Figure 11. Biomass growth in biodegradation of phenolic compounds accumulated during 2,4-DCPh ozonation

Finally, Figures 12 and 13 show the UV spectra obtained after the combined treatment. These treatments were developed according to conditions presented in table 2.

Major elimination by combined treatment was obtained when most of the not identified compounds were decomposed, respectively (4-CPh ozonated during 30 min at pH7). This result is in agreement to the higher biomass accumulation observed in Figure 10.

Pre-ozonation conditions had an important influence on the overall degradation. When most of the no identified compounds were decomposed during ozonation (in the subsequent biodegradation step) the UV spectra is very close to the control case (mineral media, absence of contaminant).

It can be noticed that ozonation by-products were not as toxic as the original ones. This is explained because they were simultaneous consumed and corresponding to the microbial growth. In all the studied combined treatments, organic acids were the preferred substrates,

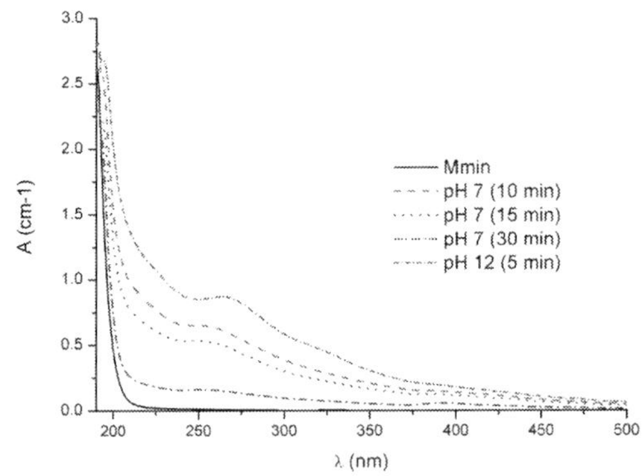

Figure 12. Effect of combined treatment in UV spectra of 4-CPh.

as they were assimilated faster than phenolic compounds. Indeed, phenolic derivatives have shown to serve as inhibitors of the biomass growing.

When 2,4-DCPh was ozonated, a similar condition to the previous one is recognized. The correlation between the biomass growing (showed in the Figure 11) and the phenolics concentrations was confirmed. Moreover, if pH was fixed to 7 and ozonation time was 8 minutes, an important organic matter decrease was observed (Figure 13).

A slighty difference between this case and the previous one shoukd be remarked: the higher organic matter decomposition is gotten. This is explained by the reaction mechanism that has been identified in preliminar studies. In this case, toxicity of byproducts can have a remarkable role on the biomass growing.

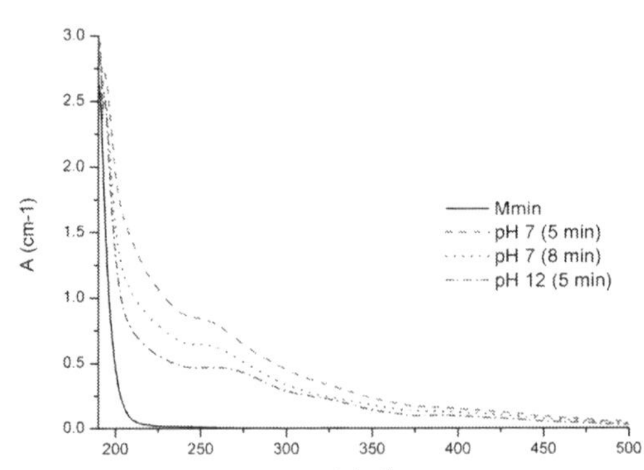

Figure 13. Effect of combined treatment in UV spectra 2,4-DCPh.

3.3.2. *Biological treatment of real residual water after ozonation step*

Two important aspects were observed to improve biodegradation in combined treatment of model solution: 1) phenolic compounds must be eliminated to the lowest achievable value and 2) short chain organic acids should be accumulated as much as possible. Considering these facts and the real residual water ozonation study, two ozonation conditions were chosen to test the suggested combined treatment: 30 and 60 minutes under pH 7. These conditions were selected from the study regarding model sample.

Figures 14 and 15 showed the global biodegradation behavior in combined treatment for real water ozonated during 30 and 60 minutes. In a similar fashion to the model solution, substrate elimination and microbial growth are parallel. To compare the results obtained for the model sample and the real wastewater, biomass growing and substrate are presented in normalized way. They are presented as diminution of UV spectra signal and optical density increases correspondingly.

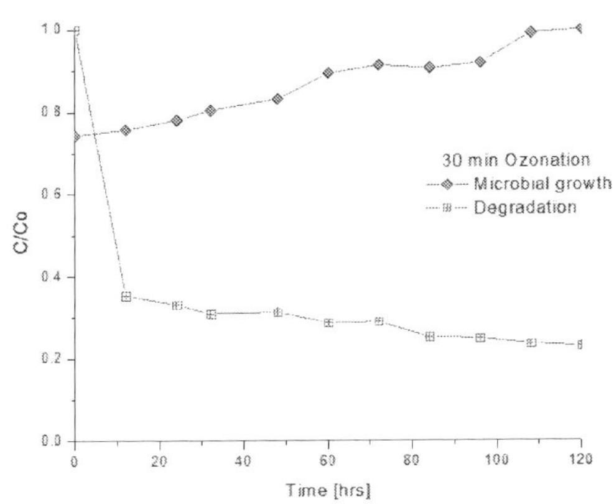

Figure 14. Growth dynamics of biomass and decomposition of organic matter after 30 minutes of ozonation.

During the first 12 hours of biodegradation for real water previously ozonated 30 minutes, more than 60% of organic matter has been metabolized by microorganisms. Moreover, after 5 days, 82% of the initial substrate was removed. In the same way, to the real water, a second sample was ozonated 60 minutes. During the first 12 hours, 78% of the substrate was eliminated and after 5 days, 83% of substrate was metabolized. As it can be observed, the degradation of all compounds that started from the first minutes.

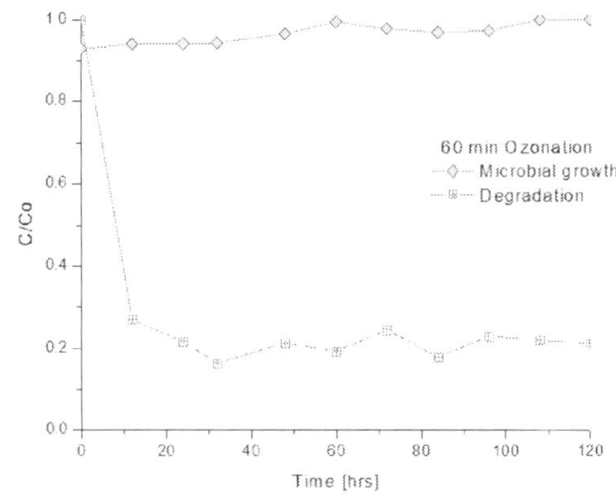

Figure 15. Growth dynamics of biomass and decomposition of organic matter after 60 minutes of ozonation.

This behavior is explained by transformation of pollutants performed by microorganisms. In particular, less toxic substrate as short organic acids were again preferred by them. A decomposition degree of 72%, after 120 hours was gotten.

In the same way for the sample ozonated by 60 minutes, after the first 12 hours, initial substrate was eliminated 78%. When the biodegradation was stopped, an organic matter decrease of 83% was obtained. On the other hand, it is necessary to pay special attention to the almost unchangeable behavior after 30 hours. This is attributed to the formation of some specific products of biodegradation, which can perhaps be subsequently consumed by microorganisms (tendency to continue decreasing). Remembering that most of the ozonated real water is composed of oxalic acid and similar short chain acids. Therefore, the substrate consumption in real water samples is having a similar portrait to organic acids consumption in model solution, it means, since the biodegradation begins, all these acids are metabolized.

Figure 16 shows the dynamics of the three systems after 5 days of biodegradation. This time was selected as the maximum time for the bioprocess. This study presented a removal of organics of 85% for sample ozonated by 30 minutes and 89% for the sample ozonated by 60 minutes.

Finally, in a quantitative way the global effect of biodegradation is the elimination of oxalic acid that was previously formed during ozonation. It is clearly observed that there is a decrease in the concentration of oxalic acid during biodegradation.

In the same way, 30 minutes of ozonation as pre-treatment is more efficient than the other one. They both have similar effects but the suggested one is using half of time for treatment with ozone. As result, final costs of combined treatment are reduced.

http://dx.doi.org/10.5772/56011

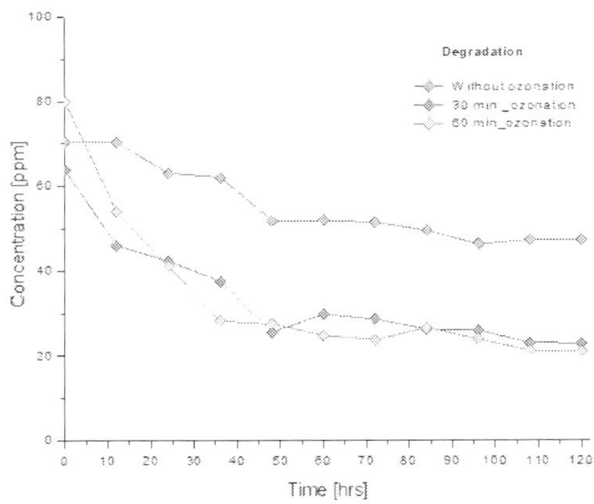

Figure 16. Decomposition dynamics of the three systems (without ozonation, 30 and 60 min of ozonation) after 5 days of biodegradation.

4. Conclusions

The combine residual water treatment using ozone before biodegradation seems to be an interesting option to eliminate more complex and toxic contaminants. The combined treatment may handle the aforementioned type of organics mixtures but with less cost than the pure chemical method and with a shorter treatment period than the biological procedure. For the model residual water, the preliminary ozonation decompose organics in the complex mixture and produce more biodegradable species like organic acids. Longer ozonation times are better if one takes into account the decomposition and accumulation dynamics of both, phenolic compounds and organic acids. When most of the no identified compounds were decomposed during ozonation, after the biodegradation step, the UV spectra was very close to the mineral media one in absence of contaminant. Then a major grade of mineralization after combined treatment is obtained. The previous acclimation of the consortium also showed an improvement of the complete treatment scheme for model residual water. The mineralization of ozonation by-products was confirmed by the microbial growth (in aerobic biodegradation, microbial growth is always accompanied by CO2 production). In the real residual water, the obtained results confirmed the completely decomposition of toxic residues after 60 minutes of ozonation. The decomposition dynamics of lignin derivatives and chlorinated phenols are proportional to the formation of the oxalic acid. This is the main final product of ozonation. Degradation dynamics of these compounds are shown as well as the formation of the oxalic acid.

Acknowledgements

This work was supported through funding provided by CONACYT grants 49367, 60976. Also, authors thanks to the IPN project SIP-20120406.

Author details

P. Guerra[1], J. Amacosta[1], T. Poznyak[1], S. Siles[1], A. García[2] and I. Chairez[3*]

*Address all correspondence to: isaac_chairez@yahoo.com

1 Superior School of Chemical Engineering, National Polytechnic Institute (ESIQIE-IPN), Mexico

2 Instituto Tecnológico de Estudios Superiores de Monterrey, Campus Guadalajara (ITESM), Mexico

3 Professional Interdisciplinary Unit of Biotechnology, National Polytechnic Institute (UPI-BI-IPN), Mexico

References

[1] Grau, P. (1991). Textile industry wastewater's treatment. Water Sci. Technol. , 24, 97-103.

[2] Beltrán, F. J, Encinar, J. M, & González, J. F. (1997). Industrial wastewater advanced oxidation. Ozone combined with hydrogen peroxide or UV radiation, Water Res. , 31, 2415-2428.

[3] Banerjee, A, & Ghosha, A. K. (2010). Isolation and characterization of hyper phenol tolerant Bacillus sp from oil refinery and exploration sites. J. Hazard. Mater. , 173, 783-788.

[4] Epa, U. S. http://www.epa.gov/waterscience/methods/pollutants.htm, last accessed (2010).

[5] Poznyak, T. I, & Vivero, J. L. (2005). Degradation of aqueous phenol and chlorinated phenols by ozone. Ozone: Sci. Eng. , 27, 447-458.

[6] Saravanan, P, Pakshirajan, K, & Saha, P. (2008). Growth kinetics of an indigenous mixed microbial consortium during phenol degradation in a batch reactor. Bioresour. Technol. , 99, 205-209.

http://dx.doi.org/10.5772/56011

[7] Wei, G, Yu, J, Zhu, Y, Chen, W, & Wang, L. (2008). Characterization of phenol degradation by Rhizobium sp. CCNWTB 701 isolated from Astragalus chrysopteru in mining tailing region. J. Hazard. Mater. , 151, 111-117.

[8] Abuhamed, T, Bayraktar, E, Mehmetoglu, T, & Mehmetoglu, U. (2004). Kinetics model for the growth of Pseudomonas putida F1 during benzene, toluene and phenol biodegradation. Process Biochem. , 39, 983-988.

[9] Kumar, A, Kumar, S, & Kumar, S. (2005). Biodegradation kinetics of phenol and catechol using Pseudomonas putida MTCC 1194. Biochem. Eng. J. , 22, 151-159.

[10] Essam, T, Amin, M. A, Tayeb, O. E, Mattiasson, B, & Guieysse, B. (2010). Kinetics and metabolic versatility of highly tolerant phenol degrading Alcaligenes strain TW1. J. Hazard. Mater. , 173, 783-788.

[11] Bajaj, M, Gallert, C, & Winter, J. (2009). Phenol degradation kinetics of an aerobic mixed culture. Bioch. Eng. J. , 46, 205-209.

[12] Nam, K, & Kukor, J. (2000). Combined ozonation and biodegradation for remediation of mixtures of polycyclic aromatic hydrocarbons in soil. Biodegrad. , 11, 1-9.

[13] Beltran-heredia, J, Torregrosa, J, Dominguez, J. R, & Garcia, J. (2000). Aerobic biological treatment of black table olive washing wastewaters: effect of an ozonation stage. Process Biochem. , 35, 1183-1190.

[14] Zhao, G, Zhou, L, Li, Y, Liu, X, Ren, X, & Liu, X. (2009). Enhancement of phenol degradation using immobilized microorganisms and organic modified montmorillonite in a two-phase partitioning bioreactor. J. Hazard. Mater. , 169, 402-410.

[15] Chen, K. C, Lin, Y. H, Chen, W. H, & Liu, Y. C. (2002). Degradation of phenol by PAA-immobilized Candida tropicalis. Enzyme Microb. Technol. , 31, 490-497.

[16] Hong PKAZeng Y. ((2002). Degradation of pentachlorophenol by ozonation and biodegradability of intermediates. Water Res. , 36, 4243-4254.

[17] Khokhawala, I. M, & Gogate, P. R. (2010). Degradation of phenol using a combination of ultrasonic and UV irradiations at pilot scale operation. Ultrason. Sonochem. , 17, 833-838.

[18] Chaichanawong, J, Yamamoto, T, & Ohmori, T. (2010). Enhancement effect of carbon adsorbent on ozonation of aqueous phenol. J. Hazard. Mater. , 175, 673-679.

[19] Derudi, M, Venturini, G, Lombardi, G, Nano, G, & Rota, R. (2007). Biodegradation combined with ozone for the remediation of contaminated soils. Eur. J. Soil Biol. , 43, 297-303.

[20] Benitez, F J, Real, F. J, Acero, J. L, Garcia, J, & Sanchez, M. (2003). Kinetics of the ozonation and aerobic biodegradation of wine vinasses in discontinuous and continuous processes. J. Hazard. Mater. B, 101, 203-218.

[21] Aparicio, M. A, Eiroa, M, Kennes, C, & Veiga, M. C. (2007). Combined post-ozonation and biological treatment of recalcitrant wastewater from a resin-producing factory. J. Hazar. Mater. , 143, 285-290.

[22] García-peña, E. I, Zarate-segura, P, Guerra-blanco, P, Poznyak, T, & Chairez, I. (2012). Enhanced Phenol and chlorinated phenols removal by combining ozonation and biodegradation, , 223, 4047-4064.

[23] Nam, K, Rodríguez, W, & Kukor, J. (2001). Enhanced degradation of polycyclic aromatic hydrocarbons by biodegradation combined with a modified Fenton reaction. Chemosphere , 45, 11-20.

[24] Contreras, S, & Rodriguez, M. Al Momani F, Sans C, Esplugas S. ((2003). Contribution of the ozonation pre-treatment to the biodegradation of aqueous solutions of 2,4-dichlorophenol. Water Res. , 7, 3164-317.

[25] El-Naas, M H, Al-zuhair, S, & Makhlouf, S. (2010). Batch degradation of phenol in a spouted bed bioreactor system. J. Ind. Eng. Chem. , 16, 267-272.

[26] Adav, S. S, Chen, M Y, Lee, D. J, & Ren, N. Q. (2007). Degradation phenol by Acinetobacter strain isolated from aerobic granules. Chemosphere , 67, 1566-1572.

[27] Godjevargova, T, Ivanova, D, Aleksieva, Z, & Dimova, N. (2003). Biodegradation of toxic organic components from industrial phenol producing wastewater by free and immobilized Trichospora cutaneum R 57. Process Biochem. , 38, 915-920.

[28] Edalatmanesh, M, Mehrvar, M, & Dhib, R. (2008). Optimization of phenol degradation in a combined photochemical-biological wastewater treatment system. Chem. Eng. Res. Des. , 86, 1243-1252.

[29] Dong, Y, Yang, H, He, K, Wu, X, & Zhang, A. (2008). Catalytic activity and stability of Y zeolite for phenol degradation in the presence of ozone. Appl. Catal. B: Environ. , 82, 163-168.

[30] Ali, M, & Sreekrishna, T. R. (2001). Aquatic toxicity from pulp and paper mill effluents: A Review. Advances in Environmental Research , 5, 175-196.

[31] Amat, A. M, Arques, A, Miranda, M. A, & López, F. (2005). Use of ozone and/or UV in the treatment of effluents from board paper industry. Chemosphere 60 (8), 1111-1117.

[32] Bijan, L, & Mohseni, M. (2005). Integrated ozone and biotreatment of pulp mill effluent and changes in biodegradability and molecular weight distribution of organic compounds. Water Research , 39, 3763-3772.

[33] Industry Profile sponsored by Contaminated Land and Liabilities Division(1996). Pulp and Paper Manufacturing Works. Department of the Environment Industry Profile, UK

[34] Fengel, D, & Wegener, G. Wood: Chemistry, ultraestructure and reactions". Walter de Gruyter ((1984). Berlin/New York.

[35] Freudenberg, K, & Neish, A. C. (1968). Constitution and Biosynthesis of Lignin. Berlin: Springer-Verlag, 129.

[36] HillCallum A.S. ((2006). Wood modification. Chemical, Thermal and Other Processes. John Wiley & Sons, Ltd, EEUU

[37] AdlerErich. ((1977). Lignin chemistry-past, present and future. Wood Science and Technology, 11(3), 169-218

[38] Demin, V. A, Shereshovets, V, & Monakov, J. B. (1999). Reactivity of Lignin and Problems of its Oxidative Destruction with Peroxy Reagents. Russian Chemical Reviews, 68(11), 937-356.

[39] Peng, G, & Roberts, J. (2000). Solubility And Toxicity Of Resin Acids. Water Research, 34(10), 2779-2785.

[40] RowellRoger. ((2012). Handbook of wood chemistry and wood composites. 2nd Edition. CRC Press Taylor & Francis Group, EEUU.

[41] Ruiz-dueñas, F. J, & Martínez, A. T. (2009). Microbial degradation of lignin: how a bulky recalcitrant polymer is efficiently recycled in nature and how we can take advantage of this. Microbial Biotechnology 2(2), 164-177

[42] Buitron, G, & Gonzalez, A. (1996). Characterization of the microorganisms from an acclimated activated sludge degrading phenolic compounds, Water Sci. Technol. , 34, 289-294.

[43] Kimet., al., ((2002). Biodegradation of phenol and chlorophenols with defined mixed culture in shake-flasks and a packed bed reactor, Process Biochem., , 37, 1367-1373.

[44] Nay, O, Erdeml, E, Kabdali, I, & Lmez, T. (2008). Advanced treatment by chemical oxidation of pulp and paper effluent from a plant manufacturing hardboard from waste paper. Environmental Technology. , 29, 1045-1051.

[45] García-peña, E. I, Zarate-segura, P, Guerra-blanco, P, Poznyak, T, & Chairez, I. (2012). Enhanced phenol and chlorinated phenols removal by combining ozonation and biodegradation, Water, Air, and Soil Pollution, 223 (7), 4047-4064.

[46] Poznyak, T. I, & Vivero, E. J. L. (2005). Degradation of aqueous phenol and chlorinated phenols by ozone, Ozone Science & Engineering, 27 (6), 447- 458.

Chapter 15

Emerging Trend in Natural Resource Utilization for Bioremediation of Oil — Based Drilling Wastes in Nigeria

Iheoma M. Adekunle, Augustine O. O. Igbuku,
Oke Oguns and Philip D. Shekwolo

Additional information is available at the end of the chapter

http://dx.doi.org/10.5772/56526

1. Introduction

1.1. Background

Nigeria is a country endowed with diverse mineral and natural resources among which is petroleum, a pivot to the national economy and sustainable development. In the past five decades, petroleum exploration and production activities have brought national economic boom but not without some aches. Acts of sabotage such as crude oil theft, pipeline bunkering and artisanal refining added to accidental spills and operational failures all combine to aggravate the oil-related aches. Oil spill into the environment, stemming from either acts of sabotage or operational failures, ultimately lead to environmental pollution with petroleum hydrocarbons [1, 2]. Petroleum mining or drilling is another factor to petroleum hydrocarbons in the environment. Most of the adverse impacts of oil spill/ petroleum hydrocarbons in the environment are experienced in the oil bearing communities, located in the Niger Delta region of the country; prominent among them being the Ogoni land pollution incidence reported by United Nations Environment Programme [1]. Petroleum exploration and production activities are strongly associated with drilling operations for oil mining. Accordingly, the extraction of petroleum resources from the earth is achieved by drilling activities. A developed drilling concept, irrespective of technological advancement, has its technical challenges, process requirements and environmental issues [3]. Drilling fluids, also referred to as drilling muds are used to enhance drilling activities via suspension of cuttings, pressure control, stabilization of exposed rocks, provision of buoyancy, cooling and lubricating.

Types of drilling fluids (muds): There are basically two categories of drilling fluids namely (i) aqueous drilling muds or water based muds (WBMs), which consist of fresh or salt water containing a weighting agent, usually barite ($BaSO_4$), clay or organic polymers and various inorganic salts, inert solids, and organic additives to modify the physical properties of the mud so that it functions optimally and (ii) non-aqueous drilling fluids (NADFs), which comprise all non-water dispersible base fluids such as oil based muds (OBMs) and synthetic based muds (SBMs) [2]. Comparative evaluation of oil based muds and water based muds shows that OBMs offer advantages over WBMs for the reasons that [3]:

- OBMs are more suitable to drill sensitive shells, allowing drilling faster than the WBMs, providing excellent shale stability
- they are more adequate to drill formulations where bottom hole temperatures exceed WBMs tolerance, especially in the presence of contaminants such as water, gases, cement, salt and temperature up to 550F
- OBMs resist formation salt leach out
- they are characterized by thin filter cakes and the friction between the pipe and wellbore is minimized, thus, reducing the risk of differential sticking and are especially suited for highly deviated and horizontal wells
- the drill of low pore pressure formations is easily accomplished, since mud weight can be maintained at a weight less than that of water (as low as 7.5 ppg)
- corrosion of pipe is controlled since oil is the external phase and coats the pipe. The oils are non-conductors and the additives are thermally stable, hence, do not form corrosive products
- bacteria do not thrive long in OBMs
- there is the possibility of using OBMs over and over again and can be stored over long periods of time since bacterial growth is suppressed
- OBM packer fluids are designed to be stable over long periods of time even when exposed to high temperature and provide long-term stable packers since additives are extremely temperature stable. Properly designed, such packer fluids can suspend weighting materials over long periods of times.

In other words, regarding shale stability, penetration rate, high temperatures, drilling salts, lubrication, low pore pressure formations, corrosion control, re-use and packer fluids, OBMs offer advantages over WBMs. It is therefore, obvious that though WBMs are more environmentally benign, they are only satisfactory for less demanding drilling of conventional vertical wells at medium depths, whereas OBMs are more suited for greater depths or in directional or horizontal drillings, which exert greater stress on drilling apparatus. As a result, OBMs are more frequently used in petroleum industries for drilling purposes. The composition of OBMs include: petroleum base fluid, weighting agent and other chemical additives.

Drill cuttings: During drilling, particles of crushed rocks produced by the grinding action of the drill bit as it penetrates the earth are referred to as drill cuttings (DC). DCs are, therefore, a mixture of rocks and particulates released from geological formulations in the drill holes made for crude oil drilling and are usually coated with the drilling fluid. Consequently, DCs are largely influenced by the chemical composition of drilling muds [2, 4].

The resultant spent OBM and drill cuttings (drilling wastes) consist of hydrocarbons, water, soils, heavy metals and water soluble salts such as chlorides and sulphates [3, 4]. Drilling wastes, which are toxic due to the presence of hydrocarbons, heavy metals and other chemical additives, if not properly treated before disposal, pose serious environmental hazards and risk to public health. Sequel to these, best practices in the management of drilling wastes cannot be over emphasized.

1.2. Health and environmental effects associated with drilling wastes

Health effects linked to drilling wastes are traceable to the basic components such as the drilling fluid and additives:

Health effects associated with drilling fluids: These health effects are attributed to the physical and chemical properties of the drilling fluids. In oil based drilling wastes, the base oil stem from petroleum stream such as crude oil, diesel (gasoil) and kerosene, which cause skin irritation. Consequently, the most commonly observed health effect associated with drilling fluids is skin irritation. Other effects include headache, nausea, eye irritation and coughing. Routes of exposure in human are dermal, inhalation, oral and some other miscellaneous routes. On exposure to drilling fluid, petroleum hydrocarbons tend to remove natural fat from the skin, which results in skin drying and cracking. These conditions allow compounds to permeate through the skin leading to irritation and dermatitis. Susceptibility to these health effects varies with individual resistance capacity and conditions of poor personal/environmental hygiene. High aromatic content fluids, especially diesel fuel contain significant levels of carcinogenic polynuclear aromatic hydrocarbons (PAHs). Diesel fuels may also be genotoxic due to high proportions of 3-7 ring PAH [2]. Skin-painting studies in mice showed that, irrespective of the level of PAH, long-term dermal exposure to diesel fuels can cause skin tumours, an effect attributed to chronic skin irritation. In humans, chronic irritation may cause small areas of the skin to thicken, eventually forming rough wart-like growths, which may become malignant. Health effects from chronic exposure to PAHs may include cataracts, kidney damage, liver damage and jaundice. Naphthalene, a specific PAH, can cause the breakdown of red blood cells, if inhaled or ingested in large amounts. Animals exposed to levels of some PAHs over long periods in laboratory studies, developed lung cancer from inhalation and stomach cancer from ingesting PAHs in food [2].

Other hydrocarbon constituents of drilling fluids are the mono-aromatics popularly referred to as BTEX (benzene, toluene, ethylbenzene and xylene). BTEX compounds are very volatile, hence, will readily evaporate in warm/hot climates of tropical regions, resulting in higher concentrations in the vapor phase. As a result, there is the possibility of exposure to human via inhalation. Exposure to high concentrations of these hydrocarbons via inhalation may result in hydrocarbon induced neurotoxicity, a non-specific effect resulting in headache, nausea, dizziness, fatigue, lack of coordination, problems with attention and memory, gait disturbances and narcosis [2].

Health effects associated with additives: In addition to the irritancy of the drilling fluid hydrocarbon constituents, several drilling fluid additives may also have irritant, corrosive or sensitizing properties. Various additives include emulsion stabilizers, pH adjusters, wetting agents, viscosifiers and fluid-loss reducing agents. For instance, calcium chloride ($CaCl_2$) has irritant properties and emulsifiers (such as polyamine) have been associated with sensitizing properties [3]. Specific chemical additives vary with locations.

1.2.1. *Environmental effects associated with drilling wastes*

Apart from health effects, environmental hazards associated with drilling wastes include land, water and air pollution [5]:

i. **Land pollution:** Farming is the major land use system in Nigeria, especially in the Niger Delta region [1]. The most significant in this aspect of environmental pollution in Nigeria is thus farmland pollution. Consequences include alteration in soil physical, biological and chemical properties, loss of soil fertility, stunted plant growth and reduced crop productivity. These lead to reduced food security and compromised food safety.

ii. **Aquatic pollution:** Large percentage of the oil spill gets spread over the surface of the aquatic system resulting in anaerobic environment in the water, below the surface. This leads to death of the natural flora and fauna where oxygen is the key element for their respiration; adversely affecting fishing profession [1]

iii. **Air pollution:** volatile organics such as benzene, toluene, ethylbenzene and xylene could have elevated concentrations in the air, leading to atmospheric pollution and consequent adverse environmental and health impacts.

Oil well drilling processes generate large volumes of drill cuttings and spent mud in the country. Drilling wastes, therefore, add to hazardous petroleum waste materials released in the environments of the Niger Delta region of the country [1, 6] and the management of drilling wastes is quite tasking. An environmentally friendly technique for the management of drilling wastes is necessary in all offshore and onshore operations; from seismic surveys, drilling operations, field development and production to decommissioning. The physical and chemical properties of the drilling wastes influence their hazardous characteristics and environmental impact abilities, which in turn depend primarily on: (i) nature of impacted material, (ii) concentration of pollutant /amount of waste material after release (iii) recipient biotic community and (iv) exposure duration. Exposure that causes an immediate effect is called acute exposure while long-term exposure is called chronic exposure. Either acute or chronic exposure has negative impacts.

1.3. Contemporary treatment of drilling waste materials

Worldwide, contemporary drilling waste management options include re-use, offshore discharge, re-injection and onshore treatment and/or disposal [7]. Each treatment and or disposal option has its pros and cons as highlighted in the options (thermal technologies and bioremediation techniques) discussed.

1.3.1. Thermal treatment

As the name suggests, thermal technologies involve the use of high temperatures to reclaim hydrocarbon contaminated materials [8]. Thermal treatment is mostly used in treating organic compounds. Additional treatment may be necessary for metals and salts depending on the final fate of the wastes. Thermal treatment technologies are designed for a fixed land based installation; however, a few mobile units exist. Two commonly practiced thermal treatment technologies are thermal desorption and incineration methods.

1.3.1.1. Thermal desorption method

Thermal desorption is an environmental remediation process that uses heat to increase the volatility of contaminants by the use of a series of equipment (desorber and oxidizer) such that the hydrocarbons and water are separated or removed from the solid matrix. It is normally carried out between the temperature range of 250-650°C. At these temperatures both the lighter and heavier hydrocarbons are removed and collected or thermally oxidized by further heating to a temperature of over 850°C. The resulting solid residue has essentially no residual hydrocarbons (having been oxidized), but does concentrate salts and heavy metals. Depending upon the success of process used, recovered hydrocarbons can be used as fuel or re-used as base fluid in the drilling fluid system and the resulting solid can be disposed of in a landfill or may be used in construction (of roads and bricks). Economical, operational and environmental implications of thermal desorption include:

1. Effective removal and recovery of hydrocarbons from solids
2. Possibility of recovering base fluid and end - product could be used for brick making
3. Low potential for future liability
4. Requires short time
5. High cost of handling environmental issues
6. Large volume of wastes is required to justify the cost of operation
7. Requires tightly controlled process parameters
8. High operating temperatures can lead to safety risks
9. Requires several operators
10. Heavy metals and salts are concentrated in residual solids
11. Process water contains some emulsified oil
12. Residue ash requires further treatment before disposal
13. End product is sterile and can no longer support plant Life.

1.3.1.2. Incineration method

Incineration involves (i) heating oil based mud and drill cuttings to a higher temperature range (1200-1500°C) in direct contact with combustion gases and (ii) oxidizing the hydrocarbons [8]. Solid/ash and vapor phases are generated. The gases produced from this operation may be passed through an oxidizer, wet scrubber, and bag house before being vented to the atmosphere. Stabilization of residual materials may be required prior to disposal to prevent constituents from leaching into the environment. Incineration of drilling wastes occurs in rotary kilns, which incinerate any waste regardless of size and composition. Incineration systems are designed to destroy only organic components of waste; however, most drilling wastes are non-exclusive in their content and therefore will contain both combustible organics and non-combustible inorganic materials. By destroying the organic fraction and converting it to carbon (IV) oxide and water vapor, incineration reduces waste volume. Inorganic components of wastes fed to an incinerator cannot be destroyed, only oxidized. The major inorganic materials are chemically classified as metals. Generally, these metals will exit the combustion process as oxides of the metals that enter. Economical, operational and environmental implications of incineration are as listed:

1. Low potential for future liability
2. High cost per volume
3. Heat produced could be used for energy generation
4. High energy cost
5. Requires air pollution control equipment because of safety concerns
6. At high temperatures, salts can form acid components
7. Air emissions pose environmental concerns.

In line with best practices, for thermal technologies, there is need for proper placement of end product. Demonstration of sufficient compliance with current regulations and adequate safety measures to cater for the potential risks of exposure to high temperatures.

1.3.2. Bioremediation technique

Bioremediation technique relies on the ability of microorganisms (mostly combination of bacteria) to feed on the hydrocarbons (HCs) as substrate, converting them into carbon dioxide, water and harmless clean solids; and the ability of some of the HCs to biodegrade over time. But in most cases, the native microorganisms are often overwhelmed by the extent of the hydrocarbon contamination and thus would require external nutrients to boost (bio-stimulation) their activity and ability to take up the HCs at a faster rate. In other cases, the native microorganisms may be needing help from their kind or other species of micro-organism which are grown or inoculated (bio-augmentation) in the laboratory and then introduced in the habitat of the native micro-organisms. Bioremediation could be carried out at the site of contamination (in-situ bioremediation technique) or off the site of contamination (ex-situ bioremediation technique). Bioremediation technologies include land farming, use of bioreactors, biopiles and compost- based technologies. Economical, operational and environmental implications of conventional bioremediation technique [9, 10, 11, 12, 13, 14] include:

1. Relatively inexpensive
2. Requires simple equipments and eliminates transportation cost as drill wastes could be treated on site
3. Less capital but may be labour-intensive.
4. Low maintenance cost; being a simple technology process that requires few machines, there are few delays due to equipment down-time
5. Process is fairly flexible and can be used for most drill wastes including OBM, NADFs, previously extracted materials and newly drilled cuttings
6. Proven technology
7. Requires a considerable period of time to complete a process
8. Appropriate bacteria and nutrient selection could be a daunting task
9. In cases where bacteria are inoculated and brought on site, adaptability to their new environment may hamper their performance
10. Minimal operation hazards
11. Environmentally friendly: once the contaminants have been degraded, the microbial population reduces considerably as they have used up their food source
12. Less impact on the environment as residue from process (TPH $< 1\%$) may require no further treatment and could be used for agricultural purposes.

Recommended best practices for bioremediation technology include ensuring (i) proper initial physical, biological and chemical characterizations to determine extent of organic and inorganic contamination, (ii) required skill and persistence for the selection of several combinations of bacteria and nutrients that can provide the desired result (iii) proper periodic tillage to provide for proper aeration that facilitates degradation of the HCs and (iv) an accurate and appropriate TPH level check in between treatment process in order to monitor progress of the remediation process. Choice of waste management options typically considers local regulations, environmental assessment, cost/benefit analysis and the composition of the drilling wastes. The Department of Petroleum Resources [15] via the Environmental Guidelines and Standards for the Petroleum Industry in Nigeria (EGSPIN) stipulated guidelines on drill cuttings discharge for inland / near-shore and offshore deep water in order to minimize the adverse impact on the surrounding environment. These requirements call for an appropriate drill cuttings treatment prior to disposal in order to meet the stipulated conditions.

1.4. Review of emerging trend in the treatment of drilling waste materials in Nigeria

There are scientific evidences showing that drilling wastes generated in the country contain toxicants that are of environmental concerns. For instance, the reports of [16] on the determination of selected physical and chemical parameters including metals concentrations in a certain drill cutting dump site in the country. Results from their study showed that oil and grease on the surface and 20 feet around the waste dump area were above the specified limit [15]. There was also lack of plant growth noticed in the study, attributed to depletion of nitrogen, phosphorus and potassium values below threshold levels for plant growth. The reports of [4] on hydrocarbon and some metal contents of drilling muds and cuttings generated during the drilling of Igbokoda onshore oil wells gave total petroleum hydrocarbon (TPH), aliphatic hydrocarbon (AH) and polycyclic aromatic hydrocarbon (PAH) as generally exceeding stipulated limits by both national and international agencies. The studies of [17] on the compositional distribution and sources of polynuclear aromatic hydrocarbons (PAHs) in Nigerian oil-based drill-cuttings, showed that the total initial PAHs concentration of the drill cuttings was 223.52 mg/kg while the initial individual PAHs concentrations ranged from 1.67 to 70.7 mg/kg, dry weight, with a 90% predominance of the combustion-specific 3-ring PAHs.

The commonly employed remediation techniques for drilling wastes in Nigeria appear to be thermal technologies. However, due to economical, operational and environmental implications of these thermal technologies; search for more acceptable techniques commenced. There is scarcity of literature on the use of natural resource materials for the remediation of drilling wastes in Nigeria. The few literature resources showed that a large percentage is still at the bench-scale platform. For instance, [18] isolated *Staphylococcus sp.* from oil-contaminated soil that was treated with 1% drilling fluid base oil (HDF-2000). Their study revealed that *Staphylococcus sp.*, is a strong primary utilizer of the base oil and has potential for application in bioremediation processes involving oil-based drilling fluids. On the other hand, the effectiveness of 2 bacterial isolates (*Bacillus subtilis* and *Pseudomonas aeruginosa*) in the restoration of oil-field drill-cuttings contaminated with polynuclear aromatic hydrocarbons was studied by [19]. In that study, a mixture of 4 kg of drill cuttings and 0.67 kg of top-soil were fed into triplicate plastic reactors labeled A1 to A3, B1 to B3, C1 to C3 and O1 to O3. These were left quiescent for 7 days under ambient conditions, followed by the addition of 20 mL working solution of pure cultures of *Bacillus* sp and *Pseudomonas* sp (each of cell density 7.6 x 10^{11} cfu/mL) to reactors A1 - A3 and B1 - B3 respectively. Another 20 mL working solution containing both cultures at cell density 1.5 x 10^{12} cfu/mL was added to reactors C1 - C3. The working solution was added to each reactor (excluding the controls, O1 - O3) every 2 weeks. Mixing and watering of the set-ups were carried out at 3 days interval under ambient temperature of 30°C for a period of 6 weeks. Results showed that the predominant 3-ring PAHs, which made up 90% w/w of the total PAHs concentration of 223.52 mg/kg, were degraded below detection and the 4-ring PAHs were reduced from 4 to 0.6% by *Pseudomonas* while *Bacillus* reduced 3 and 4-ring PAHs respectively to 0.2 and 0.8%. Their works revealed that Pseudomonas degraded 3 and 4-ring PAHs relatively better than *Bacillus*. Both strains of bacteria degraded 5 and 6-ring PAHs below detection limits. Furthermore within the 3-ring PAHs, each of the strains of bacteria reduced phenanthrene to approximately 0.2%, whereas both degraded homologues acenaphthylene, acenaphthene and fluorene as well as anthracene below detection limits. For 4-ring PAHs, *Pseudomonas* degraded fluoranthene and benzo[a]anthracene. *Bacillus* also degraded benzo[a]anthracene below detection limits. *Pseudomonas* was able to reduce pyrene and chrysene to 0.3 and 0.2% respectively; whereas *Bacillus* reduced fluoranthene, pyrene and chrysene to 0.1, 0.01 and 0.4% respectively. However, treatment with the mixed culture resulted in limited degradation of 5-ring PAHs particularly in the fourth week, which was attributed to the phenomena of co-metabolism and inhibition.

The works of [20] compared the potentials of bio-augmentation and conventional composting as bioremediation technologies for the removal of PAHs from oil-field drill-cuttings. From a mud-pit, close to a just-completed crude-oil well in the Niger Delta region of Nigeria, 4000 g of drill cuttings was obtained and homogenized with 667 g of top-soil (to serve as microbes carrier) in three separate reactors (A, B and C). The bio-augmentation of indigenous bacteria in the mix was done by adding to reactors A and B a 20-mL working solution (containing $7.6x10^{11}$ cfu/mL) of pure culture of *Bacillus* and *Pseudomonas,* respectively, while a 20-mL working solution (containing $1.5x10^{12}$ cfu/mL) of the mixed culture of *Bacillus* and *Pseudomonas* was added to reactor C. The bio-preparation was added to each reactor (excluding the control) every two weeks for six weeks. The composting experiment was conducted in a 10-litre reactor in which 4000 g of drill cuttings, 920 g of topsoil and 154 g of farmyard manure and poultry droppings were homogenized. Mixing and watering of the set-ups were carried out at 3 days interval under ambient temperature over a period of six weeks. Results showed that initial individual PAHs concentrations in the drill cuttings ranged from 1.67 to 70.7 mg/kg dry weight, with a predominance of combustion-specific 3-ring PAHs (representing 90% of a total initial PAHs. After the bioremediation exercise that lasted for 42 days, total PAHs in the drill cuttings were reduced from 223.52 to 4.25 mg/kg, representing a 98.1% reduction. Away from the use of microbial strains in the treatment of drilling wastes, a bench-scale investigation was carried out by [21] to demonstrate the efficacy of technique referred to as 'Dispersion by Chemical Reaction (DCR) technology".This particular method involved the use of hydrophobized calcium oxide (CaO) to form a dry, soil-like material that could be useful in construction works.

On the other hand, after the study on the response of four phytoplankton species in some sections of Nigeria coastal waters to crude oil in controlled ecosystem [22], that revealed the adverse impacts; a multidisciplinary environmental remediation research group (ERRG) was inaugurated with the mandate to embark on innovative, cutting-edge research and development (R & D) initiative, aimed at the development of an indigenous technology for an eco-friendly technique in the treatment of soils, sediments, sludge and drilling wastes polluted by petroleum hydrocarbons, using natural products of Nigeria origin. The goal of ERRG is to translate the technology from bench-scale to field scale and come out with on- the - shelf products that will find use for both onshore and offshore remediation works. The first phase of the R & D initiative was the exploration of the remediation potential of conventional composting technology based on the results from the works of [23]. A good start was the production of a scientifically formulated and classified compost bulk [24] that are potentially viable for environmental remediation projects [25] and able to biodegrade petroleum hydrocarbons embedded in soil and related matrices [26]. The next phase was to assess public acceptance of the principles of this technology, which culminated to the reports of [27] on population perception impact on value-added solid waste disposal in developing countries, a case study of Port Harcourt City. The feedstock utilized in product formulations in this emerging, indigenous and innovative technology is 100% biodegradable and very abundant in the Nigerian environment. Consequently, the technology has been categorized by stakeholders [27] as:

i. eco-friendly environmental remediation technique

ii. waste to wealth initiative

iii. waste to resource initiative

iv. value-added waste management option

v. a contribution to the promotion of local material development that has the potential for:

- wealth creation
- job creation
- poverty alleviation
- sound environmental management of hydrocarbon polluted wastes from the petroleum industries.

ERRG observed that either conventional composting technology or bioremediation via utilization of pure microbial isolates/strains has limitations in terms of serving the practical needs of the petroleum industry in Nigeria with regards to meeting (i) regulatory remediation targets at close – out of project and (ii) project delivery time. Subsequently, through series of bench-scale and screen house remediation investigations, products were formulated to enhance the speed of bioremediation process using nano-scale green catalysts, a technique that matured into Compost - based Nanotechnology in Bioremediation (CNB-Tech). The research group then subjected the CNB-Tech products to different scientific evaluations in order to ascertain (i) efficiency on biodegradation of petroleum hydrocarbons in oily wastes such as crude oil impacted soils, sludge and drilling wastes (drill cuttings and oil-based mud) and (ii) environmental impacts with emphasis on soil quality. Published works on assessment and prognosis of products' impact on soil quality include:

a. Assessing the effect of bioremediation agent from local resource materials in Nigeria on soil pH [28]

b. Impact of bioremediation formulation from Nigeria local resource materials on moisture contents for soils contaminated with petroleum [29]

c. Assessing and forecasting the impact of bioremediation product derived from Nigeria local raw materials on electrical conductivity of soils contaminated with petroleum products [30]

d. Soil temperature dynamics during bioremediation of petroleum products using remediation agent from Nigerian local resource materials [31].

Other works on CNB-Tech products' evaluations including (i) effect on soil heavy metal dynamics and (ii) impact on soil microbial species population and diversity are being considered elsewhere for publication. Having recorded a huge success during the laboratory scale investigations where maximum of 4000g of sample bulk and freshly hydrocarbon contaminated soils (similar to the quantities used by other investigators) [19, 20] were treated, it became necessary to assess the efficiency of CNB-Tech products on waste materials with complex nature and higher degree of hydrocarbon pollution. This aspiration was realized in collaboration with the Remediation Department of Shell Petroleum Development Company (SPDC), Port Harcourt, Nigeria through the University Liaison Team of SPDC. Sequel to this, pilot-scale projects were commissioned to evaluate the efficiency of CNB-Tech products on the degradation of hydrocarbon compounds in the following petroleum impacted materials:

i. Hydrocarbon polluted clay soils from Ejama-Ebubu legacy site of SPDC

ii. Hydrocarbon polluted carbonized soil from Ejama-Ebubu legacy site of SPDC

iii. Hydrocarbon polluted sludge from Ejama-Ebubu legacy site of SPDC

iv. Oil-based mud and drill cuttings generated from SPDC operations.

Ejama Ebubu is one of SPDC's legacy sites of up to 42 year long pollution as at the time of study in 2011 [1]. In this chapter, the efficacy of CNB-Tech products in the biodegradation of petroleum hydrocarbons in oil-based drilling wastes (OBM-DC) is presented.

1.5. Research justification

The treatment of drilling wastes, especially OBM-DC in an environmentally sound manner is a challenging task due to the complex nature of the wastes. The most popular technique adopted for the treatment of OBM-DC, thermal desorption [15] has its accompanying environmental concerns. For instance, thermal treatment technologies are associated with prohibitive capital and operational cost implications, threatening environmental consequences in addition to high occupational hazards and generation of secondary waste stream that has to be treated at extra high cost before final disposal. Consequently, there is need for a pragmatic shift to seek alternative techniques that will address the need of the oil and gas sector in the management of drilling wastes in terms of remediation target delivery time and compliance to regulatory standards in Nigeria. Regulatory standards for close-out of remediation projects vary from one country to another and success factors of a given technology are dependent on indices such as:

a. climatic conditions

b. geographical characteristics of the location

c. nature and complexity of contamination

d. expected utility of the end-products of the remediation exercise

It then becomes evident that a successful remediation technology in one part of the globe may not necessarily be efficient in another region, pointing to the need to look inward for a more practical approach to solving the environmental challenges posed by petroleum hydrocarbon polluted waste streams in Nigeria [1]. Having run laboratory, bench- scale and screen-house remediation works using CNB-Tech products on fresh hydrocarbon contaminated soils, it became necessary to conduct pilot scale remediation works on more challenging waste streams such as weathered petroleum impacted soils, sludge, sediment, oil- based drilling mud and drill cuttings, hence this project.

1.6. Research objectives

The current study comprised three major objectives:

i. to conduct a review on the emerging trends in the treatment and related studies for drilling wastes in Nigeria,

ii. to assess the efficiency of an indigenous and innovative application of compost - based nanotechnology in bioremediation (CNB-Tech) in biodegradation of hydrocarbons found in oil-based mud and drill cuttings; generated by a petroleum industry in Nigeria

iii. to investigate the beneficial utility of the remediation end-product for agricultural purpose (crop production), which is a major land use system in Nigeria.

2. Research methodology

The research methodologies employed in this study were:

i. Literature review to provide an insight to the current and emerging trend in the treatment of drilling waste materials in the country and

ii. Practical, ex-situ, pilot scale execution of biodegradation of hydrocarbon compounds in oil-based mud and drill cuttings generated by an oil company in Nigeria using an indigenous and innovative biotechnological (CNB-Tech) approach anchored on the use of natural resource materials of Nigeria origin.

2.1. Pilot-scale remediation of oil-based mud and cuttings using CNB-Tech method

This study was carried out during the 2010/ 2011 Sabbatical Programme of the University Liaison Team of Shell Petroleum Development Company (SPDC); in conjunction with the Remediation Department of SPDC, Port-Harcourt, Nigeria. The indigenous remediation products (CNB-Tech products) prepared from cellulosic natural resource materials and biogenic nanopolymers of Nigeria origin used for this pilot remediation study, were denoted as (i) Ecorem, (ii) Bioprimer and (iii) Biozator. The last two products are solids that are transformed to the aqueous form before use while the first product is used in the solid form.

2.1.1. Project site description

The present pilot-scale project, for the purposes of adequate monitoring and efficient execution, was carried out in the Industrial Area of Shell Petroleum Development Company, Port

Harcourt, Rivers State; known as "Shell IA". The earmarked project area was a relatively isolated open green field within Shell IA and according to design, a temporary sheltered facility constructed to suit the project design was erected at the site and all necessary health and safety issues were taken into consideration. The sheltered project facility comprised of three major units:

- Remediation execution section: where actual remediation took place
- Phyto-analytical section: where effects on plant life were investigated
- Mini- chemical laboratory: where necessary onsite chemical evaluations were conducted.

2.1.2. Pilot scale remediation procedure

The batch of oil-based mud and drill cuttings (OBM-DC) used in this study was generated from SPDC's operations and supplied by one of the company's certified vendors. During the conveyance procedure for OBM-DC, chain of custody document and waste stream tracking manifest was observed. Basic highlights for CNB-Tech application mode are outlined in Figure 1. Pretreatment involved recovery of free phase base fluid and stabilization involved modification of viscosity parameter.

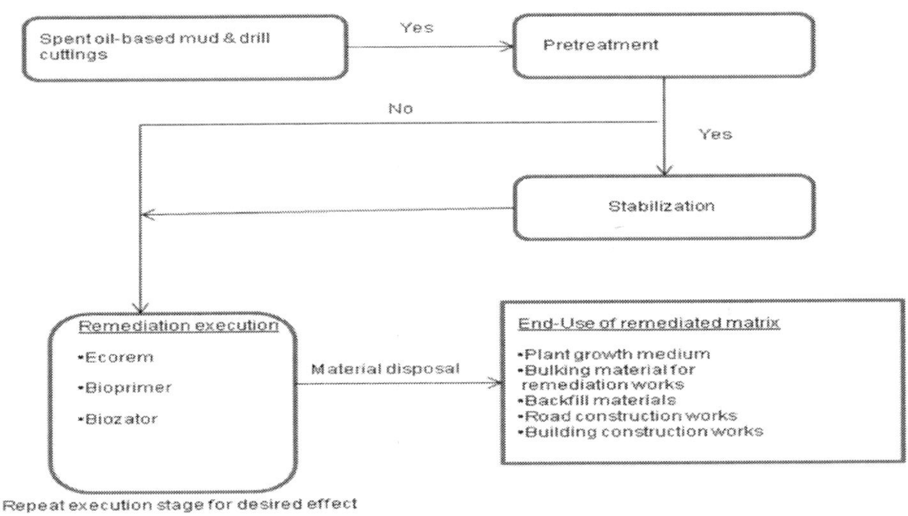

Figure 1. Application model of CNB-Tech remediation method

The biocell utilized for the remediation execution was designed by the research group, locally fabricated and lined with appropriate PVC materials. The procedures involved in the pilot remediation exercise are described as follows: A biocell of total dimension 15 m^3 was subdivided to smaller units of 3 m x 1 m x 1 m to allow for five times replication. Ecorem (a CNB-Tech product) was placed in the cells prior to loading of oil-based drilling mud and cutting

(OBM-DC) that have been previously conditioned using intervention CNB-Tech products. As the initial microbial population in OBM-DC was less than 2.0 x 10^3 cfu/mL, Ecorem was introduced at 10% by weight of waste materials. Using mechanical means, OBM-DC and Ecorem were homogenized and allowed to incubate for about 12 to 24 hours in order to trigger and stimulate natural microbial activities. CNB-Tech products (Bioprimer and Biozator) were then applied to saturate the contents in the biocells, which was followed by homogenization using mechanical devices. A CNB-Tech product was added to the leachate (process fluid) to immobilize inorganic constituents (especially metals) before recycling the leachate into the treatment network in such a manner that no leachate was produced as a by-product for discharge into the environment. OBM-DC that received no treatment served as control. Both controls and test units were subjected to the same environmental conditions.

System maintenance and monitoring: During remediation, the system was monitored for relevant environmental factors such as moisture content (I), pH (II), nitrogen content (III) and temperature (IV) using standard procedures of gravimetry for I, probe method via a calibrated pH meter for II, Kjedahl method for III and calibrated mercury in glass thermometer for IV. These environmental factors were maintained at the required range. Remediation lasted for 33 days: 6 days for actual treatment and 27 days for material fallow and recovery periods during which the treated materials were conditioned with a CNB-Tech product (Ecorem) for use as plant growth medium.

In order to validate the efficacy of this technology, representative composites were sent to an International Laboratory (RespirTeK Consulting Laboratory and affiliate Laboratories based in the United States of America) for physical, chemical and microbial assessments. RespirTek is ISO/EC accredited and certified. Three other laboratories that are based in Nigeria (certified by national regulatory bodies) were also involved in sample collection and analyses. Laboratories that participated in this study were:

1. Technology Partners International Nigeria Limited, Port Harcourt - Nigeria
2. Laser Engineering and Resources Consultants Limited, Port Harcourt- Nigeria
3. Fugro Nigeria Limited, Port Harcourt, Nigeria
4. RespirTek Consulting Laboratory - United States of America

2.2. Sample collection

At the end of the pilot remediation project using CNB-Tech products, treated materials were moved from the biocells and spread out on PVC impermeable membranes (each of dimension 650 cm for length and 248 cm for width), homogenized using mechanical means and air-dried with occasional homogenization of samples. The dry samples were returned into the biocells where further homogenization procedure was carried out. Sampling containers were sent by RespirTEK Consulting Laboratory, USA for their own use.

General sample collection: Using mechanical means, treated and dried samples in the cells were thoroughly homogenized for one week. In order to collect sample from a particular replicate, each replicate was subdivided into 4 equal parts; representative fractions were collected from the different parts and recombined to give a composite sample of 1kg.

BTEX sampling: Standard sampling kit for BTEX, sent by RespirTEK Consulting Laboratory, was utilized for the purpose. In this procedure, homogenized samples were collected from the cells using "Terra Core" sampling device. Using a 40 mL glass VOA vial containing appropriate preservatives and with the plunger seated in the handle, the Terra Core was pushed into freshly homogenized sample until the sample chamber was filled to the capacity of 5g. All sample particulates (debris) were removed from the outside of the Terra Core sampler and the sample plug was pushed into the mouth of the sampler. Excess soil that extended beyond the mouth of the sampler was removed. The plunger was then seated in the handle and rotated until it aligned with the slots in the body. The mouth of the sampler was placed into the 40 mL VOA vial containing the preservatives and sample extruded by pushing the plunger down. The lid was quickly placed back on the 40 mL VOA vial. It was ensured that when capping the 40 mL VOA vial, sample debris was removed from the top of the vial.

All samples were appropriately labeled and recorded in the chain of custody form before shipping to the USA laboratory by courier. Two Laboratories in Nigeria also collected samples for analyses, following standard procedures. The third laboratory in Nigeria was only involved in the analysis of materials using infrared and UV-absorption spectroscopic methods.

2.3. Physicochemical analysis and microbial assessment

Statement from quality control and quality assurance unit (QA/QC) of RespirTek Laboratory, USA showed that all analyses were conducted following procedures set forth by the ISO/IEC 17025:2005 accreditation program standards for which the laboratory holds certification. Quality assurance systems and quality control criteria were strictly followed. The following parameters were determined:

- Total petroleum hydrocarbons (TPH)
- Monoaromatic hydrocarbons: benzene, toluene, ethylbenzene and xylene (BTEX). For xylene, ortho -, meta - and para- derivatives were assessed
- PAHs: a total of 17 PAH compounds: (i) naphthalene, (ii) acenaphthylene, (iii) acenaphthene, (iv) fluorene, (v) phenanthrene, (vi) anthracene, (vii) fluoranthene, (viii) pyrene, (ix) benzo (a) pyrene, (x) chrysene, (xi) benzo (b) fluoranthene, (xii) benzo (k)fluoranthene, (xiii) benzo (a) pyrene, (xiv) dibenz(a,b) anthracene, (xv) benzo (ghi)perylene, (xvi) 2-methylnaphthalene and (xvii) indeno (1,2,3-cd) pyrene
- Metals: barium (Ba), calcium (Ca), copper (Cu), lead (Pb), mercury (Hg), Nickel (Ni), Sodium (Na), Potassium (K), cadmium (Cd), zinc (Zn) and arsenic (As), a metalloid
- Miscellaneous parameters: pH, salinity, nitrogen, phosphorus, total organic carbon and electrical conductivity.
- Microbial activity: assessment of 48 hr and 96 hr microbial activities of both remediation end-product and contaminated material (control) was conducted by the USA based

laboratory. Total hydrocarbon utilizing bacteria as well as total microbial count were assessed by the Nigerian based laboratories.

Hydrocarbon compounds were analyzed using Gas chromatographic method, microbial assessment was carried out using heterotrophic plate count method and metals were determined using atomic absorption spectroscopic technique. All the other parameters were carried out using standard procedures such as described in [24, 25, 32]. The CNB-Tech products (Bioprimer and (Biozator) were characterized using infrared and UV-visible spectroscopic methods. The basic characteristics of Ecorem have already been reported in [24, 25] but was slightly enhanced, in this study, for case specificity.

2.4. Assessment of seed germination potential of treated samples

The remediated materials used in this evaluation were not mixed with external soil and no external fertilizer material was added to the remediated soil. Seed germination potential (SGP) of treated samples were assessed and only viable maize seedlings were used for this purpose. In a remediated material matrix (4kg material contained in an experimental plastic pot), 6 seedlings of maize were sown. This was replicated three times. All together, 18 (6 x 3) seedlings were used to evaluate this effect. Similar set- ups were also established for the untreated oil – based mud and cuttings, which served as control systems. This gave a total of 18 (6 x 3) seeds tested for germination potential for the test systems and 18 seedlings for the control media. This phase of the evaluation lasted for 7 days.

2.5. Assessment of process fluid (leachate) effect on plant growth

Adequate leachate (process fluid) management strategy was put in place as leachate generated during remediation was recycled into the remediation process. However, this evaluation was to ensure or to prove that in the event of any leachate seepage there would be reduced environmental risk. This phytotoxicity assessment was carried out using a cereal (corn: Zea mays L.,) as an indicator crop and indices of toxicity were (i) root length and (ii) plant height. Experimental systems constituted of the following set-ups, where FS is dilution factor and SF stands for farm soil:

i. Farm soil + tap water (Code: FS + water). This served as control system for (ii) and (iii)

ii. Farm soil + stock leachate (Code: FS + LDF-0). This served as control system for (iii)

iii. Farm soil + diluted leachate series:

a. Farm soil + leachate DF-1 (Code: FS + LDF-1)

b. Farm soil + leachate DF-2 (Code: FS + LDF-2)

c. Farm soil + leachate DF-3 (Code: FS + LDF-3)

d. Farm soil + leachate DF-4 (Code: FS + LDF-4)

http://dx.doi.org/10.5772/56526

For this assessment, bulk farm soil sample, obtained from a village (K-dere, part of Ogoniland) in Rivers State, was used. Soil was sieved through a mesh and transferred at 1.5 kg per pot and designated pots were treated to 70% approximate field capacity (determined against gravity) using equal volume of appropriate fluid (water, stock leachate or diluted leachate). The systems were allowed to stabilize for 2 weeks after which viable maize seedlings were sown at 3 per pot. As the plants grew, the soil systems were treated with equal volumes of the appropriate fluid to maintain appropriate moisture level, as required by plant. Experiment lasted for 2 weeks, at the end of which the heights were recorded and plants harvested. Caution was exercised to ensure that roots were not destroyed during harvest. Root lengths were then recorded and mean values per pot calculated for each parameter.

2.6. Evaluation of beneficial utilization of end-product

Similar to the case in Section 2.4, in this evaluation, the remediated matrix was not mixed with any type of soil, neither was any external fertilizer administered. At close - out of the pilot-scale remediation project, the remediated materials were air dried, primed with one of CNB-Tech products (Ecorem) at a specified loading scheme and then utilized as a growth media. Primed end-products were transferred at 4 kg per pot of 4 liter capacity. Three indicator crops used for this project were:

- Corn (*Zea mays L.,*)
- Green leafy vegetable (Fluted pumpkin: *Telfairia Occidentalis*)
- Cassava (*Manihot esculenta Crantz*)

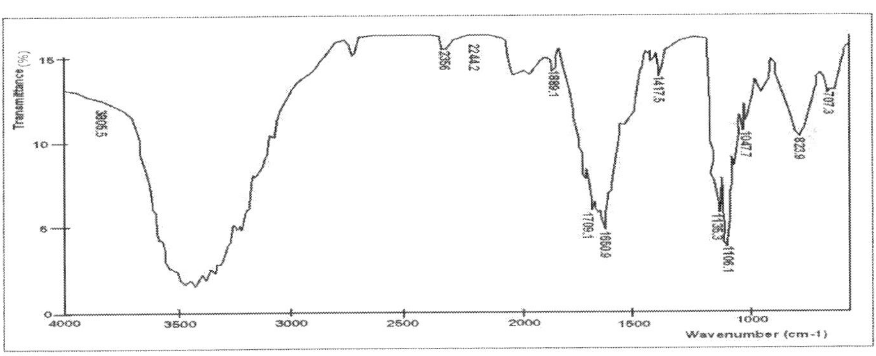

Figure 2. Infrared spectrum of Bioprimer, a CNB-Tech remediation product

The crops were used because they are commonly grown and consumed in the Niger Delta region of the country. Due to time constraint, duration of investigation varied for the crops, the longest being up to 130 days for green leafy vegetable (Fluted pumpkin: *Telfairia Occidentalis*) while corn (*Zea mays L.,)* and cassava (*Manihot esculenta Crantz*) were grown for 2 and 3 weeks respectively. Untreated OBM-DC served as a control and farm soil served as a second control.

2.7. Statistical analysis

Data generated in this study were subjected to statistical evaluations using SPSS software for Windows, version 17.0. Descriptive statistics were applied to evaluate mean and standard deviation. Paired sample T-Test and One-way analysis of variance (ANOVA) were applied to identify significant variations among treatments as appropriate. Pearson correlation was used to ascertain significant relationships.

3. Results

3.1. Typical infrared spectra of two CNB- Tech remediation products

The infrared absorption spectra of two CNB-Tech products (Bioprimer and Biozator) utilized in this pilot scale study are presented in Figures 2 and 3. Both spectra showed absorption peaks in the region of 4000 to 600 cm^{-1}.

Major information from the infrared spectra were: strong, broad absorption band of oxygen-hydrogen (O-H) of an alcohol (aryl/aliphatic) and N-H absorption bonds around 3500 - 3300 cm^{-1}; carbon-oxygen double bond (C=O) absorption band found around 1750 – 1500cm^{-1}. This could be carbonyls of ester (RCOOR), aldehyde (RCHO), ketone (RCOR) and acid (RCOOH). C-N bond of nitrogenous matter falls in the end of the range; C-O bond around 1200 – 1000 cm^{-1} and of carbon-hydrogen (C-H) bond for aromatic moieties found below 1000cm^{-1} [33].

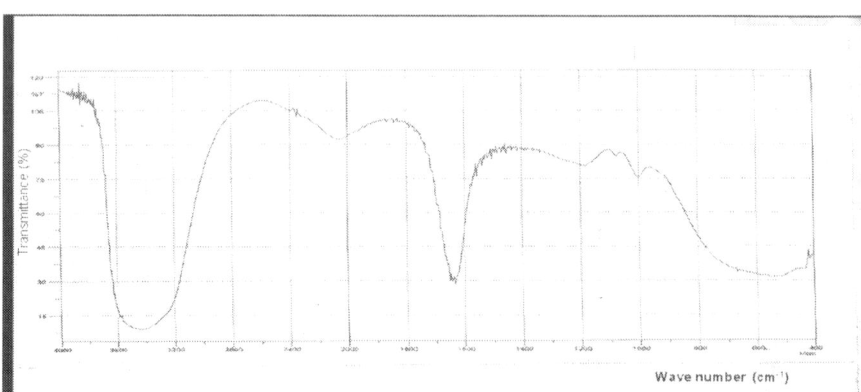

Figure 3. Infrared spectrum of Biozator, a CNB-Tech remediation product

3.2. Initial characteristics of the drilling wastes

The results presented in this paper were largely those obtained from the International laboratory. Table 1 contains the initial characteristics of the drilling wastes (oil-based mud and cuttings).

S/N	Parameter	Concentration
Inorganics		
1.	Arsenic (mg/kg)	6.69
*2.	Cadmium	Not determined
3.	Barium(mg/kg)	765
4.	Calcium(mg/kg)	87300
5.	Copper(mg/kg)	35.90
6.	Lead(mg/kg)	161
7.	Mercury(mg/kg)	0.036
8.	Nickel(mg/kg)	12.3
9.	Sodium(mg/kg)	493
10.	Potassium(mg/kg)	1930
11.	Zinc(mg/kg)	144
12.	TKN (%)	0.0357
13.	Phosphorus (%)	0.0291
*14.	pH	10.2
*15.	Electrical conductivity (mSm^{-1})	Not determined
16	Total organic carbon (%)	Not determined
17..	Salinity (mg/kg)	4300
BTEX compounds		
1.	Benzene	0.0198
2.	Ethylbenzene	0.827
3.	m- and p-xylene	0.532
4.	o-xylene	0.924
5.	toluene	1.910
PAH Compounds		
1.	Naphthalene(mg/kg)	1.94
2.	Acenaphthylene(mg/kg)	BDL
3.	Acenaphthene(mg/kg)	BDL

S/N	Parameter	Concentration
4.	Fluorene(mg/kg)	2.54
5.	Phenanthrene(mg/kg)	0.78
6.	Anthracene(mg/kg)	BDL
7.	Fluoranthene(mg/kg)	BDL
8.	Pyrene(mg/kg)	BDL
9.	Benzo (a) anthracene(mg/kg)	BDL
10.	Chrysene(mg/kg)	BDL
11.	Benzo(b)fluoranthene(mg/kg)	BDL
12.	Benzo (k)fluoranthene(mg/kg)	BDL
13.	Benzo(a)pyrene(mg/kg)	BDL
14.	Dibenz(a,h)anthracene(mg/kg)	BDL
15.	Benzo(g,h)perylene(mg/kg)	BDL
16.	2-methylnapthalene(mg/kg)	5.39
17.	Indeno(1,23-cd)pyrene(mg/kg)	BDL
	Total PAH(mg/kg)	10.65
Total petroleum hydrocarbon		
1.	TPH (mg/kg)	79 200

*Parameters not determined by the USA laboratory but quantified by Nigerian based laboratories

Table 1. Initial characteristics of the oil -based drilling mud and cuttings used in this pilot scale study

Results indicated the presence of inorganic constituents and organics (hydrocarbons compounds). Regarding inorganics, soft metal contents increased in the order: Na (493 mg/kg) < K (1930 mg/Kg) < Ca (87, 300 mg/kg). The elemental ratios were 177 for Ca/Na, 45 for Ca/K and 4 for K/Na. Heavy metal concentrations increased in the order: Hg < As < Ni < Zn < Cu < Pb < Ba. In terms of hydrocarbon contents, total concentrations of polynuclear aromatic hydrocarbon (PAH) compounds was 10.65 mg/kg with concentrations of the individual components (Figure 4) increasing as phenanthrene (0.78 mg/Kg: 7%) < naphthalene (1.94 mg/kg; 18%) < fluorene (2.54mg/kg; 24%) < 2-methylnapthalene (5.39 mg/kg; 51%). Results on monoaromatics (BTEX), shown in Figure 5, gave a total concentration of 4.213 mg/kg out of which toluene constituted the highest fraction (45.34%), followed by xylene (34.56%), ethylbenzene (19.63%) and benzene (0.47%). Total xylene concentration was 1.456 mg/kg out of which ortho-xylene constituted 63.46% while meta- and para-xylenes gave 36.54% of the total (1.456 mg/kg).

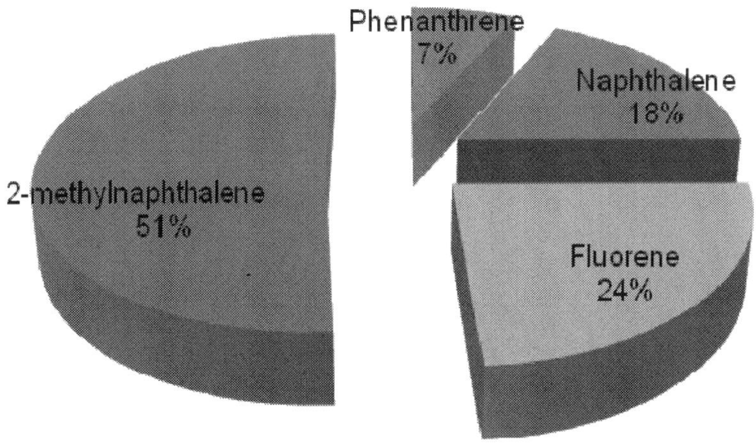

Figure 4. Percentage distribution of individual components of PAH relative to the total concentration

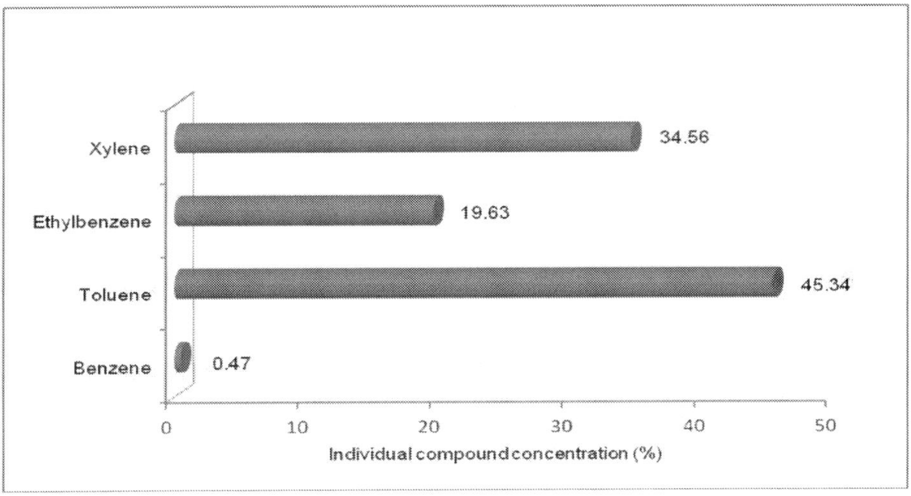

Figure 5. Percentage distribution of individual components relative to the total BTEX concentration

3.3. Results on petroleum hydrocarbon degradation

By application of CNB-Tech products, the initial TPH concentration of 79, 200 mg/kg decreased to 1888.67 ±161. 20 mg/kg. The difference in these two values was a mean TPH concentration of 77 311.33 ± 161.20 mg/kg. This difference corresponds to the total concentration of hydrocarbon compounds degraded or destroyed by the applied treatment. The initial concentration (79, 200 mg/kg) and the degraded fractions (in replicates of three) are presented in Figure 6. Specifically, results on hydrocarbon degradation (Figure 7) revealed 98% degradation for TPH, 100% degradation for BTEX and 100% degradation for PAH. Reduction in TPH level by 99% was obtained by the Nigerian laboratories.

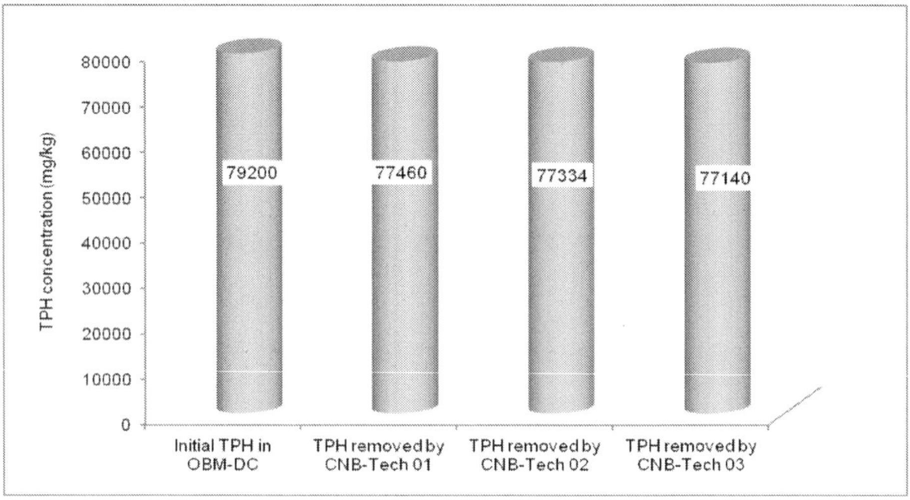

Figure 6. Graph showing concentrations of degraded TPH relative to the initial concentration

S/N	Parameter	Remarks for contaminated medium	Remarks for remediated medium
1.	Appearance	Viscous, pasty and solid interfaced in oil suspension	Transformed to non-viscous, non-sticky crumby humus soil appearance
2.	Color	Light brown	Treated matrix had characteristic dark color of humus soil
3.	Odor	Presence of strong hydrocarbon odor	Complete disappearance of hydrocarbon odor in all the treated media and all treated samples exhibited clean earthy smell
4.	Sheen test	Strong oil sheen in water suspension	Complete disappearance of oil sheen in water suspension

Table 2. Qualitative results for the remediated media

Results on qualitative assessments of the untreated OBM-DC and remediated material in terms of appearance, odor, color and sheen test are contained in Table 2 and Figure 8 depicts the materials' appearances before and after remediation.

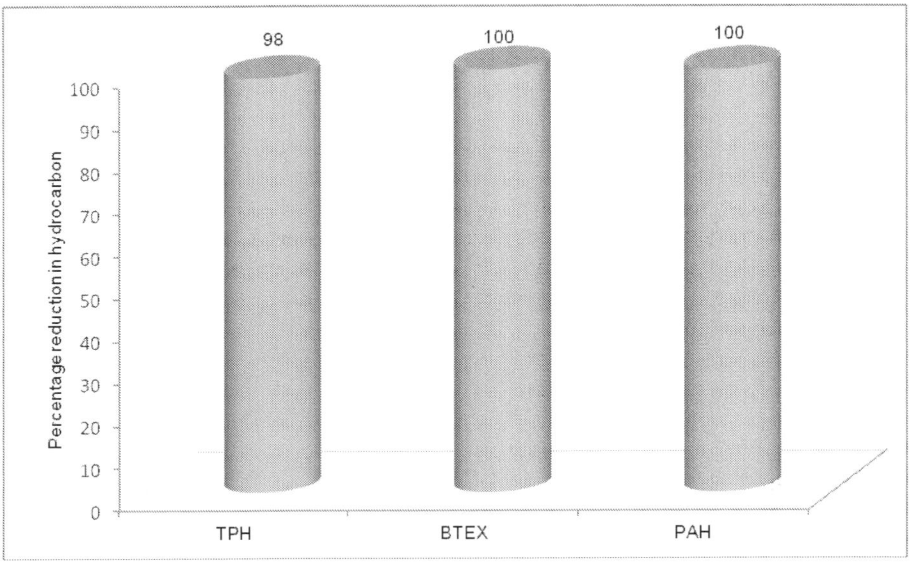

Figure 7. Percentage degradation of hydrocarbon compounds in the drilling wastes by applied CNB-Tech products

Figure 8. Photographs showing the materials before and after bioremediation by the application of CNB-Tech products

3.4. Results on inorganic constituents of the CNB -Tech treated materials

Descriptive statistics of selected inorganic constituents found in the treated media are presented in Table 3. Changes in their concentrations relative to the initial values are presented in Figure 9. For instance, the initial pH value was reduced to 7.90 from 10.20, corresponding to 23% reduction. Likewise, the following reductions were obtained: 62% for Ca, 46% for As, 44% for Cu, 70% for Pb, 100% for Hg, 57% for Ni and 37% for Zn. The concentrations of some elements such as nitrogen, phosphorus and potassium were elevated. The nitrogen-phosphorus-potassium (NPK) status, as affected by treatment, is presented in Figure 10. Nigerian laboratories obtained the same trend for NPK status. Based on the results from USA, CNB-Tech remediation option applied in this study raised the nitrogen level from 0.036% to 0.096%, raised phosphorus level from 0.0291% to 0.312%, increased potassium by 1.4 fold (Figure 10) and sodium by 3 folds. The USA based laboratory did not analyze for total organic carbon and electrical conductivity but the Nigerian based laboratory did and recorded electrical conductivity in the range of 1956 to 2063 mSm^{-1} with a mean value of 2003 ± 54 mSm^{-1} before treatment. After remediation, the electrical conductivity of the end products ranged from 594 to 696 mSm^{-1} and a mean value of 640± 52 $mSm^{-1.}$ From the mean values, there was a 68% reduction in electrical conductivity.

S/N	Element	Minimum	Maximum	Mean	Standard error	Standard deviation	Sample population
1.	pH	7.70	8.20	7.90	0.15	0.26	3
2.	Nitrogen (%)	0.070	0.130	0.096	0.016	0.028	3
3.	Phosphorus (%)	0.280	0.360	0.312	0.026	0.046	3
4.	Potassium (%)	0.50	0.77	0.61	0.08	0.14	3
5.	Copper (mg/kg)	18.10	21.70	20.10	1.06	1.83	3
6.	Zinc (mg/kg)	79.30	110	92.67	9.08	15.73	3
7.	Nickel (mg/kg)	3.99	7.05	5.29	0.92	1.59	3
8.	Calcium (mg/kg)	28900	39200	33466	3030	5248	3
9.	Arsenic (mg/kg)	2.50	4.85	3.59	0.68	1.18	3
10.	Lead (mg/kg)	5.87	54.80	27.06	14.50	25.12	3

Table 3. Concentrations of some inorganic parameters in the treated materials

Total organic carbon ranged from 2.95 to 3.06% with a mean of 2.99± 0.06% before remediation and increased to 3.84 to 3.93% with a mean of 3.88 ± 0.05%; corresponding to an increase by 23%. Before remediation, Cd concentration varied from 6.70 to 7.60 mg/kg, with a mean value of 7.03± 0.49 mg/kg. After treatment, the metal concentration ranged from 0 to 1.80 mg/kg with an average of 1.05 ± 0.94 mg/kg. By the two mean values, cadmium level was reduced by 85% due to applied CNB-Tech products.

http://dx.doi.org/10.5772/56526

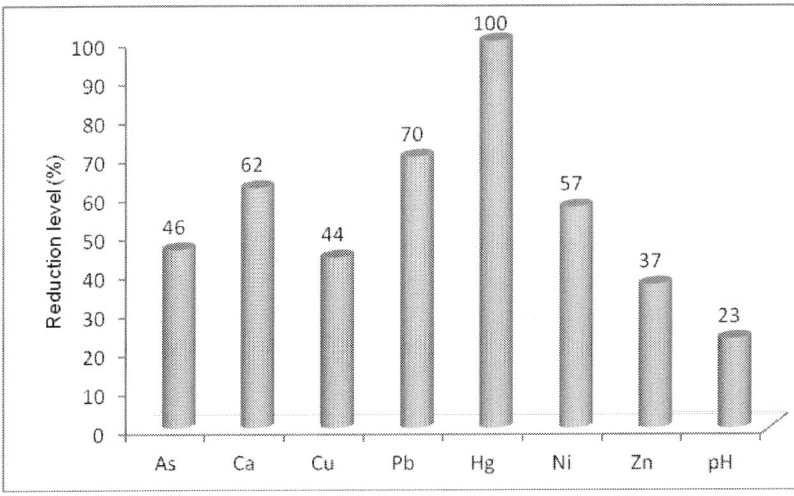

Figure 9. Reductions in some inorganic constituents of the drilling materials treated by CNB-Tech

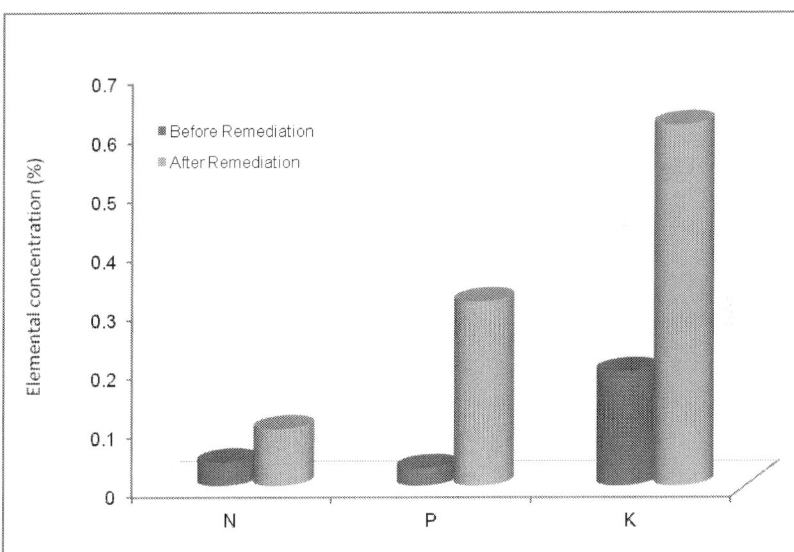

Figure 10. Nitrogen-phosphorus-potassium status before and after treatment as obtained by the USA based laboratory

3.5. Results on microbial activity

The digital photographs of heterotrophic plate count results are shown in Figure 11. Microbial activities assessed on the untreated and treated samples revealed that the contaminated oil-based mud and cuttings (no. 1 in Figure 11), contained some indigenous microorganisms of up to 1.9×10^3 (cfu/mL) while the CNB-Tech remediated samples recorded up to a maximum of 3.15×10^7 cfu/mL. An illustration of microbial enumeration for 48-hr and 96 hr counts are presented in Figure 12.

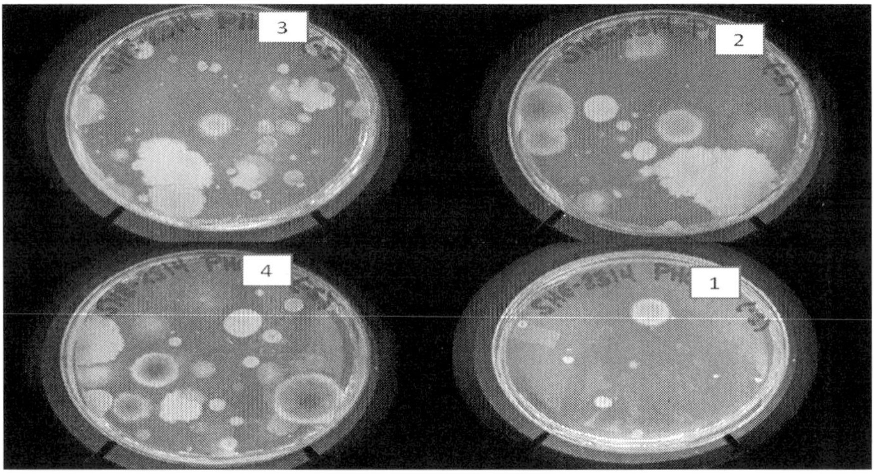

Figure 11. Heterotrophic plate count digital photographs for untreated OBM-DC (1) (before remediation) and replicates (2, 3, 4), after remediation using CNB-Tech method

At 48 hr microbial activity assessment, maximum total microbial population of 1.9×10^3 cfu/mL was obtained for untreated OBM-DC and in the materials remediated by the application of CNB-Tech products, it was 1.45×10^7 cfu/mL. These two values were significantly different at $p \leq 0.05$. At 96 hr microbial activity assessment, a total microbial population of 2.4×10^3 cfu/mL was obtained for untreated OBM-DC and 3.15×10^7 cfu/mL for the remediated matrices. Results showed that within 48 hours, the microbial activity of the remediated matrices excelled over the untreated by over 7,000 folds and at 96 hours, it excelled by over 13, 000 folds, indicating rapid multiplication of microbial activity by CNB-Tech products which also increased with time.

3.6. Results on phytotoxicity assessment of remediated samples

3.6.1. *Toxicity on seed germination potential*

The contaminated OBM-DC did not allow the germination of maize seedlings. Out of the sown 18 seedlings, none germinated. The untreated OBM-DC therefore, gave 100% toxicity to seed germination potential (SGP) of maize. On the contrary, all the 18 maize seedlings sown in the CNB-Tech remediated matrices germinated (Figure 13). Hence, resulting in 100% positive effect on SGP, indicating that the treated matrices exhibited 0% toxicity to seed germination.

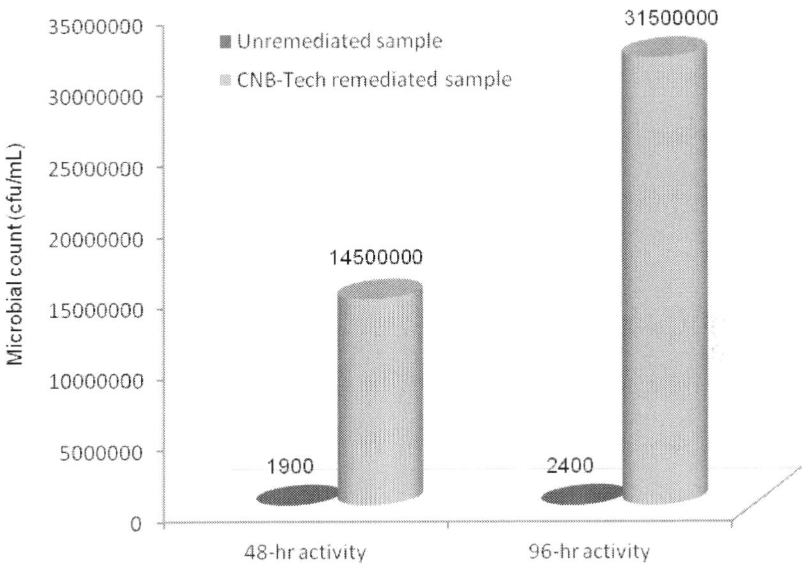

Figure 12. Microbial activity at 48 –hr and 96-hr counts for untreated oil-based drilling wastes and CNB-Tech remediated samples

Figure 13. Germinated maize seedlings growing in treated media with picture taken on day 4 of growth

3.7. Results on beneficial use of remediation end product

Figure 14, shows a cross-section of the treated materials (during recovery period) being aerated in preparation for use as plant growth media.

Figure 14. A cross section of project technical staff preparing the treated drilling wastes (OBM-DC) for use as plant growth media

During the recovery phase of the remediated end-product, treated materials were allowed to lie fallow in order to establish natural processes as a sign of wellbeing and restoration. In this project, after the fallow period, early indications of material restoration were:

- spontaneous vegetative growth,
- the presence of larva within the spontaneously grown green vegetation,
- butterflies and small birds perching on the surface of the material, which could not take place before treatment

Remediated materials supported the growth of fluted pumpkin (*Telfairia occidentalis*). A cross-section of the green leafy vegetable at over 100 days of growth and that of cassava, at one week of growth, growing in the treated materials are shown in Figure 15. Narrowing to the height of *Telfairia occidentalis*, the mean height for crops grown in the untreated OBM-DC was 0 cm as there was complete inhibition to both germination and growth. The mean height for crops grown in CNB-Tech remediated media was 217± 25 cm, a value higher than the mean height (187± 40 cm) of the vegetable crops grown in farm soil collected from the region. The difference in the two mean values was significant at $p = 0.14$. Correlation for the heights of the vegetables grown in the treated media and those grown in the farm soil gave a coefficient of 0.95 ($p = 0.204$).

Figure 15. Remediated drilling wastes as plant growth medium for Fluted pumpkin (*Telfairia occidentalis*) and cassava (*Manihot esculenta Crantz*)

3.8. Results on the impact of remediation leachate on plant life

Comparative evaluations of control system (soil treated with water only), stock leachate system (soil treated with leachate without any form of dilution) and systems treated with serial dilutions of the leachate (soil treated with leachate diluted with water by factors 1, 2, 3 and 4) are presented in Table 4.

S/N	System Code	Leachate effect of on vegetative growth relative to control (%)		Effect of serial dilution on plant using stock (undiluted leachate) as reference (%)	
		Height	Root length	Height	Root length
1.	FS+ Water (Control)	Reference	Reference	Not applicable	Not applicable
2.	FS + DF-0	-1.50	-23.45	Reference	Reference
3.	FS + DF-1	32.60	1.12	34.62	32.20
4.	FS + DF-2	45.01	16.37	42.22	50.02
5.	FS + DF-3	66.86	21.37	69.41	58.55
6.	FS + DF- 4	75.39	24.51	78.07	62.66

Negative sign stands for decrease. The other positive values stand for increase, FS = farm soil and DF = dilution factor

Table 4. Impact of leachate generated at the close-out of project on the root length and height of maize

Pictorial and graphical representations of leachate impact on plant height and root length are presented in Figures 16 and 17. Relative to the control system (soil treated with water only), leachate diluted with water by a factor of 4 improved plant height by 75.39% and root length by 24.51%. Figures16 and 17 gave all the systems at a glance, relating the control (FS + Water), system SF+LDF-0 (DF-0) and serial dilutions (DF-1 = FS+ LDF-1, DF-2 = FS+ LDF-2, DF-3 = FS + LDF-3 and DF-4 = FS + LDF - 4) for plant height and root length. Evaluating the effect of leachate dilution relative to the stock (undiluted) leachate, a 4-fold dilution excelled over the stock by 78.0% for plant height and 62.66% for root length. The relationships between plant height or root length and dilution factors are given in Figure 18. Pearson correlations gave strong coefficients: plant height versus dilution factor, r = 0.979 (p = 0.004), root length versus dilution factor, r = 0.932 (p = 0.021) and plant height versus root length, r = 0.972 (p = 0.006). From the results, plant vegetative growth increased with increasing dilution of leachate.

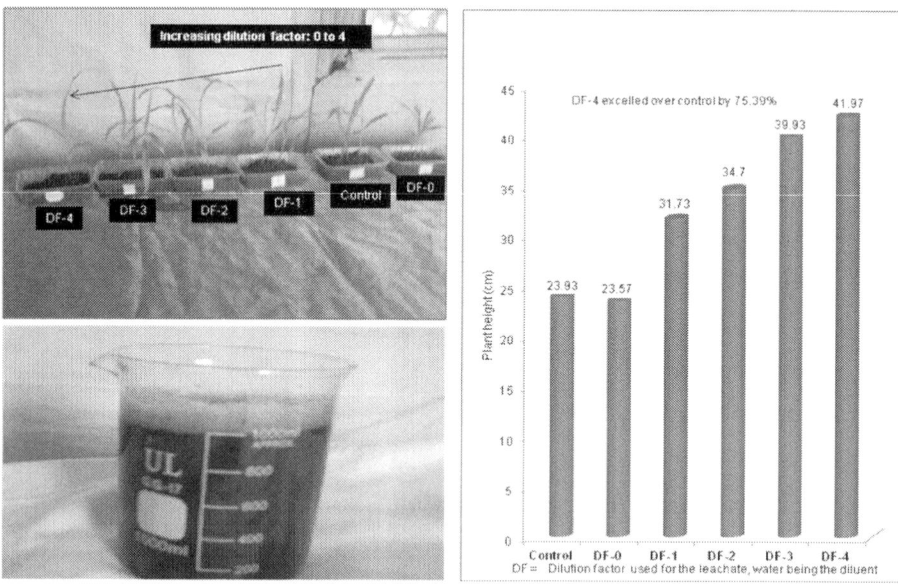

Figure 16. Pictorial and graphical representations of leachate impact on height of maize, including a picture of the stock leachate contained in a beaker

4. Discussion

The type of inorganic constituents and hydrocarbons found in the drilling wasting used in this study were consistent with the reports of [4, 17] but varied in concentrations. This confirms that the OBM-DC used in this study was toxic [2]. The remediation products of CNB-Tech

http://dx.doi.org/10.5772/56526

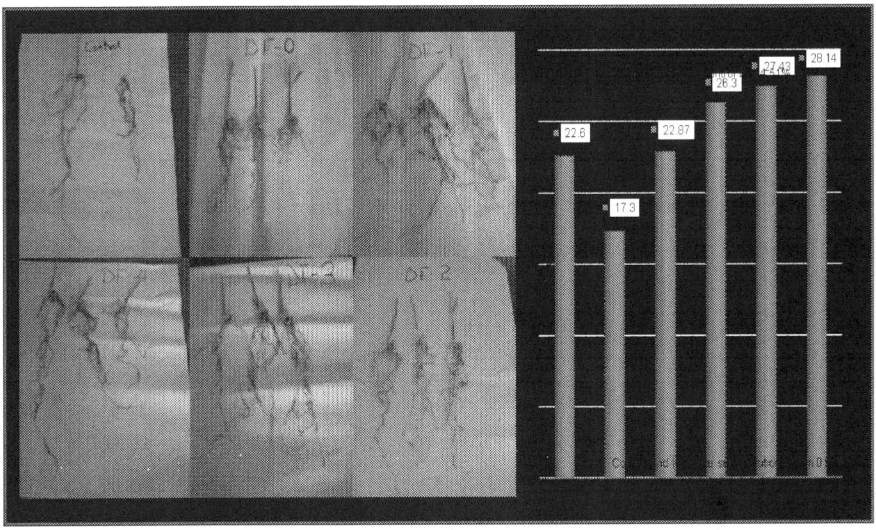

Figure 17. Pictorial and graphical representations of leachate impact on root length of maize

series used in this study demonstrated a high (98 to 100%) degradation potential for the different constituents of hydrocarbon compounds found in the drilling wastes, within a short period of 6 days. This excellent performance was attributed to the chemistry, nature and operation mechanisms of the CNB-Tech formulations.

An infrared spectrum is primarily used to identify functional groups present in a molecular fragment [33]. The infrared spectra obtained for CNB-Tech products (Biozator and Bioprimer) revealed enrichment of the molecular structure of the two products with oxo- groups, indicating oxidizing functionality. The presence of C-H of aromatic nature and the O-H stretching absorption indicate the presence of both hydrophobic and hydrophilic properties, respectively, in their molecular fragments. By implication, the remediation products are naturally endowed with:

- oxidizing ability
- polar (hydrophilic: water loving) molecular fragment
- non-polar fragment (hydrophobic: water insoluble, oil soluble) molecular fragment.

These natural endowments permit the dissolution of the products' active ingredients (solids) in water, making water the carrier medium for CNB-Tech liquid formulations. Consequently, Biozator and bioprimer are water based technical grade products. By the mentioned characteristics, the two products perform reduction and oxidation (Redox) reaction mechanisms, resulting in the degradation/ destruction of hydrocarbons compounds, without recombination

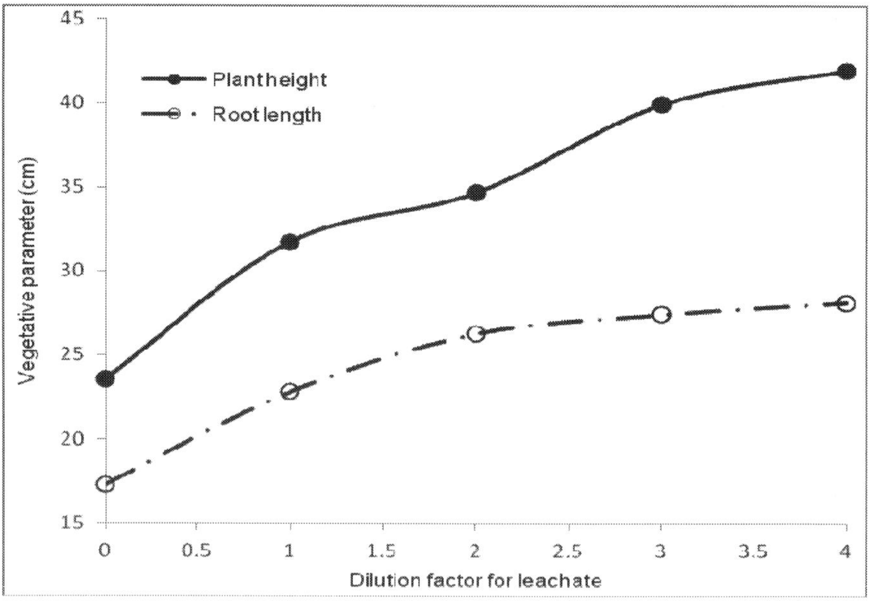

Figure 18. Relationship between plant vegetative growth and serial dilution of process fluid (leachate) generated during the remediation project

to form new hydrocarbons. These absorption peaks in the infrared spectra further reveal that CNB – Tech products are natural hydrocarbon biodegradation catalysts for the following reasons:

- enhaced water solubility of hydrocarbons via sorption, hydrolysis and oxidation mechanisms
- enhanced bioavailability of hydrocarbon pollutants for microbial degradation
- increased supply of oxygen [O] molecules required for enhanced reduction –oxidation reactions in the hydrocarbon degradation process.
- surfactant property
- emulsification of hydrocarbons

The combined actions of hydrophobic molecular fragment, hydrolysis, oxidation and surfactant property of CNB-Tech products render hydrocarbons more water soluble and subsequently more available for biodegradation. Bioprimer and Biozator also emulsify hydrocarbons into droplets that can be easily assimilated by microorganisms. By these properties, the products reduce oil-water surface tension; enhance water solubility of petroleum hydrocarbons thereby enhancing the bioavailability of the contaminants (hydrocarbons) to microorganisms for both extracellular and intracellular decompositions. The two products are 100% biodegradable. The third CNB-Tech product used in this study (Ecorem: a black amorphous solid material, also 100% biodegradable) contains major and minor plant nutrient elements and via hydro-activation, naturally generates mixed consortia of microorganisms, which multiplies with time to facilitate the destruction of hydrocarbons. No engineered microorganism or externally imported microorganism was used in this study. This technology, therefore, saves time and eliminates the daunting task of isolating pure microbial strains and associated adaptability challenges linked with conventional bioremediation techniques [7, 8, 18, 19, 20].

The microorganisms from Ecorem product perform the following functions:

- extracellular decomposition in which the naturally produced microorganisms secrete enzymes to breakdown large organic compounds (such as hydrocarbons) into smaller forms for easier absorption into the micro-organisms. Once the smaller compounds have been absorbed by the microorganisms, intracellular decomposition takes place
- increased microbial activity facilitated by Ecorem, results in thermophilic temperature modulations in the range of 55 to 60°C, a process that accelerates degradation of hydrocarbons, especially polynuclear aromatic aromatic hydrocarbons (PAHs).Thermophilic temperature modulations also controls thermo-sensitive pathogen to crops animals and man; killing off weeds and seeds that will be detrimental to land use of end products.

By the above described mechanisms, the CNB-Tech products were able to biodegrade petroleum hydrocarbon compounds with high efficiency (98% degradation for TPH and 100% degradation for PAHs and BTEX) within a short period of time of 6 days, relative to previous works on bioremediation. For instance, in a study of in-situ bioremediation of oily sludge via biostimultaion of indigenous microbes, conducted by [34], through the addition of manure at the Shengli oilfield in Northern China for 360 days, 58.2% reduction in TPH was achieved in test plots and 15.5% reduction in control plot. By treating 2 kg of drill cuttings with initial TPH of 806.36 mg/kg for 56 days under the conditions of composting of spent oyster mushroom (*P.ostreatus*) substrate, [35] recorded overall degradation of PAHs in the range of 80.25 to 92.38%. In this present study, OBM-DC used had initial TPH of 79, 200 mg/kg and was degraded by 98% within the stated short period of 6 days. In a field trial biopile composting method [36] for drilling mud polluted sites in the Southeast of Mexico with comparable TPH level of 99 300 ± 23000 mg/kg, after 180 days, TPH concentrations decreased from 99 300 ± 23000 mg/Kg to 5500 ± 700 mg/kg, corresponding to 94% degradation for amended biopile and to 22900 ±7800 mg/kg, representing 77% decrease for unamended biopile. The mean residual value of TPH (5500 ± 700 mg/kg) left in the treated matrix in their study was higher than the mean residual value (1888± 161 mg/kg) obtained in this present study.

By conducting an investigation on two bioremediation technologies (bioremediation by augmentation and conventional composting using crude manure and straw) as treatment options for oily sludge and oil polluted soil in China [12] in which the total hydrocarbon content (THC) varied from 327.7 to 371.2 g/kg (327700 to 371200 mg/kg) for dry sludge and 151.0 g/kg (151000 mg/kg) for soil for a period of 56 days; after three times of bio-preparation application, THC decreased by 46 to 53% in the oily sludge and soil. The results (98 -100% degradation) obtained from this present study was from only one dose application of CNB-Tech products. Repeated application of CNB-Tech products by two to three dose applications will achieve 100% degradation of TPH. In another instance, a 5- month field scale bioremediation of sludge matrix via the utilization of organic matter such as bark chips via conventional composting, mineral oil (equivalent to total hydrocarbons) decreased from 2400 to 700 mg/kg (70% decrease) for sludge matrix and from 700 to 200 mg/kg, corresponding to 71% decrease [14]. In treating oil sludge using composting technology in semiarid conditions for 3 months, hydrocarbons were reduced from 250 to 300g/kg (250000 to 300 000 m/kg) by 60% against reduction by 32% recorded in the control [37]. The treatment applied by [37] and consequent reduction of 60% implies that the residual hydrocarbons in the treated samples would be between 100 000 and 180 000 mg/kg unlike the results obtained in this present study that gave residual hydrocarbon of 1888.67 ±161.20 mg/kg. In a study carried out by [38], sand samples contaminated with oil spill were collected from Pensacola beach (Gulf of Mexico) and tested to isolate fungal diversity associated with beach sands and investigate the ability of isolated fungi for crude oil biodegradation. From their results, 4.7 to 7.9% biodegradation was recorded.

Elsewhere in India, Abu Dhabi and Kuwait [39], bioremediation technology was applied in field-scale degradation of hydrocarbons in different oil wastes for a period of 12 months. Table 5 illustrates different reductions in total petroleum hydrocarbons obtained in these field case studies. TPH reductions in drilling wastes were obtained in the range of 90.85 to 95.48% with residual TPH in treated samples in the range of 2600 to 10 900 mg/kg (0.26 to 1.09%).

Name of the oil Installation / type of oily waste	**Quantity of oily waste (cubic meter)**	**Number of batches**	**TPH Content (%) in oily waste before and after bioremediation**		**% Reduction in TPH**	**Residual TPH in treated material (%)**
			Before	**After**		
Abu Dhabi National Oil Company (ADNOC), Abu Dhabi / Oil contaminated drill cuttings	200	1	17.26	0.98	94.32	0.98
BG Exploration and Production India Limited (BGEPIL), India / Oil based mud (OBM)	2,428	3	5.75 – 6.23	0.26 - 0.57	95.48-90.85	0.26 – 0.57
Bharat Petroleum Corporation Limited (BPCL), India / Oily sludge	5,000	1	19. 30 – 26.5	0.26 - 0.57	98.65-97.85	0.26 -0.57

Name of the oil Installation / type of oily waste	Quantity of oily waste (cubic meter)	Number of batches	TPH Content (%) in oily waste before and after bioremediation		% Reduction in TPH	Residual TPH in treated material (%)
Cairn Energy Pty. India Limited, India / Oil contaminated drill cuttings	567	2	14.93 – 18.81	0.82 – 1.09	94.51-94.21	1.09
Chennai Petroleum Corporation Limited (CPCL), India / Oily sludge	4,444	2	26.12	0.89	96.59	0.89
Hindustan Petroleum Corporation Limited (HPCL), India / Oily sludge	5,010	3	16.70 – 52.81	0.90 – 1.60	94.61-96.97	0.90-1.60
Indian Oil Corporation Limited (IOCL) Refineries in India / Oily sludge (acidic + non acidic)	75,412	48	9.6 – 38.4	0.37 – 0.95	96.15-97.53	0.37-0.95
Kuwait Oil Company (KOC), Kuwait / Oil contaminated soil	778	1	4.6 – 12.75	0.09 – 0.10	98.04-99.21	0.09-0.10
Mangalore Refinery and Petrochemicals Limited (MRPL), India. / Oily sludge	2,222	2	8.35 – 19.86	0.84 – 0.97	89.84-95.12	0.84-0.97
Oil and Natural Gas Corporation Limited (ONGC) installations in India / Oily sludge & oil contaminated soil	95,499	145	12.0 – 51.5	0.5 – 1.2	95.83-97.67	0.50--1.20
Oil India Limited (OIL) , Assam / Oily sludge & oil contaminated soil	15,921	14	21.6 – 37.7	0.49 – 0.53	97.73-98.59	0.49-0.53

Name of the oil Installation / type of oily waste	Quantity of oily waste (cubic meter)	Number of batches	TPH Content (%) in oily waste before and after bioremediation		% Reduction in TPH	Residual TPH in treated material (%)
Reliance Energy Limited (RIL), India / Oily sludge	611	2	19.15	0.5	97.39	0.50

Table 5. Reductions in TPH levels obtained in field case studies of different types of petroleum impacted wastes (soils, drill cuttings and oil-based mud) in Abu Dhabi, Kuwait and India [39].

The residual TPH level (1888.67 ± 161.20 mg/kg) obtained in this present study was below the Environmental Guidelines and standards for the Petroleum Industry in Nigeria (EGASPIN) intervention value for mineral oil (petroleum hydrocarbon) of 5000 mg/kg [15]. By repeated application of CNB-Tech products, it is possible to meet a very strict regulatory standard for residual TPH level of less than 50 mg/kg. The changes in metal concentrations found in this study were attributed to (i) immobilization via chelate formation (ii) preferential supplementation of trace plant nutrient elements using the three products, (iii) natural electrochemical process whereby the positively or negatively charged organic molecules (generated during the natural transformation process occurring when the products were in use) bond with their counterparts in organic matter. These processes include oxidation, methylation, hydroxylation, carboxylation, coupling and polymerization [40] thereby enhancing bioavailability of the metals to microorganisms that utilize the organic matter supplied by the CNB –Tech products as energy source.

Microbial population found in a typical tropical soil under Nigerian climate is in the neighborhood of 8.19×10^6 cfu/mL [41]. Relative to this value, the population found in the contaminated OBM-DC (1.9 to 2.4×10^3 cfu/mL) showed suppressed microbial population, attributed to strong hydrocarbon (TPH level of 79, 200mg/kg) pollution. This is in agreement with the reports of [3]. The microbial population (1.45 to 3.15×10^7 cfu/mL) found in treated samples revealed restoration of soil microbial population using CNB-Tech products. It excelled over the value recorded in polluted material by over 7000 folds and higher than the value reported by [34], where TPH degraders and PAH degraders increased by one to two orders of magnitude via the addition of manure. Furthermore, the use of CNB-Tech products modified the pH value of the drilling wastes, transforming it from strongly alkaline (pH of 10) medium to pH of 7.90 medium; comparable to the 7.3±0.1 obtained by [34] for bioremediated soils. The very high pH of the untreated drilling waste materials could be attributed to some of the additives in the drilling fluid. Drilling fluids contain an internal phase of brine such as calcium salts [3]. This was confirmed by the high content of Ca (87 300 mg/kg) obtained in this study for the untreated material. One dose application of CNB-Tech products reduced this concentration by up to 62%, repeated dose application would definitely bring Ca level to any desired value.

Observations made during the recovery /fallow period were signs of drastic positive change in toxicity conditions, implying reduced toxicity. Reduction of soil toxicity by bioremediation, evidenced by increase in EC50 of the soil was reported by [34]. In this study, bioremediation using CNB-Tech products reduced toxicity in treated materials relative to untreated OBM-DC, evidenced by 100% positive effect on seedling germination potential and improved crop vegetative growth. Reduced material toxicity also explains the increased microbial activity of the treated matrices in comparison to the untreated drilling wastes, obtained in this study. The agricultural potential for the remediation end-products was also manifested by:

- increased microbial activities
- increased nitrogen-phosphorus-potassium (NPK) status
- increased soil crumby nature as against very viscous and pasty characteristics of untreated drilling wastes.

These nutrient elements (NPK) enhance microbial growth, microbial population, microbial activity and consequently increase soil fertility [41]. By these, CNB-Tech products could overcome the extreme phytotoxicity [100% toxicity to seedling germination potential of maize and 100% inhibition to vegetative growth for three different types of plant (maize, fluted pumpkin and cassava)], caused by the untreated drilling waste. CNB-Tech products transformed oil-based drilling mud/cuttings to arable soil; capable of supporting seed germination and plant growth; excelling the performance of a control (farm soil apparently not impacted by drilling waste or crude oil) by 14%.

Electrical conductivity, a measure of dissolved ions in solution, is influenced by several soil physical and chemical properties such as salinity, saturation percentage, water content, bulk density, organic matter content, temperature and cation exchange capacity of the soil matrix. Impact of these influencing factors must be reflected in interpreting electrical conductivity effect on plant growth. Generally, elevated electrical conductivity and high salinity levels in agricultural soils may result in reduced plant growth and productivity or in extreme cases, the elimination of crops and native vegetation [42]. The reduction of electrical conductivity by 68% is a positive development because it demonstrates that the products could also modify the salinity of the material. In situations of very high initial electrical conductivity, there is a step-down CNB-Tech product as was carried out in this study and in situations of very low electrical conductivity, there is also a step-up CNB-Tech product as reported in a previous publication [30].Results in this present study on excellent growth of crops planted in the remediated matrices were indicators of acceptable soil salinity level for plant growth. The beneficial use of the end-products obtained in this study for crop production were attributed to postulations based on findings from this study and previous works on this subject matter, which include:

a. stimulation of beneficial microorganisms in soil, which enhances soil fertility [25]

b. possible increased photosynthetic rate in plants evidenced by increased photosynthetic pigments (chlorophylls a and b) [40]

c. increase in soil buffering capacity [28]

d. increased soil moisture retention capacity by reducing hydrophobicity tendency [29]

e. positive soil temperature modifications that enhance soil nutrient bioavailability to plants [31, 40]

f. formation of stable chelates with toxic metals such as Pb, Cu and Cd in order to reduce their bioavailability to plants [40]

g. preferential exclusion of the chelated toxic metals from soil solution, allowing the plant nutrient elements to be assimilated into plant cells

h. improvement of soil physicochemical properties via:

i. increased aeration and water retention [29]

j. activation of the macro and micro nutrients in soil in forms readily assimilated by plants [30, 40]

k. improvement of plant root development and growth

l. improvement of seed sprout of plants and subsequent shoot growth

m. improved plant biomass production [26]

n. enhanced soil nitrogen, phosphorus and potassium status for improved soil fertility

o. acting as plant growth hormone, having positive stimulant action for plant growth [25, 26]

p. improvement of soil permeability, promoting plant drought resistance [29]

q. promotion of increased soil porosity and organic matter content, hence greatly promoting the microorganism activity and improving soil fertility.

Regarding leachate generation and management during the remediation exercise; fluid (leachate) produced as remediation progressed was recycled by incorporation into the biocell and used to regulate moisture content, thereby reducing water usage and conserving water resources. Expertise applied during the project ensured that at remediation project close-out, no isolated fluid system was actually produced. Nonetheless, the assessment of leachate effect on plant growth carried out in this work was to establish the fact that even in the event of accidental release of some fluid into the environment, there would be minimal risk to the receptor biotic community. More evaluations are still ongoing in this regard. Results from this study revealed that the leachate generated, though a concentrate, supported plant growth and when diluted with ordinary tap water gave a better support; reasons being that:

- toxic petroleum hydrocarbons in the contaminated drilling wastes have been destroyed to an acceptable level, evidenced by natural foamability of the concentrated leachate. Foamability would hardly occur if oil was still present

- leachate is also enriched with plant nutrients such as nitrogen, phosphorus and potassium

The process fluid, therefore, had some fertilizer value. The percentage decreases (1.50% and 23.45%) obtained for plant height and root length respectively, for the stock leachate was attributed to concentrated level of nutrients, confirmed by better performance of dilute leachate series. Naturally, in any formulated fertilizer, plant nutrients are applied at specified concentrations otherwise may hinder plant growth. Comparative evaluations of control system (soil treated with water only), stock leachate system (soil treated with leachate without

http://dx.doi.org/10.5772/56526

any form of dilution) and systems with serial dilutions of the leachate (soil treated with leachate diluted with water by factors 1, 2, 3 and 4) revealed that the leachates were not toxic to receptor plants. The implication of this is that in the event of occasional spill of the leachate to the adjacent environment; dilution with water is, therefore, an adequate safety measure.

The ability of the end products to sustain the growth of green leafy vegetable: fluted pumpkin (*Telfairia ocidentis*) and root tuber crop, cassava (*Manihot esculenta Crantz*) and cereal crop (maize) is a demonstration of the utility of the remediation end product. It therefore stands that the use of CNB-Tech products as a biotechnological tool for hydrocarbon degradation in drilling waste converts these waste materials into non-toxic and potentially useful end products. In addition to the beneficial use of the remediation end-product for agricultural purposes, other possible utility options, shown in Figure 19, include:

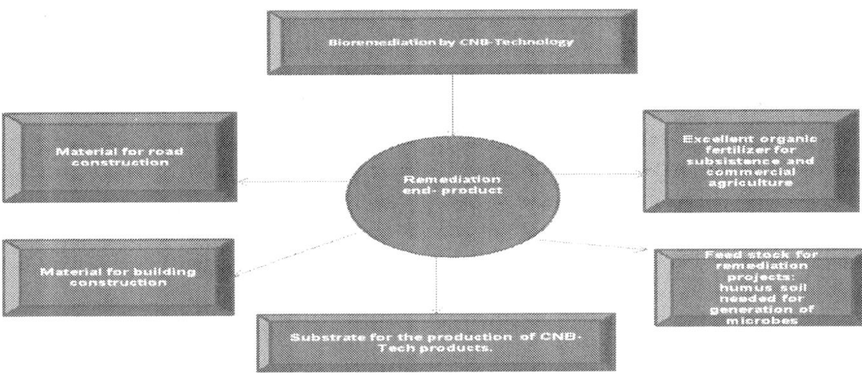

Figure 19. Potential utility of end - products from bioremediation using CNB-Tech products

- material for road construction
- material for building construction
- substrate for the production CNB-Tech bioremediation agents
- excellent organic fertilizer for subsistence and commercial agriculture
- feedstock for bioremediation projects

Table 6 is a comparative evaluation of economic, operational and environmental implications of thermal technologies as reported by [3] and CNB-Technology based on the results and learning from this study.

S/N	Thermal Technology	CNB-Tech
1.	Effective removal and recovery of hydrocarbons from solids	Effective removal of hydrocarbons from solid

S/N	Thermal Technology	CNB-Tech
2.	Possibility of recovering base fluid and end - product could be used for brick making	Effective recovery of free phase oil and end product has other uses apart from brick making
3.	Low potential for future liability	No future liability
4.	Requires short time	Time is relatively short
5.	High cost of handling environmental issues, since end-product dispersion would be below organic layer where vegetation growth is desired	Very minimized environmental issues
6.	Large volume of wastes is required to justify the cost of operation	Cost-effective for either small or large volume of wastes
7.	Requires tightly controlled process parameters	Does not require tightly controlled process parameters
8.	Heavy metals and salts are concentrated in processed solids	Reduces heavy metals and salts concentrations in process solid
9.	High operating temperatures can lead to safety risks	Low operating temperature. Operates at ambient temperature; modulation does not exceed 60°C.
10.	Requires several operators	Does not require several operators
11.	Process water contains some emulsified oils	Process water does not contain some emulsified oils
12.	Residue ash requires further treatment	No residue ash. End-product is clean soil
13.	End product is sterile and can no longer support plant Life	End product is fertile and can support microbial and plant Life

Table 6. Comparative evaluation between thermal technology and CNB-technology

5. Conclusions and recommendations

This study revealed that it is possible to harness natural, biodegradable and local resource materials of Nigeria origin; translate them to scientifically formulated products that can be utilized for efficient biodegradation of hydrocarbon polluted matrices such as oil-based mud and drill cuttings within a reasonable short period of 6 days. This technology thus converts hydrocarbon polluted oil-based mud and drill cuttings to beneficial end-products of high order reuse such as soil amendment, without the generation of secondary waste materials. Field-scale trial adopting CNB-Technology is recommended.

Acknowledgements

This project was carried out under full financial support of the Remediation Department, Shell Petroleum Development Company (SPDC), Port Harcourt, Nigeria through the University

Liaison Team of the company. The support of the Oil well Team of SPDC that facilitated the procurement of oil- based mud and drill cuttings is also acknowledged.

Author details

Iheoma M. Adekunle[1*], Augustine O. O. Igbuku[2], Oke Oguns[3] and Philip D. Shekwolo[2]

*Address all correspondence to: imkunle@yahoo.com

1 Environmental Remediation Research Group, Department of Chemical Sciences (Chemistry), Federal University Otuoke, Bayelsa State, Nigeria

2 Restoration of Ogoniland Project Team, Shell Petroleum Development Company, Port Harcourt, Nigeria

3 Remediation Team, Shell Petroleum Development Company, Port Harcourt, Nigeria

References

[1] United Nations Environmental Programme (UNEP), 2011. Environmental Assessment of Ogoniland. P.1-262. ISBN:978-92-807-9 Available on line at: http://postconflict.unep.ch/publications/OEA /UNEP_OEA.pdf

[2] Department of Health, Government of South Australia (DHGSA). Public Health Fact Sheet on Polycyclic Aromatic Hydrocarbons (PAHs): Health effects 2009 http://www.dh.sa.gov.au/pehs/PDF-files/ph-factsheet-PAHs-health.pdf

[3] Neff, M.M and Duxbury, MA. Composition, environmental fates, and biological effects of water based drilling muds and cuttings discharged to the marine environment: A Synthesis and Annotated Bibliography. Prepared for Petroleum Environmental Research Forum (PERF) and American Petroleum Institute. 2005. http://perf.org/pdf/APIPERFreport.pdf

[4] Gbadebo, A.M., Taiwo, A.M. and U. Eghele, U Environmental impacts of drilling mud and cutting wastes from the Igbokoda onshore oil wells, Southwestern Nigeria. Indian Journal of Science and Technology, 2010; 3(5), 504 -510.

[5] Environmental Protection Agency (EPA). An Assessment of the Environmental Implications of Oil and Gas Production: A Regional Case Study, 2008

[6] Osuji, L.C., Erondu, E.S and Ogali, R.E Upstream petroleum degradation of mangroves and intertidal shores: The Niger Delta Experience. Chemistry and Biodiversity, 2010: 7, 116 -128.

[7] Knez, D., Jerzy, A, G and Czekaj Trends in the drilling waste management. Acts Montanistica Rocnlk, 2006:11, 80-83.

[8] Morillon, A., Vidalie, J.F., hamzah, U.S., Suripno and Hadinota, E.K "Drilling and Waste management", SPE 73931, Intenationa; Conference on Health, Safety and Environment in oil and gas exploration and production, 2002: March 20-22

[9] Zimmerman, P.K. and Rober, J.D Oil-based drill cuttings treated by landfarming. Oil and Gas J, 1991: 12, 81-84

[10] Rojas-Avelizapa, N.G., Roldan-carrillo, T., Zegarra-Martinez, H., Munez-Colunga, A.M and Fernandez-Linares A field trial for an ext-situ bioremediation of a drilling mud-polluted site. Chemosphere 2007: 66, 1595-1600.

[11] Frydda, S and Randle, J.B Case study: Biological treatement of Geothermal drilling cuttings. Proceedings World Geothermal Congress, Bali, Indonesia, 25-29, 2010: 1-3.

[12] Ouyang, W., Liu, H., Murygina, V., Yu, Y., Xiu, Z and Kalyuzhnyi, S Comparison of bio-augmentation and composting for remediation of oily sludge: A field-scale study in China. Process Biochemistry, 2005: 40, 3763 -3768.

[13] Vidali, M. Bioremediation: An overview. Pure and Applied Chemistry, 2001: 73(7), 1163-1173

[14] Jorgensen, K.S., Puutstinen, J and Suortt, A. –M Bioremediation of petroleum hydrocarbon-contaminated soil by composting in biopiles. Environmental Pollution, 2000: 107, 245-254.

[15] Department of Petroleum Resources. Environmental Guidelines and Standard for the Petroleum Industry in Nigeria, 2002

[16] Joel, O.F and Amajuoyi, C.A Determination of selected physicochemical parameters and heavy metals in a drilling cutting dump site at Ezeogwu–Owaza, Nigeria. J. Appl. Sci. Environ. Manage, 2009: 13(2), 27- 31.

[17] Okparanma, R.N., Ayotamuno, J. M Polycyclic aromatic hydrocarbons in Nigerian oil-based drill-cuttings; evidence of petrogenic and pyrogenic effects. World Applied Sciences Journal 2010; 11 (4): 394-400, ISSN 1818-4952.

[18] Nweke, C.O and Okpokwasili, G. C Drilling fluid base oil biodegradation potential of a soil Staphylococcus species. African Journal of Biotechnology 2003; 2 (9), pp. 293-295. http://www.academicjournals.org/AJB

[19] Ayotamuno, J.M., Okparanma, R, N and Araka, P.P Bioaugmentation and composting of oil-field drill-cuttings containing polycyclic aromatic hydrocarbons (PAHs). Journal of Food, Agriculture & Environment 2009; 1.7 (2): 6 5 8 - 664. www.world-food.net

[20] Okparanma, R.N Ayotamuno, J.M and Araka, P.P Bioremediation of hydrocarbon contaminated-oil field drill-cuttings with bacterial isolates. African Journal of Envi-

ronmental Science and Technology 2009 3 (5), pp. 131-140. Available online at http://www.academicjournals.org/AJEST

[21] Ifeadi, C.N The treatment of drill cuttings using dispersion by chemical reaction (DCR). A paper prepared for presentation at the DPR Health, Safety & Environment (HSE) International Conference on Oil and Gas Industry in Port Harcourt, Nigeria. 2004.

[22] Adekunle, I.M., Ajijo, M.R., Omoniyi, I.T and Adeofun, C.O Response of four phytoplankton species in some sections of Nigeria coastal waters to crude oil in controlled ecosystem. Int. J. Environ., Res., Iran, 2009; 4 (1): 65 -74 http://ijer.ut.ac.ir

[23] Adekunle, I.M and Onianwa, P.C Functional group characteristics of humic acid and fulvic acid extracted from some agricultural wastes. Nigerian Journal of Science, Nigeria, 2001: 35 (1), 15 – 19.

[24] Adekunle, I.M Evaluating environmental impact from utilization of bulk composted wastes of Nigerian origin using laboratory extraction test. Environmental Engineering and Management Journal 2010; 9 (5): 721 -729.: http://omicron.ch.tuiasi.ro/EEMJ/

[25] Adekunle I.M., Adekunle, A.A., Akintokun, A.K., Akintokun, P and Arowolo,T.A Recycling of organic wastes through composting for land applications: a Nigerian experience. Waste Management & Research 2011; 29 (6): 582 – 593. DOI: 10.1177/ http://wmr.sagepub.com/content/29/6/582.abstract

[26] Adekunle, I.M Bioremediation of soils contaminated with Nigerian petroleum products using composted municipal wastes. Bioremediation Journal, 2011; 15 (4): 230-241, DOI: 10.1080/10889868.2011.624137. http://dx.doi.org/10.1080/10889868.2011.624137

[27] Adekunle I.M., Oguns, O., Shekwolo, P.D., Igbuku, O.O and Ogunkoya, O.O Assessment of population perception impact on value-added solid waste disposal in developing countries, a case study of Port Harcourt City, Nigeria. In: Xiao-Ying, Y (Ed) Municipal and Industrial Waste Disposal. Intech; 2012, p177-206.

[28] Adekunle A. A., Adekunle, I.M., Igba, T. O Assessing the effect of bioremediation agent from local resource materials in Nigeria on soil pH. Journal of Emerging Trends in Engineering and Applied Sciences 2012; 3 (3) 526-532. http://jeteas.scholarlinkresearch.org/articles/Assessing%20the%20Effect%20of%20Bioremediation%20Agent.pdf

[29] Adekunle A.A., I.M. Adekunle and Igba, T.O Impact of bioremediation formulation from Nigeria local resource materials on moisture contents for soils contaminated with petroleum products. International Journal of Engineering Research and Development 2012; 2(4) 40-45 http://www.ijerd.com/paper/vol2-issue4/F02044045.pdf

[30] Adekunle A.A, Adekunle, I.M. and Igba, T.O Assessing and forecasting the impact of bioremediation product derived from Nigeria local raw materials on electrical conductivity of soils contaminated with petroleum products. Journal of Applied technol-

ogy in Environmental Sanitation 2012; 2 (1) 57 -66. http://www.trisanita.org/jates/atespaper2012/ates09v2n1y2012.pdf

[31] Adekunle A.A., I. M. Adekunle and Igba T. O Soil temperature dynamics during bioremediation of petroleum products using remediation agent for Nigerian local resource materials. International Journal of Engineering Science and Technology 2012; 1 (4): 1-8. http://www.ijert.org/browse/june-2012-edition

[32] Association of Official Analytical Chemists (AOAC), Official Method and Analysis of The Association oh The Official Analytical Chemists 11th Edition Washington D C, 1970.

[33] Finar, I.L Organic Chemistry, volume I The Fundamental principles. 6th Ed, Longman, 1973.

[34] Liu, W., Luo, Y and Teng, Y Bioremediation of oily sludge-contaminated soil by stimulating indigenous microbes. Environ Geochem health 2010: 32, 23 -29.

[35] Ayotamuno, J.M., Okparanma, R.N., Davis, DD and allagoa, M. PAH removal from Nigerian oil-based drill-cuttings with spent oyster mushroom (Pleurotus ostretus) substrate. Journal of Food, Agriculture and Environment 2010: 8 (3 &4), 914 -919.

[36] Rojas-Avelizapa, N.G., Roldan-Carrillo, T., Zegarra-Martinez, H., Munoz-Colunga, A.M and Fernadez-Linares A field trail for an ex-situ bioremediation of a drilling mud-polluted site. Chemospher, 2007: 66, 1595 – 1600.

[37] Martin, J.A., Moreno, J.L., Hernandez, T and Garcia, C Bioremediation by composting of heavy oil refinery sludge in semiarid conditions. Biodegradation, 17:, 251 – 261.

[38] Al-Nasrawi, H Biodegradation of Crude Oil by Fungi Isolated from Gulf of Mexico. J Bioremed Biodegrad 2012, 3:4

[39] Mandal, A.K., Sarma, P. M., Singh, B., Jeyaseelan, C.P., Channasshettar, V.A., Lal, B and Datta, J bioremediation : an environment friendly sustainable biotechnological solution for remediation of petroleum hydrocarbon contaminated waste. ARPN Journal of Science and Technology, 2012: 2, 1-12

[40] Stevenson, F.J Humus Chemistry, 2004. Wiley & Sons

[41] Obayori, O.S., Ilori, M.O., Adebusoye, S.A., Amund, O.O and Oyetibo, G.O Microbial population changes in tropical agricultural soil experimentally contaminated with crude petroleum. African Journal of Biotechnology, 2008: 7 (24), 4512-4520.

[42] Corwin, D.L and Lesch, S.M. Apparent soil electrical conductivity measurements in agriculture. Computers and Electronics in Agriculture, 2005: 46, 11–4

Chapter 16

Biocomposites: Influence of Matrix Nature and Additives on the Properties and Biodegradation Behaviour

Derval dos Santos Rosa and Denise Maria Lenz

Additional information is available at the end of the chapter

http://dx.doi.org/10.5772/56290

1. Introduction

Composite materials are material systems which consist of one or more discontinuous phases embedded in a continuous phase. Thus, at least two distinct materials that are completely immiscible are combined to form a composite. The continuous phase are termed matrix and the discontinuous phase can be a reinforcement (reinforcing agent) or filler. Also, other additives as plasticizers, pigments, heat and light stabilizers are frequently added in order to provide certain properties. The type and reinforcement geometry impart strength to the matrix and the resultant composite shows optimized properties such as high specific strength, stiffness and hardness with respect to the specific components [1].

As conventional plastics are resistant to biodegradation, the concept of using biobased plastics (biodegradable polymers or biopolymers) as reinforced matrices for biocomposites is gaining more and more approval day by day [2]. A variety of natural and synthetic biodegradable polymers that can be used as biocomposite matrix are commercially avaiable. These biocomposite materials are designed to have a better environmental impact than conventional plastics as well as to promote an improvement in their mechanical properties so that their applications can be expanded. By embedding natural fibers with renewable resource-based biopolymers such as cellulosic plastics; polylactides; starch plastics; polyhydroxyalkanoates (bacterial polyesters); soy-based plastics, the so-called green biocomposites could soon be the future [3].

Biocomposites are composites that present natural reinforcements (like vegetable fibers) in their composition and can be: (i) partial biodegradable with non-biodegradable polymers matrices such as thermoplastic polymers (e.g., polypropylene, polyethylene) and thermoset polymers (e.g., epoxy, polyester) or (ii) fully biodegradable with biodegradable polymers matrices such as renewable biopolymer matrices (e.g., soy plastic, starch plastic, cellulosic plastic) and petrobased biodegradable polymer matrices (e.g., aliphatic co-polyester, polyesteramides). The fully biodegradable ones are 100% biobased materials and show biodegradability and/or compostability properties [2, 4, 5]. For the purpose of this chapter, only fully biodegradable biocomposites are the subject considered.

Natural fiber reinforced plastics by using biodegradable polymers as matrices are the most environmental friendly materials which can be composted at the end of their life cycle. Unfortunately, the overall physical properties of those composites are far away from glass-fiber reinforced thermoplastics. Further, a balance between life performance and biodegradation has to be developed [6].

Hybrid composites are resulted from the incorporation of several types of reinforcing agents with the purpose of tailoring the properties of the obtained composite according to engineering requirements. A synergistic effect between the different kinds of reinforcements enhances the overall performance of the composite. Bionanocomposites are a emerging class of nanostructured biohybrid material which exhibit a singular combination of structural and functional properties together with biocompatibility and biodegradability that was not found in nature. These hybrid materials consist mainly in the assembly of biopolymers and silicates from clay mineral family that have shown extraordinary potential to be used in many applications [7].

In the present chapter, an overview of the current biodegradable polymer matrices and some of the most used reinforcements is described as well as the properties and applications of the obtained biocomposites are dicussed.

2. Biodegradable polymer matrices

There are various ways that biodegradable polymers can be adressed. Depending on their origin, they may be divided as: natural, synthetic or microbial polymers.

2.1. Natural biodegradable polymers

Natural biodegradable polymers are polymers formed naturally during the growth cycle of living organisms. Their synthesis generally involves enzyme-catalyzed reactions and reactions of chain growth from activated monomers which are formed inside the cells by complex metabolic processes. Natural polymers such as proteins (collagen, silk and keratin), carbohydrates (starch, glycogen) are widely used materials for conventional and novel pharmaceutical dosage forms [8]. These materials are chemically inert, nontoxic, less expensive than the synthetic ones, eco-friendly and widely avaiable [8,9]. The families of natural polymers are low-cost materials along with some disavantages such as inferior thermal and mechanical properties. The natural polymers here described are from two groups, i.e., those obtained from vegetable and those from animal sources, as shown in Table 1.

Plant Source	Carbohydrates	Polysaccharides	Cellulose
			Starch
			Pectin
	Proteins		Soy derivatives
			Polypeptides
	Lignins		Polyphenols
Animal Source	Proteins		Silk
			Wool
			Polypeptides
	Polysaccharides		Chitin
			Chitosan
			Glycogen

Table 1. Classification of natural polymers based on their sources.

There are several types of carbohydrates: monosaccharides, disaccharides, oligosaccharides and polysaccharides. The latter ones, of particular interest, are comprised of hundreds or thousands of monosaccharides, commonly glucose, forming linear chains, such as cellulose, or branched chains, as in starch and glycogen. For this chapter, cellulose and its derivatives, starch and chitosan will be presented as natural biodegradable polymers [10].

2.1.1. Cellulose derivatives

Cellulose acetate (CA), universally recognized as the most important organic ester of cellulose because of its extensive applications in fibres, plastics and coatings, is prepared by reacting cellulose with acetic anhydride using acetic acid as a solvent and perchloric acid or sulphuric acid as a catalyst. CA is a carbohydrate composed of β-glucose molecules that are covalently linked through β-1,4-glycosidic bonds, widely found in nature in algae and land plants which has been valued as a functional material. CA comes to meeting the diverse needs of today's society including biodegradability characteristics, its hydrophilic behaviour and biocompatibility [11].

Several applications for cellulose and its derivatives have been shown, for example: in paints, textiles, pharmaceuticals and beauty, fibers, ionic liquids, construction technology and so on [12, 13]. Cellulose esters for coating applications are nearly always used as miscible blends with acrylics, polyesters and other polymers. This is possible because of their ability to form hydrogen bonds through the presence of hydroxyl groups and the carboxyl groups of the ester. An increase in ester molecular weight increases the toughness and melting point but decreases the compatibility and solubility, whereas hardness and density are unaffected. Compatibility, solubility and the maximum non-volatile content all decrease as the ester molecular weight increases. The hydroxyl group content inversely affects the moisture resistance and toughness [11].

Ignácio et al. [14] evaluated the production of cellulosic polymer membranes based on cellulose acetate and thus advanced technology was brougth to be used in membranes for separation processes (ultrafiltration, microfiltration, reverse osmosis, nanofiltration, gas separation, etc.). The use of these membranes has been shown to be effective for water treatment in chemical industries and pharmaceutical processes. Mulinari et al. [15] studied the preparation and characterization of a hybrid composite composed by bleached cellulose and hydrous zirconium oxide. Authors showed that these cellulose composites obtained by the crushed sugarcane combined with an inorganic material has intrinsic advantages such as low cost, biodegradability and simplicity in preparation and handling.

2.1.2. *Starch*

Starch, a low-cost biodegradable polymer, is abundant in plants, where it is stored in granule form and acts as an energy reserve [16]. Starch is composed of two polymers: amylose and amylopectin, both of which contain α-D-glucose units. Amylose is mostly a linear molecule of $\alpha(1\rightarrow4)$-linked-D-glucopyranosyl units with the ring oxygen atoms all on the same side. Amylopectin is the major branched component of starch and presents a $(1\rightarrow6)$ linkage that forms branch points. The hydrophilicity of these polymers is responsible for their incompatibility with most hydrophobic polymers [17]. When exposed to a soil environment, the starch component is easily consumed by microorganisms, leading to increase its porosity by void formation and the loss of integrity of the plastic matrix. The plastic matrix will be broken down into smaller particles.

Addition of a plasticizer like glycerin can further improve the ductility of starch, forming a polymer that is known as thermoplastic starch (TPS) which is capable of flowing easily. This plastifying agent lowers the glass transition temperature of starch as well as the melting temperature of the mixture by the introduction of mechanical and heat energy. The starch plastification is commonly carried out by extrusion at temperatures close to 120 °C. The mixtures of TPS with other polymers have the potential to behave in a similar manner to more conventional polymer-polymer blends. This would allow greater control of the dispersed phase morphology since the TPS should undergo deformation, disintegration and coalescence [18].

The crystalline nature of starch granules reflects the organization of amylopectin molecules within the granules whereas amylose is the most constituent of the amorphous portion that is randomly distributed among the amylopectin clusters.The conversion of starch into a thermoplastic material by extrusion or by gel casting into films results in the loss of the natural organization of the chains [19]. Figure 1 shows granular starch (a) and pregelatinized starch (b).

Blends of starch with synthetic polymers such as ethylene–vinyl alcohol copolymer, starch/poly(ethylene-co-vinyl alcohol), copolymers of ethylene with vinyl acetate, vinyl alcohol, acrylic acid, cellulose derivatives and other natural polymers, recycled high density polyethylene (HDPE) and other polyethylenes (PE) as well as compounds with a mixture of glycerin as plasticizer have been studied. Among the environmentally friendly starch-synthetic polymer products currently marketed on a commercial scale are Mater-Bi TM (Novamont, Italy), Bioplast (Biotech, Germany), Biopar (Biop Biopolymer Technologies AG, Germany), and NovonTM (produced by Chisso in Japan and Warner Lambert in the USA [20].

http://dx.doi.org/10.5772/56290

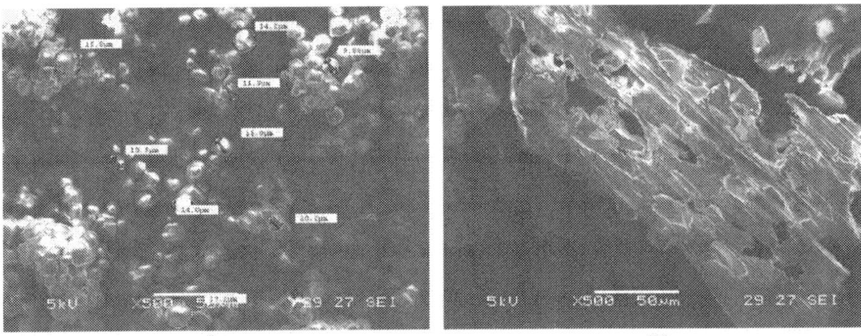

(a) (b)

Figure 1. Scanning electron microscopy (SEM) photomicrographs of (a) granular starch and (b) pregelatinized starch. Reprinted from Carbohydrate Polymers, 59, Pedroso A. G. and Rosa D. S., Mechanical, thermal and morphological characterization of recycled LDPE/corn starch blends, 1–9, Copyright (2005) [19] with permission from Elsevier.

The blending of biodegradable starch with inert polymers, such as polyethylene (PE), has received considerable attention currently. The reasoning behind this approach is the possibility of disintegration and disappearing of the all plastic films in the waste disposal environment if the biodegradable component is present in sufficient amounts and can be removed by microorganisms.

Pedroso and Rosa [19] studied blends with recycled low density polyethylene (LDPE) and corn starch containing 30, 40 and 50 wt% starch. The blends were prepared by extrusion and characterized by the melt flow index (MFI), tensile test, dynamic mechanical thermal analysis (DMTA) and scanning electron microscopy (SEM). For comparison, virgin LDPE/corn starch blends were prepared and characterized under the same conditions. The addition of starch to LDPE reduced the MFI values, the tensile strength and the elongation at break whereas the modulus increased. The decreases in the MFI and tensile properties were most evident when 40 and 50 wt% were added. SEM images showed that the interfacial interaction was weak for blends containing virgin and recycled LDPE. Blends prepared with recycled LDPE showed the same behavior as those blends prepared with virgin LDPE, indicating that starch was the main factor that influenced the blend.

In other work [21], the same authors blended high density polyethylene (HDPE) and polypropylene (PP), both post-consumer polymers, with thermoplastic starch (TPS). Corn starch plastification was carried out by extrusion with glycerin addition. The processing, thermal and mechanical behaviours of the produced TPS were investigated as well as the morphology characterization of post-consumer HDPE/PP blends (100/0, 75/25, and 0/100 wt.%) in different proportions of TPS (30%, 40% and 50% wt.%). In conclusion, the addition of TPS to recycled PP reduces its melting flow index (MFI) whereas the MFI of HDPE and HDPE/PP blends increases. TPS also decreases the tensile strength and increases the rigidity of the polymers. The incorporation of TPS in polyolefin matrices results in the separation of phases and a disintegration of the starch granules.

2.1.3. Chitosan

Chitosan (CS) is a biopolymer (poly-β-1,4-glucosamine) having immense structural possibilities for chemical and mechanical modifications to generate novel properties, functions and applications, especially in biomedical area. Chitosan is no longer just a waste by-product from the seafood processing industry. This material is now being utilized by industry to solve problems and to improve existing products as well as to create new ones. CS is composed by linear nitrogenous polysaccharides - a basic polysaccharide homopolymer from natural sources, biodegradable, biocompatible and non toxic. Chitosan is produced commercially by deacetylation of chitin, naturally occurring polysaccharides which is the structural element in the exoskeleton of crustaceans (crabs, shrimp, etc.). Due to its variable and incomplete deacetylation process, it acts as a copolymer of varying amounts of N-acetyl glucosamine and N-glucosamine repeated units. The presence of reactive primary amino groups renders special property that makes CS very useful in pharmaceutical applications [22].

CS has three types of reactive functional groups, an amino group as well as both primary and secondary hydroxyl groups. Chemical modifications of these groups have provided numerous useful materials in diferent fields of application. Chitosan oligomers as well as chitosan have been shown to inhibit growth of several fungi and bacteria, especially pathogens. Hirano and Nagao [23] have studied the relationship between the degree of polymerization of chitosan and the inhibition efect.

At room temperature, chitosan forms aldimines and ketimines with aldehydes and ketones, respectively. Reaction with ketoacids followed by reaction with sodium borohydride produces glucans carrying proteic and nonproteic amino groups. N-Carboxymethyl chitosan is obtained from glyoxylic acid and its potential uses are in chromatographic media and metal ion collection [24].

2.2. Biodegradable polymers of microbiological origin

Polymers of microbial origin are produced as intracellular reserve material for a variety of bacteria and have gained prominence due to their possible applications as well as their biodegradable and renewable characteristics.

In the last three decades, the polymers, especially polysaccharides, have acquired great importance in a wide range of industrial processes [25]. Several species of fungi and yeasts produce polymers of commercial interest; however, polymers from bacterial origin are those with greater viability in terms of industrialization and commercialization since they present quality and constant supply. Among these polymers, we highlight the PHB and the PHBV which comprise the group of polyhydroxyalkanoates whose classification is presented in Table 2.

Polysaccharides	Polyhydroxyalkanoates	Poly(3-hydroxy-butyrate) - PHB
		Poly(β-hydroxybutyrate-co-valerate) PHB-V

Table 2. Examples of polyhydroxyalkanoates.

The polyhydroxyalkanoates (PHAs) are thermoplastic polyesters which degrade completely into microbiologically active environments in addition to being biocompatible and may be biosynthesized by a large number of Gram negative and Gram positive bacteria, from different carbon sources or made from renewable and non-renewable genetically modified (GM) plants. Examples of pure cultures used for industrial production of PHAs include *Ralstonia eutropha, Alcaligenes lotus, Azotobacter vineland* and various Pseudomonas species [26-32].

Genetically modified plants, such as potatoes (*Solanum tuberosum*) and tobacco (*Nicotiana tabacum*) produce cereals such as sunflower and soybean that can provide other ways of producing PHAs. However, the yield (4% of the weight of the plant) is much less than the one obtained by bacteria which reduces the production of PHAs by this method [26-32].

2.2.1. *Poly(3-Hydroxy-Butyrate) (PHB)*

Poly(3-hydroxy-butyrate), PHB, which is a PHA produced by the *Alcaligens eutrophorus* bacteria, is one of the most interesting biodegradable polymers because it is obtained by bacterial fermentation from renewable resources. PHB can also be synthesized by ring-opening polymerization of β-butyrolactone using distannoxane derivatives as catalysts, such as zinc and alluminium [33]. PHB is linear, homochiral, thermoplastic polyester produced by micro-organisms as intracellular fat deposits in response to limited nutrient availability. PHB belongs to a polyhydroxyalkanoate class of shorter pendant groups that confers a high degree of cristalinity [34].

However, PHB presents some drawbacks like thermal instability at temperatures close to its melting point and a relatively low impact resistance [16]. PHB molar mass decreases proportionately with some processing parameters like time and temperature. In spite of its narrow processing window, PHB with high molar mass can be processed like other thermoplastics if adequate processing parameters are used.

Two main efforts have been used to change PHB properties: biosynthesis and blending. Since blends are a cheaper and faster method to improve polymer properties than synthesis, blends have often been used to improve mechanical properties and processability of PHB. [16, 35].

The biosynthesis of this polymer allows a cyclical process through sustainable renewable sources by replacing cutting edge technologies related to the production and use of synthetic polymeric materials. Among the microorganisms that produce PHB, bacteria like *Alcaligenes eutrophus, Azotobacter vinelandii* and *Ralstonia eutropha* can be detached. [36].

According to Lenz et al. [31], the chemical structure of the polyester is an important factor in determining its physical properties and determining the activity of the enzymes involved in their biosynthesis and biodegradation. PHB is a saturated linear polyester, behaving like conventional thermoplastic materials. It has high crystallinity and melting temperature of approximately 176°C. Its glass transition temperature (Tg) is below 5°C and its properties resemble those of polypropylene (PP).

Comparing to polymer commodities, conventional PHB and its copolymers have the advantage of biodegradability and biocompatibility. In contrast, presents the disadvantage of having a poor thermal stability and impact resistance relatively low. Its use spans several segments, such as applications in biomedical areas, agriculture, food packaging and pharmaceutical products, as well as the segments of packaging and agricultural films strongly highlighted. The combination of high temperature and crystallinity provides shine to the films, whereas the rigidity and low impact resistance presented by PHB hinder their use. PHB copolymers have better mechanical properties. The copolymer PHB-V, for example, provide an improvement in ductility and impact resistance, making it more interesting from the point of view of application and end products compared to PHB [30, 32, 37-40].

2.3. Synthetic biodegradable polymers

This class of polymers has been widely used in biomedical uses, such as controlled-release capsules of drugs in living organisms, fasteners surgery (sutures, implants for bone pins) and special packaging. Polymers of this class that have been studied more recently are poly(lactic acid) (PLA), polyglycolic acid (PGA), poly (glycolic acid-lactic acid) (PGLA) and poly(e-caprolactone) (PCL) [35].

For greater understanding, synthetic biodegradable polymers are separated into classes. Table 3 shows the classification of non-natural synthetic biodegradable polymers.

	Aliphatic	- PLA
		Poly(glycolic acid) - PGA
		Poly(e-caprolactone) - PCL
	Aliphatic Aromatics (PAA)	Polytrimethylene terephthalate - PTT
		Poly(butylene terephthalate) -PBT
		Poly (butylene succinate) - PBS

Table 3. Classification of non-natural synthetic biodegradable polymers.

The polyesters compete an important position among the group of biodegradable plastics and some biodegradable polyesters are already commercially available.The main biodegradable polyesters are those based on hydroxy-carbonic acids. The biodegradable polyesters still have high cost, but they have aroused great interest due to their accessible production by fermentation or synthetic routes [35].

During the last two decades, aliphatic polyesters such as poly(ε-caprolactone) (PCL) and poly (L-lactic acid) (PLLA) have been extensively studied due to their ability to undergo hydrolysis in the human body as well as in natural circumstances [37, 41, 42].

2.3.1. Poly(Lactic Acid) (PLA)

Poly(lactic acid) (PLA) is a hydrolytically degradable aliphatic polyester which presents water vapor permeability that may have a significant influence on its rate of degradation. The poly(lactic acid) (PLA) is an aliphatic polyester obtained by polymerization of lactic acid. This can be found in the form of two optical isomers: L-and D-lactide. PLA has potential for applications in the medical, pharmaceutical and packaging, mainly as implantable devices temporarily (sutures, staples, nano-reservoirs for drugs etc). Other applications involve the sectors of textiles and fibers, agriculture, electronics, appliances and housewares [43, 44, 45].

PLA presents some advantages like biocompatibility, has better thermal processibility compared to other biopolymers such as poly(hydroxyalkanoates) (PHAs), shows eco-friendly characteristics and requires 25–55% less energy to produce than petroleum-based polymers. Nevertheless, PLA is a very brittle material and chemically inert with no reactive side-chain groups making its surface and bulk modifications a challenging task. Besides, PLA shows low degradation rates and is hydrophobic [46].

Henry et al. [47] investigated systems including poly (lactic acid) (PLA). The thermal analysis showed that the glass transition temperature (Tg) of the polymer is about 320 K. The β relaxation was observed between -150 °C and -30 °C, depending on the measurement frequency (1 Hz - 100 kHz) and was determined as secondary relaxation in the glassy state. The authors studied the changes that are associated with water penetration into the polymer which directly affect the relaxation process. Water molecules confined (outlined / permeated) and the polymer chains in polymer networks represent an important function in matrix degradation and, thus, the authors were able to observe the evolution of degradation for a few weeks in an environment with controlled humidity. It is accepted that water penetrates preferentially in amorphous areas, but also affects the crystalline regions. It is a clear evolution of the observed activation energy of relaxation during polymer degradation. The resulting dielectric relaxations are complemented with measures of molecular weight during degradation with time.

2.3.2. PCL

Poly (e-caprolactone) (PCL) is a synthetic aliphatic polyester made from ring opening polymerization. This biodegradable polyester presents good mechanical properties that is compatible with many types of polymers and is one of the most hydrophobic biodegradable polymers currently available. PCL has been widely studied for use in drug release systems [48]. Extracellular enzymes present in soil can cleave the extensive chains of PCL before assimilation of the polymer by microorganisms. However, the high cost of PCL has prevented its widespread industrial use. PCL has been thoroughly examined as a biodegradable medium and as a matrix in controlled drug-release systems [14, 49].

The main limitation of PCL is its low melting temperature (Tm 65°C) and also has some drawbacks, including a poor, long-term stability caused by water absorption, poor mechanical and processing properties. Some of these problems can be overcome by physical or chemical modifications, including the blending of these polymers. [49]

PCL/CA blends are generally incompatible, immiscible and show poor interpolymeric adhesion [14, 49]. Rosa et al. [11] reported miscibility between several CAs and aliphatic polyesters. The miscibility of the cellulose polymer with a polymeric plasticizer is important in order to maintain the already complex mixture as homogenous as possible. The use of coupling agents usually improves the elongation of composites, but frequently results in a decrease in strength. One approach to improve the compatibility between the constituent polymers in PCL/CA mixtures is to incorporate a compatibilizer into the mixture. The chemical modification of aliphatic polyesters by grafting is another way of improving the compatibility between starch and aliphatic polyesters in polymeric blends. The effects of polyethylene grafted with maleic anhydride (PE-g-MA) on the thermal and mechanical properties, as well as on the morphology of blends of low-density polyethylene (LDPE) and corn starch have been studied using differential scanning calorimetry (DSC), tensile strength measurements and scanning electron microscopy [14, 49-53].

3. Natural reinforcement agents as additives for biocomposites

Polymer reinforcements are generally used in order to provide stiffness and strength to the polymer matrix resulting in improved mechanical properties for the obtained composites Besides, properties like water and gas barrier as well as fire resistance and flame retardant properties and so on can be enhanced by the employ of reinforcements in polymer matrices [54-56].

The present review focuses on vegetable fibers (also reported as natural or plant fibers), nanofibers extracted from them and nanoclays in particular mineral silicates as reinforcement agents for biobased polymer matrices. Instead of being a natural non-renewable source, nanoclays are abundantly available and improve mechanical properties at lower loadings [57].

3.1. Natural or vegetable fibers

The interest in the use of vegetable fibers as reinforcement agents in polymeric composites is growing currently owing to environmental regulations and ecological concerns of the actual society.

Vegetable fibers are abundantly available, fully and easily recyclable, non-toxic, biodegradable, non-abrasive to the molding machinery, easily colored as well as have lower cost, lower density and lower energy consumption in producing step with respect to synthetic fibers as glass and carbon fibers [58,59]. In addition to having lower processing energy requirements and more shatter resistant when compared with synthetic fibers, vegetable fibers have good sound abatement capability, non-brittle fracture on impact, high specific tensile modulus and tensile strength, low thermal expansion coefficient and low mold shrinkage [59].

There are thousands of different fibers in the world and a few of them have been studied. All vegetable fibers (wood or non-wood fibers) are constituted by cellulose; hemicellulose and lignin combined to some extent as major constituents [6]. In fact, the so-called lignocellulosic fibers have cellulose as the main fraction of the fibers. Cellulose is a semicrystalline polysaccharide made up of D-glucosidic bonds. A large amount of hydroxyl groups in cellulose (three in each repeating unit) imparts hydrophilic properties to the natural fibers [60]. Thus, they are hydrophilic in nature. Cellulose forms slender rodlike crystalline microfibrils that are embedded in a network of hemicellulose and lignin, i. e., the microfibrils are bonded together through an amorphous and complex lignin/hemicellulose matrix that acts as a cementing material.

Hemicellulose is a polysaccharide with lower molecular weight than cellulose.The main difference between cellulose and hemicellulose is that hemicellulose has much shorter chains and also has branches with short lateral chains consisting of different sugars while cellulose is a linear macromolecule [52]. Both are easily hydrolyzed by acids, but only hemicellulose is soluble in alkali solutions as well as lignin. Lignin is a hydrocarbon polymer with a complex composition that presents hydroxyl, methoxyl and carbonyl functional groups [4].

Lignocellulosic fibers may be found in different parts of the plant like leaf, bast, seed and fruit. Some fibers derived from leaf part - leaf fibers: abaca (Manila hemp), sisal, curauá, banana leaf fiber, pineapple leaf fiber (PALF) and henequen; fibers derived from the inner bark part - bast fibers: flax, ramie, kenaf/mesta, hemp, piaçava and jute; fibers derived from plant seed - seed fibers: cotton and kapok and fruit fibers: coconut husk, i.e., coir and luffa. Climatic conditions, age of plant and the digestion process influence not only the structure of fibers but also their chemical composition [56, 61]. Plant fibers from wheat straw, rice straw, oat straw, esparto, elephant grass, bamboo, bagasse (sugar cane) are classified as grass and reed fibers [56] Some of these non-wood fibers were been studied as raw material source (pulp) for papermaking in many developing countries and for biocomposites manufacture whose composites can be applied mainly for food or non-food packaging, automobile parts and biomedical engineering in repairing or restoring tissues and implants as well as drug/gene delivery [62, 63].

Wood fibers have numerous types distributed in softwoods and hardwoods. Hardwoods are, in general, more complex and heterogeneous in structure than softwoods having a characteristic type of cell called vessel element (or pore) for water transport [64].

Table 4 shows the chemical composition of some non-wood vegetable fibers. The concentration of cellulose and other components of lignocellulosic fibers exhibit a considerable variation even for the same fiber. The references therein indicate concentration values all along the presented concentration range. The spiral angle of the cellulose microfibrils and the content of cellulose, determines generally the mechanical properties of the cellulose-based natural fibers [6]. For instance, these two structure parameters were used to calculate the Young's modulus of the fibers through models developed by Hearle et al [65] cited by Bledzski and Gassan [6].

As natural materials, vegetable fibers have nonuniformity such in dimensions as in mechanical properties when compared to synthetic fibers. Other drawbacks for the use of vegetable fibers in biocomposites are: (i) the lower processing temperature (limited to approximately 200°C) due to fiber degradation and/or volatile emissions; (ii) the high moisture absorption due to fiber hydrophilic nature and (iii) incompatibility with most hydrophobic polymers. These problems are well known and countless research has been developed to reduce them with reasonable success [66, 67]. Nevertheless, vegetable fibers (as fillers or reinforcements) are the latest growing type of polymer additives [68].

Because of the low interfacial properties between vegetable fiber and polymer matrix which often reduce their potential as reinforcing agents due to fiber hydrophilic nature, chemical modifications are considered to optimize the interface of fibers. Chemicals may activate hydroxyl groups or introduce new moieties that can effectively interlock with the matrix [69].

Chemical Composition						
Fiber	**Cellulose (wt%)**	**Lignin (wt%)**	**Hemicellulose (wt%)**	**Ash (wt%)**	**Microfibrilar/spiral angle (Deg.)**	**References**
Abaca	56-63	7-13	15-25	5	--------	[2, 56, 68, 69]
Curauá	70.7–73.6	7.5-11.1	9.9	0.9	--------	[2, 66, 67]
Flax	64–71	2–5	18.6–20.6	5	5-10	[2, 56, 68, 69]
Hemp	57-77	3.7-13	14-22.4	--------	2-6.2	[2, 56, 68, 69]
Henequen	77.6	13.1	4-8	--------	--------	[68, 69]
Jute	45–72	12–26	12–21	0.5–2	8.0	[2, 6, 56, 68, 69]
Kenaf	31–72	8–21	22–24	2–5	--------	[2, 56, 68, 69]
PALF	70-82	5-12.7	--------	--------	14	[2, 68]
Ramie	68.6–91	0.6–0.7	5–16.7	--------	7.5	[2, 6, 56, 68, 69]
Sisal	47–78	8–13	10–24	0.6–1	10-22	[2, 6, 56, 68, 69]

Table 4. Chemical composition of some common vegetable fibers.

Over the last decade, many approaches towards enhancing interfacial adhesion have been pursued. Generally improvements can be accomplished, but there must be a critical cost-benefit evaluation of using the added interfacial agents or processing steps [63].

Alkaline treatment or mercerization is one of the most used chemical treatments of natural fiber. The important modification done is the disruption of hydrogen bonding in the fiber network structure, increasing surface roughness. This treatment removes a certain amount of lignin, wax and oils covering the external surface of the fiber cell wall, depolymerizes cellulose and exposes the short length crystallites [69, 70]. As a result; the adhesive characteristics of the fiber surface are enhanced [71]. Figure 2 shows the aspect of curauá vegetable fiber before and after treatment of NaOH solution.

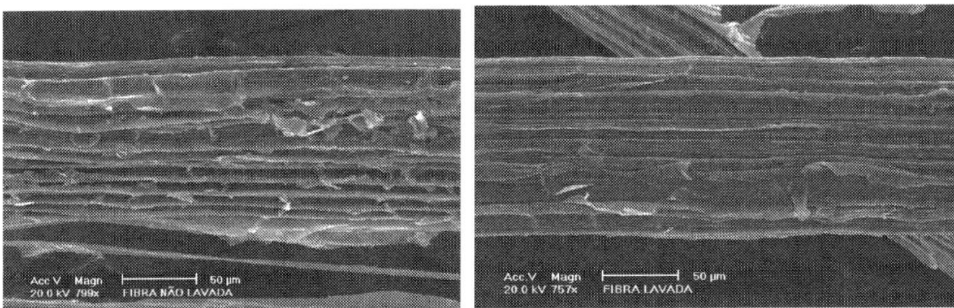

(a) (b)

Figure 2. SEM micrographs of curauá fiber: (a) as received (b) washed with 0.1 M NaOH solution 24 h at room temperature. Source: Authors

The efficiency of the alkali treatment depends on the type and concentration of the alkaline solution as well as time and temperature of the treatment. Different conditions for alkali treatment of vegetable fibers can be found in literature as well as combinations with other treatments [6, 72].

Authors reported that alkali concentration and reaction time of mercerization has a significant effect on the surface modification [73]. C. indica vegetable fibers were immersed firstly in 2% NaOH for the different time intervals at room temperature to optimize the mercerization time. Afterwards, the mercerization of C. indica fiber was also carried out in 4, 6, 8, 10, 12, and 14% NaOH solutions to study the effect of different concentrations of NaOH on the mercerization of the fibers. Maximum mercerization observed in terms of weight loss of fiber polymer backbone was observed at 210 min. With respect to the concentration of NaOH solution, the weight loss increases with the increase in alkali concentration and shows maximum weight loss at 10% alkali concentration. This happens due to the removal of lignin, hemicelluloses, pectin and other surface impurities with NaOH.

Campos et al. [74] reported the development of biocomposites of thermoplastic starch and polycaprolactone (PCL) with sisal fibers as reinforcement agent. Sisal fibers were treated with sodium hydroxide solution (NaOH 5% (w/v) at 90ºC under agitation for 60 min. After that, sisal fibers were bleached with a blend solution of peroxide hydrogen (H_2O_2 16%) and sodium hydroxide (NaOH 5%) at 55 ºC for 90 min. The authors observed strong adhesion fiber-matrix and interaction between carboxyl groups in PCL-starch and hydroxyl groups in sisal fibers.

Nevertheless, alkaline treatment or other chemical/physical treatment may damage vegetable fiber surface structure, reducing its strength [75, 76]. When a chemical treatment is applied on synthetic fibers like glass fibers only fiber surface is modified. On the contrary, chemical treatments applied on vegetable fibers can produce important chemical and structural changes not only at fiber surface but also on the interphase between elementary fibers [66]. Furthermore, the orientation of microfibrils of cellulose within each elementary fiber plays an important role because it changes the crystallinity of the natural fiber [77]. A different variety of chemical treatments applied on sisal fibers resulted in greater extensibility and lower modulus. These phenomena must be related to the structural variation in the ultimate cells, that is, swelling and partial removal of lignin and hemicellulose [78].

Moraes et al. [76] showed the use of sodium borohydride ($NaBH_4$) (1% wt/vol) as protective agent for vegetable sisal fibers under alkaline treatment with sodium hydroxide (NaOH). The authors reported that the effectiveness of hydride ions (H^-) to protect the sisal fiber was more pronounced in moderate NaOH concentrations (5 wt/vol %) at room temperature or higher (10 wt/vol %) for shorter alkaline treatment times.

Acetylation of natural fibers is a well-known esterification method causing plasticization of cellulosic fibers. Acetylation reduces the hygroscopic nature of natural fibers and increases the dimensional stability of composites [54]. Acetylation is based on the reaction of cell wall hydroxyl groups of lignocellulosic materials with acetic or propionic anhydride at elevated temperature [70]. Other chemical treatments that have already used for fiber treatment are mainly benzoylation treatment, permanganate treatment, isocyanate treatment and peroxide treatment [69].

The use of coupling agents is also extensively used for chemical modification of synthetic and vegetable fibers. Organosilanes and maleic anhydride are both coupling agents that not only produce surface modification but also can produce grafting polymers [63, 79]. Acrylonitrile grafting has also been reported as fiber treatment for glass fibers as well as for vegetable fibers [69]. Coupling agents can be found inserted in polymer matrices (grafted polymer matrices) or in vegetable fibers or even introduced during reactive melt processing of the biocomposite.

In work of Chang et al. [80], kenaf fiber dust was added to a previous maleated polycaprolactone/thermoplastic sago starch blend used as biocomposite matrix. The addition of Kenaf fiber up to 30 phr decreased the water absorption capacity of the maleated treated biocomposites with respect to non-treated biocomposites. The decrease in water absorption was due to the enhanced adhesion between the Kenaf fiber dust and the matrix through grafting which led to decrease of voids between fiber/matrix interfaces. Besides, Kenaf fiber addition improved the mechanical properties of the maleated and non-maleated biocomposites. Nevertheless, tensile strength and modulus reached higher values for maleated biocomposites with higher Kenaf fiber loadings. The effective coupling mechanism of maleic anhydride between polymer matrix and Kenaf has been attributed to esterification reaction between the hydroxyl groups of the Kenaf and anhydride group to form ester linkages [69, 80].

Different authors have applied different methods for silane treatment and have studied the effect of silane treatment on surface morphological and hygroscopic character of the natural fibers. Most of the silane groups have the following formula: $R_{(4-n)} - Si -(R'X)_n$ $(n = 1,2)$ where R is alkoxy, X represents an organofunctionality, and R′ is an alkyl bridge connecting the silicon atom and the organofunctionality [81].

Some authors prepared bamboo fiber-reinforced polylactic acid (PLA) biocomposites using a film-stacking process [71]. Bamboo fibers were subjected to three different silane treatments: direct silane coupling, silane coupling after plasma treatment and silane coupling during UV irradiation. Biocomposites with silane coupling after plasma-treated fibers presented the highest increase in tensile strength with respect to biocomposites with untreated fibers and among all tested fiber treatments, showing a close adhesion between the PLA matrix and fibers. Fiber surface modifications was related to the silane that should have two functional groups to effectively couple fiber and matrix: a hydrolyzable alkoxy group to condense with hydroxyls on the surface of bamboo fibers and an organofunctional group capable of interacting with the PLA matrix that can result in a copolymerization (grafting) and/or formation of a interpenetrating network.

Other works [81, 82] also reported that in general the interaction of the silane coupling agent with vegetable fibers involves four steps: (i) hydrolysis of silane monomers in presence of water to yield reactive silanol (–Si-OH), (ii) self-condensation of silanol, (iii) The silanol monomers or oligomers are physically adsorbed to hydroxyl groups of fibers by hydrogen bonds on the fiber surfaces and/or in the cell walls. The free silanols also adsorb and may react with each other forming rigid polysiloxane structures linked with a stable –Si-O-Si– bond and (iv) grafting under heating conditions since the hydrogen bonds between the silanols and the hydroxyl groups of fibers can be converted into the covalent –Si-O-C– bonds and liberating water.

In order to enhance the behavior of Kenaf/PLA biocomposites, authors [43] treated kenaf fibers with sodium hydroxide and 3-aminopropyltriethoxysilane (APS) coupling agent. The authors described the hypothetical reaction of silanol and the fiber: the ethoxy groups of APS hydrolyze in water or a solvent producing a silanol and next the silanol reacts with the OH group of the kenaf fiber which forms stable covalent bonds to the cell wall that are chemisorbed onto the fiber surface. In other work [83], ramie fibers were treated with permanganate acetone solution and with permanganate acetone solution followed by silane acetone solution to produce biocomposites with poly(L-lactic acid) PLLA matrix by hot press molding. The fiber surface-treatment with permanganate acetone solution followed by silane acetone solution improved the interfacial adhesion with PLLA matrix. Both treatments accelerate the water permeation rate in PLLA biocomposites, which plays a critical role in the decline of interfacial adhesion strength.

Also, physical treatments have been used. These treatments change structural and surface properties of the fiber and thus influence the mechanical bonding with the polymer matrix. Some pf these treatments envolve fibrillation and electric discharge (Corona, cold plasma, sputtering) and so on [72]. Cold plasma treatment causes chemical implantion, etching, polymerization, free radical formation and crystallization whereas sputtering promotes physical changes such as fiber surface roughness that leads to fiber/matrix interface adhesion [71, 84].

Nevertheless, the hydrophilic character of natural (biobased) polymers has contributed to the successful development of environmentally friendly composites, as most natural fibers and nanoclays are also hydrophilic in nature [85]. Most of the published studies on biocomposites with biodegradable polymers are with polyester matrix, such PHA, due to its polar character that provides better adhesion to lignocellusic fibers [86].

Authors [87] showed that curauá vegetable fibers have good interfacial adhesion to a polyester-based matrix even without coupling agent addition. In this work coupling agent was added during reactive extrusion at the same time with the neat matrix and a masterbatch containing curauá fiber and the blend matrix. The authors reported the importance of the coupling agent addition, beside the NaOH treatment of the fiber, for improved interfacial fiber/matrix adhesion. Figure 3 shows SEM analysis of tensile fracture cross-section samples of polyester blend/curauá fiber biocomposite. Figure 3a revealed a weak fiber/matrix interface with numerous irregularly shaped microvoids and some de-bondings for composites in the absence of coupling agent, which could be responsible for deterioration of the stress transfer from the matrix to the fibers having an adverse effect on the mechanical properties. On the other hand, composites with coupling agent showed an improvement in polymer/fiber adhesion, avoiding fiber pull-out that leads to voids emerging. In this case, curauá fibers were broken under tension (Figure 3b).

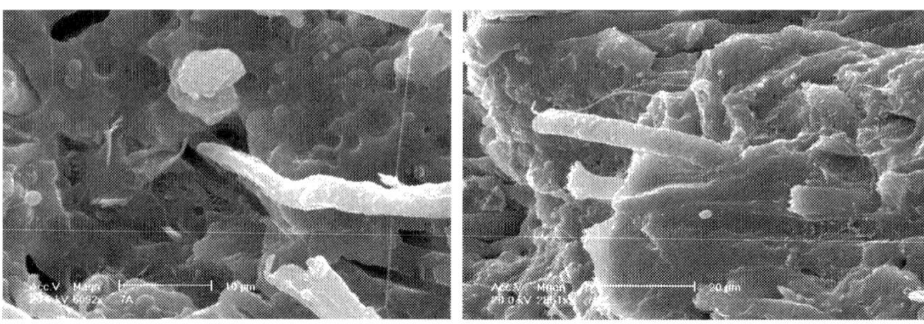

(a) (b)

Figure 3. SEM micrographs of fracture cross section of polyester blend/curauá fibers: (a) without coupling agent and (b) with coupling agent. Reprinted with kind permission from Springer Science and Business Media: Journal of Polymers and the Environment Biodegradable Polyester-Based Blend Reinforced with Curauá Fiber: Thermal, Mechanical and Biodegradation Behaviour 20, 2012, 237-244, Harnnecker F., Rosa, D. S., Lenz, D. M., Figure 3a and 3b [87].

3.2. Cellulose nanofibers from vegetable fibers

Cellulose is the most abundant renewable carbon resource on Earth. Thus, it can be obtained from many natural sources. Aside from occurring in wood, cotton and other plant-based materials derived from agricultural crops and by-products, cellulose is also produced by algae, some bacteria and tunics of marine animals – tunicates. [88, 89]. The main difference between cellulose obtained by plants and bacteria is that plant-synthesised cellulose usually also contains hemicellulose, lignin and pectin while cellulose produced by bacteria on the other hand, is pure cellulose without foreign substances [90]. Also, highly crystalline cellulose in the native state can be extracted from tunicates which shows high aspect ratio (length/diameter ratio) as well as allows better matrix-to-filler stress transfer [91].

Nanofibers are fibers that have at least one of its linear dimensions smaller than 100 nm. One of the more significant characteristics of nanofibers is the enormous availability of surface area per unit mass - 1 m^2 of them weighs only 0.1 - 1 gram [3, 92]. Cellulose nanofibers are one class of natural fibers that have resulted in structures with remarkable mechanical properties. These nanofibers have received an increasing interest in the bio-based materials community since nanocellulose reinforced biopolymers will be less expensive than many common plastics derived from petroleum resources if processing costs can be kept to between $0.20–$0.25/lb [93]. However, the full reinforcing potential of nanofibers has yet to be realized partly because of issues related to scaling manufacturing processes [94].

Cellulose nanofibers are nano-reinforcements from biomass that have been improved the biobased polymers properties such as thermal stability, mechanical toughness and barrier properties at much lower fiber fractions than those required in conventional vegetable fiber composites. Biocomposite materials have been showed potential to be used in packaging with PLA matrix [95] and medical applications using polyurethane - PU - matrix [96].

There are many different methods to obtain nanofibres from vegetable fibres. Cellulose nanocrystals, also reported in the literature as nanowhiskers (or just simply "whiskers"), nanofibers, cellulose crystallites or crystals, are the crystalline domains of cellulosic fibers, isolated mainly by acid hydrolysis [97].

Cellulosic materials intended for use as nano-reinforcements in biocomposites are usually subjected to hydrolysis by strong acids such as sulfuric or hydrochloric acid, yielding in a selective degradation of amorphous regions of cellulose and, consequently, the splitting of micro-fibril beams. As a result of cellulose hydrolysis, the disintegration of its hierarchical structure takes place to form crystalline nanofibers [89]. Usually the acid hydrolysis is combined with sonication [88]. The source of cellulose and hydrolysis conditions (acid concentration, acid to cellulose ratio, temperature and reaction time directly affect the morphology of the nanocrystals [89, 98]. The length of the so-produced nanocrystals generally ranges between 100 and 300 nm and width of 5-20 nm [88, 99]. Invariably these nanocrystals from plant fibers present a rod-like structure [91].

Cellulose nanoparticles are obtained as stable aqueous suspensions and thus the processing of cellulose nanocomposites was first limited to using hydrosoluble (or at least hydrodispersible) or latex-form polymers as nanocomposite matrices. After dissolution of the hydrosoluble (or hydrodispersible) polymer, the aqueous solution was mixed with the aqueous suspension of cellulosic nanoparticles to form a mixture that was cast and evaporated to obtain a solid nanocomposite film. The use of the extrusion processing technique was hampered due to the hydrophilic nature of cellulose which causes irreversible agglomeration of the nanofibers in polymer matrices [3]. The development of newer industrially viable processing techniques as melt compounding is the focus currently. PLA nanocomposites reinforced by cellulose nanofibers separated from kenaf pulp were obtained using a two-step process: masterbatch preparation using a solvent mixture of acetone and chloroform followed by extrusion process and injection molding. The tensile modulus and the tensile strength of the PLA nanocomposite using 5 wt% of nanofiber showed an increase of 24% and 21%, respectively [100].

Cellulose nanocrystals can also be produced by submitting vegetable fibres to high mechanical shearing forces, disintegration of the fibres occurs, leading to a material called microfibrillated cellulose (MFC) [88, 101]. However, depending upon the raw material and the degree of processing, chemical treatments (alkaline, enzimatic or oxidation treatments) may be applied prior to mechanical fibrillation which aim to produce purified cellulose, such as bleached cellulose pulp, which can then be further processed [101]. These nanofibrils ideally consist of individual nanoparticles with a lateral dimension around 5 nm, but MFC generally consists of nanofibril aggregates, whose lateral dimensions range between 10 and 30 nm or more [88].

The major obstacle when producing cellulose based nanocomposites is to disperse the hydrophilic reinforcement in the hydrophobic polymer matrix without degradation of the biopolymer or the reinforcing phase. This can be addressed by improving the interaction between cellulose nanofibers and the matrix and/or by using suitable processing methods [102]. Jute nanofibers submitted to alkali, dimethyl sulfoxide (DMSO) and acid hydrolysis treatments were incorporated into the biocopolyester matrix by melt mixing in varying weight percentages ranging from 0% to 15%. The enhancement in properties was highest for 10 wt % jute nanofiber loaded composites, indicating the most uniform dispersion in this material [103]. In work of Wang and Drzal [104], the solvent evaporation technique (commonly used for drug microencapsulation) was employed to suspend PLA in water as microparticles. The suspension of the PLA microparticles was mixed with high pressure homogenized cellulose nanofibers, producing nanocomposites with good fiber dispersion after water removal by membrane filtration followed by compression molding. Tensile modulus and strength increased up to 58% and 210%, respectively, with respect to neat PLA.

In other work, a hybrid multi-scale biocomposite composed by microfibrillated cellulose (MFC) and bamboo fiber bundles in a polylactic acid (PLA) matrix were successfully processed by extrusion using a surfactant which favoured the dispersion of nanowhiskers in PLA matrix [105]. A hierarchy structure of reinforcement was created with bamboo fiber as the primary reinforcement and cellulose creates an interphase in the PLA matrix around the bamboo fiber that prevents sudden crack growth.

In work of Cherian et al. [106], the nanodimensional cellulose embedded in pineapple fibers was extracted applying acid coupled steam treatment. This treatment was found to be effective in the depolymerization and defibrillation of the fiber to produce nanofibrils of these fibers. Figure 4 shows the cellulose nanofibers extracted through this treatment. These nanofibrils were used to reinforce the polyurethane (PU) by compression moulding [96]. The addition of 5 wt% of cellulose nanofibrils to PU increased the strength nearly 300% and the stiffness by 2600%. The developed composites were utilized to fabricate various versatile medical implants.

A new type of modification of vegetable fibers which consists in the deposition of a nanosized cellulose coating onto natural fibers or the dispersion of nanosized cellulose in natural fiber reinforced composites has been studied in order to develop hierarchical structures. This fiber modification has great potential to improve the fiber-matrix interface and the overall mechanical performances of such composites. Nevertheless, the aspect ratio and alignment of the cellulose nanofiller need optimization as well as novel processing techniques need to be developed to take advantage of the potential use of cellulose nanocrystals [107].

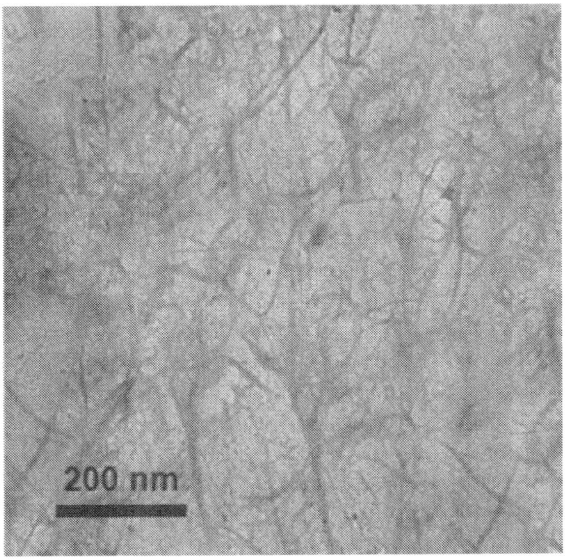

Figure 4. Transmission electron micrograph of cellulose nanofibers from pineapple fibers. Reprinted from Carbohydrate Polymers, 81, Bibin Mathew Cherian, Alcides Lopes Leão, Sivoney Ferreira de Souza, Sabu Thomas, Laly A. Pothan, M. Kottaisamy Isolation of nanocellulose from pineapple leaf fibers by steam explosion, 720–725, Copyright (2010) [106] with permission from Elsevier.

3.3. Nanoclays

Various inorganic nano-particles have been recognized as possible additives to enhance the polymer performance such as polymer nanofibers, the cellulose whiskers and the carbon nanotube. Among these, up to now only the layered inorganic solids like nanoclay have attracted some attention by the packaging industry. This is not only due to their availability and low cost but also due to their relative simple processability and significant improvements in some properties of the resulting polymer composites that include [108, 109]:

- Mechanical properties;
- Decreased permeability to gases, water and hydrocarbons;
- Thermal stability and heat distortion temperature;
- Flame retardancy and reduced smoke emissions;
- Chemical resistance;
- Surface appearance;
- Electrical and thermal conductivity;
- Optical clarity in comparison to conventionally filled polymers.

Most of synthetic bionanocomposites result from the assembly of biopolymers and silicates belonging to the clay mineral family. The effect of nanoclay minerals on polymer properties is mainly attributed to their high surface area and high aspect ratio as well as the combination of singular properties such as chemical inertness, low or null toxicity, good biocompatibility with high adsorption ability and cation exchange capacity [110]. Nanoreinforcement of biobased polymers with nanoclays can thus create new value-added applications of "green" polymers in the materials world [111].

Montmorillonite (MMT) clays, part of the smectite family clays, are the clay minerals most used as fillers in polymer nanocomposites due to environmental and economic criteria [112]. The chemical structure of MMT clays consist of two fused silica tetrahedral sheets sandwiching an edge-shared octahedral sheet of either magnesium or aluminum hydroxide establishing a nanometer scale platelets of magnesium aluminum silicate [113]. Each platelet of MMT is about 1 nm in thickness and varies in lateral dimension from 50 nm to several micrometers, showing high aspect ratio. Also, the platelet has a negative charge arising from isomorphous substitution in the lattice structure, which is compensated by naturally occurring cations that are located within the gallery (or interlayer) regions between the platelets [8]. Clay structure is formed by hundreds of layered platelets stacked into particles or tactoids of approximately 8 to 10 μm in diameter [114, 115].

MMT clays have hydrophilic nature due to the presence of inorganic cations on the basal planar surface of montmorillite layer [116]. The hydrophilicity of the surface of MMT clays makes their dispersion in organic matrices difficult [117]. Thus, MMT clays must be submitted to treatments which play an important role in the preparation of nanocomposites since it can affect their final properties. The most widely used treatments are the diverse functionalizations of clay by various organic cations through ion exchange where the inorganic cations are replaced by organic cations intercalated into the silicate layers. Its hydrophilic nature and ionic exchange capacity allow the silicate mineral to be intercalated by organic cations, which in most cases are alkylammonium ions, to make the clay organophilic and compatible with polymer matrices, preferably with polymers with polar groups which exhibit a higher affinity towards the alkylammonium ion-modified clays [118]. Functionalization of MMT clay by means of the silylation reaction with 3-aminopropyltriethoxysilane and *N*-[3-(trimethoxysilyl)propyl]ethylene-diamine was also reported [119].

There are three possible morphologies for polymer-clay nanocomposites that include: (i) immiscible, (ii) intercalated and (iii) exfoliated structures [115, 120]. In the immiscible structure the polymer does not penetrate between the clay platelets and the interlayer space of the clay gallery does not expand due to its poor affinity with the polymer, so this structure is also known as phase separated morphology or tactoid morphology. Intercalation is attained when polymer chains slightly penetrate within the gallery space and induce moderate expansion of the clay platelets. Exfoliation is characterized by a random distribution of the clay platelets due to extensive penetration of the polymer chains, resulting in the delamination of the clay platelets and the loss of the crystalline structure of the clay. This is due to a high affinity between polymer and clay.

There are three main processing routes for the development of well dispersed clay/biobased nanocomposites [108, 121]: (i) the solvent route which consists in swelling the layered silicates in a polymer solvent, (ii) the *in-situ* polymerization route for which the layered silicates are swollen in the monomer or monomer solution so as the polymer formation can occur between the intercalated sheets and (iii) the melt processing route which is based on polymer processing in the molten state (extrusion, injection molding, etc) which is highly preferred in the context of sustainable development since it avoids the use of organic solvents.

4. Biocomposites of biobased polymers and natural reinforcement agents: Properties and applications

The development of biocomposites started in the late 1980s and most of the biodegradable polymers which are now available in the market do not yet satisfy each of the requirements for bio-composites. Although promising results were obtained, development of biocomposites is still in its preliminary stage. More data on properties of biocomposites are required to establish confidence in their use [122]. Nanotechnologies promise many stimulating changes in composite materials in order to enhance health, wealth and quality of life, while reducing the environmental impact [108]. Thus, many researches in the biocomposite area can be found in literature. Some of them are reported in the following items.

4.1. PLA based biocomposites

One of the most studied biocomposites is PLA (polylactide) based biocomposite since PLA was the first commodity plastic produced from annually renewable resources [123]. Lactid acid based polymers (polylactides) are polyesters made from lactic acid. PLLA (poly-L-lactide) is a polymer built with only repeating units of L-stereoisomer configuration. The general term PLA (polylactide) is used for polymers without isomer specification.

PLA is brittle, so it needs modification for pratical applications. Bledzki and Jaszkiewicz [124] reported that one of the main drawbacks concerning technical applications of biodegradable polymers, especially for PLA polymers, is their low impact strength. Most research on PLA biocomposite ultimately seeks to improve the mechanical properties to a level that satisfies a particular application [125]. The mechanical properties of biocomposites depend on a number of parameters such as percentage of fiber content, interfacial characteristics between fiber and matrix, fiber aspect ratio, surface modification of fibers and addition of various additives (coupling agents) to enhance the compatibility between fiber and matrix [126].

Huda et al. [82] studied the addition of alkali and/or silane treated Kenaf fibers in PLA matrix through compression molding using the film-stacking method with a fiber content of 40 wt%. Although the introduction of treated kenaf fibers significantly improves flexural modulus compared to the neat PLA matrix, the flexural strength of the PLA composites decreases with the addition of Kenaf fibers. The composite with silane-treated fibers showed an increase of 69% in modulus than that of alkali treated fibers. The notched Izod impact strength of surface-treated composites was higher than those of the neat PLA. The impact strength of neat PLA improved almost 45% with the addition of 40 wt% untreated fiber and 90% with alkali treated Kenaf fibers with the same content. The high toughness of this natural fiber laminated biocomposite places it in the category of tough engineering materials. Other authors [63] used a carding process that provided a uniform blend of PLA fiber and Kenaf fiber that was followed by needle punching, pre-pressing and further hot-pressing in presence of silane coupling agent to form the biocomposite material. The flexural modulus and flexural strength of the treated fiber biocomposites increased with respect to neat PLA and untreated fiber biocomposites.

In other work, tensile strength and Charpy notched strength were evaluated for PLA biocomposites with a variety of types of natural fiber: abaca fibers, man-made cellulose, jute and flax fibers. Authors observed that increasing the content of fibers up to 30 wt% the composite's stiffness significantly increases as well as tensile and impact strengths with respect to neat PLA [127]. The same improvement in mechanical properties was reported by Choie and Lee [128] using ramie fibers and PLA resin as matrix.

Tensile strength, Young's modulus and impact strength of short hemp fibre reinforced PLA biocomposites increased with increased fibre content (10–30 wt.%) as well as with the application of surface fiber treatments like alkali and silane treatments. It was found that PLA could be reinforced with a maximum of 30 wt.% fibres using conventional injection moulding, but could not be processed at higher fibre contents due to poor melt flow of the compounded materials [123]. In Table 5 the best results of each reference for some mechanical properties of PLA biocomposites with vegetable fiber are summarized.

As shown in Table 5, PLA biocomposites have shown different mechanical properties. Kenaf and hemp fiber PLA biocomposites showed a significantly increase in tensile strength and Young's modulus while a decrease in impact strength with respect to neat PLA was also reported [129]. In this work, neat PLA showed a tensile strength of 30.1 MPa, Young Modulus of 3.6 GPa and 24.4 kJ/m^2 for unnotched Charpy impact strength. The same observation was achieved by Oksman et al. [130] for unnotched Charpy impact strength of PLA biocomposite (12 kJ/m^2) with respect to neat PLA (15 kJ/m^2). Different values for neat PLA mechanical properties were reported and they depend mainly on inherent PLA properties (average molar mass, density, etc.) as well as the manufacturing process. Nevertheless, some authors have already observed an increase from a notched impact test for PLA biocomposites [82, 123, 124, 131] for different types of vegetable fibers.

Biodegradable composites have showed insufficient impact strength, preventing a broader field of application of these materials in automotive sector and in electronic devices. However, PLA reinforced with a man-made cellulose (Cordenka®) produced a biocomposite which have met performance requirements, especially for impact properties (72 kJ/m^2 for unnotched Charpy impact strength), that can be used in automotive and electronic industry [132]. Authors [129] also reported PLA biocomposites with man-made cellulose that have shown good tensile and impact properties and they can be used in different fields of application like household appliances and in bumpers in the automobile industry.

Fiber and Content (wt%)	Interface Treatment	Manufacturing Process	Tensile strength (MPa)	Young's modulus, (GPa)	Impact strength (kJ/m²)	Reference
Abaca (30)	Untreated fibers	Extrusion followed by injection molding	74.0	8.0	5.0 (notched Charpy)	124
Bamboo (20)	Plasma and silane coupling	a filmstacking procedure	90	1.8	-	71
Flax (30)	Enzime retting of fiber	Extrusion followed by compression molding	53	8.3	12 (unnotched Charpy)	130
Hemp (30)	Mercerized fiber	Extrusion followed by injection molding	75.5	8.2	2.64 (notched Charpy)	123
Hemp (40)	Untreated fibers	Roller carding with PLA followed by compression molding	57.5	8	9.5 (unnotched Charpy)	129
Jute (30)	Untreated fibers	Extrusion followed by injection molding	81.9	9.6	4.8 (notched Charpy)	124
Kenaf (40)	Untreated fibers	Roller carding with PLA followed by compression molding	52.9	7.1	9.0 (unnotched Charpy)	129
Kenaf (30)	5 wt% Coupling agent (maleic anhydride grafted PLA)	Internal mixing followed by compression molding	-	-	3.46 ± 0.13 (notched Charpy)	131
Man-made cellulose (Lyocell) (40)	Untreated fibers	Roller carding with PLA followed by compression molding	81.8	6.8	39.7 (unnotched Charpy)	129
Man-made cellulose (30)	Untreated fibers	Extrusion followed by injection molding	92	5.8	8.0 (notched Charpy)	124

Table 5. Tensile strength, Young's modulus and impact strength (room temperature) of PLA-based biocomposites with vegetable fibers.

Biocomposites that show high tensile strength and stiffness as well as low impact strength could be used in manufacture of furniture, boardings or holders for grinding discs and so on which are not subjected to high impact stress. Biocomposites that show the combination of properties as low tensile strength with high impact strength leads to application of these materials in interior parts in cars or safety helmets [129]. Also, kenaf fiber–reinforced PLA matrix biocomposites which the processing is based on injection molding have been used for spare tire covers and circuit boards [133] and these biocomposites were proposed to be used in an automotive headliner made from a 50/50 PLA/Kenaf fiber using a carding process [63].

The mechanical properties are thus among the most widely tested properties of natural fiber reinforced composites [2]. Compared with widespread research on mechanical properties of biocomposites, there are few reports on flame retardancy of biopolymers and biocomposites [134, 135]. The flame retardancy of ramie fiber reinforced PLA biocomposites was tested using halogen-free ammonium polyphosphate (APP). PLA biocomposites using flame-retardant treatment of ramie fibers have demonstrated a certain flame retardancy but cannot be classified by UL94 testing (Test for flammability of plastic materials for parts in devices and appliances) because of low APP loading (4.5 wt%). When PLA matrix is mixed in a extruded with APP, biocomposites with treated or non-treated ramie fibers and having the same APP loading (10.5 wt%) achieved V-0 rating (short burning time, no dripping; self-extinguishing). Low loading of APP does not adversely affect the mechanical performance of PLA/ramie biocomposites [136]. Other authors [137] also studied PLA biocomposites using plasma-treated coconnut fiber and prepared using the commingled yarn method. As expected, plasma-treated coconut fibers improved mechanical properties like tensile strength and modulus of biocomposites compared to neat PLA, but no significant changes on the fire retardant properties was achieved for the biocomposites with respect to neat PLA, according to the limiting oxygen index (LOI) value: around 25 for neat PLA and 10 wt% treated coconut fiber biocomposite. Generally, when the LOI value is greater than 26, materials can be considered to have flame retardancy [134].

Nanoreinforcements were also tested in fully biodegradable biocomposites of PLA matrix. These biocomposites help to provide new food packaging materials with improved mechanical, barrier, antioxidant and antimicrobial properties [138]. The addition of cellulose nanowhiskers to PLA matrix reduced the water permeability by up to 82% and the oxygen permeability by up to 90% with only 3 wt% of nanofiller content [139]. Moreover, the incorporation of organomodified mica-based clay to PLA matrix enhanced barrier properties to UV light; besides other barrier properties.This property is highly important for food packaging as protection against light which is a basic requirement to preserve the quality of many food products [140].

In previous research, PLA matrix was reinforced by 5wt% microcrystalline cellulose or 5wt% commercial organically modified bentonite (layered silicate) [141]. The bionanocomposite reinforced by bentonite showed great improvements in tensile modulus and strength as well as a decrease in oxygen permeability whereas the bionanocomposite reinforced with microcrystalline cellulose only showed a tendency to improve strength as well as a reduction in elongation at break. No changes for oxygen permeability were observed. This was attributed to the larger surface area of bentonite that allows interaction with a larger amount of PLA chains.

http://dx.doi.org/10.5772/56290

In other work, the presence of a surfactant favoured the dispersion of cellulose nanocrystals in the PLA matrix, yielding bionanocomposites with higher tensile modulus and strength. The addition of silver nanoparticles to the bionanocomposite did not enhance these mechanical properties. Besides, an antibacterial activity against *Staphylococcus aureus* and *Escherichia coli* cells was detected for ternary systems, indicating that these bionanocomposites have great potencial to be applied in food packaging when an antibacterial effect is required [95].

Polylactides and their copolymers were been widely reported to be used in the fields of orthopedic and reconstructive surgery due to its biodegradability and better features for use in the human body (nontoxicity) [142, 143]. According to Walker et al. [144], polylactides degrade in vivo by hydrolytic mechanisms of the ester bonds into lactic acid which is processed through metabolic pathways and is eliminated from the body through the renal and/or respiratory mechanisms. PLLA constructs have a longer degradation time when compared to other polymers, having shown to be present at 3 years after implantation. Its structural characteristics have proven useful for the construction of orthopedic hardware.

Bionanocomposites of hydroxyapatite (HPA) nanospheres which is the main inorganic constituent of natural bone and PLLA microspheres were tested for biomedical application to produce scaffolds using a laser sintering process [145]. HPA particles can reinforce polymer matrices and decrease the degradation rate of PLA [146]. Also, other work showed that PLA/organoclay bionanocomposites have enhanced their thermomechanical properties and gas barrier properties with respect to neat PLA and their biodegradation rate depends on the organoclay nature, organoclay content, organoclay dispersion as well as the organic modifier used to treat the nanoclay [147]. The relative hydrophilicity of the clay layers has been shown to play a key role in the hydrolytic degradation of the PLA chains [148].

Biodegradability of flax fiber reinforced PLA based biocomposites in presence of amphiphilic additives like benzilic acid, mandelic acid, dicumyl peroxide (DCP) and zein protein was investigated by soil burial test with farmland soil. Authors reported that neat PLA films degraded rapidly compared to natural fiber reinforced biocomposites. But, regarding the use of amphiphilic additives, the higher loss in weight is obtained for flax reinforced PLA biocomposites in the presence of mandelic acid. In the presence of DCP, the biodegradability of the biocomposites was comparatively delayed. Depending on the end-uses of the biocomposites, suitable amphiphilic additives can be used as triggers for inducing controlled biodegradation [149].

The aerobic biodegradation of biocomposites of PLA, thermoplastic starch (TPS) and a blend of 75 wt% of PLA and 25 wt% of TPS with short natural fiber (coir) with and without the addition of maleic anhydride (MA) coupling agent were investigated under controlled composting conditions. TPS showed higher biodegradation rates than PLA, probably due to the TPS domains preferentially attacked by microorganisms. Besides, authors ascertained that coir fibers probably have no influence in the biodegradation process due to the slight differences in carbon dioxide produced for neat polymers and their biocomposites with coir fiber. Also, the presence of coupling agent decreased the percentage of evolved CO_2 compared to biocomposites without coupling agent [150].

In other work, bacterial (*Burkholderia cepacia* bacteria) biodegradation studies were performed for biocomposites of PLA and mercerized banana fiber (BF) produced by melt blending followed by compression molding. Banana fibers were also treated with various silanes to improve their compatibility with PLA matrix. Authors reported improvements in tensile and impact strength of the biocomposites with respect to neat PLA. Weight loss experiments showed that PLA had 60% of degradation within a period of 25 days and all biocomposites showed higher degradation rates (80–100%). While biocomposites with untreated and alkaline-treated BF degraded almost completely, silane-treated biocomposites degraded at lower rates. Water absorption studies supported this evidence [151, 152].

4.2. PHBV biocomposites

Poly(hydroxyl-alkanoates) (PHAs).are a family of bacterial polyesters which poly(hydroxy-butyrate) (PHB) and its copolymer poly (3-hydroxybutyrate-co-3-valerate) (PHBV) make part. According to Bledzki and Jaszkiewicz [124], PHBV has been technologically developed to improve the known weaknesses of PHB like brittleness and poor processability.

Biocomposites of PHBV with wood and bamboo fibers were fabricated using extrusion followed by injection molding. Tensile and flexural modulus increased with fiber loading for biocomposites with the two kinds of fiber and no appreciable difference among the two fiber loadings (30 and 40 wt% fiber) was noticed. However, notch impact strength of PHBV decreased with the fiber addition and the reduction was greater in case of bamboo fiber biocomposites [153]. However, in other work biocomposites of PHBV and bamboo pulp fibers which were prepared by melt compounding and injection molding showed substantially increase of the impact strength by the addition of bamboo pulp fiber as well as increased tensile strength and modulus and flexural strength and modulus. The maleic anhydride grafted PHBV used as coupling agent improved polymer/fiber interactions and therefore resulted in increased strength and modulus. However, the toughness of the composites was substantially reduced due to the hindrance to fiber pullout [154]. Also, authors [124] reported an increase of the impact strength for PHBV biocomposites using 30 wt% of man-made cellulose, abaca and jute fibers at 23ºC and also at -30 ºC. The most pronounced results were obtained with man-made cellulose. PHBV was blended with 27.6 wt% of poly (butylene adipate-co-butylene terephtalate) (PBAT) and 2.4 wt% of processing aids. Moreover, tensile strength and modulus were increased.

In recent work, PHBV was blended with PBAT using extrusion (in a twin-screw extruder) followed by injection molding. Biocomposites were performed with 20–40wt% switchgrass and the compatibilizer pMDI. With the addition of 25wt% switchgrass the tensile and flexural strengths of the biocomposite have improved. On increasing the fiber content to 30wt% and further to 40wt%, both tensile and flexural strength dropped but the modulus of the composites increased progressively with increasing fiber content. With regard to uncompatibilized composites, impact strength of 53 J/m was achieved for composites with 25wt% switchgrass because of the proper wetting achieved between the fiber and the matrix. Impact strength reduced with increase in fiber content. The use of the pMDI compatibilizerer in biocomposites with 30 wt% switchgrass promoted interfacial interactions between the matrix and the fiber and significantly improved the mechanical properties of the biocomposites. The addition of pMDI significantly increased the impact strength of the composites. The notched impact strength increased 80% compared to the uncompatibilized composite owing to the enhanced interfacial adhesion [155]. Also, by incorporation of biomass fiber reinforcement like corn straw, soy stalk and wheat straw into the PHBV by melt mixing technique, authors showed that the alkali treatment of wheat straw fibers enhanced strain, break and impact strength of PHBV composites by 35%, hardly increasing strength and modulus compared to their untreated counterparts. Authors also showed that the tensile and storage modulus of PHBV were improved by maximum 256% and 308%, respectively, with 30 wt% of the biomass and these values were much higher than the corresponding polypropylene (PP) composites [156].

Nanoparticles also have already been incorporated into PHBV matrix. Well-dispersed cellulose nanocrystals into PHBV matrix were obtained with simultaneous enhancements on the mechanical property and thermal stability of PHBV. Compared to neat PHBV, a 149% improvement in tensile strength and 250% increase in Young's modulus were obtained for the resulting nanocomposites with 10 wt% of cellulose nanocrystals [157]. Lower concentrations of cellulose nanowhiskers (0–4.6 wt%) were used to prepare PHBV bionanocomposites by solution casting [158]. The mechanical properties of the films increased with increasing cellulose nanowhiskers content until the content reached 2.3 wt %. Real permittivity of the composites also peaked at 2.3 wt % cellulose nanowhiskers over a wide spectrum of frequencies (0.01–10^6 Hz). These property transitions at 2.3% cellulose nanowhiskers content were due to the transition of cellulose nanowhiskers dispersion from homogeneous dispersion to agglomeration. Nevertheless, rheological results of the bionanocomposites indicated a transition point lower than 2.3% due to the formation of a biopolymer-fiber network in the composite melt.

Some authors [159] showed that the incorporation of low concentrations of nanoclays (5 wt%) and cellulose nanowhiskers (3 wt%) into PHBV matrix and other biodegradable matrices like PLA and polycaprolactone (PCL) resulted in improvements in oxygen permeability that can be very useful for food packaging. With respect to water permeability, authors showed that PHBV films with 1 wt% alpha cellulose fiber content had a water permeability drop of 71% compared to the unfilled material, whereas PHBV films with a fiber content of 10 wt% showed a water permeability reduction of around 52% due to fiber agglomeration. However, the lowest water and limonene permeability coefficient values were obtained for the bionanocomposites containing 5 wt% of clay due to the good morphology for these nanocomposites. The same work also reported that mica-based nanoclays exerted certain UV/visible light blocking action in PLA and PHBV matrices. The blocking effect of PHBV in the UV-Vis region was higher than that of PLA since PHBV is a translucent material. Moreover, greater reductions in vapour permeability were attained for PHBV bionanocomposites with clay contents of 1 wt% [94]. Furthermore, the PHBV processing behavior could be improved with addition of montmorillonite nanoclay since the processing temperature range enlarged by lowering melting temperature with the increasing clay content. The tensile properties of the corresponding materials were improved by incorporation of 3wt% of clay [160].

Thus, in general many properties have been improved with the incorporation of fibers and mainly nanofibers and nanoclays into PHBV which are helpful to overcome many obstacles and enhance the efficiency in a diverse number of applications. In this way, it is found that nanofibers can induce fast regeneration of many tissues/organs in medical applications and improve the efficiency of many chemical and electronic applications [161].

PHA's family was related to be used in numerous biomedical applications, such as sutures, cardiovascular patches, wound dressings, scaffolds in tissue engineering, tissue repair/ regeneration devices, drug carriers and so on, but much deep studies [162]. PHBV bionano-composites were manufactured with various calcium phosphate-reinforcing phases for bone tissue regeneration while inducing a minimal inflammatory response. Authors showed that the addition of a mineral nano-sized reinforcing phase to PHBV reduced the proinflammatory response and also improved osteogenic properties with respect to pure PHBV [163].

With respect to biodegradation behaviour, biocomposites of PHBV matrix and 10, 20 and 25 wt% of peach palm particles were investigated [164]. Soil biodegradation tests were carried out according to ASTM G160-98 with test exposures of up to 5 months. The addition of peach palm particles reduced the maximum strength but improved the Young's modulus and also soil biodegradation tests indicated that the biocomposites degraded faster than the neat polymer due to the presence of cavities that resulted from introduction of the peach palm particles and that degradation increased with increasing particles content. These voids allowed for enhanced water adsorption and greater internal access to the soil-borne degrader micro-organisms. Similarly, other authors found that biocomposites with PHBV and wood fiber have higher degradation rates than the neat polymer [165]. On the other hand, some authors reported no significant difference between the degradability of PHBV and its composite with wheat straw using either Sturm tests or soil burial tests [166].

5. Conclusion

Due to the high demand for environmental sustainable products, researchers continue to seek materials derived from renewable resources that can be applied in a wide range of applications. This overview provided a survey of some of the current researches on the biocomposites area. Within this context, this chapter showed that there have been many attempts to produce biocomposites using natural reinforcements and biobased polymers since improvements in their mechanical, barrier and other properties can be accomplished through the use of reinforcement agents like vegetable fibers and nanoparticles (cellulose nanofiber or nano-clays). Vegetable fibers are generally submitted to chemical treatments, mostly alkaline and acid treatments in order to favour interfacial adhesion between polymer matrices and the fiber. Also, the use of coupling agents enhance adhesion by surface modification as well as they can produce grafting reactions between matrix and fiber. Moreover, the presence of polar groups in most biobased poymers contributes to better affinity to cellulosic groups of vegetable fibers. All these issues dramatically influence the mechanical properties of the biocomposites. With respect to nanoreinforcements, cellulose nanofibers and organic functionalized clays (orga-noclays) are the most used as fillers in bionanocomposites.

PLA based biocomposites are one of the most studied biocomposites and some researches showed that the use of vegetable fiber can improve the impact strengh of the PLA matrix, but insufficient strength values were found to enable their application in automotive sector and in electronic devices. PLA biocomposites with a man-made cellulose fiber that fulfill the requirements for mechanical properties were already reported and their use can be extended to diferent fields of application. The use of nanoreinforcements in PLA matrices produced bionanocomposites with remarkable mechanical, thermal, barrier, antioxidant and antimicrobial properties, presenting a new material with potential for food packaging application. The biodegradability of PLA biocomposites with vegetable fibers showed to be sensitive to the additives used in biocomposite processing. The presence of coupling agents provides lower degradation times than neat PLA. Also, depending on the nature of the amphilic additives, they may speed up or delay the biodegradation process. Researches with organoclay in bionanocomposites showed that their biodegradation rate depends on the nature, the content and the dispersion level of organoclay in the bionanocomposite as well as the nature of organic modifier of the clay.

PHBV based biocomposites also showed an increase in mechanical properties in presence of treated vegetable fibers and coupling agents. However, the incorporation of cellulose nanofibers and organoclays in PHBV matrix promoted greater improvements not only in mechanical properties but also in oxygen and water permeability. The bionanocomposites produced can be used in medical applications due to the faster regeneration of many tissues/organs and in many chemical and electronic applications. The specific use of organoclays also produced UV-Vis blocking effects and greater reductions in vapour permeability as well as processing behaviour improvements. The biodegradability of these bionanocomposites showed to be similar or faster than the neat PHBV matrix.

Therefore, bionanocomposites arised as a promissing area that can overcome some of the drawbacks of biobased polymers and their biocomposites since the use of nanoparticles generally promotes greater improvements in many properties with respect to biocomposites. However developments must be performed on processing techniques and key research callenges like nanoparticles dispersion into biopolymers. Thus, the construction of a biocomposite/bionanocomposite is not a simple process and it needs the knowledge of the real contribution of each composite phase for property tuning. Moreover, biocomposites/bionanocomposites will be only attractive if material and process costs are competitive compared to conventional composites which use petrochemical resources.

Acknowledgements

The author Derval dos Santos Rosa thanks FAPESP – Process no 2012/13445-8 and UFABC for support.

Author details

Derval dos Santos Rosa[1] and Denise Maria Lenz[2*]

1 Universidade Federal do ABC, SP, Brazil

2 Universidade Luterana do Brasil, RS, Brazil

References

[1] Chand, N and Fahim, M Tribology of natural fiber polymer composites. Chapter 1, 1-58, Woodhead Publishing and CRC Press (2008).

[2] Faruk, O, Andrzej, K, Bledzki, H-P, Fink, M S Biocomposites reinforced with natural fibers: 2000–2010. Progress in Polymer Science, 37(11)1552-1596 (2012).

[3] Kalia, S, Dufresne, A, Cherian, B M, Kaith, B S, Avérous, L, Njuguna, J, Nassiopoulos, E Cellulose-Based Bio- and nanocomposites: a review. International Journal of Polymer Science (2011) Article ID 837875, 35 pages doi:10.1155/2011/837875.

[4] John, M J, Thomas, S Biofibres and biocomposites. Carbohydrate Polymers, 71: 343-364 (2008).

[5] Fowler, P A, Hughes, J M, Elias, R M Review Biocomposites: technology, environmental credentials and market forces. Journal of the Science of Food and Agriculture, 86(12):1781-1789 (2006).

[6] Bledzki, A K, Gassan, J Composites reinforced with cellulose-based fibres, Progress in Polymer Science, 24(2): 221-274 (1999).

[7] Ruiz-Hitzky, E, Darder, M, Aranda, P Progress in bionanocomposite materials. In: Annual Review Nano Research Volume 3, Chapter 3, 149-189, Guozhong Cao, Qifeng Zhang and C. Jeffrey Brinker(Ed) World Scientific Publishing, Singapore (2010).

[8] Malviya, R, Srivastava, P, Bansal, V and Sharma, P K Formulation, evaluation and comparison of sustained release matrix tablets of diclofenac sodium using natural polymers as release modifier, International Journal of Pharma and Biosciences,1(2): 1-8 (2010).

[9] Malviya, R, Srivastava, P, Bansal M and Sharma P K Preparation and evaluation of disintegrating properties of Cucurbita maxima pulp powder. International J. Pharmaceutical Sci., 2(1):395-399 (2010).

[10] Šimkovic, I What could be greener than composites made from polysaccharides? Carbohydrate Polymers, 74 (4): 759-762 (2008).

[11] Rosa, D S, Guedes, C G F, Casarin, F, Braganca, F C The effect of the Mw of PEG in PCL/CA blends. Polymer Testing, 24(5): 542–548 (2005).

[12] Klemm, D , Heublein, B, Fink, H-P, Bohn, A Cellulose: fascinating biopolymer and sustainable raw material Angewandte Chemie International Edition, 44 (22): 3358-3393 (2005).

[13] Edgar, K J , Buchanan, C M, Debenham, J S, Rundquist, P A, Seiler, B D, Shelton, M C, Tindall, D Advances in cellulose ester performance and application Progress in Polymer Science, 26 (9):1605-1688 (2001).

[14] Ignácio C and Barros D M Membranas celulósicas – efeitos da concentração de polímero-solvente-não solvente na morfologia da membrana. 8º Brazillian Polymer Congress, Águas de Lindóia, 624-625 (2005).

[15] Mulinari, D R, Silva M L C P, Silva G L J P Preparação e caracterização dos compósitos celulose branqueada/ZrO2nH2O preparados pelos métodos da precipitação convencional e precipitação em solução homogênea. 8º Brazillian Polymer Congress, Águas de Lindóia, 245-246 (2005).

[16] Vroman, I and Tighzert, L Biodegradable Polymers, Materials, 2: 307-344 (2009) doi: 10.3390/ma2020307.

[17] Rosa D S, Carvalho C L, Gaboardi, M L F, Rezende, M I, Tavares B, Petro M S M, Calil M R Evaluation of enzymatic degradation based on the quantification of glucose in thermoplastic starch and its characterization by mechanical and morphological properties and NMR measurements. Polymer Testing, 27(7): 827-834 (2008).

[18] Rosa, D S, Guedes, C G F, Carvalho C L Processing and thermal, mechanical and morphological characterization of post-consumer polyolefins/thermoplastic starch blends. Journal of Materials Science, 42(2): 551–557 (2007).

[19] Pedroso, A G and Rosa, D S Mechanical, thermal and morphological characterization of recycled LDPE/corn starch blends. Carbohydrate Polymers, 59 (1): 1-9 (2005).

[20] Rosa, D S, Guedes, C G F, Volponi, J E Biodegradation and dynamic mechanical properties of starch gelatinization in poly(-caprolactone)/corn starch blends. Journal of Applied Polymer Science, 102 (1): 825-832 (2006).

[21] Rosa, D S, Guedes C G F, Carvalho, C L Processing and thermal, mechanical and morphological characterization of post-consumer polyolefins/thermoplastic starch blends. Journal of Materials Science, 42(2): 551–557 (2007).

[22] Furusaki E., Ueno Y., Sakairi N., Nishi N. and Tokura, S. Facile Preparation and inclusion ability of a chitosan derivative bearing carboxymethyl-b-cyclodextrin. Carbohydrate Polymers, 9: 29-34 (1996),

[23] Hirano, S and Nagano, N Effects of chitosan, pectic acid, lysozyme and chitinase on the growth of several phytopathogens. Agricultural and Biological Chemistry, 53: 3065-3066 (1989).

[24] Kumar, M N V R A review of chitin and chitosan applications. Reactive & Functional Polymers, 46 (1): 1-27 (2000).

[25] Piermaria, J A, Pinotti, A, Garcia, M A, Abraham, A G Films based on kefiran, an exopolysaccharide obtained from kefir grain: development and characterization. Food Hydrocoll., 23: 684-690 (2009).

[26] Chandra, R, Rustgi, R Biodegradable polymers. Progress in Polymer Science, 23: 1273-1335 (1998).

[27] Andrade, C T, Lopes, L Polímeros de origem microbiana: polissacarídeos bacterianos. Revista de Química Industrial, 703:19-23 (1995).

[28] Serafim S L, Lemos P C, Reis, M A M Produção de bioplásticos por culturas microbianas mistas. Boletim de Biotecnologia, 76:15-20 (2003).

[29] Pradella, J G C Biopolímeros e intermediários químicos. Technical Report nº 84396-205, Centro de Tecnologia de Processos e Produtos, Laboratório de Biotecnologia Industrial – LBI/CTPP (2006).

[30] Coutinho, B C, Miranda, G B, Sampaio, G R, Souza, L B S, Santana, W J, Coutinho, H D M A importância e as vantagens do polihidroxibutirato (plástico biodegradável). Holos 3: 76-81 (2004).

[31] Lenz, R W, Marchessault, R H Bacterial polyesters: biosynthesis, biodegradable plastics and biotechnology. Biomacromolecules 6 (1):1-8 (2005).

[32] Rosa, D S, Penteado, D F, Calil, M R Propriedades térmicas e biodegradabilidade de PCL e PHB em um pool de fungos. Revista Ciência & Tecnologia, 15: 75-80 (2000).

[33] Corrêa, M C S, Rezende, M L, Rosa, D S, Agnelli, J A M, Nascente, P A P Surface composition and morphology of poly(3-hydroxybutyrate) exposed to biodegradation. Polymer Testing 27 (4): 447-452, 2008.

[34] Quental, A C, de Carvalho, F P, Rezende, M I, Rosa, D S, Felisberti M I Aromatic/aliphatic polyester blends. Journal of Polymers and the Environment, 18 (3): 308-317 (2010).

[35] Luckachan, G. E., Pillai C. K. S. Biodegradable polymers- a review on recent trends and emerging perspectives Journal of Polymers and the Environment , 19: 637–676 (2011).

[36] Keshavarz, T and Roy, I Polyhydroxyalkanoates: bioplastics with a green agenda Current Opinion in Microbiology, 13:321-326 (2010).

[37] Rosa, D S and Pântano Filho, R Biodegradação: um ensaio com polímeros. Moara (Ed.), Itatiba, S.P; Editora Universitária São Francisco (Ed.), Bragança Paulista, S.P. (2003).

[38] Lotto, N T Dimensionamento da degradação dos polímeros PHB e PHB-V através da variação da rugosidade. (Master Dissertation), 78 p. Universidade São Francisco, S. P., Brazil (2003).

[39] Rosa, D S, Franco, B L M, Calil, M R Biodegradabilidade e propriedades mecânicas de novas misturas poliméricas. Polímeros: Ciência e Tecnologia, 2 (11): 82-88 (2001).

[40] Lotto, N T, Calil, M R, Guedes. C G F, Rosa, D S The effect of temperature on the biodegradation test. Materials Science and Engineering: C, 24(5): 659-662 (2004).

[41] Vogelsanger, N, Formolo, M C, Pezzin, A P T, Schneider, A L S, Furlan, A A, Bernardo, H P, Pezzin S H, Pires, A T N, Duek, E A R Blendas biodegradáveis de poli(3-hidroxibutirato)/poli(-caprolactona): obtenção e estudo da miscibilidade. Materials Research, 6 (3):359-365 (2003).

[42] Raghavan, A D Characterization of Biodegradable Plastics. Polymer-Plastics Technology and Engineering, 34(1): 41-63 (1995).

[43] Zenkiewicz, M, Richert, J, Różański, A Effect of blow moulding ratio on barrier properties of polylactide nanocomposite films. Polymer Testing, 29(2): 251-257 (2010).

[44] Bhatia, A, Gupta, R, Bhattacharya, S, Choi, H. Effect of clay on thermal, mechanical and gas barrier properties of biodegradable poly(lactic acid)/poly(butylene succinate) (PLA/PBS) nanocomposites. International Polymer Processing, 25 (1):5-14 (2010).

[45] Martino, V P, Ruseckaite, R A, Jiménez, A, Averous, L Correlation between composition, structure and properties of poly(lactic acid)/polyadipate-based nano-Biocomposites. Macromolecular Materials and Engineering, 295(6): 551-558 (2010).

[46] Rasala, R M, Janorkarc, A V, Hirta, D E Poly(lactic acid) modifications Progress in Polymer Science, 35:338–356 (2010).

[47] Henry, F, Costa, L C, Devassine, M The evolution of poly(lactic acid) degradability by dielectric spectroscopy measurements European Polymer Journal, 41(9): 2122–2126 (2005).

[48] Ulery, B D, Nair, L S, Laurencin, C T Biomedical applications of biodegradable polymers Journal of Polymer Science Part B: Polymer Physics, 49: 832–864 (2011).

[49] Bragança, F C and Rosa, D S Thermal, mechanical and morphological analysis of poly(3-caprolactone), cellulose acetate and their blends, Polymers for Advances Technology 14(10):669–675 (2003).

[50] Landry, C J T, Lum, K K, O'Reilly, J M Physical aging of blends of cellulose acetate polymers with dyes and plasticizers, Polymer, 42 (13):5781-5792 (2001).

[51] Wang, Z, Qu, B, Fan, W, Hu, Y, Shen, X Effects of PE-g-DBM as a compatibilizer on mechanical properties and crystallization behaviors of magnesium hydroxide- based LLDPE blends. Polymer Degradation and Stability, 76(1): 123-128 (2002).

[52] Kim, C-H, Cho, K Y, Park, .J-K Grafting of glycidil methacrylate onto polycaprolactone: preparation and characterization. Polymer, 42(12): 5135-5142 (2001).

[53] Liu, W, Wang, Y L, Sun, Z Effects of polyethylene-grafted maleic anhydride (PE-g-MA) on thermal properties, morphology, and tensile properties of low-density polyethylene (LDPE) and corn starch blends, Journal of Applied Polymer Science, 88 (13): 2904–2911 (2003),.

[54] Mucci, V, Pérez, J, Vallo, C I Preparation and characterization of light-cured methacrylate/montmorillonite nanocomposites Polymer International, 60 (2): 247-254 (2011).

[55] Sreekumar, P A and Thomas, S Matrices for natural-fibre reinforced composites. In: Properties and performance of natural-fibre composites. Chapter 2, 67-126, Kim L. Pickering (Ed) Woodhead Publishing Limited and CRC Press (2008).

[56] Kozlowski, R and Wladyka-Przybylak, M Uses of natural fiber reinforced plastics. In: Natural Fibers, Plastics and Composites. Chapter 14, 249-271, Wallenberger FT, Weston NE (Ed). Kluwer Academic Publishers: Dordrecht (2004).

[57] Yang, L, Hu, Y, Lu, H, Song, L Morphology, thermal, and mechanical properties of flame-retardant silicone rubber/montmorillonite nanocomposites. Journal of Applied Polymer Science, 99(6): 3275–3280 (2006).

[58] Duhovic, M, Peterson, S, Jayaraman, K Natural-fibre-biodegradable polymer composites for packaging In: Properties and performance of natural-fibre composites. Chapter 9, 301-329, Kim L. Pickering (Ed), Woodhead Publishing Limited and CRC Press (2008).

[59] Satyanarayana, K G, Arizaga, G G C, Wypych, F Biodegradable composites based on lignocellulosic fibers- an overview. Progress in Polymer Science, 34(9): 982-1021 (2009),.

[60] Spinacé, M A S, Lambert, C S, Fermoselli, K K G, De Paoli, M-A. Characterization of lignocellulosic curaua fibres. Carbohydrate Polymers, 77(1): 47-53 (2009).

[61]

[62] Sánchez, C Lignocellulosic residues: Biodegradation and bioconversion by fungi. Biotechnology Advances, 27(2):185-194 (2009).

[63] Pilla, S Engineering applications of bioplastics and biocomposites – an overview. In: Handbook of Bioplastics and Biocomposites - Engineering Applications, chapter 1, 1-14, Srikanth Pilla (Ed), Scrivener Publishing LCC and John Wiley & Sons, Massachusetts and New Jersey (2011).

[64] Lee, B-H, Kim, H-S, Lee, S, Kim, H-J, Dorgan, J R Bio-composites of kenaf fibers in polylactide: Role of improved interfacial adhesion in the carding process. Composites Science and Technology, 69(15-16): 2573–2579 (2009).

[65] Stelte, W, Sanadi, R A Preparation and characterization of cellulose nanofibers from two commercial hardwood and softwood pulps. Industrial and Engineering Chemical Research, 48 (24):11211–11219 (2009).

[66] Hearle, J W S and Sparrow, J T Mechanics of the extension of cotton fibers. I. Experimental studies of the effect of convolutions. Journal of Applied Polymer Science, 24 (6): 1465-1477 (1979).

[67] Silva, R V and Aquino, E M F Curaua Fiber: a new alternative to polymeric composites Journal of Reinforced Plastics and Composites, 27(1): 103-112 (2008).

[68] Santos, P A, Spinacé, M A S, Fermoselli, K K G, De Paoli, M.-A Polyamide-6/vegetable fiber composite prepared by extrusion and injection molding. Composites Part A: Applied Science and Manufacturing , 38 (12): 2404–2411 (2007).

[69] Bismark, A, Mishra, S, Lampke,, T. Plant Fibers as reinforcement for green composites. In: Natural fibers, biopolymers and biocomposites, Chapter 2, 37-108, A. K. Mohanty, M. Misra and L. T. Drzal (Eds) Taylor & Francis, CRC Press (2005).

[70] Li. X, Tabil, L G, Panigrahi, A S Chemical Treatments of Natural Fiber for Use in Natural Fiber-Reinforced Composites: A Review. Journal of Polymer and the Environment,15 (1): 25–33 (2007).

[71] Kalia, S, Kaith, B S, Kaur, I. Pretreatments of natural fibers and their application as reinforcing material in polymer composites—a review. Polymer Engineering & Science, 49 (7): 1253–1272 (2009).

[72] Ma, H and Joo, C W Influence of surface treatments on structural and mechanical properties of bamboo fiber-reinforced poly(lactic acid) biocomposites. Journal of Composite Materials, 45(23): 2455-2463 (2011).

[73] Franco, P J H, Valadez-González, A Fiber-matrix adhesion in natural fiber composites In: Natural fibers, biopolymers and biocomposites, Chapter 6, 177-230, A. K. Mohanty, M. Misra and L. T. Drzal (Eds) Taylor & Francis, CRC Press (2005).

[74] Singha, A S and Rana, A K Effect of silane treatment on physicochemical properties of lignocellulosic C. indica fiber Journal of Applied Polymer Science, 124(3): 2473–2484 (2012).

[75] Campos, A, Marconcini, J M, Martins-Franchetti, S M, Mattoso, L H C The influence of UV-C irradiation on the properties of thermoplastic starch and polycaprolactone biocomposite with sisal bleached fibers. Polymer Degradation and Stability, 97 (10): 1948-1955 (2012).

[76] Stocchi, A, Bernal, C, Vazquez, A, Biagotti, J, Kenny, J A silicone treatment compared to traditional natural fiber treatments: effect on the mechanical and viscoelastic prop-

erties of jute–vinylester laminates. Journal of Composite Materials, 41 (16): 2005-2024 (2007).

[77] Moraes, A G O, Sierakowski, M R, Amico, S C The novel use of sodium borohydride as a protective agent for the chemical treatment of vegetable fibers Fibers and Polymers. 13 (5): 641-646 (2012) DOI: 10.1007/s12221-012-0641-7.

[78] Mukherjee, A, Ganguly P K, Sur, D J Structural mechanics of jute: the effects of hemicellulose or lignin removal. Journal of the Textile Institute. 84 (3):348-353 (1993).

[79] Rong, M Z, Zhang, M Q, Liu, Y., Yang, G C, Zeng, H M The effect of fiber treatment on the mechanical properties of unidirectional sisal-reinforced epoxy composites. Composites Science and Technology, 61 (10): 1437–1447 (2001).

[80] Kim, H-S, Lee, B-H, Choi, S-W, Kim, S, Kim, H-J The effect of types of maleic anhydride-grafted polypropylene (MAPP) on the interfacial adhesion properties of bio-flour-filled polypropylene composites. Composites: Part A, 38 (6): 1473–1482 (2007).

[81] Chang, S Y, Ismail, H, Ashan, Q Effect of maleic anhydride on kenaf dust filled polycaprolactone/ thermoplastic sago starch composites. Bioressurces, 7(2):1594-1616 (2012).

[82] Xie, Y, Hill, C A S, Xiao, Z, Militz, H, Carsten, M Silane coupling agents used for natural fiber/polymer composites: A review. Composites: Part A 41(7): 806–819 (2010),.

[83] Huda, M S, Drzal, L T, Mohanty, A K, Misra, M Effect of fiber surface-treatments on the properties of laminated biocomposites from poly(lactic acid) (PLA) and kenaf fibers. Composites Science and Technology 68 (2): 424–432 (2008)..

[84] Chen, D, Li, J, Ren, J Influence of fiber surface-treatment on interfacial property of poly(l-lactic acid)/ramie fabric biocomposites under UV-irradiation hydrothermal aging. Materials Chemistry and Physics, 126 (3): 524–531 (2011).

[85] Jayamol, G, Sreekala, M S, Thomas, S A review on interface modification and characterization of natural fiber reinforced plastic composites. Polymer Engineering & Science 41 (9): 1471–1485 (2001)..

[86] Yu, L, Dean, K, Li, L Polymer blends and composites from renewable resources. Progress in Polymer Science 31(6): 576–602 (2006).

[87] Avérous, L Biocomposites based on biodegradable thermoplastic polyester and lignocellulose fibers In: Cellulose Fibers: Bio- and Nano-Polymer Composites: Green Chemistry and Technology. Chapter 17,453- 478, Susheel Kalia, B. S. Kaith, Inderjeet Kaur(Eds), Springer, Heidelberg (2011).

[88] Harnnecker, F, Rosa, D S, Lenz, D M Biodegradable Polyester-Based Blend Reinforced with Curaua Fiber: Thermal, Mechanical and Biodegradation Behaviour Journal of Polymers and the Environment, 20(1): 237-244 (2012).

[89] Eichhorn, S J, Dufresne, A, Aranguren, M, Marcovich, N E, Capadona, J R, Rowan, S J, Weder, C, Thielemans, W, Roman, M, Renneckar, S, Gindl, W, Veigel, S, Keckes, J,

Yano, H, Abe, K, Nogi, M, Nakagaito, A N, Mangalam, A, Simonsen, J, Benight, A S, Bismarck, A, Berglund, L A, Peijs, T Review: current international research into cellulose nanofibres and nanocomposites Journal of Materials Science, 45(1):1–33 (2010).

[90] Szczęsna-Antczak, M, Kazimierczak, J, Antczak, T Nanotechnology - Methods of Manufacturing Cellulose Nanofibres. Fibers & Textiles in Eastern Europe, 20, 2(91): 8-12 (2012).

[91] Iguchi, M, Yamanaka S, Budhiono A Bacterial cellulose - a masterpiece of nature's arts. Journal of Materials Science, 35(2):261-270 (2000).

[92] Rusli R., Shanmuganathan K., Rowan S. J., Weder C, Eichhorn S J. Stress Transfer in Cellulose Nanowhisker Composites - Influence of Whisker Aspect Ratio and Surface Charge. Nature 472: 334-337 (2011).

[93] European Commission, Nanotechnology Research needs on nanoparticles. Proceedings of the workshop held in Brussels, 25-26.01.2005.

[94] Halley, P J and Dorgan, J R Next-generation biopolymers: Advanced functionality and improved sustainability MRS Bulletin, 36: 687-691 (2011).

[95] Masoodi, R, Hajjar, R E, Pillai, K M, Sabo, R Mechanical characterization of cellulose nanofiber and bio-based epoxy composite. Materials and Design, 36: 570-576 (2011).

[96] Fortunati, E, Armentano, I, Zhou, Q, Iannoni, A, Saino, E, Visai, L, Berglund, L A, Kenny, J M Multifunctional bionanocomposite films of poly(lactic acid), cellulose nanocrystals and silver nanoparticles. Carbohydrate Polymers, 87(2): 1596–1605 (2012).

[97] Cherian, B M, Leão, A L, de Souza, S F, Costa, L M M, Olyveira, G M, Kottaisamy, M, Nagarajan, E R, Thomas, S Cellulose nanocomposites with nanofibers isolated from pineapple leaf fibers for medical applications Carbohydrate Polymers, 86(4): 1790–1798 (2011).

[98] Souza Lima, M M and Borsali, R. Rodlike Cellulose Microcrystals: Structure, Properties, and Applications. Macromolecular Rapid Communications, 25 (7): 771-787 (2004).

[99] Habibi, Y, Lucia, L A, Rojas, O J Cellulose Nanocrystals: Chemistry, Self-Assembly, and Applications Chemical Reviews, 110: 3479–3500 (2010).

[100] George, J, Ramana, K V, Bawa, A S and Siddaramaiah. Bacterial cellulose nanocrystals exhibiting high thermal stability and their polymer nanocomposites. International Journal of Biological Macromolecules, 48(1): 50-57 (2011).

[101] Jonoobi, M, Harun, J, Mathew, A P, Oksman, K Mechanical properties of cellulose nanofiber (CNF) reinforced polylactic acid (PLA) prepared by twin screw extrusion, Composites Science and Technology, 70(12): 1742-1747 (2010).

[102] Siró, I and Plackett, D Microfibrillated cellulose and new nanocomposite materials: a review. Cellulose, 17 (3): 459–494 (2010).

[103] Oksman, K, Mathew, A P, Sain, M Novel bionanocomposites: processing, properties and potential applications. Plastics, Rubber and Composites, 38(9-10): 396-405 (2009).

[104] Das, K, Ray, D, Banerjee, C, Bandyopadhyay, N R, Sahoo, S, Mohanty, A K, Misra, M Physicomechanical and Thermal Properties of Jute-Nanofiber-Reinforced Biocopolyester Composites. Industrial & Engineering Chemical Research, 49(6): 2775–2782 (2010).

[105] Wang, T and Drzal, L T Cellulose-nanofiber-reinforced poly(lactic acid) composites prepared by a water-based approach. Applied Materials & Interfaces, 4(10): 5079–5085 (2012).

[106] Okubo, K, Fujii, T, Thostenson, E T Multi-scale hybrid biocomposite: processing and mechanical characterization of bamboo fiber reinforced PLA with microfibrillated cellulose Composites Part A, 40(4): 469-475 (2009).

[107] Cherian, B M, Leão, A L, de Souza, S F, Thomas, S, Pothan, L A, Kottaisamy, M Isolation of nanocellulose from pineapple leaf fibers by steam explosion. Carbohydrate Polymers, 81(3): 720–725, (2010).

[108] Blaker, J J, Lee, K-Y, Bismarck, A. Hierarchical Composites Made Entirely from Renewable Resources Journal of Biobased. Materials and Bioenergy, 5(1): 1-16 (2011).

[109] Sorrentino, A, Gorrasi, G, Vittoria, V Potential Perspectives of Bio-Nanocomposites for Food Packaging Applications, Trends in Food Science & Technology, 18(2): 84-95 (2007).

[110] Chrissafis, K and Bikiaris, D Can nanoparticles really enhance thermal stability of polymers? Part I: An overview on thermal decomposition of addition polymers. Thermochimica Acta, 523(1): 1–24 (2011).

[111] Choy, J H, Choi, S J, Oh, J M, Park, T Clay minerals and layered double hydroxides for novel biological applications. Applied Clay Science, 35(1-3): 122-132 (2007).

[112] Liu, Z and Erhan, S Z "Green" composites and nanocomposites from soybean oil. Materials Science and Engineering A, 483–484: 708–711 (2008).

[113] Annabi-Bergaya, F Layered clay minerals. Basic research and innovative composite applications. Microporous and Mesoporous Materials, 107(1-2): 141–148 (2008).

[114] Fu, H-K, Huang. C-F, Huang, J-M, Chang, F-C Studies on thermal properties of PS nanocomposites for the effect of intercalated agent with side groups. Polymer, 49(5): 1305-1311 (2008).

[115] Rodriguez, F, Sepulveda, H M, Bruna, J, Guarda, A, Galotto, M J Development of Cellulose Eco-nanocomposites with Antimicrobial Properties Oriented for Food

Packaging. Packaging Technology and Science, 2012 DOI: 10.1002/pts.1980. http://onlinelibrary.wiley.com/doi/10.1002/pts.1980/full (acessed 20 July 2012).

[116] Arora, A and Padua, G W Review: Nanocomposites in Food Packaging Journal of Food Science, 75(1): R43–R49 (2010).

[117] Rajkiran, R, Tiwari, K C, Khilar, U N Synthesis and characterization of novel organo-montmorillonites. Applied Clay Science, 38(3–4): 203–208 (2008).

[118] Kim, J-T, Lee, D-Y, Oh, T-S, Lee, D-H Characteristics of nitrile–butadiene rubber layered silicate nanocomposites with silane coupling agent Journal of Applied Polymer Science, 89(10): 2633–2640 (2003).

[119] Panagiotis, I, Xidas, K, Triantafyllidis, S Effect of the type of alkylammonium ion clay modifier on the structure and thermal/mechanical properties of glassy and rubbery epoxy–clay nanocomposites European Polymer Journal, 46(3): 404–417 (2010).

[120] Piscitelli, F, Scamardella, A M, Valentina, R, Lavorgna, M, Barra, G, Amendola, E Epoxy composites based on amino-silylated MMT: The role of interfaces and clay morphology Journal of Applied Polymer Science, 124(1): 616–628 (2012).

[121] Tjong, S C Structural and mechanical properties of polymer nanocomposites, Materials Science and Engineering: R: Reports, 53(3–4): 73–197 (2006).

[122] Bordes, P, Pollet, E, Avérous, L Nano-biocomposites: Biodegradable polyester/nanoclay systems Progress in Polymer Science, 34(2): 125–155 (2009)

[123] Kumar, K A A, Sreekala, M S, Arun, S Studies on Properties of Bio-Composites from Ecoflex/Ramie Fabric-Mechanical and Barrier Properties Journal of Biomaterials and Nanobiotechnology, 3: 396-404 (2012).

[124] Sawpan, M A, Pickering, K L, Fernyhough, A Improvement of mechanical performance of industrial hemp fiber reinforced polylactide biocomposites. Composites Part A: Applied Science and Manufacturing, 42(3): 310-319 (2011).

[125] Bledzki, A K, Jaszkiewicz, A Mechanical performance of biocomposites based on PLA and PHBV reinforced with natural fibres – A comparative study to PP. Composites Science and Technology, 70 (12): 1687–1696 (2010).

[126] Huda, M S, Drzal, L T, Mohanty, A K, Misra, M. Chopped glass and recycled newspaper as reinforcement fibers in injection molded poly(lactic acid) (PLA) composites: A comparative study. Composites Science and Technology, 66 (11–12): 1813–1824 (2006).

[127] Bajpai, P K, Singh, I, Madaan, J Development and characterization of PLA-based green composites: A review. Journal of Thermoplastic Composite Materials 2012, 1–30, DOI: 10.1177/0892705712439571, http://jtc.sagepub.com/content/early/2012/03/21/0892705712439571.full.pdf+html (accessed 10 August 2012).

[128] Bledzki, A K, Jaszkiewicz, A, Murr, M, Sperber, V E, Lützkendorf, R, Reußmann, T Processing techniques for natural and wood–fibre composites. In: Properties and per-

formance of natural-fiber composites. Chapter 4, 163–92, Pickering KL (Ed) Cambridge, UK: Woodhead Publishing (2008).

[129] Choi, H-Y and Lee, J- S Effects of surface treatment of ramie fibers in a ramie/poly(lactic acid) composite. Fibers and Polymers, 13(2): 217-223 (2012).

[130] Graupner, N, Herrmann, A S, Müssig, J Natural and man-made cellulose fibre-reinforced poly(lactic acid) (PLA) composites: An overview about mechanical characteristics and application areas. Composites: Part A, 40(6-7): 810–821 (2009).

[131] Oksman, K, Skrifvars, M, Selin, J F. Natural fibres as reinforcement in polylactic acid (PLA) composites. Composites Science and Technology, 63(9):1317–24 (2003).

[132] Avella, M, Bogoeva-Gaceva, G, Bużarovska, A, Errico, M E, Gentile, G, Grozdanov, A. Poly(lactic acid)-based biocomposites reinforced with Kenaf fibers. Journal of Applied Polymer Science, 108(6): 3542–3551 (2008).

[133] Bax, B and Mussig, J. Impact and tensile properties of PLA/cordenka and PLA/flax composites. Composites Science and Technology, 68(7-8): 1601–1607 (2008).

[134] Nakamura, R, Goda, K, Noda, J, Ohgi, J High temperature tensile properties and deep drawing of fully green composites. Express Polymer Letters; 3(1): 19–24 (2009).

[135] Matkó Sz., Toldy, A, Keszei, S, Anna, P, Bertalan, Gy, Marosi, Gy. Flame retardancy of biodegradable polymers and bio-composites. Polymer Degradation and Stability, 88(1): 138–145 (2005).

[136] Bourbigot, S, Fontaine, G, Duquesne, S, Delobe, R. PLA nanocomposites: quantification of clay nanodispersion and reaction to fire. International Journal of Nanotechnology, 5(6-8): 683-692 (2008).

[137] Shumao, L, Jie, R, Hua, Y, Tao, Y, Weizhong, Y. Influence of ammonium polyphosphate on the flame retardancy and mechanical properties of ramie fiber-reinforced poly(lactic acid) biocomposites Polymer International, 59(2): 242–248 (2010).

[138] Jangm, J Y, Jeong, T K , Oh, H J, Youn, J R, Song, Y S. Thermal stability and flammability of coconut fiber reinforced poly(lactic acid) composites. Composites Part B: Engineering , 43(5): 2434–2438 (2012).

[139] Jiménez, A and Ruseckaite, R A. Nano-Biocomposites for Food Packaging In: Environmental Silicate Nano-Biocomposites Green Energy and Technology, Chapter 15, 393-408, L. Avérous and E. Pollet (Eds) Springer-Verlag, London (2012).

[140] Sanchez-Garcia, M D and Lagaron, J M. On the use of plant cellulose nanowhiskers to enhance the barrier properties of polylactic acid. Cellulose, 17(5): 987-1004 (2010).

[141] Sanchez-Garcia, M D and Lagaron, J M Novel clay-based nanobiocomposites of biopolyesters with synergistic barrier to UV light, gas and vapour. Journal of Applied Polymer Science, 118(1): 188-199 (2010).

[142] Petersson, L and Oksman, K. Biopolymer based nanocomposites: Comparing layered silicates and microcrystalline cellulose as nanoreinforcement. Composites Science and Technology, 66 (13): 2187-2196 (2006).

[143] Lasprilla, A J R, Martinez, G A R, Lunelli, B H, Jardini, A L, Maciel Filho, R. Polylactic acid synthesis for application in biomedical devices - A review. Biotechnology Advances, 30(1): 321-328 (2012).

[144] Chen, L, Yang, J, Wang, K, Chen, F, Fu, Q. Largely improved tensile extensibility of poly(L-lactic acid) by adding poly (ε-caprolactone). Polymer International, 59(8): 1154-1161 (2010).

[145] Walker, P A, Aroom, K R, Jimenez, F, Shah, S K, Harting, M T, Gill, B S, Cox Jr, C S. Advances in progenitor cell therapy using scaffolding constructs for central nervous system injury Stem Cell Reviews and Reports, 5(3): 283–300 (2009).

[146] Zhou, W Y, Lee, S H, Wang, M, Cheung, W L, Ip, W Y. Selective laser sintering of porous tissue engineering scaffolds from poly (L-lactide)/carbonated hydroxyapatite nanocomposite microspheres Journal of Materials Science-Materials in Medicine, 19: 2535-2540 (2008).

[147] Hong, Z K, Qiu, XY, Sun, J R, Deng, M X, Chen, X S, Jing, X B. Grafting polymerization of l-lactide on the surface of hydroxyapatite nano-crystals. Polymer, 45(19): 6699-6706 (2004).

[148] Ray, S S and Okamoto, M. Biodegradable Polylactide and Its Nanocomposites: Opening a new dimension for plastics and composites. Macromolecular Rapid Communications, 24(14): 815-840 (2003).

[149] Paul, M-A, Delcourt, C, Alexandre, M, Degée, Ph., Monteverde, F, Dubois, Ph. Polylactide/montmorillonite nanocomposites: study of thehydrolyticdegradation Polymer Degradation and Stability, 87(3): 535–542 (2005).

[150] Kumar, R, Yakubu, M K, Anandjiwala, R D. Biodegradation of flax fiber reinforced poly lactic acid. Polymer Letters, 4(7): 423–430 (2010).

[151] Iovino, R, Zullo, R, Rao, M A, Cassar, L, Gianfreda, L. Biodegradation of poly(lactic acid)/starch/coir biocomposites under controlled composting conditions. Polymer Degradation and Stability, 93(1): 147–157 (2008).

[152] Jandas, P J, Mohanty, S, Nayak, S K, Srivastava, H. Effect of surface treatments of banana fiber on mechanical, thermal, and biodegradability properties of PLA/banana fiber biocomposites. Polymer Composites, 32(11): 1689–1700 (2011).

[153] Jandas, P J, Mohanty S, Nayak S K. Renewable Resource-Based Biocomposites of various surface treated banana fiber and poly lactic acid: Characterization and Biodegradability. Journal of Polymers and the Environment, 20(2): 583-595 (2012).

[154] Singh, S, Mohanty, A K, Sugie, T, Takai, Y, Hamada, H. Renewable resource based biocomposites from natural fiber and polyhydroxybutyrate-co-valerate (PHBV) bioplastic. Composites: Part A, 39(5): 875–886 (2008).

[155] Jiang, L, Huang, J, Qian, J, Chen, F, Zhang, J, Wolcott, M P, Zhu, Y. Study of Poly(3-hydroxybutyrate-co-3-hydroxyvalerate) (PHBV)/bamboo pulp fiber composites: Effects of nucleation agent and compatibilizer. Journal of Polymers and the Environment, 16(2): 83-93 (2008).

[156] Nagarajan, V, Misra, M, Mohanty, A K New engineered biocomposites from poly(3-hydroxybutyrate-co-3-hydroxyvalerate) (PHBV)/poly(butylene adipate-co-terephthalate) (PBAT) blends and switchgrass: Fabrication and performance evaluation. Industrial Crops and Products, 42: 461–468 (2013)

[157] Ahankari, S S, Mohanty, A K, Misra, M. Mechanical behaviour of agro-residue reinforced poly(3-hydroxybutyrate-co-3-hydroxyvalerate), (PHBV) green composites: A comparison with traditional polypropylene composites. Composites Science and Technology, 71(5): 653–657 (2011).

[158] Yu, H-Y, Qin, Z-Y, Liu, Y-N, Chen, L, Liu, N, Zhou, Z. Simultaneous improvement of mechanical properties and thermal stability of bacterial polyester by cellulose nanocrystals Carbohydrate Polymers, 89(3): 971–978 (2012).

[159] Ten, E, Bahr, D F, Li, B, Jiang, L, Wolcott, M P. Effects of Cellulose Nanowhiskers on mechanical, dielectric and rheological properties of poly(3-hydroxybutyrate-co-3-hydroxyvalerate)/cellulose nanowhisker. Composites Industrial & Engineering Chemical Research, 51 (7): 2941–2951 (2012).

[160] Sanchez-Garcia, M D, Lopez-Rubio, A, Lagaron, J M. Natural micro and nanobiocomposites with enhanced barrier properties and novel functionalities for food biopackaging applications. Trends in Food Science & Technology, 21(11): 528-536 (2010).

[161] Chen, G X, Hao, G J, Guo, T Y, Song, M D, Zhang, B H. Structure and mechanical properties of poly(3-hydroxybutyrate-co-3-hydroxyvalerate) (PHBV)/clay nanocomposites. Journal of Materials Science Letters, 21(20):1587–1589 (2002).

[162] Nguyen L. T. H., Chen S., Elumalai N. K., Prabhakaran M. P., Zong Y., Vijila C., Allakhverdiev S. I., Ramakrishna S. Biological, chemical, and electronic applications of nanofibers. Macromolecular Materials and Engineering, 297(11): 1035–1123 (2012).

[163] Hazer, D B, Kılıçay, E, Hazer, B. Poly(3-hydroxyalkanoate)s: Diversification and biomedical applications. A state of the art review. Materials Science and Engineering C, 32(4): 637–647 (2012).

[164] Cool, S M, Kenny, B, Wu, A. Nurcombe V., Trau M., Cassady A. I., Grøndahl L. Poly(3-hydroxybutyrate-co-3-hydroxyvalerate) composite biomaterials for bone tissue regeneration: In vitro performance assessed by osteoblast proliferation, osteoclast

adhesion and resorption and macrophage proinflammatory response. Journal of Biomedical Materials Research Part A, 82: 599-610 (2007).

[165] Batista, K C, Silva, D A K, Coelho, L A F, Pezzin, S H, Pezzin, A P T. Soil biodegradation of PHBV/peach palm particles biocomposites Journal of Polymers and the Environment, 18(3): 346-354 (2010).

[166] Petersson, S, Jayaraman, K, Bhattacharyya, D. Forming performance and biodegradability of wood-fibre BiopolTM composites. Composites Part A, 33: 1123-1134 (2002).

[167] Avella, M, La Rota, G, Martuscelli, E, Raimo, M, Sadocco, P, Elegir, G, Riva, R. Poly (3-hydroxybutyrate-co-3-hydroxyvalerate) and wheat straw fibre composites: thermal, mechanical properties and biodegradation behaviour. Journal of Materials Science, 35(4): 829-836 (2000).